LIFE CYCLE RELIABILITY ENGINEERING

LIFE CYCLE
RELIABILITY
ENGINEERING

Guangbin Yang
Ford Motor Company

JOHN WILEY & SONS, INC.

Published by John Wiley & Sons, Inc., Hoboken, New Jersey
Published simultaneously in Canada.

For general information on our other products and services or for technical support, please contact our Customer Care Department within the United States at (800) 762-2974, outside the United States at (317) 572-3993 or fax (317) 572-4002.

Wiley also publishes its books in a variety of electronic formats. Some content that appears in print may not be available in electronic formats. For more information about Wiley products, visit our web site at www.wiley.com.

Library of Congress Cataloging-in-Publication Data:

Yang, Guangbin, 1964-
   Life cycle reliability engineering / Guangbin Yang.
      p. cm.
   Includes bibliographical references and index.
   ISBN-13: 978-0-471-71529-0 (cloth)
   ISBN-10: 0-471-71529-8 (cloth)
1. Reliability (Engineering) 2. Product life cycle. I. Title.
   TS173.Y36 2007
   620'.00452—dc22

                                                    2006019150

Printed in the United States of America

10 9 8 7 6 5 4 3 2 1

To Ling, Benjamin, and Laurence

CONTENTS

PREFACE

This decade has witnessed a rapid development of economic globalization characterized by international free trade and free flow of capital across countries. It is now common to see similar products from different countries in the same market at the same time. This competition is further intensified by customers' immediate access to detailed information about products through the ubiquitous Internet. With little time and effort, customers can compare competing products in terms of features, cost, reliability, service, and many other factors. It is not surprising that the best-informed customers are the picky ones; they always choose the products that work best and cost least. To survive and grow in such a competitive business environment, manufacturers must deliver reliable products with more features, at lower cost, and in less time. In response to these market forces, manufacturers are challenging reliability professionals as well as providing opportunities to improve reliability, shorten design cycle, reduce production and warranty costs, and increase customer satisfaction. To meet these challenges, reliability professionals need more effective techniques to assure product reliability throughout the product life cycle.

This book is designed to provide useful, pragmatic, and up-to-date reliability techniques to assure product reliability throughout the product life cycle, which includes product planning; design and development; design verification and process validation; production; field deployment; and disposal. In particular, we discuss techniques for understanding customer expectations, for building reliability into products at the design and development stage, for testing products more efficiently before design release, for screening out defective products at the production stage, and for analyzing warranty data and monitoring reliability performance in the field. The book is comprised of 11 chapters organized according to the sequence of the product life cycle stages. In Chapter 1 we describe briefly the concepts of reliability engineering and product life cycle and the integration of reliability techniques into each stage of the life cycle. Chapter 2 delineates the reliability definition, metrics, and product life distributions. In Chapter 3 we present techniques for analyzing customer expectations, establishing reliability targets, and developing effective reliability programs to be executed throughout the product life cycle. Chapter 4 covers methodologies and practical applications of system reliability modeling and allocation. Confidence intervals for system reliability are also addressed in this chapter. Chapter 5 is one of the most important chapters in the book, presenting robust reliability design techniques aimed at building reliability and robustness into products in the design and development phase. In Chapter 6 we describe reliability tools used to detect, assess, and eradicate design mistakes. Chapter 7 covers accelerated life test methods, models, plans, and data analysis techniques illustrated with many industrial examples. In Chapter 8 we discuss degradation testing and data analysis methods that cover both destructive and nondestructive inspections. In Chapter 9 we present reliability techniques for design verification and process validation and in Chapter 10 address stress screening topics and describe more advanced methods for degradation screening. The last chapter, Chapter 11, is dedicated to warranty analysis, which is important for manufacturers in estimating field reliability and warranty repairs and costs.

The book has the following distinct features:

- It covers many new and practical reliability techniques, including customer-driven reliability target setting, customer-driven reliability allocation, reliability design using design for six sigma, robust reliability design, accelerated life tests with higher usage rates, accelerated life tests with tightened thresholds, destructive degradation testing and analysis, sample size reduction based on physical characteristics, degradation screening, two-dimensional warranty analysis, and many other techniques.

- Pragmatism is emphasized throughout the book. All reliability techniques described are immediately applicable to product planning, design, testing, screening, and warranty analysis.

- Examples and exercises deal with real-world applications. Although many problems have been from the automotive industry, other industries have essentially the same problems.

- The book closely relates reliability to customer satisfaction and presents reliability and quality techniques, such as quality function deployment and customer-driven reliability allocation, for improving customer satisfaction.

- We provide and review recent advances in important reliability techniques which researchers will find useful in new developments.

- Some 300 references, representing helpful resources for pursuing further study of the topics, are cited.

The book is designed to serve engineers working in the field of reliability and quality for the development of effective reliability programs and implementation of programs throughout the product life cycle. It can be used as a textbook for students in industrial engineering departments or reliability programs but will also be useful for industry seminars or training courses in reliability planning, design, testing, screening, and warranty analysis. In all cases, readers need to know basic statistics.

I am indebted to a number of people who contributed to the book. Mr. Z. Zaghati, Ford Motor Company, encouraged, stimulated, and assisted me in completing the book. I am most grateful to him for his continuing support.

I would like to specially acknowledge Dr. Wayne Nelson, a consultant and teacher in reliability and statistics. He gave me detailed feedback on parts of this book. In addition, Dr. Nelson generously shared with me some of his unpublished thoughts and his effective and valuable book-writing skills.

A number of people provided helpful suggestions and comments on parts of the book. In particular, I am pleased to acknowledge Prof. Thad Regulinski, University of Arizona; Dr. Joel Nachlas, Virginia Polytechnic Institute and State University; Dr. Ming-Wei Lu, DaimlerChrysler Corporation; Prof. Fabrice Guerin and Prof. Abdessamad Kobi, University of Angers, France; and Dr. Loon-Ching Tang, National University of Singapore. I am also grateful for contributions from Dr. Vasiliy Krivtsov, Ford Motor Company. Over the years I have benefited from numerous technical discussions with him.

I would also like to thank Prof. Hoang Pham, Rutgers University; Dr. Greg Hobbs, Hobbs Engineering Corporation; and Prof. Dimitri Kececioglu, University of Arizona, who all generously reviewed parts of the manuscript and offered comments.

Finally, I would like to express my deep appreciation and gratitude to my wife, Ling, and sons Benjamin and Laurence. Their support was essential to the successful completion of the book.

GUANGBIN YANG

Dearborn, Michigan
May 2006

1

RELIABILITY ENGINEERING
AND PRODUCT LIFE CYCLE

1.1 RELIABILITY ENGINEERING

Reliability has a broad meaning in our daily life. In technical terms, *reliability* is defined as the probability that a product performs its intended function without failure under specified conditions for a specified period of time. The definition, which is elaborated on in Chapter 2, contains three important elements: intended function, specified period of time, and specified conditions. As reliability is quantified by probability, any attempts to measure it involve the use of probabilistic and statistical methods. Hence, probability theory and statistics are important mathematical tools for reliability engineering.

Reliability engineering is the discipline of ensuring that a product will be reliable when operated in a specified manner. In other words, the function of reliability engineering is to avoid failures. In reality, failures are inevitable; a product will fail sooner or later. Reliability engineering is implemented by taking structured and feasible actions that maximize reliability and minimize the effects of failures. In general, three steps are necessary to accomplish this objective. The first step is to build maximum reliability into a product during the design and development stage. This step is most critical in that it determines the inherent reliability. The second step is to minimize production process variation to assure that the process does not appreciably degrade the inherent reliability. Once a product is deployed, appropriate maintenance operations should be initiated to alleviate performance degradation and prolong product life. The three steps employ a large variety of reliability techniques, including, for example, reliability planning and specification, allocation, prediction, robust reliability design, failure mode and effects analysis (FMEA), fault tree analysis (FTA), accelerated life testing, degradation testing, reliability verification testing, stress screening, and warranty analysis. To live up to the greatest potential inherent to these reliability techniques for specific products, we must develop and implement appropriate and adequate reliability programs that synthesize these individual reliability techniques. In particular, such programs include the tasks of specifying the reliability requirements, customizing and sequencing the reliability techniques, orchestrating the implementation, and documenting the results. In subsequent chapters we describe in detail reliability programs and individual reliability techniques.

1.2 PRODUCT LIFE CYCLE

Product life cycle refers to sequential phases from product planning to disposal. Generally, it comprises six main stages, as shown in Figure 1.1. The stages, from product planning to production, take place during creation of the product and thus are collectively called the *product realization process*. The tasks in each stage are described briefly below.

Product Planning Phase   Product planning is to identify customer needs, analyze business trends and market competition, and develop product proposals. In the beginning of this phase, a cross-functional team should be established that represents different functions within an organization, including marketing, financing, research, design, testing, manufacturing, service, and other roles. Sometimes supplier representatives and consultants are hired to participate in some of the planning work. In this phase, the team is chartered to conduct a number of tasks, including business trends analysis, understanding customer expectations, competitive analysis, and market projection. If the initial planning justifies further development of the product, the team will outline the benefits of the product to customers, determine product features, establish product performances, develop product proposals, and set the time to market and time lines for the completion of such tasks as design, validation, and production.

Design and Development Phase   This phase usually begins with preparation of detailed product specifications on reliability, features, functionalities, economics, ergonomics, and legality. The specifications must meet the requirements defined in the product planning phase, ensure that the product will satisfy customer expectations, comply with governmental regulations, and establish strong competitive position in the marketplace. The next step is to carry out the concept design. The starting point of developing a concept is the design of a functional structure that determines the flow of energy and information and the physical interactions. The functions of subsystems within a product need to be clearly defined; the requirements regarding these functions arise from the product specifications. Functional block diagrams are always useful in this step. Once the architecture is complete,

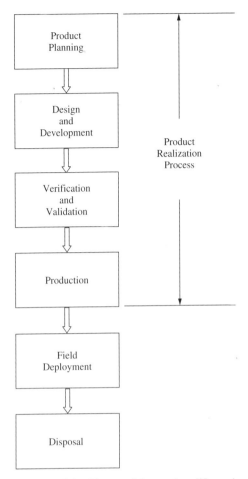

FIGURE 1.1 Phases of the product life cycle

the physical conception begins to determine how the functions of each subsystem can be fulfilled. This step benefits from the use of advanced design techniques such as TRIZ and axiomatic design (Suh, 2001; K. Yang and El-Haik, 2003) and may result in innovations in technology. Concept design is a fundamental stage that largely determines reliability, robustness, cost, and other competitive potentials.

Concept design is followed by detailed design. This step begins with the development of detailed design specifications which assure that the subsystem requirements are satisfied. Then physical details are devised to fulfill the functions of each subsystem within the product structure. The details may include physical linkage, electrical connection, nominal values, and tolerances of the functional parameters. Materials and components are also selected in this step. It is worth noting that design and development is essentially an iterative task as a result of design review and analysis. The implementation of effective reliability programs will reduce repetition.

Verification and Validation Phase   This phase consists of two major steps: design verification (DV) and process validation (PV). Once a design is completed successfully, a small number of prototypes are built for DV testing to prove that the design achieves the functional, environmental, reliability, regulatory, and other requirements concerning the product as stipulated in the product specifications. Prior to DV testing, a test plan must be developed that specifies the test conditions, sample sizes, acceptance criteria, test operation procedures, and other elements. The test conditions should reflect the real-world use that the product will encounter when deployed in the field. A large sample size in DV testing is often unaffordable; however, it should be large enough so that the evidence to confirm the design achievement is statistically valid. If functional nonconformance or failure occurs, the root causes must be identified for potential design changes. The redesign must undergo DV testing until all acceptance criteria have been met completely.

Parallel to DV testing, production process planning may be initiated so that pilot production can begin once the design is verified. Process planning involves the determination of methods for manufacturing a product. In particular, we choose the steps required to manufacture the product, tooling processes, process checkpoints and control plans, machines, tools, and other requirements. A computer simulation is helpful in creating a stable and productive production process.

The next step is PV testing, whose purpose is to validate the capability of the production process. The process must not degrade the inherent reliability to an unacceptable level and must be capable of manufacturing products that meet all specifications with minimum variation. By this time, the process has been set up and is intended for production at full capacity. Thus, the test units represent the products that customers will see in the marketplace. In other words, the samples and the final products are not differentiable, because both use the same materials, components, production processes, and process monitoring and measuring techniques. The sample size may be larger than that for DV testing, due to the need to evaluate process variation. The test conditions and acceptance criteria are the same as those for DV testing.

Production Phase   Once the design is verified and the process is validated, full capacity production may begin. This phase includes a series of interrelated activities such as materials handling, production of parts, assembly, and quality control and management. The end products are subject to final test and then shipped to customers.

Field Deployment Phase   In this phase, products are sold to customers and realize the values built in during the product realization process. This phase involves marketing advertisement, sales service, technical support, field performance monitoring, and continuous improvement.

Disposal   This is the terminal phase of a product in the life cycle. A product is discarded, scraped, or recycled when it is unable to continue service or is not cost-effective. A nonrepairable product is discarded once it fails; a repairable product may be discarded because it is not worthy of repair. The service of some repairable products is discontinued because their performance does not meet customer demands. The manufacturer must provide technical support to dispose of, dismantle, and recycle the product to minimize the associated costs and the adverse impact on the environment.

1.3   INTEGRATION OF RELIABILITY ENGINEERING INTO THE PRODUCT LIFE CYCLE

From a manufacturer's perspective in gaining a competitive advantage, the product realization process should be minimized with respect to time and cost. On the other hand, once they take ownership, customers expect products to operate reliably and to incur little maintenance cost. The conflicting interests have motivated manufacturers to integrate reliability programs into the product life cycle. As described in Chapter 3, a reliability program consists of a series of reliability tasks that are well sequenced to achieve the reliability target and customer satisfaction. The reliability tasks are customized to fit the needs of specific products and implemented throughout the product life cycle. In the product realization process, reliability tasks are especially important because of the amount of value that they can add to products. In particular, reliability techniques are aimed at building reliability into products and reducing the cost and time associated with the process. To maximize efficiency, reliability tasks should be incorporated into the engineering activities that take place in this process. A comprehensive reliability program adds value to the product life cycle even after a product enters the field deployment phase.

The functions of reliability tasks are important in each phase of the product life cycle. In the product planning phase, a multidisciplinary reliability team is organized to develop a reliability program suitable for the product, to set a reliability target, to translate customer expectations into engineering requirements, and to conceive and evaluate product proposals from a reliability perspective. Whenever possible, reliability tasks should be incorporated into other planning activities. The reliability decisions made in this phase have a tremendous impact on each stage of the product life cycle. For example, setting a reliability target has strong effects on cost, time to market, and competitiveness. An overly ambitious target would incur unaffordable design and development costs and prolong the product realization process, and thus jeopardize competitive advantages. Conversely, a low reliability target certainly undermines competitiveness simply by losing customers.

Reliability tasks play an especially important role in the design and development phase; reliability activities usually add more value to a product in this phase than in any other phase. The objective of reliability tasks in this phase is to design-in the reliability of the product while designing-out potential failure modes. This may be accomplished by allocating the product reliability target to the integral subsystems or components and assuring achievement of the respective reliability goals through the implementation of reliability design techniques such as robust reliability design, FMEA, FTA, and design controls. These proactive reliability tasks are aimed at designing things right the first time. Doing so would cut off the design–test–fix loop, which was a typical design model in the old days and unfortunately is still sometimes used. Clearly, a reliability program can accelerate the design and development cycle and save the associated costs.

Reliability tasks are vital elements of DV and PV. In the DV stage, reliability verification testing is performed to demonstrate that the design meets the reliability requirements. In the PV step, the test is intended to prove the capability of the production process. The process must be capable of manufacturing final products that fulfill the reliability target that has been specified. The determination of economic and statistically significant sample size and test time is a challenge to almost all manufacturers. The reliability techniques described in Chapter 9 have the power to make the best trade-off. The function of reliability tasks in this phase is more than testing. For example, life and performance data analysis is often necessary to arrive at meaningful conclusions about the conformability of the design and process under evaluation.

The reliability tasks in the production phase are intended to assure that the process results in the manufacture of uniform and reliable products. To maintain a stable process over time, process control plans and charts are implemented to monitor the process and help identify special causes as soon as they emerge. Sometimes, reactive reliability methods are needed in the production phase. For example, acceptance sampling may determine whether or not to accept a particular production lot being concerned. Products may be subjected to an environmental stress screening to precipitate defective units before being shipped to customers.

The reliability tasks involved in the field deployment phase include the collection and analysis of warranty data, identification and projection of failure trends, customer feedback analysis, and failure analysis of warrantied parts. Six-sigma projects are often initiated in this phase to determine the causes of significant failure modes and to recommend containment and permanent corrective actions.

1.4   RELIABILITY IN THE CONCURRENT PRODUCT REALIZATION PROCESS

A conventional product realization process is serial; that is, a step starts only after the preceding step has been completed. In the sequential model, the information flows in succession from phase to phase. Design engineers in the upstream part of the process usually do not address the manufacturability, testability, and serviceability in their design adequately because of a lack of knowledge. Once the design is verified and the process fails to be validated due to inadequate manufacturability, design changes in this phase will increase cost substantially compared to making the changes in the design and development phase. In general, cost to fix a design increases an order of magnitude with each subsequent phase (Levin and Kalal, 2003).

The application of concurrent engineering to a product realization process is the solution to problems associated with the sequential model. In the framework of concurrent engineering, a cross-functional team is established representing every aspect of the product, including design, manufacturing process, reliability and quality planning, marketing and sales, purchasing, cost accounting, material handling, material control, data management and communication, service, testing, and others. The team relays information to design engineers concerning all aspects of the product so that from the very beginning the engineers will address any potential issues that would otherwise be ignored. The information flow is multidirectional between all functional areas, as stated above, and continues throughout the entire product realization process. As a result, other phases in addition to design and development also benefit from the concurrent involvement. For example, test plans can be developed in the design and development phase, with valuable input from design engineers and other team members. If a testability problem is discovered in this phase, design engineers are more likely to make design changes. Under concurrent engineering, most phases of the product realization process can take place simultaneously. The resulting benefits are twofold: maximization of the chance of doing things right the first time, and reducing the time to market. Ireson et al. (1996) and Usher et al. (1998) describe concurrent engineering in greater detail and present application examples in different industries.

In the environment of concurrent engineering, a multidisciplinary reliability team is required to perform effective reliability tasks. The team is an integral part of the engineering team and participates in decision making so that reliability objectives and constraints are considered. Because reliability tasks are incorporated into engineering activities, a concurrent product realization process entails multiple reliability tasks to be conducted simultaneously. The environment allows reliability tasks to be implemented in the upfront phases of the process to consider the potential influences that might be manifested in subsequent phases. For example, the reliability allocation performed at the beginning of the process should take into account the technological feasibility of achieving the reliability, economics of demonstrating the reliability, and manufacturability of components. Although being concurrent is always desirable, some reliability tasks must be performed sequentially. For example, a process FMEA usually starts after a design FMEA has been completed because the former utilizes the outputs of the latter. In these situations, we should understand the interrelationships between the reliability tasks and sequence the tasks to maximize the temporal overlap.

PROBLEMS

1.1 Explain the concept of reliability and the function of reliability engineer

1.2 Describe the key engineering tasks and reliability roles in each phase of a product life cycle.

1.3 Explain the important differences between serial and concurrent product realization processes.

1.4 How should a reliability program be organized in the environment of concurrent engineering?

2

RELIABILITY DEFINITION, METRICS, AND PRODUCT LIFE DISTRIBUTIONS

2.1 INTRODUCTION

The focus in this book is on the development of effective reliability programs to be implemented throughout a product life cycle. A reliability program usually consists of reliability planning, design, testing, and analysis. In reliability planning, reliability must be defined for the particular product of concern; that is, the intended function, the specified conditions and the specified period of time must adhere to and satisfy or exceed the pertinent design requirements. To make a reliability definition operational, ambiguous and qualitative terms should be avoided. Once reliability has been defined, appropriate metrics must be chosen to characterize it. Efficient metrics are those sensitive to stress and time, and most concern both the manufacturer and customers. Selection of the metrics is critical; once determined, they are used throughout the reliability program. In other words, reliability target setting, design, testing, and analysis should work consistently on the same metrics. Evaluation of the metrics may sometimes apply nonparametric approaches; however, parametric approaches are more useful and insightful, especially when inference or prediction is involved. In this book we use parametric methods. Therefore, statistical distributions are necessary in measuring reliability.

In this chapter we define reliability and elaborate on the three essential elements of the definition. Various reliability metrics and their relationships are presented. In addition, we describe the most commonly used statistical distributions, including the exponential, Weibull, mixed Weibull, smallest extreme value,

Life Cycle Reliability Engineering, by Guangbin Yang
Copyright © 2007 John Wiley & Sons, Inc.

normal, and lognormal. The reliability definition, metrics, and life distributions presented in this chapter are the basic materials for subsequent reliability design, testing, and analysis.

2.2 RELIABILITY DEFINITION

First we define terms that relate to the definition of reliability.

- *Binary state*. The function of a product is either success or failure.
- *Multistate*. The function of a product can be complete success, partial success, or failure. Performance degradation is a special case of multistate.
- *Hard failure*. This is a catastrophic failure that causes complete ceasation of a function. Such a failure mode occurs in a binary-state product.
- *Soft failure*. This is partial loss of a function. This failure mode occurs in a multistate (degradation) product.

In our daily life, reliability has a broad meaning and often means dependability. In technical terms, *reliability* is defined as the probability that a product performs its intended function without failure under specified conditions for a specified period of time. The definition contains three important elements: intended function, specified period of time, and specified conditions.

The definition above indicates that reliability depends on specification of the intended function or, complementarily, the failure criteria. For a binary-state product, the intended function is usually objective and obvious. For example, lighting is the intended function of a light bulb. A failure occurs when the light bulb is blown out. For a multistate or degradation product, the definition of an intended function is frequently subjective. For example, the remote key to a car is required to command the operations successfully at a distance up to, say, 30 meters. The specification of a threshold (30 meters in this example) is somewhat arbitrary but largely determines the level of reliability. A quantitative relationship between the life and threshold for certain products is described in G. Yang and Yang (2002). If the product is a component to be installed in a system, the intended function must be dictated by the system requirements, and thus when used in different systems, the same components may have different failure criteria. In the context of commercial products, the customer-expected intended functions often differ from the technical intended functions. This is especially true when products are in warranty period, during which customers tend to make warranty claims against products that have degraded appreciably even through they are technically unfailed.

Reliability is a function of time. In the reliability definition, the period of time specified may be the warranty length, design life, mission time, or other period of interest. The design life should reflect customers' expectations and be competitive in the marketplace. For example, in defining the reliability of a passenger automobile, the length of time specified is 10 years or 150,000 miles, which define the useful life that most customers expect from a vehicle.

It is worth noting that time has different scales, including, for example, calendar time, mileage, on–off cycles, and pages. The life of some products can be measured on more than one scale. Sometimes, it is difficult to choose from among different scales because all appear to be relevant and of interest. The guide is that the scale selected should reflect the underlying failure process. For example, the deterioration of the body paint of an automobile is more closely related to age than to mileage because the chemical reaction to the paint takes place continuously whether or not the vehicle is operating. In some cases, more than one scale is needed to characterize the life. The most common scales are age and usage. A typical example is the automobile, whose life is usually measured by both age and mileage. Modeling and evaluating reliability as a function of time and usage has been studied in, for example, Eliashberg et al. (1997), S. Yang et al. (2000), and G. Yang and Zaghati (2002).

Reliability is a function of operating conditions. The conditions may include stress types and levels, usage rates, operation profiles, and others. Mechanical, electrical, and thermal stresses are most common. Usage rate (frequency of operation) is also an important operating condition that affects the reliability of many products. Its effects on some products have been studied. Tamai et al. (1997) state that the number of cycles to failure of a microrelay is larger at a high usage rate than at a low rate. Tanner et al. (2002) report that microengines survive longer when running at high speed than when running at low speed. According to Harris (2001), rotation speed is a factor that influences revolutions to failure of a bearing. Nelson (1990, 2004) describes the effects but ignores them in accelerated life test data analyses for the sake of simplicity. In Chapter 7 we describe a model that relates usage rate to product life.

Understanding customer use of a product is a prerequisite for specifying realistic conditions. Most products are used in a wide range of conditions. The conditions specified should represent the real-world usage of most customers. In many situations it is difficult or impossible to address all operating conditions that a product will encounter during its lifetime. However, the stresses to which a product is most sensitive should be included. Chan and Englert (2001) describe most stresses that have large adverse effects on the reliability of electronic products.

The reliability definition for a specific product should be operational. In other words, the reliability, intended function, specified condition, and time must be quantitative and measurable. To achieve this, qualitative and uninformative terms should be avoided. If the product is a component to be installed in a system, the definition should be based on the system's requirements. For example, the reliability of an automobile is defined as the probability that it will meet customer expectations for 10 years or 150,000 miles under a real-world usage profile. The 10 years or 150,000 miles should be translated into the design life of the components within the automobile.

2.3   RELIABILITY METRICS

In this section we describe common metrics used to measure reliability. In practice, the appropriate and effective metrics for a specific product must be determined based on product uniqueness and use.

Probability Density Function (pdf)   The *pdf*, denoted $f(t)$, indicates the failure distribution over the entire time range and represents the absolute failure speed. The larger the value of $f(t)$, the more failures that occur in a small interval of time around t. Although $f(t)$ is rarely used to measure reliability, it is the basic tool for deriving other metrics and for conducting in-depth analytical studies.

Cumulative Distribution Function (cdf)   The *cdf*, denoted $F(t)$, is the probability that a product will fail by a specified time t. It is the probability of failure, often interpreted as the population fraction failing by time t. Mathematically, it is defined as

$$F(t) = \Pr(T \le t) = \int_{-\infty}^{t} f(t)dt. \tag{2.1}$$

Equation (2.1) is equivalent to

$$f(t) = \frac{dF(t)}{dt}. \tag{2.2}$$

For example, if the time to failure of a product is exponentially distributed with parameter λ, the pdf is

$$f(t) = \lambda \exp(-\lambda t), \qquad t \ge 0. \tag{2.3}$$

and the cdf is

$$F(t) = \int_{0}^{t} \lambda \exp(-\lambda t)dt = 1 - \exp(-\lambda t), \qquad t \ge 0. \tag{2.4}$$

Reliability   The *reliability function*, denoted $R(t)$, also called the *survival function*, is often interpreted as the population fraction surviving time t. $R(t)$ is the probability of success, which is the complement of $F(t)$. It can be written as

$$R(t) = \Pr(T \ge t) = 1 - F(t) = \int_{t}^{\infty} f(t)dt. \tag{2.5}$$

From (2.4) and (2.5), the reliability function of the exponential distribution is

$$R(t) = \exp(-\lambda t), \qquad t \ge 0. \tag{2.6}$$

Hazard Function   The *hazard function* or *hazard rate*, denoted $h(t)$ and often called the *failure rate*, measures the rate of change in the probability that a surviving product will fail in the next small interval of time. It can be written as

$$h(t) = \lim_{\Delta t \to 0} \frac{\Pr(t < T \le t + \Delta t | T > t)}{\Delta t} = \frac{1}{R(t)} \left[-\frac{dR(t)}{dt} \right] = \frac{f(t)}{R(t)}. \quad (2.7)$$

From (2.3), (2.6), and (2.7), the hazard rate of the exponential distribution is

$$h(t) = \lambda. \quad (2.8)$$

Equation (2.8) indicates that the hazard rate of the exponential distribution is a constant.

The unit of hazard rate is failures per unit time, such as failures per hour or failures per mile. In high-reliability electronics applications, FIT (failures in time) is the commonly used unit, where 1 FIT equals 10^{-9} failures per hour. In the automotive industry, the unit "failures per 1000 vehicles per month" is often used.

In contrast to $f(t)$, $h(t)$ indicates the *relative failure speed*, the propensity of a surviving product to fail in the coming small interval of time. In general, there are three types of hazard rate in terms of its trend over time: decreasing hazard rate (DFR), constant hazard rate (CFR), and increasing hazard rate (IFR). Figure 2.1 shows the classical bathtub hazard rate function. The curve represents the observation that the life span of a population of products is comprised of three distinct periods:

1. *Early failure period*: The hazard rate decreases over time.
2. *Random failure period*: The hazard rate is constant over time.
3. *Wear-out failure period*: The hazard rate increases over time.

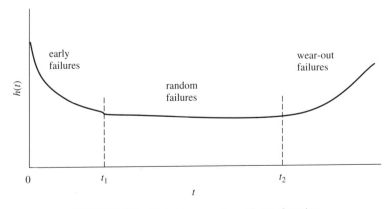

FIGURE 2.1   Bathtub curve hazard rate function

Early failures are usually caused by major latent defects, which develop into patent defects early in the service time. The latent defects may be induced by manufacturing process variations, material flaws, and design errors; customer misuse is another cause of early failures. In the automotive industry, the *infant mortality* problem is significant. It is sometimes called the *one-month effect*, meaning that early failures usually occur in the first month in service. Although a decreasing hazard rate can result from infant mortality, early failures do not necessarily lead to a decreasing hazard rate. A substandard subpopulation containing latent defects may have an increasing hazard rate, depending on the life distribution of the subpopulation. For example, if the life distribution of the substandard products is Weibull with a shape parameter of less than 1, the hazard rate decreases over time. If the shape parameter is greater than 1, the hazard rate has an increasing trend.

In the random failure period, the hazard rate remains approximately constant. During this period of time, failures do not follow a predictable pattern and occur at random due to the unexpected changes in stresses. The stresses may be higher or lower than the design specifications. Higher stresses cause overstressing, whereas lower stresses result in understressing. Both over- and understressing may produce failures. For instance, an electromagnetic relay may fail due to a high or low electric current. A high current melts the electric contacts; a low current increases the contact resistance. In the constant hazard rate region, failures may also result from minor defects that are built into products due to variations in the material or the manufacturing process. Such defects take longer than major defects to develop into failures.

In the wear-out region, the hazard rate increases with time as a result of irreversible aging effects. The failures are attributable to degradation or wear out, which accumulates and accelerates over time. As a product enters this period, a failure is imminent. To minimize the failure effects, preventive maintenance or scheduled replacement of products is often necessary.

Many products do not illustrate a complete bathtub curve. Instead, they have one or two segments of the curve. For example, most mechanical parts are dominated by the wear-out mechanism and thus have an increasing hazard rate. Some components exhibit a decreasing hazard rate in the early period, followed by an increasing hazard rate. Figure 2.2 shows the hazard rate of an automotive subsystem in the mileage domain, where the scale on the y-axis is not given here to protect proprietary information. The hazard rate decreases in the first 3000 miles, during which period the early failures took place. Then the hazard rate stays approximately constant through 80,000 miles, after which failure data are not available.

Cumulative Hazard Function   The *cumulative hazard function*, denoted $H(t)$, is defined as

$$H(t) = \int_{-\infty}^{t} h(t)dt. \qquad (2.9)$$

Handwritten annotations at top of page:

Probability of failure in given time interval
$[t_1, t_2]$
$\int_{t_1}^{t_2} f(t)\,dt = R(t_1) - R(t_2)$
$= F(t_2) - F(t_1)$

Probability of failure in small interval Δt
$\dfrac{P}{\Delta t} = \dfrac{F(t + \Delta t) - F(t)}{R(t) \cdot \Delta t}$

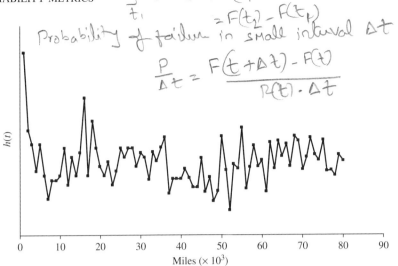

FIGURE 2.2 The hazard rate of an automotive subsystem in mileage domain

For the exponential distribution, we have

$$H(t) = \lambda t, \qquad t \geq 0. \tag{2.10}$$

From (2.7) and (2.9) the relationship between $H(t)$ and $R(t)$ can be written as

$$R(t) = \exp[-H(t)]. \tag{2.11}$$

If $H(t)$ is very small, a Taylor series expansion results in the following approximation:

$$R(t) \approx 1 - H(t). \tag{2.12}$$

$H(t)$ is a nondecreasing function. Figure 2.3 depicts the $H(t)$ associated with DFR, CFR, and IFR. The shapes of $H(t)$ for DFR, CFR, and IFR are convex,

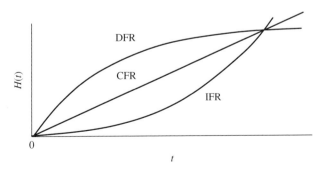

FIGURE 2.3 Cumulative hazard functions corresponding to DFR, CFR and IFR

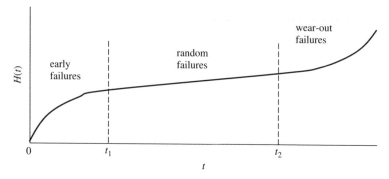

FIGURE 2.4 Cumulative hazard function corresponding to the bathtub curve hazard rate

flat, and concave, respectively. Figure 2.4 illustrates the $H(t)$ corresponding to the bathtub curve hazard rate in Figure 2.1.

Percentile The *percentile*, denoted t_p, is the time by which a specified fraction p of the population fails. It is the inverse of $F(t)$: namely,

$$t_p = F^{-1}(p). \tag{2.13}$$

For the exponential distribution, we have

$$t_p = \frac{1}{\lambda} \ln\left(\frac{1}{1-p}\right). \tag{2.14}$$

In application, t_p is sometimes used to characterize the time by which an allowable small portion of the population would fail. A special percentile is B_{10}, where $B_{10} = t_{0.1}$, the time by which 10% units of the population would fail. This special quantity is frequently used in industry; for example, it is a common characteristic of bearing life.

$R(t_p) = 1 - R(t).$

Mean Time to Failure (MTTF) MTTF is the expected life $E(T)$ of a nonrepairable product. It is defined as

expected life

$$\text{MTTF} = E(T) = \int_{-\infty}^{\infty} t f(t)\, dt. \tag{2.15}$$

If the range of T is positive, (2.15) can be written as

$$\text{MTTF} = \int_{0}^{\infty} R(t)\, dt. \tag{2.16}$$

For the exponential distribution, MTTF is

$$\text{MTTF} = \int_0^\infty \exp(-\lambda t)\, dt = \frac{1}{\lambda}. \tag{2.17}$$

MTTF is a measure of the center of a life distribution. For a symmetric distribution such as the normal distribution (Section 2.8), the MTTF is the same as the median. Otherwise, they are not equal. For a highly skewed distribution, the difference between the two is appreciable.

Variance   The *variance*, denoted Var(T), is a measure of the spread of a life distribution, defined as

$$\text{Var}(T) = \int_{-\infty}^\infty [t - E(T)]^2 f(t)\, dt. \tag{2.18}$$

For the exponential distribution, the variance is

$$\text{Var}(T) = \int_0^\infty \left(t - \frac{1}{\lambda}\right)^2 \lambda \exp(-\lambda t)\, dt = \left(\frac{1}{\lambda}\right)^2. \tag{2.19}$$

In many applications we prefer using the standard deviation, $\sqrt{\text{Var}(T)}$, since it has the same unit as T. Generally, the standard deviation is used together with MTTF.

2.4   EXPONENTIAL DISTRIBUTION

The $f(t)$, $F(t)$, $R(t)$, $h(t)$, $H(t)$, $E(T)$, and Var(T) of the exponential distribution are given in (2.3), (2.4), (2.6), (2.8), (2.10), (2.17), and (2.19), respectively. In these equations, λ is called the *hazard rate* or *failure rate*. The $f(t)$, $F(t)$, $R(t)$, $h(t)$, and $H(t)$ are shown graphically in Figure 2.5, where $\theta = 1/\lambda$ is the mean time. As shown in the figure, when $t = \theta$, $F(t) = 0.632$ and $R(t) = 0.368$. The first derivative of $R(t)$ with respect to t evaluated at time 0 is $R'(0) = -1/\theta$. This implies that the tangent of $R(t)$ at time 0 intersects the t axis at θ. The tangent is shown in the $R(t)$ plot in Figure 2.5. The slope of $H(t)$ is $1/\theta$, illustrated in the $H(t)$ plot in Figure 2.5. These properties are useful for estimating the parameter λ or θ using graphical approaches such as the probability plot (Chapter 7) or the cumulative hazard plot (Chapter 11).

The exponential distribution possesses an important property: that the hazard rate is a constant. The constant hazard rate indicates that the probability of a surviving product failing in the next small interval of time is independent of time. That is, the amount of time already exhausted for an exponential product has no effects on the remaining life of the product. Therefore, this characteristic is also called the *memoryless property*. Mathematically, the property can be expressed as

$$\Pr(T > t + t_0 | T > t_0) = \Pr(T > t), \tag{2.20}$$

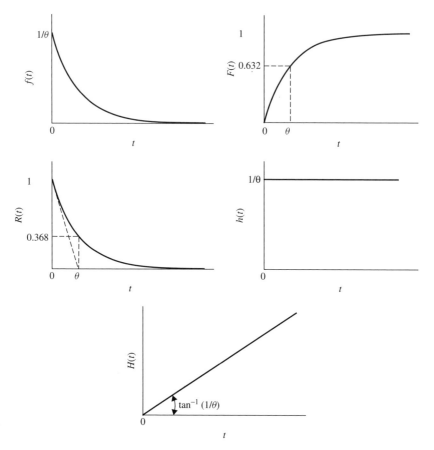

FIGURE 2.5　Exponential $f(t)$, $F(t)$, $R(t)$, $h(t)$ and $H(t)$

where t_0 is the time already exhausted. Equation (2.20) is proved as follows:

$$\Pr(T > t + t_0 | T > t_0) = \frac{\Pr(T > t + t_0 \cap T > t_0)}{\Pr(T > t_0)} = \frac{\Pr(T > t + t_0)}{\Pr(T > t_0)}$$

$$= \frac{\exp[-\lambda(t + t_0)]}{\exp(-\lambda t_0)} = \exp(-\lambda t) = \Pr(T > t).$$

The exponential distribution may be appropriate for modeling random failures. This can be explained by the following arguments. Random failures are usually caused by external shocks such as an unexpected change in load. External shocks usually can be modeled using the Poisson process. If every shock causes a failure, the product life can be approximated using the exponential distribution. The arguments suggest that the exponential distribution may be adequate for a failure process described by the threshold strength model, where a failure occurs when the stress is greater than the threshold strength. On the other hand, the arguments imply that the exponential distribution is inappropriate for failures due to degradation or wear out.

The exponential distribution is widely used and is especially popular in modeling the life of some electronic components and systems. For example, Murphy et al. (2002) indicate that the exponential distribution adequately fits the failure data of a wide variety of systems, such as radar, aircraft and spacecraft electronics, satellite constellations, communication equipment, and computer networks. The exponential distribution is also deemed appropriate for modeling the life of electron tubes, resistors, and capacitors (see, e.g., Kececioglu, 1991; Meeker and Escobar, 1998). However, the author's test data suggest that the exponential distribution does not adequately fit the life of several types of capacitors and resistors, such as electrolytic aluminum and tantalum capacitors and carbon film resistors. The Weibull distribution is more suitable. The failure of these components is driven primarily by performance degradation; for example, an electrolytic aluminum capacitor usually fails because of exhaustion of the electrolyte.

The exponential distribution is often mistakenly used because of its mathematical tractability. For example, the reliability prediction MIL-HDBK-217 series assumes that the life of electronic and electromechanical components follows an exponential distribution. Because of the exponential assumption and numerous other deficiencies, this handbook has been heavily criticized and is no longer actively maintained by the owner [U.S. Department of Defense (U.S. DoD)]. Another common misuse lies in redundant systems comprised of exponential components. Such systems are nonexponentially distributed (see, e.g., Murphy et al. 2002).

2.5 WEIBULL DISTRIBUTION

The pdf of the Weibull distribution is

$$f(t) = \frac{\beta}{\alpha^\beta} t^{\beta-1} \exp\left[-\left(\frac{t}{\alpha}\right)^\beta\right], \qquad t > 0. \tag{2.21}$$

The Weibull cdf is

$$F(t) = 1 - \exp\left[-\left(\frac{t}{\alpha}\right)^\beta\right], \qquad t > 0. \tag{2.22}$$

The hazard function is

$$h(t) = \frac{\beta}{\alpha}\left(\frac{t}{\alpha}\right)^{\beta-1}, \qquad t > 0. \tag{2.23}$$

The cumulative hazard function is

$$H(t) = \left(\frac{t}{\alpha}\right)^\beta, \qquad t > 0. \tag{2.24}$$

The 100pth percentile is

$$t_p = \alpha[-\ln(1-p)]^{1/\beta}. \qquad (2.25)$$

The mean and variance are   *[handwritten: MTTF or MTBF or Expected Life]*

$$E(T) = \alpha\Gamma\left(1+\frac{1}{\beta}\right),$$ *[handwritten: Function not Var]*

$$\mathrm{Var}(T) = \alpha^2\left[\Gamma\left(1+\frac{2}{\beta}\right) - \Gamma^2\left(1+\frac{1}{\beta}\right)\right], \qquad (2.26)$$

respectively, where $\Gamma(\cdot)$ is the gamma function, defined by

$$\Gamma(x) = \int_0^\infty z^{x-1}\exp(-z)dz.$$

In the Weibull formulas above, β is the shape parameter and α is the characteristic life; both are positive. α also called the *scale parameter*, equals the 63.2th percentile (i.e., $\alpha = t_{0.632}$). α has the same unit as t: for example, hours, miles, and cycles. The generic form of the Weibull distribution has an additional parameter, called the location parameter. Kapur and Lamberson (1977), Nelson (1982) and Lawless (2002), among others, present the three-parameter Weibull distribution.

To illustrate the Weibull distribution graphically, Figure 2.6 plots $f(t)$, $F(t)$, $h(t)$, and $H(t)$ for $\alpha = 1$ and $\beta = 0.5$, 1, 1.5, and 2. As shown in the figure, the shape parameter β determines the shape of the distribution. When $\beta < 1$ ($\beta > 1$), the Weibull distribution has a decreasing (increasing) hazard rate. When $\beta = 1$, the Weibull distribution has a constant hazard rate and is reduced to the exponential distribution. When $\beta = 2$, the hazard rate increases linearly with t, as shown in the $h(t)$ plot in Figure 2.6. In this case the Weibull distribution is called the *Rayleigh distribution*, described in, for example, Elsayed (1996). The linearly increasing hazard rate models the life of some mechanical and electromechanical components, such as valves and electromagnetic relays, whose failure is dominated by mechanical or electrical wear out.

It can be seen that the Weibull distribution is very flexible and capable of modeling each region of a bathtub curve. It is the great flexibility that makes the Weibull distribution widely applicable. Indeed, in many applications, it is the best choice for modeling not only the life but also the product's properties, such as the performance characteristics.

Example 2.1   The life of an automotive component is Weibull with $\alpha = 6.2 \times 10^5$ miles and $\beta = 1.3$. Calculate $F(t)$, $R(t)$, and $h(t)$ at the end of the warranty period (36,000 miles) and $t_{0.01}$.

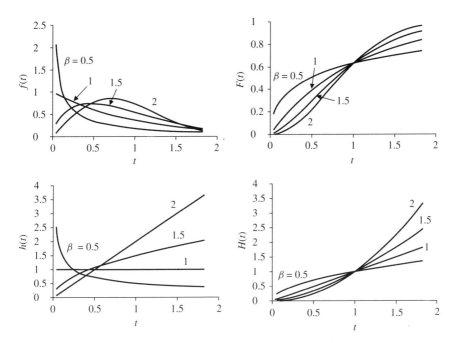

FIGURE 2.6 Weibull $f(t)$, $F(t)$, $h(t)$ and $H(t)$ for $\alpha = 1$

SOLUTION From (2.22) the probability of failure at 36,000 miles is

$$F(36,000) = 1 - \exp\left[-\left(\frac{36,000}{6.2 \times 10^5}\right)^{1.3}\right] = 0.024.$$

This indicates that 2.4% of the component population will fail by the end of warranty period. The reliability at 36,000 miles is

$$R(36,000) = 1 - 0.024 = 0.976.$$

That is, 97.6% of the component population will survive the warranty period. From (2.23), the hazard rate at 36,000 miles is

$$h(36,000) = \left(\frac{1.3}{6.2 \times 10^5}\right)\left(\frac{36,000}{6.2 \times 10^5}\right)^{1.3-1} = 0.89 \times 10^{-8} \text{ failures per mile.}$$

Because $\beta > 1$, the hazard rate increases with mileage. From (2.25), $t_{0.01}$, the mileage by which 1% of the population will fail, is given by

$$t_{0.01} = 6.2 \times 10^5 \times [-\ln(1 - 0.01)]^{1/1.3} = 18,014 \text{ miles.}$$

2.6   MIXED WEIBULL DISTRIBUTION

A *mixed distribution* comprises two or more distributions. Mixture arises when the population of interest contains two or more nonhomogeneous subpopulations. Such cases occur frequently in practice. A common example is that a good subpopulation is mixed with a substandard subpopulation due to manufacturing process variation and material flaws. The substandard subpopulation fails in early time, but the good one survives considerably longer. In addition to the mixture of good and bad products, a manufacturing process fed with components from different suppliers usually produces nonhomogeneous subpopulations. The use condition, such as environmental stresses or usage rate, is also a factor contributing to the mixture of life distributions. When a homogeneous population of products is operated at different conditions, the life of the products usually has multiple modes. This case is often seen in the warranty data analysis. The automotive industry is a frequent witness to mixed distributions, due to all the factors described above. Such distributions are especially common on new vehicle lines, of which the assembly processes may be on the learning curve early in the launch time. Then the processes become stable and stay under control as the inherent problems are corrected as a result of ongoing reliability improvement programs such as six-sigma projects, early warranty analysis, and fleet testing.

It is always desirable and valuable to distinguish between the nonhomogeneous subpopulations and analyze them individually, as discussed in Evans (2000). Sometimes, however, it is difficult or impossible to make a separation. For example, an automobile manufacturer has no way to know if a vehicle sold in a region continues operation in the same region; then the mixture of vehicles in different locations arises. In this situation, a mixed distribution is useful or essential.

A mixture of two distributions is usually of most interest. The mixed distribution is often bimodal and finds extensive applications in the development of burn-in or screen plans. Some examples are given in Jensen and Peterson (1982), Kececioglu and Sun (1995), Chan and Englert (2001), and G. Yang (2002). The bimodal Weibull distribution is perhaps the most common mixed distribution in practice because of its inherent flexibility. In Chapter 10, module-level screening models are based on the bimodal Weibull distribution. The pdf of the mixed Weibull distributions of two subpopulations $f_1(t)$ and $f_2(t)$ is

$$f(t) = \rho f_1(t) + (1 - \rho) f_2(t)$$

$$= \rho \frac{\beta_1}{\alpha_1^{\beta_1}} t^{\beta_1 - 1} \exp\left[-\left(\frac{t}{\alpha_1}\right)^{\beta_1} \right]$$

$$+ (1 - \rho) \frac{\beta_2}{\alpha_2^{\beta_2}} t^{\beta_2 - 1} \exp\left[-\left(\frac{t}{\alpha_2}\right)^{\beta_2} \right], \qquad t > 0, \qquad (2.27)$$

$R(t) = \rho R_1(t) + (1-\rho) R_2 t$

$h(t) = \omega(t) h_1(t) + (1 - \omega(t)) h_2(t)$ $\left[\omega(t) = \rho \dfrac{R_1(t)}{R_t(t)} \right.$

where ρ is the fraction of the first subpopulation accounting for the entire population and β_i and α_i $(i = 1, 2)$ are the shape parameter and characteristic life of subpopulation i. The associated cdf is

$$F(t) = \rho F_1(t) + (1 - \rho) F_2(t)$$

$$= 1 - \rho \exp\left[-\left(\frac{t}{\alpha_1}\right)^{\beta_1}\right] - (1 - \rho) \exp\left[-\left(\frac{t}{\alpha_2}\right)^{\beta_2}\right], \quad t > 0. \quad (2.28)$$

Example 2.2   An automotive component population produced in the first two months of production contains 8% defective units. The mileage to failure of the components has a bimodal Weibull distribution with $\beta_1 = 1.3$, $\alpha_1 = 12{,}000$ miles, $\beta_2 = 2.8$, and $\alpha_2 = 72{,}000$ miles. Plot $f(t)$ and $F(t)$, and calculate the probability of failure at the end of the warranty period (36,000 miles).

SOLUTION $f(t)$ is obtained by substituting the data into (2.27), plotted in Figure 2.7. $F(t)$ is calculated from (2.28) and shown in Figure 2.8. The probability of failure at 36,000 miles is

$$F(36{,}000) = 1 - 0.08 \times \exp\left[-\left(\frac{36{,}000}{12{,}000}\right)^{1.3}\right]$$

$$- 0.92 \times \exp\left[-\left(\frac{36{,}000}{72{,}000}\right)^{2.8}\right] = 0.202.$$

This indicates that 20.2% of the component population produced in the first two months of production will fail by 36,000 miles.

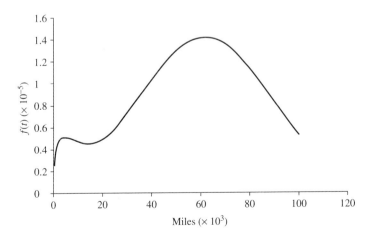

FIGURE 2.7 $f(t)$ of the mixed Weibull distribution

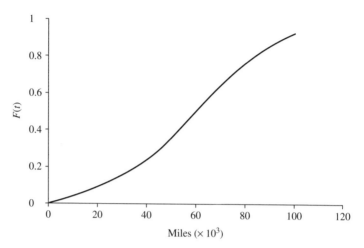

FIGURE 2.8 $F(t)$ of the mixed Weibull distribution

identical components (handwritten)

2.7   SMALLEST EXTREME VALUE DISTRIBUTION

In terms of mechanical, electrical, or thermal structure, some products, such as cable insulators and printed circuitry, can be considered to consist of a series of identical components. These products fail when the weakest components do not function; the life of the weakest components determines the product life. The smallest extreme value distribution may be suitable for modeling the life of such products. The pdf of this distribution is

$$f(t) = \frac{1}{\sigma} \exp\left(\frac{t-\mu}{\sigma}\right) \exp\left[-\exp\left(\frac{t-\mu}{\sigma}\right)\right], \qquad -\infty < t < \infty. \quad (2.29)$$

The cdf is

$$F(t) = 1 - \exp\left[-\exp\left(\frac{t-\mu}{\sigma}\right)\right], \qquad -\infty < t < \infty. \quad (2.30)$$

The hazard function is

$$h(t) = \frac{1}{\sigma} \exp\left(\frac{t-\mu}{\sigma}\right), \qquad -\infty < t < \infty. \quad (2.31)$$

The cumulative hazard function is

$$H(t) = \exp\left(\frac{t-\mu}{\sigma}\right), \qquad -\infty < t < \infty. \quad (2.32)$$

μ = location parameter (handwritten)
σ = scale parameter (handwritten)

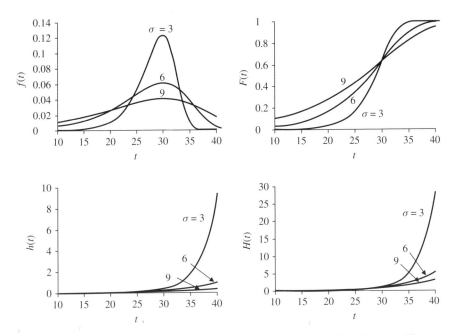

FIGURE 2.9 Smallest extreme value $f(t)$, $F(t)$, $h(t)$ and $H(t)$ for $\mu = 30$

The cumulative hazard function is simply the hazard function times σ. The mean, variance, and $100p$th percentile are

$$E(T) = \mu - 0.5772\sigma, \qquad \text{Var}(T) = 1.645\sigma^2, \qquad t_p = \mu + u_p\sigma, \qquad (2.33)$$

respectively, where $u_p = \ln[-\ln(1 - p)]$.

In the formulas above, μ is the location parameter and σ is the scale parameter; $-\infty < \mu < \infty$ and $\sigma > 0$. They have the same unit as t. When $\mu = 0$ and $\sigma = 1$, the distribution is called the *standard smallest extreme value distribution*, and the hazard rate equals the cumulative hazard rate. Figure 2.9 plots the distribution for $\mu = 30$ and various values of σ. The hazard rate plot indicates that the distribution may be suitable for products whose hazard rate increases rapidly with increased time in service, due to excessive degradation or wear out. However, the distribution is seldom used to model product life in reality because it allows the life to be negative, and the probability of failure is greater than zero when the time is zero. Instead, the distribution is very useful in analytical studies when the Weibull distribution is involved because of their relation. If y is Weibull with shape parameter β and characteristic life α, $t = \ln(y)$ has the smallest extreme value distribution with $\sigma = 1/\beta$ and $\mu = \ln(\alpha)$. This relationship is used in Chapter 7 to develop accelerated life test plans.

2.8 NORMAL DISTRIBUTION

The pdf of the normal distribution is

$$f(t) = \frac{1}{\sqrt{2\pi}\sigma} \exp\left[-\frac{(t-\mu)^2}{2\sigma^2}\right], \qquad -\infty < t < \infty. \tag{2.34}$$

The normal cdf is

$$F(t) = \int_{-\infty}^{t} \frac{1}{\sqrt{2\pi}\sigma} \exp\left[-\frac{(y-\mu)^2}{2\sigma^2}\right] dy, \qquad -\infty < t < \infty. \tag{2.35}$$

The mean and variance are

$$E(T) = \mu \quad \text{and} \quad \text{Var}(T) = \sigma^2, \tag{2.36}$$

respectively. The hazard function and cumulative hazard function can be obtained from (2.7) and (2.9), respectively. They cannot be simplified and thus are not given here.

When T has a normal distribution, it is usually indicated by $T \sim N(\mu, \sigma^2)$. μ is the location parameter and σ is the scale parameter. They are also the population mean and standard deviation as shown in (2.36) and have the same unit as t, where $-\infty < \mu < \infty$ and $\sigma > 0$. Figure 2.10 plots the normal distribution for $\mu = 15$ and various values of σ.

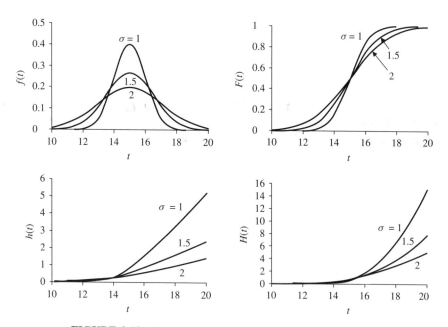

FIGURE 2.10 Normal $f(t)$, $F(t)$, $h(t)$ and $H(t)$ for $\mu = 15$

$$h(t) = \frac{f(t)}{R(t)} = \frac{\phi\left(\frac{t-\mu}{\sigma}\right)/\sigma}{1 - \Phi\left(\frac{t-\mu}{\sigma}\right)}$$

When $\mu = 0$ and $\sigma = 1$, the normal distribution is called the *standard normal distribution*. Then the pdf becomes

$z =$ slaidard normal variance

$$f(t) = \phi(z) = \frac{1}{\sqrt{2\pi}} \exp\left(-\frac{z^2}{2}\right), \qquad -\infty < z < \infty. \qquad (2.37)$$

The cdf of the standard normal distribution is

$$F(t) = \Phi(z) = \int_{-\infty}^{z} \frac{1}{\sqrt{2\pi}} \exp\left(-\frac{y^2}{2}\right) dy, \qquad -\infty < z < \infty. \qquad (2.38)$$

$\Phi(z)$ is tabulated in, for example, Lewis (1987) and Nelson (1990, 2004). Many commercial software packages such as Minitab and Microsoft Excel are capable of doing the calculation. With the convenience of $\Phi(z)$, (2.35) can be written as

$$F(t) = \Phi\left(\frac{t-\mu}{\sigma}\right), \qquad -\infty < t < \infty. \quad = \Phi\left(\frac{t-\mu}{\sigma}\right) \qquad (2.39)$$

$f(t) = \phi\left(\frac{t-\mu}{\sigma}\right)/\sigma$

The 100pth percentile is

$$t_p = \mu + z_p \sigma, \qquad z_p = \Phi^{-1}(p) \qquad (2.40)$$

where z_p is the 100pth percentile of the standard normal distribution [i.e., $z_p = \Phi^{-1}(p)$]. As a special case, $t_{0.5} = \mu$; that is, the median equals the mean.

The normal distribution has a long history in use because of its simplicity and symmetry. The symmetric bell shape describes many natural phenomena, such as the height and weight of newborn babies. The distribution is considerably less common in modeling life because it allows the random variable to be negative. It may be suitable for some product properties if the coefficient of variation (σ/μ) is small. The normal distribution is very useful in statistical analysis. For example, the analysis of variance presented in Chapter 5 assumes that the data are normally distributed. In analytic reliability studies, the normal distribution is often needed when the lognormal distribution is involved. The relationship between the normal and lognormal distribution is described in Section 2.9.

The normal distribution has an important property frequently utilized in reliability design. If $X_1, X_2, \ldots, X_n$ are independent random variables and normally distributed with (μ_i, σ_i^2) for $i = 1, 2, \ldots, n$, then $X = X_1 + X_2 + \cdots + X_n$ has a normal distribution with mean and variance

$$\mu = \sum_{i=1}^{n} \mu_i \quad \text{and} \quad \sigma^2 = \sum_{i=1}^{n} \sigma_i^2.$$

Example 2.3 An electronic circuit contains three resistors in series. The resistances (say, R_1, R_2, R_3) in ohms of the three resistors can be modeled with the normal distributions $R_1 \sim N(10, 0.3^2)$, $R_2 \sim N(15, 0.5^2)$, and $R_3 \sim N(50, 1.8^2)$.

Calculate the mean and standard deviation of the total resistance and the probability of the total resistance being within the tolerance range $75 \pm 5\%$.

SOLUTION   The total resistance R_0 of the three resistors is $R_0 = R_1 + R_2 + R_3$. The mean of the total resistance is $\mu = 10 + 15 + 50 = 75\ \Omega$. The standard deviation of the total resistance is $\sigma = (0.3^2 + 0.5^2 + 1.8^2)^{1/2} = 1.89\ \Omega$. The probability of the total resistance being within $75 \pm 5\%$ is

$$\Pr(71.25 \le R_0 \le 78.75) = \Phi\left(\frac{78.75 - 75}{1.89}\right) - \Phi\left(\frac{71.25 - 75}{1.89}\right)$$

$$= 0.976 - 0.024 = 0.952.$$

2.9 LOGNORMAL DISTRIBUTION

The pdf of the lognormal distribution is

$$f(t) = \frac{1}{\sqrt{2\pi}\,\sigma t} \exp\left\{-\frac{[\ln(t) - \mu]^2}{2\sigma^2}\right\} = \frac{1}{\sigma t}\phi\left[\frac{\ln(t) - \mu}{\sigma}\right], \qquad t > 0, \quad (2.41)$$

where $\phi(\cdot)$ is the standard normal pdf. The lognormal cdf is

$$F(t) = \int_0^t \frac{1}{\sqrt{2\pi}\,\sigma y} \exp\left\{-\frac{[\ln(y) - \mu]^2}{2\sigma^2}\right\} dy = \Phi\left[\frac{\ln(t) - \mu}{\sigma}\right], \qquad t > 0, \tag{2.42}$$

where $\Phi(\cdot)$ is the standard normal cdf. The $100p$th percentile is

$$t_p = \exp(\mu + z_p\sigma), \tag{2.43}$$

where z_p is the $100p$th percentile of the standard normal distribution. An important special case of (2.43) is the median $t_{0.5} = \exp(\mu)$. The mean and variance of T are, respectively,

$$E(T) = \exp(\mu + 0.5\sigma^2) \quad \text{and} \quad \text{Var}(T) = \exp(2\mu + \sigma^2)[\exp(\sigma^2) - 1]. \tag{2.44}$$

When T has a lognormal distribution, it is usually indicated by $\text{LN}(\mu,\sigma^2)$. μ is the scale parameter and σ is the shape parameter; $-\infty < \mu < \infty$ and $\sigma > 0$. It is important to note that unlike the normal distribution, μ and σ here are not the mean and standard deviation of T. The relationships between the parameters and the mean and standard deviation are given in (2.44). However, μ and σ are

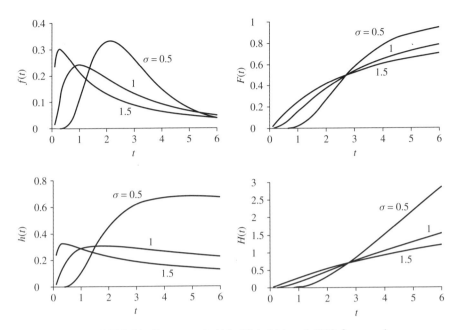

FIGURE 2.11 Lognormal $f(t)$, $F(t)$, $h(t)$ and $H(t)$ for $\mu = 1$

the mean and standard deviation of $\ln(T)$ because $\ln(T)$ has a normal distribution when T is lognormal.

The lognormal distribution is plotted in Figure 2.11 for $\mu = 1$ and various values of σ. As the $h(t)$ plot in Figure 2.11 shows, the hazard rate in not monotone. It increases and then decreases with time. The value of t at which the hazard rate is the maximum is derived below.

The lognormal hazard rate is given by

$$h(t) = \frac{\phi}{\sigma t (1 - \Phi)}, \tag{2.45}$$

where ϕ and Φ are the function of $[\ln(t) - \mu]/\sigma$. Taking the natural logarithm on both sides of (2.45) gives

$$\ln[h(t)] = \ln(\phi) - \ln(\sigma) - \ln(t) - \ln(1 - \Phi). \tag{2.46}$$

Equate to zero the first derivative of (2.46) with respect to t. Then

$$\frac{\ln(t) - \mu}{\sigma} + \sigma - \frac{\phi}{1 - \Phi} = 0. \tag{2.47}$$

The solution of t, say t^*, to (2.47) is the value at which the hazard rate is maximum. Before t^*, the hazard increases; after t^*, the hazard rate decreases.

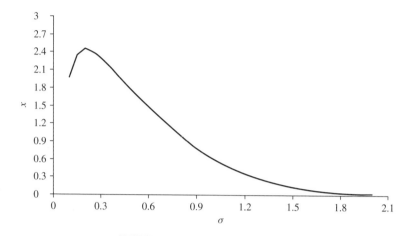

FIGURE 2.12   Plot of x versus σ

There is no closed form for t^*; it is calculated with a numerical method. Let

$$t^* = x \exp(\mu). \tag{2.48}$$

Figure 2.12 plots x versus σ. From the chart the value of x is read for a specific σ value. Then it is substituted into (2.48) to calculate t^*. It can be seen that the hazard rate decreases in practically an entire lifetime when $\sigma > 2$. The hazard rate is in the increasing trend for a long period of time when σ is small, especially at the neighbor of 0.2. The time at which the hazard rate begins to decrease frequently interests design engineers and warranty analysts. If this time occurs after a very long time in service, the failure process during the useful life is dominated by degradation or wear out.

The lognormal distribution is useful in modeling the life of some electronic products and metals due to fatigue or crack. The increasing hazard rate in early time usually fits the life of the freak subpopulation, and the decreasing hazard rate describes the main subpopulation. The lognormal distribution is also frequently used to model the use of products. For example, as shown in Lawless et al. (1995), M. Lu (1998), and Krivtsov and Frankstein (2004), the accumulated mileages of an automobile population at a given time in service can be approximated using a lognormal distribution. This is illustrated in the following example and studied further in Chapter 11.

Example 2.4   The warranty plan of a car population covers 36 months in service or 36,000 miles, whichever comes first. The accumulated mileage U of the car population by a given month in service can be modeled with the lognormal distribution with scale parameter $6.5 + \ln(t)$ and shape parameter 0.68, where t is the months in service of the vehicles. Calculate the population fraction exceeding 36,000 miles by 36 months.

SOLUTION   From (2.42) the probability that a car exceeds 36,000 miles by 36 months is

$$\Pr(U \geq 36{,}000) = 1 - \Phi\left[\frac{\ln(36{,}000) - 6.5 - \ln(36)}{0.68}\right] = 0.437.$$

That is, by the end of warranty time, 43.7% of the vehicles will be leaving the warranty coverage by exceeding the warranty mileage limit.

In this section, lognormal distribution is defined in terms of the natural logarithm (base e). Base 10 logarithm can be used, but its use is less common in practice, especially in analytical studies.

PROBLEMS

2.1 Define the reliability of a product of your choice, and state the failure criteria for the product.

2.2 Explain the concepts of hard failure and soft failure. Give three examples of each.

2.3 Select the metrics to measure the reliability of the product chosen in Problem 2.1, and justify your selection.

2.4 Explain the causes for early failures, random failures, and wear-out failures. Give examples of methods for reducing or eliminating early failures.

2.5 The life of an airborne electronic subsystem can be modeled using an exponential distribution with a mean time of 32,000 hours.

   (a) Calculate the hazard rate of the subsystem.
   (b) Compute the standard deviation of life.
   (c) What is the probability that the subsystem will fail in 16,000 hours?
   (d) Calculate the 10th percentile.
   (e) If the subsystem survives 800 hours, what is the probability that it will fail in the following hour? If it survives 8000 hours, compute this probability. What can you conclude from these results?

2.6 The water pump of a car can be described by a Weibull distribution with shape parameter 1.7 and characteristic life 265,000 miles.

   (a) Calculate the population fraction failing by the end of the warranty mileage limit (36,000 miles).
   (b) Derive the hazard function.
   (c) If a vehicle survives 36,000 miles, compute the probability of failure in the following 1000 miles.
   (d) Calculate B_{10}.

2.7 An electronic circuit has four capacitors connected in parallel. The nominal capacitances of the four are 20, 80, 30, and 15 μF. The tolerance of each capacitor is ± 10%. The capacitance can be approximated using a normal distribution with mean equal to the nominal value and the standard deviation equal to one-sixth of the two-side tolerance. The total capacitance is the sum of the four. Calculate the following:

(a) The mean and standard deviation of the total capacitance.

(b) The probability of the total capacitance being greater than 150 μF.

(c) The probability of the total capacitance being within the range 146 ± 10%.

2.8 The time to failure (in hours) of a light-emitting diode can be approximated by a lognormal distribution with $\mu = 12.3$ and $\sigma = 1.2$.

(a) Plot the hazard function.

(b) Determine the time at which the hazard rate begins to decrease.

(c) Compute the MTTF and standard deviation.

(d) Calculate the reliability at 15,000 hours.

(e) Estimate the population fraction failing in 50,000 hours.

(f) Compute the cumulative hazard rate up to 50,000 hours.

3

RELIABILITY PLANNING AND SPECIFICATION

3.1 INTRODUCTION

Today's competitive business environment requires manufacturers to design, develop, test, manufacture, and deploy higher-reliability products in less time at lower cost. Reliability, time to market, and cost are three critical factors that determine if a product is successful in the marketplace. To maintain and increase market share and profitability, manufacturers in various industrial sectors have been making every effort to improve competitiveness in terms of these three factors. Reliability techniques are known powerful tools for meeting the challenges. To maximize effectiveness, individual reliability techniques are orchestrated to create reliability programs, which are implemented throughout the product life cycle. Recognizing the benefits from reliability programs, many large-scale manufacturers, such as Ford Motor Company, General Electric, and IBM, have established and continue to enhance reliability programs suitable for their products. The past few years have also witnessed the development and implementation of various programs in a large number of midsized and small-scale organizations.

A reliability program should begin at the beginning of a product life cycle, preferably in the product planning stage. Planning reliability simultaneously with product features, functionality, cost, and other factors ensures that customer expectations for reliability are addressed in the first place. Concurrent planning enables reliability to be considered under constraints of other requirements, and vice versa. Reliability planning includes understanding customer expectations

Life Cycle Reliability Engineering, by Guangbin Yang
Copyright © 2007 John Wiley & Sons, Inc.

and competitors' positions, specification of the reliability target, determination of reliability tasks to achieve the target, and assurance of resources needed by the tasks.

In this chapter we describe the quality function deployment method for understanding customer needs and translating the needs into engineering design requirements. Then we present methods for specifying a reliability target based on customer needs, warranty cost objectives, and total cost minimization. Also discussed is the development of reliability programs for achieving the reliability target.

3.2 CUSTOMER EXPECTATIONS AND SATISFACTION

Customers' choice determines the market share of a manufacturer. Loss of customers indicates erosion of market share and profitability. Thus, customer satisfaction has become a critical business objective and the starting point of product design. To satisfy customer demands to the greatest extent, manufacturers must understand what customers expect and how to address their expectations throughout the product life cycle. As an integral part of the design process, reliability planning and specification should be driven by customer needs. In this section we present the quality function deployment method, which evaluates customer needs and translates the needs into engineering design requirements and production control plans.

3.2.1 Levels of Expectation and Satisfaction

Customer expectations for a product can be classified into three types: basic wants, performance wants, and excitement wants. *Basic wants* describe customers' most fundamental expectations for the functionality of a product. Customers usually do not express these wants and assume that the needs will be satisfied automatically. Failure to meet these needs will cause customers to be seriously dissatisfied. Examples of such wants are cars being able to start and heaters being able to heat. *Performance wants* are customers' spoken expectations. Customers usually speak out for the needs and are willing to pay more to meet the expectations. A product that better satisfies the performance wants will achieve a higher degree of customer satisfaction. For example, good fuel economy and responsive steering are two typical performance wants on cars. Reliability is also a performance want; meeting customer expectation for reliability increases customer satisfaction significantly. *Excitement wants* represent potential needs whose satisfaction will surprise and delight customers. Customers do not realize that they have such needs, so these wants are unspoken. Examples of such needs related to a gasoline car are fuel economy of 45 miles per gallon and acceleration from 0 to 60 miles in 6 seconds.

Meeting the three types of customer expectations can lead to different levels of customer satisfaction. Satisfying basic wants does not appreciably increase

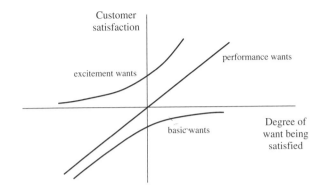

FIGURE 3.1   Customer expectations described by the Kano model

customer satisfaction; however, failure to do so greatly dissatisfies customers. Meeting performance wants increases customer satisfaction linearly. If these wants are not satisfied, customer satisfaction decreases linearly. When excitement wants are fulfilled, customer satisfaction increases exponentially. When they are not met, customer satisfaction stays flat. The relationships between customer satisfaction and the three types of customer expectations can be described by the Kano model, shown in Figure 3.1. The relationships indicate that manufacturers must satisfy all three types of wants to maximize customer satisfaction. It is worth noting that a higher-level customer want can degrade to a lower one as time goes. In particular, an excitement want may become a performance or a basic want, and a performance want may turn into a basic want. For example, an antilock braking system, commonly known as ABS, installed in automobiles was an excitement want in the early 1990s and is a performance want today. We can expect that it will be a standard feature and thus become a basic want in the near future.

3.2.2   QFD Process and Reliability Deployment

Global competition has pressed manufacturers to deliver products that meet customer demands, including reliability expectation, to the greatest extent. In Section 3.2.1 we discussed customer expectations and satisfaction that must be addressed in each phase of the product life cycle. A powerful method for doing this is *quality function deployment* (QFD), a structured tool that identifies important customer expectations and translates them into appropriate technical characteristics which are operational in design, verification, and production. Akao (1990) describes QFD in detail. QFD enables resources to be focused on meeting major customer demands. Figure 3.2 shows the structure of a QFD, which is often called a *house of quality* because of its shape. A house of quality contains a customer (horizontal) axis and a technical (vertical) axis. The *customer axis* describes what customers want, the importance of the wants, and the competitive performance. The customer wants component is often referred to as WHAT,

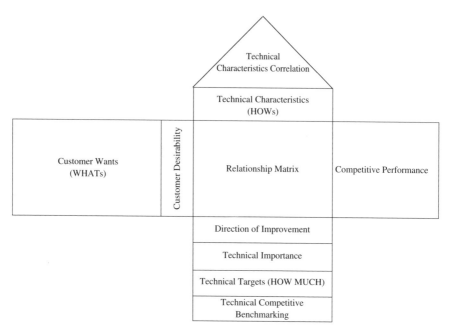

FIGURE 3.2   Typical house of quality

meaning what is to be addressed. The *technical axis* explains the technical characteristics that affect customer satisfaction directly for one or more customer expectations. Also on the technical axis are the correlations, importance, and targets of the technical characteristics and technical competitive benchmarking. The technical characteristics component is often referred to as HOW, meaning how to address WHAT; then technical targets are accordingly called HOW MUCH. The interrelationships between customer wants and technical characteristics are evaluated in the relationship matrix.

The objective of QFD is to translate customer wants, including reliability expectation, into operational design characteristics and production control variables. This can be done by deploying the houses of quality in increasing detail. In particular, customer wants and reliability demands are converted to technical characteristics through the first house of quality. Reliability is usually an important customer need and receives a high importance rating. The important technical characteristics, which are highly correlated to reliability demand, among others, are cascaded to design parameters at the part level through the second house of quality. The design parameters from this step of deployment should be closely related to reliability and can be used in subsequent robust design (Chapter 5) and performance degradation analysis (Chapter 8). Critical design parameters are then deployed in process planning to determine process parameters through the third house of quality. Control of the process parameters identified directly minimizes unit-to-unit variation (an important noise factor in robust design) and reduces

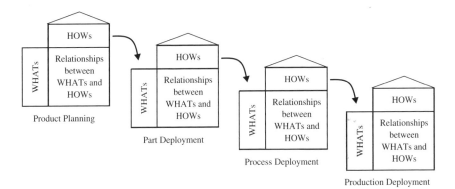

FIGURE 3.3 Quality function deployment process

infant mortality and variation in degradation rates. The fourth house of quality is then used to translate the process parameters into production requirements. The deployment process is illustrated in Figure 3.3. A complete QFD process consists of four phases: (1) product planning, (2) part deployment, (3) process deployment, and (4) production deployment. The four phases are described in detail in the following subsections.

3.2.3 Product Planning Phase

In this phase, customer expectations collected from different sources, which may include market research, customer complaints, and comparison with competitors, are translated into specified technical characteristics using the first house of quality. The steps in developing the house of quality are described below and illustrated with an example of an automobile windshield wiper system. The example is only typical for such a system and is not intended to be exhaustive.

1. State what customers want in the WHAT entries of the first house of quality. These customer expectations are usually nontechnical and fuzzy expressions. For example, customers may state their reliability wants as "long life," "never fail," and "very dependable." Technical tools such as affinity and tree diagrams may be used to group various customer requirements (Bossert, 1991). For an automobile windshield wiper system, customer expectations include high reliability, minimal operation noise, no residual water traces, no water film, and large wiping area, among others. These wants are listed in the WHAT entries of the quality house, as shown in Figure 3.4.

2. Determine customer desirability, which rates the desirability for each customer want relative to every other want. Various scaling approaches are used in practice, but none of them is theoretically sound. In the windshield wiper example, we use the analytic hierarchy process approach (Armacost et al., 1994), which rates importance levels on a scale of 1 to 9, where 9 is given to the

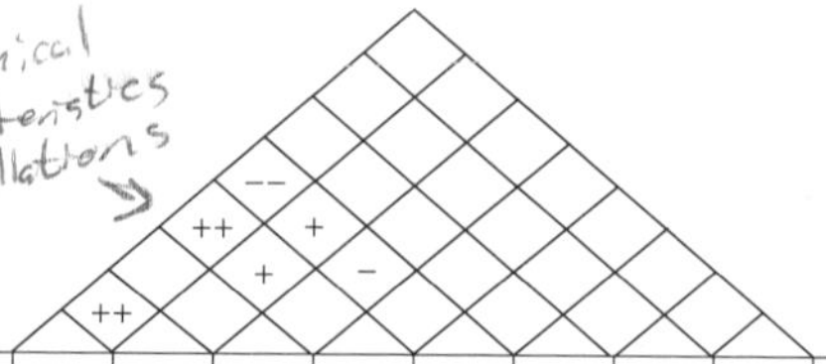

(Handwritten annotations:) Technical Characteristics Correlations →; Technical Characteristics (How's to address Whats); Customer Needs (Whats)

Legend (Technical Characteristics Correlations roof):
++ : strongly positive
+ : positive
− : negative
−− : strongly negative

	Customer Desirability	Arm Rotation Torque	Motor Load	Motor Operating Angle	Blade-to-Glass Friction Coefficient	Arm Length	Low Motor Speed	High Motor Speed	...	Competitive Performance 1	2	3	4	5
High Reliability	9	3	9	1	3		1	1	...		A	B D	C	
Minimal Noise	5	9	3		3	3	1	3				A	B	C D
No Residual Water Traces	7	3	3		3	3	3	3		A	C	B D		
No Water Film	9		3		3	3	3	3			A	C D	B	
Large Wiping Area	3			9		3						C D	A	B
... ...	...	...			...			...	...			...		...
Direction of Improvement		o	o	o	o	o	o	o	...					
Technical Importance		93	144	36	90	24	62	72	...					
Technical Targets		0.16	1.2	115	1.7	0.35	40	58	...					
Technical Competitive Benchmarking — A		0.21	1.5	113	1.6	0.35	45	62						
B		0.19	1.3	115	1.6	0.38	39	62						
C		0.15	1.2	110	1.8	0.34	42	56						
D		0.14	1.3	110	1.6	0.34	42	56						
Unit		Nm	Nm	deg		m	rpm	rpm	...					

A, B, C: Competing products

D: Prior-generation product

FIGURE 3.4 Example QFD for automobile windshield wiper system planning

extremely important level, 7 to strongly important, 5 to very important, 3 to important, and 1 to not important. Even numbers 2, 4, 6, and 8 are assigned to the importance levels in between. It is recommended that customer demand as to reliability receive a high importance rating (8 or higher), especially when products are safety related. In the wiper system example, customer expectations for high reliability and no water film reflect customer safety concern and thus are assigned the highest ratings. The customer desirability column of Figure 3.4 shows the ratings for all customer wants listed.

3. **Evaluate the competitive performance for major competing products and the prior-generation product.** The evaluation is accomplished by rating customer satisfaction as to each of the customer wants on a scale of 1 to 5, where 5 is assigned to the state of being completely satisfied, 4 to very satisfied, 3 to fairly well satisfied, 2 to somewhat dissatisfied, and 1 to very dissatisfied. The objective

of the evaluation is to assess the strengthes and weaknesses of the product being designed and to identify areas for improvement. Shown on the right-hand side of Figure 3.4 are the competitive performance ratings for three competitors (denoted A, B, and C) and the predecessor of this wiper system (denoted D).

4. List the technical characteristics that directly affect one or more customer wants on the customer axis. These characteristics should be measurable and controllable and define technically the performance of the product being designed. The characteristics will be deployed selectively to the other three houses of quality in subsequent phases of deployment. In this step, fault tree analysis, cause-and-effect diagrams, and test data analysis of similar products are helpful because the technical characteristics that strongly influence reliability may not be obvious. In the wiper system example, we have identified the technical characteristics that describe the motor, arm, blade, linkage, and other components. Some of them are listed in Figure 3.4.

5. Identify the interrelationships between customer wants and technical characteristics. The strength of relationship may be classified into three levels, where a rating of 9 is assigned to a strong relation, 3 to a medium relation, and 1 to a weak relation. Each technical characteristic must be interrelated to at least one customer want; one customer want must also be addressed by at least one technical characteristic. This ensures that all customer wants are concerned in the product planning, and all technical characteristics are established properly. The ratings of the relation strength for the wiper system are entered in the relationship matrix entries of the quality house, shown in Figure 3.4. It can be seen that the motor load is one of the technical characteristics that strongly affect system reliability.

6. Develop the correlations between technical characteristics and indicate them in the roof of the house of quality. The technical characteristics can have a positive correlation, meaning that the change of one technical characteristic in a direction affects another characteristic in the same direction. A negative correlation means otherwise. Four levels of correlation are used: a strongly positive correlation, represented graphically by ++; positive by +; negative by −; and strongly negative by −−. Correlations usually add complexity to product design and would result in trade-off decisions in selecting technical targets if the correlations are negative. Correlations among the technical characteristics of the wiper system appear in the roof of the quality house, as shown in Figure 3.4.

7. Determine the direction of improvement for each technical characteristic. There are three types of characteristics: larger-the-better, nominal-the-best, and smaller-the-better (Chapter 5), which are represented graphically by ↑, ○, and ↓, respectively, in a house of quality. The direction is to maximize, set to target, or minimize the technical characteristic, depending on its type. The technical characteristics listed in Figure 3.4 are all nominal-the-best type.

8. Calculate ratings of technical importance. For a given technical characteristic, the values of the customer desirability index are multiplied by the corresponding strength ratings. The sum of the products is the importance rating of

the technical characteristic. The importance ratings allow the technical characteristics to be prioritized and thus indicate the significant characteristics that should be selected for further deployment. Characteristics with low values of rating may not need deployment to subsequent QFD phases. In the wiper system example, the importance rating of the motor load is $9 \times 9 + 5 \times 3 + 7 \times 3 + 9 \times 3 = 144$. Ratings of the listed technical characteristics are in the technical importance row, shown in Figure 3.4. The ratings indicate that the motor load is an important characteristic and should be deployed to lower levels.

9. Perform technical competitive benchmarking. Determine the measurement of each technical characteristic of the predecessor of the product as well as the competing products evaluated on the customer axis. The measurements should correlate strongly with the competitive performance ratings. Lack of correlation signifies inadequacy of the technical characteristics in addressing customer expectations. This benchmarking allows evaluation of the position of the predecessor relative to competitors from a technical perspective and assists in the development of technical targets. In the wiper system example, the measurements and units of the technical characteristics of the products under comparison are shown in Figure 3.4.

10. Determine a measurable target for each technical characteristic with inputs from the technical competitive benchmarking. The targets are established so that identified customer wants are fulfilled and the product being planned will be highly competitive in the marketplace. The targets of the technical characteristics listed for the wiper system are shown in Figure 3.4.

3.2.4   Part Deployment Phase

The product planning phase translates customer wants into technical characteristics, determines their target values that will make the product competitive in the marketplace, and identifies the significant technical characteristics that need to be expanded forward to the part deployment phase. To perform part deployment, outputs (i.e., the significant technical characteristics) from the product planning phase are carried over and become the WHATs of the second house of quality. In this phase, part characteristics are identified to address the WHATs. The steps for developing the second house of quality are similar to these for the first. Outputs from this phase include important part characteristics and their target values. These part characteristics are highly correlated to customer wants and are indicators of product reliability. These characteristics should be deployed to the next phase and serve as the control factors for robust design (Chapter 5). ReVelle et al. (1998) describe the use of QFD results for robust design.

3.2.5   Process Deployment Phase

The significant part characteristics identified in the part deployment phase are expanded further in the process deployment phase. In this phase the third house of quality is developed, where the WHATs are carried from the significant HOWs of the second house of quality, and the new HOWs are the process parameters to produce the WHATs at the target values. Outputs from this phase include critical process parameters and their target values, which should be deployed to the next phase for developing control plans. Deployment is critical in this phase, not only because this step materializes customer wants in production, but also because the process parameters and target values determined in this step have strong impacts on productivity, yield, cost, quality, and reliability.

3.2.6 Production Deployment Phase

The purpose of production deployment is to develop control plans to ensure that the target values of critical process parameters are achieved in production with minimum variation. To fulfill this purpose, the critical process parameters and their target values are expanded further through the fourth house of quality. Outputs from this phase include process control charts and quality control checkpoints for each process parameter. The requirements and instructions for implementing control plans should also be specified in this phase. Readers interested in statistical quality control may refer to Montgomery (2001a). From the reliability perspective, this phase is a critical step in the QFD process because effective control plans are needed to minimize infant mortality and unit-to-unit variation of parts and to improve field reliability and robustness.

3.3 RELIABILITY REQUIREMENTS

For most products, reliability is a performance need for which customers are willing to pay more. Meeting this expectation linearly increases customer satisfaction, which decreases linearly with failure to do so, as depicted in Figure 3.1. To win the war of sustaining and expanding market share, it is vital to establish competitive reliability requirements, which serve as the minimum goals and must be satisfied or exceeded through design and production. In this section we describe three methods of setting reliability requirements, which are driven by customer satisfaction, warranty cost objectives, and total cost minimization. Lu and Rudy (2000) describe a method for deriving reliability requirement from warranty repair objectives.

3.3.1 Statement of the Requirements

As defined in Chapter 2, reliability is the probability that a product performs its intended function without failure under specified conditions for a specified period of time. The definition consists of three essential elements: intended function, specified period of time, and specified conditions. Apparently, reliability changes as any of them varies. To be definitive, quantitative, and measurable, the reliability requirements must contain the three elements. We should avoid vague or incomplete requirements such as "no failure is allowed" or "the reliability goal is 95%."

Reliability requirements must define what constitutes a failure (i.e., the failure criteria). The definition may be obvious for a hard-failure product whose failure is the complete termination of function. For a soft-failure product, failure is defined in terms of performance characteristics crossing specified thresholds. As pointed out in Chapter 2, the thresholds are more or less subjective and often arguable. It thus is important to have all relevant parties involved in the specification process and concurring as to the thresholds. In a customer-driven market, the thresholds should closely reflect customer expectations. For example, a refrigerator may be said to have failed if it generates, say, 50 dB of audible noise, at which level 90% of customers are dissatisfied.

As addressed earlier, life can be measured in calendar time, usage, or other scales. The most appropriate life scale should be dictated by the underlying failure mechanism that governs the product failure process. For example, mechanical wear out is the dominant failure mechanism of a bearing, and the number of revolutions is the most suitable life measure because wear out develops only by rotation. The period of time specified should be stated on such a life scale. As discussed before, reliability is a function of time (e.g., calendar age and usage). Reliability requirements should define the time at which the reliability level is specified. For many commercial products, the specified time is the design life. Manufacturers may also stipulate other times of interest, such as warranty lengths and mission times.

Reliability is influenced largely by the use environment. For example, a resistor would fail much sooner at a high temperature than at ambient temperature. Reliability requirements should include the operating conditions under which the product must achieve the reliability specified. The conditions specified for a product should represent the customer use environment, which is known as the *real-world usage profile*. In designing subsystems within a product, this profile is translated into the local operating conditions, which in turn become the environmental requirements for the subsystems. Verification and validation tests intended to demonstrate reliability must correlate the test environments to the use conditions specified; otherwise, the test results will be unrealistic.

3.3.2 Customer-Driven Reliability Requirements

In a competitive business climate, meeting customer expectations is the starting point and driving force of all design, verification, and production activities. In the context of reliability planning, customer needs should be analyzed and further correlated with reliability requirements. Satisfying these requirements should lead to customer satisfaction. In this section we present a method of specifying reliability requirements based on customer expectations.

Suppose that n important customer wants, denoted $E_1, E_2, \ldots, E_n$, are linked to m independent critical performance characteristics, denoted $Y_1, Y_2, \ldots, Y_m$. The thresholds of $Y_1, Y_2, \ldots, Y_m$ are $D_1, D_2, \ldots, D_m$, respectively. The performance characteristics may be identified through QFD analysis. Each customer want is strongly interrelated to at least one performance characteristic; the strength

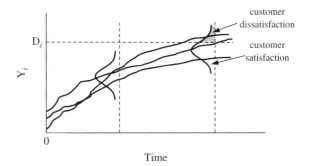

FIGURE 3.5 Correlation between performance degradation and customer dissatisfaction

of the relationship has a rating of 8 or 9, as designated in Section 3.2.3. We further assume that the performance characteristics degrade monotonically and that customers are dissatisfied when any of the characteristics crosses the threshold. The thresholds may or may not be the critical values that are specified from an engineering perspective. Often, customers have tighter values than those deemed functionally operational. The probability that a performance characteristic lies within the threshold measures the degree of customer satisfaction as to performance. The complement of the probability is the degree of customer dissatisfaction. Without loss of generality, Figure 3.5 shows the correlation between performance degradation and customer dissatisfaction for a smaller-the-better characteristic.

The degree of customer satisfaction on E_i can be written as

$$S_i = \prod_{j}^{m_i} \Pr(Y_j \leq D_j), \qquad i = 1, 2, \ldots, n, \tag{3.1}$$

where S_i is the degree of customer satisfaction on E_i, Y_j a performance characteristic that is interrelated to E_i with a strength rating of 8 or 9, and m_i the total number of such performance characteristics. Note that the index j may not be numerically consecutive.

If the minimum allowable customer satisfaction on E_i is S_i^*, we have

$$\prod_{j}^{m_i} \Pr(Y_j \leq D_j) = S_i^*, \qquad i = 1, 2, \ldots, n. \tag{3.2}$$

When the number of important customer wants equals the number of critical performance characteristics (i.e., $n = m$), (3.2) is a system containing m equations with m unknowns. Solving the equation system gives unique solutions of the probabilities, denoted p_i $(i = 1, 2, \ldots, m)$. If the two numbers are unequal, unique solutions may be obtained by adopting or dropping less important customer wants, which have lower values of customer desirability.

Because the product is said to have failed if one of the m independent performance characteristics crosses the threshold, the reliability target R^* of the product can be written as

$$R^* = \Pr(Y_1 \leq D_1)\Pr(Y_1 \leq D_1)\cdots\Pr(Y_m \leq D_m) = p_1 p_2 \ldots p_m. \quad (3.3)$$

It is worth noting that meeting the minimum reliability level is a necessary and not a sufficient condition to achieve all specified customer satisfactions simultaneously, because the reliability depends only on the product of p_i ($i = 1, 2, \ldots, m$). This is illustrated in the following example. To fulfill all customer satisfactions, it is important to ensure that $\Pr(Y_i \leq D_i) \geq p_i (i = 1, 2, \ldots, m)$ for each performance characteristic in product design.

Example 3.1   Customers have highly desirable wants E_1, E_2, and E_3 on a product. QFD indicates strong correlations of E_1 to Y_1 and Y_2, E_2 to Y_1 and Y_3, and E_3 to Y_2, where Y_1, Y_2, and Y_3 are the independent performance characteristics of the product. Customer satisfactions on E_1, E_2, and E_3 are required to be, respectively, greater than or equal to 88%, 90%, and 95% at the design life. Determine the minimum reliability at this time.

SOLUTION   From (3.2) we have $p_1 p_2 = 0.88$, $p_1 p_3 = 0.9$, and $p_2 = 0.95$. Solving this equation system gives $p_1 = 0.93$, $p_2 = 0.95$, and $p_3 = 0.97$. Then from (3.3), the reliability target is

$$R^* = p_1 p_2 p_3 = 0.93 \times 0.95 \times 0.97 = 0.857.$$

Note that meeting this overall reliability target does not guarantee all customer satisfactions. For instance, $R^* = 0.857$ may result in $p_1 = 0.98$, $p_2 = 0.92$, and $p_3 = 0.95$. Then E_3 is not satisfied.

3.3.3   Warranty Cost-Driven Reliability Requirements

Although meeting customer demands is the objective of commercial business, some organizations may not have enough market information about customer expectations and would base reliability requirements on warranty cost objectives. Let C_w^* be the maximum allowable warranty cost, t_0 be the warranty period, c_0 be the average cost per repair, and n be the production volume. If the product is subject to a free replacement warranty policy (Chapter 11), the expected warranty cost C_w of n units is

$$C_w = c_0 n W(t_0), \quad (3.4)$$

where $W(t_0)$ is the expected number of repairs per unit by t_0. If the repair is a minimal repair (i.e., the failure rate of the product immediately after repair equals that right before failure), $W(t_0)$ can be written as

$$W(t_0) = \ln\left[\frac{1}{R(t_0)}\right]. \quad (3.5)$$

product is same as before failure after repair.

In Chapter 11 we describe the concept of minimal repair and gives (3.5).
Using (3.5), we can rewrite (3.4) as

$$C_w = c_0 n \ln \left[\frac{1}{R(t_0)} \right].$$ (3.6)

Because the total warranty cost must not be greater than C_w^*, from (3.6) the
reliability target is

$$R^* = \exp \left(-\frac{C_w^*}{c_0 n} \right).$$ (3.7)

For a complicated product, the costs per repair and failure rates of subsystems
may be substantially different. In this situation, (3.4) does not provide a good
approximation to the total warranty cost. Suppose that the product has m subsystems connected in series and the life of the subsystems can be modeled with
the exponential distribution. Let c_{0i} and λ_i denote the cost per repair and failure
rate of subsystem i, respectively. The expected warranty cost is

$$C_w = n t_0 \sum_{i=1}^{m} c_{0i} \lambda_i.$$ (3.8)

In many applications it is reasonable to assume that the failure rate of subsystem i is proportional to its production cost: namely, $\lambda_i = K C_i$, where K is a
constant and C_i is the production cost of subsystem i. Then the failure rate of
subsystem i can be written as

$$\lambda_i = \frac{C_i}{C} \lambda,$$ (3.9)

where C and λ are, respectively, the production cost and failure rate of the
product, and $C = \sum_{i=1}^{m} C_i$ and $\lambda = \sum_{i=1}^{m} \lambda_i$. Substituting (3.9) into (3.8) gives

$$C_w = \frac{\lambda n t_0}{C} \sum_{i=1}^{m} c_{0i} C_i.$$ (3.10)

Because the total warranty cost must not exceed C_w^*, the maximum allowable
failure rate of the product can be written as

$$\lambda^* = \frac{C_w^* C}{n t_0 \sum_{i=1}^{m} c_{0i} C_i}.$$ (3.11)

Example 3.2 A product consists of five subsystems. The production cost and
cost per repair of each subsystem are shown in Table 3.1. The manufacturer plans
to produce 150,000 units of such a product and requires that the total warranty
cost be less than \$1.2 million in the warranty period of one year. Determine the
maximum allowable failure rate of the product.

TABLE 3.1   Subsystem Production Cost and Cost per Repair

Cost (dollars)	Subsystem				
	1	2	3	4	5
c_{0i}	25	41	68	35	22
C_i	38	55	103	63	42

SOLUTION   The total production cost is $C = \sum_{i=1}^{5} C_i = \$301$. From (3.11), the maximum allowable failure rate of the product is

$$\lambda^* = \frac{1,200,000 \times 301}{150,000 \times 8760 \times (25 \times 38 + \cdots + 22 \times 42)}$$

$$= 2.06 \times 10^{-5} \text{ failures per hour.}$$

Some products are subject to two-dimensional warranty coverage. In other words, the products are warranted with restrictions on time in service as well as usage. For example, the bumper-to-bumper warranty plans of cars in the United States typically cover 36 months in service or 36,000 miles, whichever occurs first. Under the two-dimensional warranty coverage, failures of the products are not reimbursed if the use at failure exceeds the warranty usage limit. This policy lightens warranty cost burden for manufacturers to some extent. The reduced warranty cost should be taken into account in the development of reliability requirements.

After such products are sold to customers, usages are accumulated over time. Because customers operate their products at different rates, the usages at the same time in service are widely distributed. The distribution and usage accumulation model may be determined by using historical data. In Chapter 11 we discuss use of the lognormal distribution and linear relationship to describe the mileage accumulation of automobiles over time. Suppose that the warranty time and usage limits are t_0 and u_0, respectively. Using customer survey data, recall data, or warranty repair data, we may compute the probability $\Pr[U(t) \leq u_0]$, where $U(t)$ is the accumulated use by time t. Then (3.8) can be modified to

$$C_w = n \int_0^{t_0} \Pr[U(t) \leq u_0] dt \sum_{i=1}^{m} c_{0i} \lambda_i. \tag{3.12}$$

Accordingly, (3.11) becomes

$$\lambda^* = \frac{C_w^* C}{n \int_0^{t_0} \Pr[U(t) \leq u_0] dt \sum_{i=1}^{m} c_{0i} C_i}. \tag{3.13}$$

Example 3.3   Refer to Example 3.2. Suppose that the product is warranted with a two-dimensional plan covering one year or 12,000 cycles, whichever comes first. The use is accumulated linearly at a constant rate (cycles per month) for a particular customer. The rate varies from customer to customer and can be modeled using the lognormal distribution with scale parameter 6.5 and shape parameter 0.8. Calculate the maximum allowable failure rate of the product.

SOLUTION The probability that a product accumulates less than 12,000 cycles by t months is

$$\Pr[U(t) \leq 12{,}000] = \Phi\left[\frac{\ln(12{,}000) - 6.5 - \ln(t)}{0.8}\right] = \Phi\left[\frac{2.893 - \ln(t)}{0.8}\right].$$

From (3.13) the maximum allowable failure rate is

$$\lambda^* = \frac{1{,}200{,}000 \times 301}{150{,}000 \int_0^{12} \Phi[[2.893 - \ln(t)]/0.8]dt\,(25 \times 38 + \cdots + 22 \times 42)}$$

$$= 0.0169 \text{ failures per month} = 2.34 \times 10^{-5} \text{ failures per hour.}$$

Comparing this result with that from Example 3.2, we note that the two-dimensional warranty plan yields a less stringent reliability requirement, which favors the manufacturer.

3.3.4 Total Cost-Driven Reliability Requirements

Reliability requirements derived from warranty cost target take into account the failure cost and do not consider the cost associated with reliability investment. In many applications it is desirable to determine a reliability target that minimizes the total cost. Conventionally, the total cost comprises the failure cost and the reliability investment cost, as depicted in Figure 3.6. If the two types of costs can be quantified, the optimal reliability level is obtained by minimizing the total.

The traditional view perceives reliability program costs as investment costs. This might have been true in the old days, when reliability efforts focused on

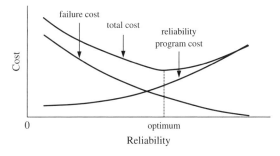

FIGURE 3.6 Costs associated with a reactive reliability program

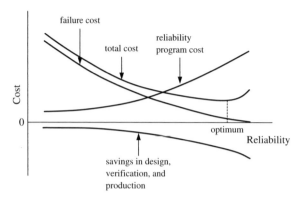

FIGURE 3.7   Costs and savings associated with a proactive reliability program

testing and were essentially reactive. Such programs do not add much value at the beginning of the design cycle. Nowadays, reliability design techniques such as robust design are being integrated into the design process to build reliability into products. The proactive methods break the design–test–fix loop, and thus greatly reduce the time to market and cost. In almost every project, reliability investment is returned with substantial savings in design, verification, and production costs. Figure 3.7 illustrates the costs and savings. As a result of the savings, the total cost is reduced, especially when the required reliability is high. If the costs and savings can be quantified, the optimal reliability level is the one that minimizes the total cost. Clearly, the optimal reliability is considerably larger than the one given by the conventional total cost model.

As we know, modeling failure cost is a relatively easy task. However, estimating the costs and savings associated with a reliability program is difficult, if not impossible. Thus, in most applications the quantitative reliability requirements cannot be obtained by minimizing the total cost. Nevertheless, the principle of total cost optimization is universally applicable and useful indeed in justifying a high-reliability target and the necessity of implementing a proactive reliability program to achieve the target.

3.4   RELIABILITY PROGRAM DEVELOPMENT

In Section 3.3 we described methods for establishing reliability requirements which have to be satisfied for products to be competitive in the marketplace. Reliability activities conducted to achieve the established targets are not free and sometimes require a large amount of investment. However, the investment costs will be paid off through the reduction in failure expenses and engineering design, verification, and production costs, as illustrated in Figure 3.7. Furthermore, the reliability tasks compress the design, verification, and production cycles and thus lead to a shorter time to market. The reliability tasks are more effective when they are well orchestrated and integrated into a reliability program. In this section we describe a generic reliability program, considerations for developing product-specific programs, and management of reliability programs.

3.4.1 Generic Reliability Program

An effective reliability program consists of a series of reliability tasks to be implemented throughout the product life cycle, including product planning; design and development; verification and validation; production; field deployment; and disposal. The reliability activities are not independent exercises; rather, they should be integrated into engineering projects in each stage of the life cycle and assist successful completion of the projects. Figure 3.8 shows the main stages of a typical product life cycle and the reliability tasks that may be implemented in each of the stages. The listed reliability tasks are not intended to be exhaustive; other reliability techniques, such as redundancy design, are not included because of fewer applications in commercial products.

In the product planning stage, reliability tasks are intended to capture customer expectations, establish competitive reliability requirements, and organize a team and secure the resources needed by the reliability program. The reliability tasks in this stage are explained briefly below.

1. *Organizing a reliability team.* A cross-functional team should be assembled at the beginning of a product planning stage so that the reliability requirements are considered in the decision-making process. Even though reliability requirements are ultimately driven by customers, the top leaders of some organizations still unfortunately perceive reliability deployment as a luxurious exercise. In these situations it is vital to seek a management champion of the team and assure that the resources needed throughout the reliability program will be in place. Outputs of the team can be maximized if the team members have diversified expertise, including reliability, market research, design, testing, manufacture, and field service.

2. *Quality function deployment (QFD).* This is a powerful tool for translating customer expectations into engineering requirements. The method was described in detail in Section 3.2.

3. *Reliability history analysis.* This task is to collect and analyze customer feedback, test data, and warranty failure data of the prior-generation product. The analysis should indicate what customer wants were not reasonably satisfied and reveal areas for improvement. Methods of reliability analysis using warranty data are described in Chapter 11.

4. *Reliability planning and specification.* The objective of this task is to establish a competitive reliability target that is economically achievable and develop an effective reliability program to reach or exceed the target. This task can be assisted by utilizing the results of QFD and reliability historical data analysis.

In the design and development stage and before prototypes are created, reliability tasks are to build reliability and robustness into products and to prevent

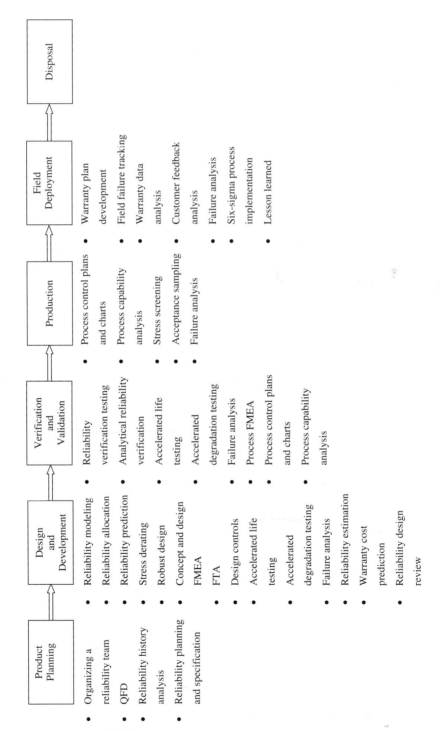

FIGURE 3.8   Reliability tasks for a typical product life cycle

potential failure modes from occurrence. The reliability tasks in this stage are described below.

1. *Reliability modeling.* This task is to model product reliability according to the architecture of the product. The architecture lays out the logic connections of components, which may be in series, parallel, or more complex configurations. Product reliability is expressed as a function of component reliabilities. This relationship is useful in reliability allocation, prediction, and analysis. We discuss this task in detail in Chapter 4.

2. *Reliability allocation.* The reliability target established in the product planning stage should be apportioned appropriately to lower-level structures (subsystems, modules, or components) of the product. The reliability allocated to a structure becomes the reliability target of that structure. Organizations responsible for lower-level structures must achieve their respective targets so that the overall reliability target is attained. Reliability allocation methods are presented in Chapter 4.

3. *Reliability prediction.* In the early design stage, it is frequently desirable to predict reliability for comparing design alternatives and components, identifying potential design issues, determining if a design meets the allocated reliability target, and projecting reliability performance in the field. Several methods are often employed for prediction in this stage. Part count and part stress analysis for electronic equipment, well documented in MIL-HDBK-217 (U.S. DoD, 1995), was a prevailing approach until the mid-1990s. The approach assumes that component lifetimes are exponentially distributed (with a constant failure rate) and that a system is in the logic series of the components. In addition to these assumptions, part stress analysis overemphasizes temperature effects and overlooks other stresses, such as thermal cycling and transient conditions, which are the primary causes of failure in many systems. It was reported repeatedly that the DoD's handbook produced overly pessimistic results, especially when used in commercial products. Unsurprisingly, the handbook was subjected to overwhelming criticism and it is no longer upgraded. A more recent prediction methodology known as PRISMPlus was developed by the Reliability Analysis Center (RAC), now the Reliability Information Analysis Center (RIAC). The methodology includes component-level reliability prediction models and a process for assessment of system reliability due to noncomponent variables such as software and process. The prediction program is comprised of RAC failure models and failure data, user-defined data, and a system failure model that applies process-grading factors. Smith and Womack (2004) report that the methodology produced a more realistic result for an airborne system in a correlation study. There are various commercial software packages that are capable of performing reliability prediction based on this methodology. Another approach to reliability prediction in the early design stage is modeling system reliability as a function of component reliabilities based on the system configuration (Chapter 4). Component reliabilities may be estimated from historical test data, warranty data, or other sources. Ideally, prediction of component reliability should be driven by

a physics-based model, which describes the underlying failure process. Unfortunately, such models are unavailable for most applications, due to the difficulty in understanding and quantifying the failure mechanisms.

4. *Stress derating.* This task is to enhance reliability by reducing stresses that may be applied to a component to levels below the specified limits. When implemented in an electronic design as it often is, derating technique lowers electrical stress and temperature versus the rated maximum values. This alleviates parameter variation and degradation and increases long-term reliability. Useful references for this technique include, for example, U.S. DoD (1998) and O'Connor (2002).

5. *Robust design.* A failure can be attributed to either a lack of robustness or the presence of mistakes induced in design or production. The purpose of robust design is to build robustness and reliability into products in the design stage through implementation of a three-stage process: concept design, parameter design, and tolerance design. In Chapter 5 we describe in detail the methodology of robust reliability design with an emphasis on parameter design. This technique can result in a great improvement in reliability and robustness but has not been implemented as extensively have as conventional reliability tools.

6. *Concept and design FMEA (failure mode and effects analysis).* As stated above, the causes of failure can be classified into two groups: lack of robustness and presence of mistakes. Concept and design FMEA are performed to uncover potential failure modes, analyze effects, and determine causes of failures. The FMEA process is intended to detect design errors that have been embedded into a design and supports recommendations for corrective actions. In Chapter 6 we describe the FMEA methodology.

7. *Fault tree analysis (FTA).* Some failure modes of a design may evoke special concerns, especially when safety is involved. In such situations, FTA is often needed to identify the root causes of the failure modes and to assess the probability of failure occurrence. We introduce the FTA technique in Chapter 6.

8. *Design controls.* This task is aimed at detecting design deficiencies before a design is prototyped. This is accomplished by analyzing product responses to stresses such as temperature, humidity, vibration, mechanical and electrical load, and electromagnetic interference. The common problems uncovered in design control include crack, fatigue, overheating, and open or short circuit, among others. Once concerns are identified and evaluated, corrective actions should be recommended. Implementation of design controls usually requires dedicated computer programs. In Chapter 6 we describe concisely several design control techniques that are applied widely in industry.

9. *Accelerated life testing.* Testing products in the design stage is essential in nearly all design programs for the purpose of comparing design options, uncovering failure modes, estimating reliability, and verifying a design. Testing a product to failure at a normal operating condition is often unfeasible economically, especially in the current competitive business environment. Instead, we conduct accelerated life tests at higher stress levels, which shortens test time and reduces test cost. This task can be a part of robust reliability design, which often

requires testing products to failure at different combinations of design settings. In Chapter 7 we present accelerated life test methods and life data analysis.

10. *Accelerated degradation testing.* Even under accelerating conditions, testing high-reliability products to failure may be too lengthy a task to be affordable. For some products a failure is said to have occurred if one of the performance characteristics crosses a specified threshold. The characteristics are the indicators of product reliability. Therefore, it is possible to estimate reliability of such products by using degradation measurements, which are recorded during testing. This type of test is more efficient than an accelerated life test in terms of test time and cost. This task can be a part of robust reliability design aimed at determining the optimal levels of design parameters. In Chapter 8 we describe accelerated degradation test methods and degradation data analysis.

11. *Failure analysis.* Accelerated life or degradation testing may produce failures. The failed units should be analyzed for failure modes, effects, and mechanisms. Failure analyses at the component or material level usually enable a deep understanding of the root causes and may lead to prevent the same failure modes. All products that fail prior to field deployment should be analyzed thoroughly for causes. Even in the field deployment phase, most warranty return parts are subjected to failure analysis to determine the failure modes and mechanisms in the real world.

12. *Reliability estimation.* This task is needed throughout the product life cycle for a variety of purposes. In many applications it is not a separate task. Rather, it is a part of, for example, reliability history analysis, accelerated testing, design comparison, and warranty analysis. In Chapters 7, 8, and 11 we present methods for reliability estimation from different types of data.

13. *Warranty cost prediction.* Warranty cost not only quantifies the revenue that would be eroded by warranty repairs but also indicates customer satisfaction and competitiveness once a product enters the marketplace. From an engineering perspective, warranty cost reflects the reliability as well as maintainability, both of which should be considered in design. Warranty cost depends on warranty policy, product reliability, sales volume, and cost per repair. In the design stage, product reliability may be estimated from test data, computer simulation, or historical data. In Chapter 11 we present methods for estimating warranty cost.

14. *Reliability design review.* A reliability program should establish several checkpoints at which reliability tasks are reviewed. The objective of the review is to audit whether the reliability program is executed as planned in terms of schedule and accuracy. Importantly, the review team should evaluate the possibility that the reliability target will be achieved through implementation of the established reliability program based on what has been accomplished. If necessary, the team should recommend actions to improve the effectiveness of the program. Whenever possible, reliability design reviews should be conducted along with engineering design reviews. The concurrent reviews enable the reliability

accomplishments to be examined by design engineers from a product design perspective, and vice versa. These interdisciplinary reviews usually identify concerns that would not be discovered in individual reviews.

In the product verification and process validation stage, reliability tasks are intended to verify that the design achieves the reliability target, to validate that the production process is capable of manufacturing products that meet the reliability requirements, and to analyze the failure modes and mechanisms of the units that fail in verification and validation tests. As presented in Chapter 1, process planning is performed in this phase to determine the methods of manufacturing the product. Thus, also needed are reliability tasks that assure process capability. The tasks that may be executed in this phase are explained below.

1. *Reliability verification testing.* This task is to demonstrate with minimum test time and sample size that a product meets the reliability target. In Chapter 9 we describe test methods, approaches to determination of sample size and test time, and techniques for sample size reduction.

2. *Analytical reliability verification.* Reliability verification through testing may be too expensive and time consuming to be affordable in some situations. When there are adequate mathematical models that relate product life to stresses, design parameters, and manufacturing variables, the product reliability may be verified by evaluating such models. This approach, often referred to as *virtual validation*, involves finite element analysis, computer simulation, and numerical calculation.

3. *Process FMEA.* This task is performed in the process planning stage to detect potential process failure modes, analyze effects, and determine causes of failure. Then actions may be recommended to correct the process steps and prevent the failure modes from occurrence in production. In Chapter 6 we present the concept, process, and design FMEA, with focus on the design FMEA.

In the production stage, the objective of reliability tasks is to assure that the production process has minimum detrimental impact on the design reliability. The tasks that may be implemented in this phase are described as follows.

1. *Process control plans and charts.* Process variation increases unit-to-unit variation and infant mortality and thus should be minimized in each step of the production process. This task is to develop and implement process control plans for the critical performance characteristics which are identified in the fourth house of quality of the QFD process. In Chapter 11 we present statistical process control charts for monitoring infant mortality using early warranty data. Montgomery (2001a) describes in detail methods for process control.

2. *Process capability analysis.* Process capability measures the uniformity of a production process. A process of low capability produces high variability in performance and low reliability. This task is to estimate the process capability

and to provide information for minimizing process variation. Process capability analysis is well described in Montgomery (2001a).

3. *Stress screening*. Some products may have latent defects due to material flaws, process variation, or inadequate design. Defective products will fail in early service time and should be eliminated before being shipped to customers. For this purpose, stress screening is often conducted. This task is covered in Chapter 10.

4. *Acceptance sampling*. This task is accomplished, as needed, to make a decision as to whether to accept a particular production lot based on measurements of samples drawn at random from the lot. Due to material defects or the process running out of control, certain lots may contain a large portion of defective units. Failure to reject such substandard lots will result in low field reliability and customer dissatisfaction. ANSI/ASQ (2003a, b) provide standard methods for acceptance sampling.

In the field deployment stage, reliability tasks are aimed at developing a warranty plan, tracking field failures, assessing field reliability performance, evaluating customer satisfaction, analyzing warrantied parts, and developing containment and permanent corrective actions as needed. The major reliability tasks are described below.

1. *Warranty plan development.* A preliminary warranty coverage may be planned in the product planning stage. The plan is finalized when the product is ready for marketing. Although a warranty plan is determined largely by market competition, the final decisions are driven by financial analysis. An important component of the financial analysis is the warranty repair cost, which may be estimated from warranty repair modeling or reliability prediction. Warranty policies and repair cost estimation are addressed in Chapter 11.

2. *Field failure tracking*. This task is intended to collect failure information from warranty repairs and customer complaints. The failure information should be as specific and accurate as possible and include failure modes, operating conditions at which the failures occur, failure time and usage (e.g., mileage), and others. Often, a computer system (i.e., a warranty database) is needed to store and retrieve these failure data. This task is covered in Chapter 11.

3. *Warranty data analysis.* This task is to estimate field reliability, project warranty repair numbers and costs, monitor field failures, and detect unexpected failure modes and patterns. Early detection of unusual failure modes and high failure probability can promote corrective actions to change ongoing manufacturing process and repair strategies. For safety-related products such as automobiles, timely warranty data analysis enables the assessment of risks associated with critical failure modes and may warrant recalls. In Chapter 11 we present methods for warranty data analysis.

4. *Customer feedback analysis*. The real-world usage profile is the ultimate environment in which a product is validated, and customer satisfaction is the

predominant factor that drives the success of a product. Customer feedback on both functional and reliability performance must be analyzed thoroughly to determine what product behaviors do and do not satisfy customers. The results are valuable inputs to the QFD development of the next-generation product. The sources for collecting feedback may include customer surveys, warranty claims, and customer complaints.

5. _Six-sigma process implementation._ Warranty data analysis of early failures may indicate unusual failure modes and failure probability. The root causes of the failures should be identified and eliminated in subsequent production. This may be accomplished by implementing the six-sigma process. The process is characterized by DMAIC (define, measure, analyze, implement, and control). The first step of the process is to define the problem and the project boundaries, followed by creating and validating the measurement system to be used for quantifying the problem. The analyze step is to identify and verify the causes of the problem, and the improve step is to determine the methods of eliminating the causes. Finally, improvement is implemented and sustained through the use of control plans. The DMAIC approach has been used extensively in industry and has generated numerous publications, including, for example, Pyzdek (2003) and Gitlow and Levine (2005). In Section 3.5 we discuss six sigma in more detail.

6. _Lessons learned._ This task documents lessons learned as well as success stories. Lessons should cover all mistakes, from missing a checkpoint to failure to meet customer expectations. The causes of mistakes must be identified, and recommendations must be made to prevent similar mishaps in the future. Success stories are processes and actions that have proved successful in improving the effectiveness of a reliability program and reducing the cost and time associated with the program. Both lessons learned and success stories should be communicated to teams that are developing other products and should be archived as reliability historical data to be reviewed in the development of next-generation products.

3.4.2 Product-Specific Reliability Program

A product-specific reliability program should serve as a road map leading product planning, design and development, testing, production, and deployment to achieve the reliability target and customer satisfaction at low cost and in a short time. The specific program may be customized starting with the generic reliability program presented in Section 3.4.1. Such a program should be suitable for the particular product, minimize the cost and time incurred by the program itself, and maximize the savings in time and costs associated with engineering design, verification, and production, as explained in Figure 3.7. Developing such a reliability program is basically the process of answering the following questions:

- What reliability tasks should be selected for the specific product?

- When should the individual reliability tasks be implemented in the product life cycle?
- How is the effectiveness of the reliability tasks assured and improved?

Selection of suitable reliability tasks requires a thorough understanding of the theoretical background, applicability, merits, and limitations of the tasks. The interrelationships, similarities, and differences between the reliability tasks under consideration need to be well understood and assessed to improve program efficiency. For example, to consider QFD as a candidate for a reliability program, we should have knowledge of the QFD process, the procedures for developing houses of quality, and the inputs and outputs of each house of quality. We should also understand that QFD is applicable to customer-driven products as well as contract-driven products. Here, the customers are not necessarily the end users; they can be internal users. Understanding the applicability allows this technique to be considered for government-contracted products. The function of QFD enables customer expectations to be linked to engineering technical characteristics so that the design activities and production controls are aligned with satisfying customers. Often, a complex system benefits more than a simple component from this technique, because technical characteristics of a simple component are relatively easily related to customer wants without using a structured tool. Having these understandings, we would consider QFD for a complex government-contracted product and not use it for a simple commercial component. The relation of QFD to other reliability tasks should also be considered and exploited to increase the efficiency of the reliability program. For example, the QFD technique is correlated with robust reliability design. The technical characteristics scoped in the houses of quality are the control factors of subsequent robust reliability design. Thus, robust reliability design will be facilitated if QFD is implemented earlier.

Cost-effectiveness is an important factor that must be considered when selecting a reliability task. Once a task is identified as be applicable to the product being planned, the effectiveness of the task is evaluated. This may be assisted by reviewing the relevant lessons learned and success stories and analyzing the pros and cons of the technique. Some reliability techniques that are powerful for a product may not be effective or are even not suitable for another product. For example, accelerated degradation testing is an efficient test method for a product whose performance degrades over time but is not suitable for a binary-state product. In addition to the effectiveness, the cost and time that a reliability task will incur and save must be taken into account. A reliability task certainly increases the expense. However, it saves the time and costs associated with engineering design, verification, and production. The essence is illustrated in Figure 3.7. A good reliability task incurs low cost and produces substantial savings.

Now let's address the second question: When should individual reliability tasks be implemented in the product life cycle? Once appropriate reliability tasks are selected, the sequence of these tasks should be optimized for maximum effectiveness. Essentially, reliability tasks are integral parts of the engineering design, verification, and production program. Thus, the time lines of reliability tasks should be aligned with those of the design, verification, and production activities. For example, QFD is a tool for translating customer wants into engineering requirements and supports product planning. Thus, it should be performed in the product planning stage. Another example is that of design controls, which are conducted only after the design schematics are completed and before a design is released for production. In situations where multiple reliability tasks are developed to support an engineering design, verification, or production activity, the sequence of these tasks should be orchestrated carefully to reduce the associated cost and time. Doing so requires fully understanding of interrelationships among the reliability tasks. If a task generates outputs that serve as the inputs of another task, it should be completed earlier. For instance, thermal analysis as a method of design control for a printed circuit board yields a temperature distribution which can be an input to reliability prediction. Thus, the reliability prediction may begin after the thermal analysis has been completed.

The time line for a reliability program should accommodate effects due to changes in design, verification, and production plans. Whenever changes take place, some reliability tasks need to be revised accordingly. For example, design changes must trigger the modification of design FMEA and FTA and the repetition of design control tasks to verify the revised design. In practice, some reliability tasks have to be performed in early stages of the life cycle with limited information. As the life cycle proceeds, the tasks may be repeated with more data and more specific product configurations. A typical example is reliability prediction, which is accomplished in the early design stage based on part count to provide inputs for comparison of design alternatives and is redone later to predict field reliability with specific product configuration, component information, stress levels, and prototype test data.

Once the reliability program and time lines are established, implementation strategies should be developed to assure and improve the effectiveness of the program. This lies in the field of reliability program management and is discussed in the next subsection.

Example 3.4   An automotive supplier was awarded a contract to produce a type of electronic module to be installed in automobiles. The supplier has developed a reliability program for the design, verification, and production of modules to meet a specified reliability target and customer satisfaction. The program was not intended to include field deployment because the original equipment manufacturer (OEM) had a comprehensive reliability program to cover the modules deployed in the field. Suitable reliability tasks were selected for the modules and integrated into the design, verification, and production activities. The time lines of the tasks were aligned with these of design, verification, and production. Figure 3.9 shows the main activities of design, verification, and production and the reliability tasks aligned with each activity. The design process is essentially iterative; design changes are mandated if the design does not pass the design reviews, design verification testing, or process validation testing. Some reliability tasks such as the FMEA and reliability prediction, are repeated whenever design changes take place.

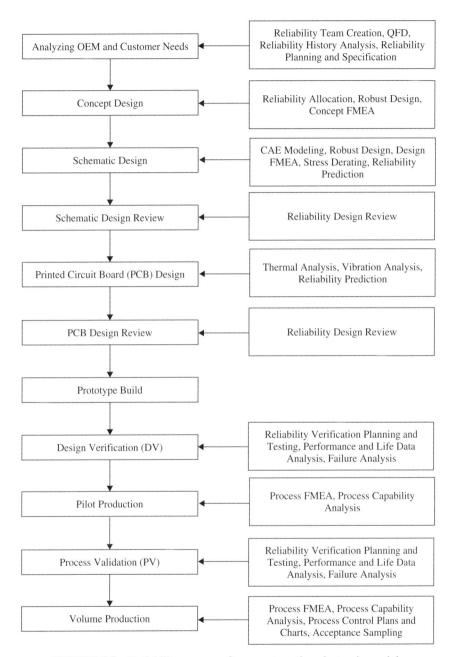

FIGURE 3.9 Reliability program for an automotive electronic module

3.4.3   Reliability Program Management

Effective management is an important dimension that assures the success of a reliability program. The tasks of management include organizing an efficient team, assigning roles and responsibilities, planning schedules, developing emergency plans, securing required resources, creating effective communication avenues and a cooperative working environment, reviewing progress, motivating smart and hard work, and so on. Although the accomplishment of a task moves one step forward to success, only the achievement of all tasks in a timely and economical manner maximizes the effectiveness of a reliability program.

The first task of a reliability program usually is to organize a reliability team for reliability planning at the beginning of the product planning stage. Since the potential reliability tasks will cover all stages of the product life cycle, the team should be cross-functional and comprised of reliability and quality engineers as well as representatives of marketing, design, testing, production, and service personnel. This diversified representation enables a reliability program to be planned and implemented concurrently with the engineering design, verification, and production activities. As discussed earlier, a reliability program is an integral part of the engineering design, verification, and production tasks. Therefore, the product manager should champion the reliability team, although he or she may not be a reliability expert. The championship is beneficial to the reliability team in that it has the authority to assure the project resources, gain the cooperation of design, verification, and production personnel, and exercise veto power over premature design release. The product manager may appoint a reliability expert as the reliability program leader; however, the manager is still responsible for the success of the reliability program.

The resources needed by a reliability program should be planned and secured as early as possible. The resources may include, for example, team member skills, time commitment, test equipment, failure analysis tools, software, measurement systems, and samples. In regard to the team member skills, many reliability engineers do not have sufficient knowledge of the physics of the products on which they are to work. In reality, it is unrealistic to expect otherwise because of their educational and industrial backgrounds. Thus, it is important for reliability engineers to receive sufficient training on the physics of products. Such training definitely increases the effectiveness of communications with design, verification, and production engineers and enables a deeper understanding of reliability issues, with higher productivity. On the other hand, team members without a reliability background should also be trained in basic reliability knowledge so that they can have reliability thinking in their design, verification, and production work and command a common language in teamwork. The hardware and software resources that will be consumed in a reliability program shall be planned for well ahead. The budget for procuring new equipment and software is a part of the product program budget and should be approved by the product manager. In planning, some outsourcing jobs, such as failure analysis using dedicated equipment, may be anticipated, and the resulting costs should be added to the total budget. In a reliability program, test samples are nearly essential for testing and analysis.

The sample size needed in design, verification, and production is determined by the reliability team. The product program planner must count in this sample size when ordering test samples. The reliability team should get its full order. It is not uncommon for the reliability team to be short of test samples, especially when functional tests require additional samples.

An effective reliability program should set up reasonable checkpoints at which progress is reviewed, potential problems are identified, corrective actions are recommended, and plans for the next steps may be adjusted. Progress review should check not only the amount of work done versus the goal, but also the accuracy of the work and the conclusions drawn from the work by looking into the methods, data, and process used to accomplish the work. Review at checkpoints usually results in identification of existing and potential problems, such as improper use of test and data analysis methods. This is followed by recommendations for correcting the problems and preventing them from reoccurring. If necessary, the plans for the next steps should be adjusted to allow the changes.

3.5 RELIABILITY DESIGN AND DESIGN FOR SIX SIGMA

Reliability design is aimed at establishing reliability into products at the design stage through the use of a matrix of reliability tools. It is perhaps the most important component in a reliability program in terms of the benefits versus costs it brings. The component consists of a series of tasks, such as QFD, robust reliability design, and FMEA. As addressed in Section 3.4.2, the sequence of these tasks should be optimized so that the effectiveness of the reliability design is maximized. This may be achieved by using the design for six sigma (DFSS) approach. In this section we describe concisely six-sigma methodologies and a reliability design process using the DFSS approach. More description of six sigma may be found in, for example, Bruce and Launsby (2003) and K. Yang and El-Haik (2003).

3.5.1 Overview of Six Sigma

As we know, *sigma* is a statistical term representing the standard deviation that quantifies a process variation. The sigma number is used to measure the capability of a process to produce defect-free products. For example, a three-sigma process yields 66,800 defects per million opportunities, whereas a six-sigma process produces only 3.4 defects per million opportunities. Most organizations in the United States are operating at three- to four-sigma quality levels. Therefore, six sigma represents a quality objective toward which an organization may strive. The approach to achieving the objective is called simply *six sigma*.

Six sigma is a highly disciplined process that helps us focus on designing, developing, and delivering near-perfect products, processes, and services. It is a structured approach composite of many different methodologies, tools, and philosophies, such as robust design, FMEA, gauge repeatability and reproducibility, and statistical process control. The essence of six sigma does not lie in the individual disciplines that it utilizes but in the synergy that all the different methodologies and tools provide in the pursuit of improvement. Currently, there are two types of six-sigma approaches: six sigma and DFSS. The six-sigma approach is aimed at resolving the problems of existing products or processes through use of the DMAIC process, introduced in Section 3.4.1. Because of its reactive nature, it is essentially a firefighting method, and thus the value it can add is limited. In contrast, DFSS is a proactive approach deployed at the beginning of a design cycle to avoid building potential failure modes into a product. Thus, DFSS is capable of preventing failure modes from occurring. DFSS is useful in the design of a new product or the redesign of an existing product.

DFSS is implemented by following the ICOV process, where I stands for *identify*, C for *characterize*, O for *optimize*, and V for *validate*. The main activities in each phase of the ICOV process are described below.

- *Identify requirements phase (I phase).* This phase involves defining the scope and objective of a project, developing a team and a team charter, assigning roles and responsibilities, gathering customer expectations, and performing competitive analysis. In this phase, customer wants are translated into technical characteristics, and the requirements of the characteristics are specified.
- *Characterize design phase (C phase).* This phase is to translate the technical characteristics into product functional characteristics, develop design concepts, and evaluate design alternatives.
- *Optimize design phase (O phase).* This phase is to optimize product performance with minimum sensitivity to use conditions and variations in both material and production process. The optimal setting of design parameters is determined, and the resulting product performance is predicted.
- *Validate design phase (V phase).* This phase is to validate the optimal design. Test plans must be devised carefully and be valid statistically. Test operation procedures must be standardized. Failed units are analyzed to determine the root causes; the analysis may lead to design changes.

3.5.2 Reliability Design by the DFSS Approach

DFSS is essentially a process that drives achievement of the six-sigma objective in an effective manner. As stated earlier, individual reliability tasks should be well sequenced to maximize effectiveness. A powerful approach to sequencing is the DFSS. To implement DFSS in a reliability program, we may orchestrate the reliability tasks for design and verification according to the ICOV process.

In the context of the ICOV process, the first phase in reliability design is "identify requirements", which begins in the product planning stage. In this phase, market research, benchmarking, and competitor analysis are performed and customer expectations are collected and analyzed. Then a QFD is constructed to determine the technical characteristics, and a reliability target is specified. Also defined in the I phase are the reliability team and the roles and responsibilities of the team members.

The second phase of reliability design in the framework of the ICOV process is "characterize design", which occurs in the early design and development stage. In this phase the technical characteristics identified in the I phase are translated further, into product functional characteristics to be used in the O phase. The translation is done by expanding the first house of quality to the second house. Reliability modeling, allocation, prediction, FMEA, and FTA may be performed to help develop design alternatives. For example, a concept FMEA may rule out potential design alternatives that have a high failure risk. Once detailed design alternatives are generated, they are evaluated with respect to reliability by applying reliability techniques such as reliability prediction, FMEA, FTA, and design control methods. Outputs from this phase are the important product characteristics and the best design alternative that has high reliability.

The next phase in reliability design by the ICOV process is "optimize design", which is implemented in the late design and development stage. The concept design has been finalized by this stage and detailed design is being performed. The purpose of reliability design in this phase is to obtain the optimal setting of design parameters that maximizes reliability and makes product performance insensitive to use condition and process variation. The main reliability tasks applied for this purpose include robust reliability design, accelerated life testing, accelerating degradation testing, and reliability estimation. Functional and reliability performance may be predicted at the optimal setting of design parameters. Design control methods such as thermal analysis and mechanical stress analysis should be implemented after the design optimization is completed to verify that the optimal design is free of critical potential failure modes.

The last phase of reliability design by the ICOV process is to validate the optimal design in the verification and validation stage. In this phase, samples are built and tested to verify that the design has achieved the reliability target. The test conditions should reflect real-world usage. For this purpose, a P-diagram (a tool for robust design, described in Chapter 5) may be employed to determine the noise factors that the product will encounter in the field. Accelerated tests may be conducted to shorten the test time; however, it should be correlated to real-world use. Failure analysis must be performed to reveal the causes of failure. This may be followed by a recommendation for design change.

In summary, the DFSS approach provides a lean and nimble process by which reliability design can be performed in a more efficient way. Although this process improves the effectiveness of reliability design, the success of reliability design relies heavily on each reliability task. Therefore, it is vital to develop suitable reliability tasks that are capable of preventing and detecting potential failure modes in the design and development stage. It is worth noting that the DFSS is a part of a reliability program. Completion of the DFSS process is not the end of the program; rather, the reliability tasks for production and field deployment should begin or continue.

PROBLEMS

3.1 Define the three types of customer expectations, and give an example of each type. Explain how customer expectation for reliability influences customer satisfaction.

3.2 Describe the QFD process and the inputs and outputs of each house of quality. Explain the roles of QFD in reliability planning and specification.

3.3 Perform a QFD analysis for a product of your choice: for example, a lawn mower, an electrical stove, or a refrigerator.

3.4 A QFD analysis indicates that the customer expectations for a product include E_1, E_2, E_3, and E_4, which have customer desirability values of 9, 9, 8, and 3, respectively. The QFD strongly links E_1 to performance characteristics Y_1 and Y_3, E_2 to Y_1 and Y_2, and both E_3 and E_4 to Y_2 and Y_3. The required customer satisfaction for E_1, E_2, E_3, and E_4 is 90%, 95%, 93%, and 90%, respectively. Calculate the reliability target.

3.5 A manufacturer is planning to produce 135,000 units of a product which are warranted for 12 months in service. The manufacturer sets the maximum allowable warranty cost to $150,000 and expects the average cost per repair to be $28. Determine the reliability target at 12 months to achieve the warranty objective.

3.6 Refer to Example 3.3. Suppose that the customers accumulate usage at higher rates, which can be modeled with the lognormal distribution with scale parameter 7.0 and shape parameter 0.8. Determine the minimum reliability requirement. Compare the result with that from Example 3.3, and comment on the difference.

3.7 Describe the roles of reliability tasks in each phase of the product life cycle and the principles for developing an effective reliability program.

3.8 Explain the process of six sigma and design for six sigma (DFSS). What are the benefits of performing reliability design through the DFSS approach?

4

SYSTEM RELIABILITY EVALUATION AND ALLOCATION

4.1 INTRODUCTION

Webster's College Dictionary (Neufeldt and Guralnik, 1997) defines a *system* as a set or arrangement of things so related or connected as to form a unity or organic whole. Technically, a system is a collection of independent and interrelated components orchestrated according to a specific design in order to achieve a specified performance and reliability target and simultaneously meet environmental, safety, and legal requirements. From the hierarchical structure point of view, a system is comprised of a number of subsystems, which may be further divided into lower-level subsystems, depending on the purpose of system analysis. Components are the lowest-level constituents of a system. For example, a car is a typical system. It consists of a powertrain, a chassis, a body, and an electrical subsystem. A powertrain subsystem contains engine, transmission, and axle, which are still tremendously complex and can be broken down further into lower-level subsystems.

In Chapter 3 we described methods for setting reliability targets and developing effective reliability programs to achieve the targets and customer satisfaction. A reliability target is usually established for an entire product, which may be considered as a system. To ensure the overall target, it is important to allocate the target to individual subsystems that constitute the product, especially when suppliers or contractors are involved in a product realization process. The apportioned reliability to a subsystem becomes its target, and the responsible organization must guarantee attainment of this target. In the car example, the

---

Life Cycle Reliability Engineering, by Guangbin Yang
Copyright © 2007 John Wiley & Sons, Inc.

overall reliability target for a car should be allocated to the powertrain, chassis, body, and electrical subsystem. The reliability allocated to the powertrain is further apportioned to the engine, transmission, and axle. The allocation process is continued until the assembly level is reached. Then the auto suppliers are obligated to achieve the reliability of the assemblies they are contracted to deliver. In this chapter we present various reliability allocation methods.

A comprehensive reliability program usually requires evaluation of system (product) reliability in the design and development stage for various purposes, including, for example, selection of materials and components, comparison of design alternatives, and reliability prediction and improvement. Once a system or subsystem design is completed, the reliability must be evaluated and compared with the reliability target that has been specified or allocated. If the target is not met, the design must be revised, which necessitates a reevaluation of reliability. This process continues until the desired reliability level is attained. In the car example, the reliability of the car should be calculated after the system configuration is completed and assembly reliabilities are available. The process typically is repeated several times and may even invoke reliability reallocation if the targets of some subsystems are unattainable.

In this chapter we describe methods for evaluating the reliability of systems with different configurations, including series, parallel, series–parallel, and k-out-of-n voting. Methods of calculating confidence intervals for system reliability are delineated. We also present measures of component importance. Because system configuration knowledge is a prerequisite to reliability allocation, it is presented first in the chapter.

4.2 RELIABILITY BLOCK DIAGRAM

A *reliability block diagram* is a graphical representation of logic connection of components within a system. The basic elements of logic connections include series and parallel, from which more complicated system configurations can be generated, such as the series–parallel and k-out-of-n voting systems. In a reliability block diagram, components are symbolized by rectangular blocks, which are connected by straight lines according to their logic relationships. Depending on the purpose of system analysis, a block may represent a lowest-level component, a module, or a subsystem. It is treated as a black box for which the physical details are not shown and may not need to be known. The reliability of the object that a block represents is the only input that concerns system reliability evaluation. The following example illustrates the construction of reliability block diagrams at different levels of a system.

Example 4.1 Figure 4.1 shows the hierarchical configuration of an automobile that consists of a body, a powertrain, and electrical and chassis subsystems. Each subsystem is broken down further into multiple lower-level subsystems. From a reliability perspective, the automobile is a series system (discussed in the next section) which fails if one or more subsystems break. Figure 4.2 shows

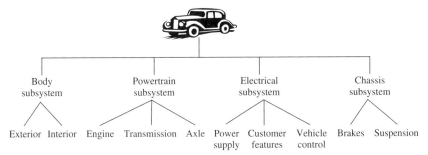

FIGURE 4.1 Hierarchical configuration of a typical automobile

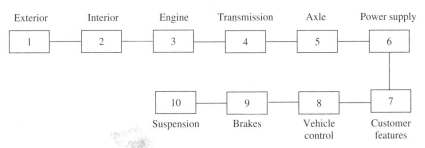

FIGURE 4.2 Reliability block diagram with blocks representing first-level subsystems

FIGURE 4.3 Reliability block diagram with blocks representing second-level subsystems

the reliability block diagram of the automobile, in which the blocks represent the first-level subsystems, assuming that their reliabilities are known. Figure 4.3 is a diagram illustrating second-level subsystems. Comparing Figure 4.2 with Figure 4.3, we see that the complexity of a reliability block diagram increases with the level of subsystem that blocks represent. The reliability block diagram of a typical automobile contains over 12,000 blocks if each block represents a component or part.

In constructing a reliability block diagram, keep in mind that physical configurations in series or parallel do not necessarily indicate the same logic relations in terms of reliability. For example, an automobile engine may have six cylinders connected in parallel mechanically. From a reliability perspective, the six cylinders are in series because the engine is said to have failed if one or more cylinders fails. Development of a reliability block diagram for a large-scale system is time consuming. Fortunately, today, the work can be facilitated by using a commercial software package, such as Reliasoft, Relex, or Item.

A reliability block diagram is a useful and basic tool for system reliability analysis. In Chapter 6 we describe application of the diagram to fault tree analysis. In the sections that follow we present methods for calculating system reliability based on the diagram. The software mentioned above is capable of doing the calculations.

4.3 SERIES SYSTEMS

A system is said to be a *series system* if the failure of one or more components within the system results in failure of the entire system. In other words, all components within a system must function for the system to succeed. Figures 4.2 and 4.3 show the automobile series systems at two hierarchical levels. The reliability of a general series system can be calculated as follows.

Suppose that a series system consists of n mutually independent components. Here, mutual independence implies that the failure of one component does not affect the life of other components. We use the following notation: E_i is the event that component i is operational, E the event that the system is operational, R_i the reliability of component i, and R the system reliability. By definition, successful operation of a system requires all components to be functional. From probability theory, the system reliability is

$$R = \Pr(E) = \Pr(E_1 \cdot E_2 \cdots E_n).$$

Because of the independence assumption, this becomes

$$R = \Pr(E_1)\Pr(E_2)\cdots\Pr(E_n) = \prod_{i=1}^{n} R_i. \tag{4.1}$$

If the n components are identical with reliability R_0, the system reliability is

$$R = R_0^n. \tag{4.2}$$

Equation (4.1) indicates that the system reliability is the product of reliabilities of components. This result is unfortunate, in that the system reliability is less than the reliability of any component. Furthermore, the system reliability decreases rapidly as the number of components in a system increases. The observations support the principle of minimizing the complexity of an engineering design.

Let's consider a simple case where the times to failure of n components in a system are modeled with the exponential distribution. The exponential reliability function for component i is $R_i(t) = \exp(-\lambda_i t)$, where λ_i is the failure rate of component i. Then from (4.1), the system reliability can be written as

$$R(t) = \exp\left(-t\sum_{i=1}^{n}\lambda_i\right) = \exp(-\lambda t), \tag{4.3}$$

where λ is the failure rate of the system and

(4.4 of)

$$\lambda = \sum_{i=1}^{n} \lambda_i.$$ (4.4)

The mean time to failure of the system is

$$\text{MTTF} = \int_0^\infty R(t)\, dt = \frac{1}{\sum_{i=1}^{n} \lambda_i}$$ (4.5)

Equation (4.3) indicates that the life of a system follows the exponential distribution if all components within the system are exponential and the failure rate of the system is the sum of all individual failure rates. Equation (4.3) is widely used, and perhaps misused, because of its simplicity. For example, MIL-HDBK-217F (U.S. DoD, 1995) assumes that all components have constant failure rates and uses (4.3) to calculate the system failure rate.

Example 4.2 Refer to Figure 4.2. Suppose that the lifetimes of the body, powertrain, and electrical and chassis subsystems are exponentially distributed with $\lambda_1 = 5.1 \times 10^{-4}$, $\lambda_2 = 6.3 \times 10^{-4}$, $\lambda_3 = 5.5 \times 10^{-5}$, and $\lambda_4 = 4.8 \times 10^{-4}$ failures per 1000 miles, respectively. Calculate the reliability of the vehicle at 36,000 miles and the mean mileage to failure.

SOLUTION Substituting the values of $\lambda_1, \lambda_2, \lambda_3$, and λ_4 into (4.4) yields

$$\lambda = 5.1 \times 10^{-4} + 6.3 \times 10^{-4} + 5.5 \times 10^{-5} + 4.8 \times 10^{-4}$$
$$= 16.75 \times 10^{-4} \text{ failures per 1000 miles.}$$

The reliability at 36,000 miles is $R(36,000) = \exp(-16.75 \times 10^{-4} \times 36) = 0.9415$. The mean mileage to failure (MMTF) is obtained from (4.5) as

$$\text{MMTF} = \frac{1}{16.75 \times 10^{-4}} = 597,000 \text{ miles.}$$

Now let's consider another case where the times to failure of n components in a system are modeled with the Weibull distribution. The Weibull reliability function for component i is

$$R_i(t) = \exp\left[-\left(\frac{t}{\alpha_i}\right)^{\beta_i} \right],$$

where β_i and α_i are, respectively, the shape parameter and the characteristic life of component i. From (4.1) the system reliability is

$$R(t) = \exp\left[-\sum_{i=1}^{n} \left(\frac{t}{\alpha_i}\right)^{\beta_i} \right].$$ (4.6)

Then the failure rate $h(t)$ of the system is

$$h(t) = \sum_{i=1}^{n} \frac{\beta_i}{\alpha_i} \left(\frac{t}{\alpha_i} \right)^{\beta_i - 1}. \tag{4.7}$$

Equation (4.7) indicates that like the exponential case, the failure rate of the system is the sum of all individual failure rates. When $\beta_i = 1$, (4.7) reduces to (4.4), where $\lambda_i = 1/\alpha_i$.

If the n components have a common shape parameter β, the mean time to failure of the system is given by

$$\text{MTTF} = \int_0^\infty R(t)\,dt = \frac{\Gamma((1/\beta) + 1)}{\left[\sum_{i=1}^{n}(1/\alpha_i)^\beta\right]^{1/\beta}}, \tag{4.8}$$

where $\Gamma(\cdot)$ is the gamma function, defined in Section 2.5.

Example 4.3 A resonating circuit consists of an alternating-current (ac) power supply, a resistor, a capacitor, and an inductor, as shown in Figure 4.4. From the reliability perspective, the circuit is in series; the reliability block diagram is in Figure 4.5. The times to failure of the components are Weibull with the

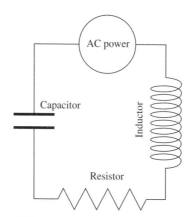

FIGURE 4.4   Resonating circuit

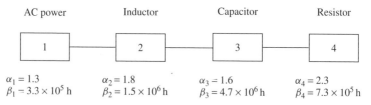

FIGURE 4.5   Reliability block diagram for the resonating circuit

parameters shown in Figure 4.5. Calculate the reliability and failure rate of the circuit at 5×10^4 hours.

SOLUTION Substituting the values of the Weibull parameters into (4.6) gives

$$
R(5 \times 10^4) = \exp\left[-\left(\frac{5 \times 10^4}{3.3 \times 10^5}\right)^{1.3} - \left(\frac{5 \times 10^4}{1.5 \times 10^6}\right)^{1.8} - \left(\frac{5 \times 10^4}{4.7 \times 10^6}\right)^{1.6} \right.
$$
$$
\left. - \left(\frac{5 \times 10^4}{7.3 \times 10^5}\right)^{2.3}\right]
$$
$$
= 0.913.
$$

The failure rate is calculated from (4.7) as

$$
h(5 \times 10^4) = \frac{1.3}{3.3 \times 10^5}\left(\frac{5 \times 10^4}{3.3 \times 10^5}\right)^{1.3-1} - \frac{1.8}{1.5 \times 10^6}\left(\frac{5 \times 10^4}{1.5 \times 10^6}\right)^{1.8-1}
$$
$$
- \frac{1.6}{4.7 \times 10^6}\left(\frac{5 \times 10^4}{4.7 \times 10^6}\right)^{1.6-1} - \frac{2.3}{7.3 \times 10^5}\left(\frac{5 \times 10^4}{7.3 \times 10^5}\right)^{2.3-1}
$$
$$
= 2.43 \times 10^{-6} \text{ failures per hour.}
$$

4.4 PARALLEL SYSTEMS

A system is said to be a *parallel system* if and only if the failure of all components within the system results in the failure of the entire system. In other words, a parallel system succeeds if one or more components are operational. For example, the lighting system that consists of three bulbs in a room is a parallel system, because room blackout occurs only when all three bulbs break. The reliability block diagram of the lighting system is shown in Figure 4.6. The reliability of a general parallel system is calculated as follows.

Suppose that a parallel system consists of n mutually independent components. We use the following notation: E_i is the event that component i is operational;

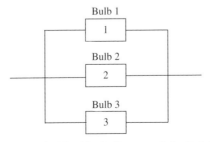

FIGURE 4.6 Reliability block diagram of the lighting system

E the event that the system is operational; $\overline{X}$ the complement of X, where X represents E_i or E; R_i the reliability of component i; F the system unreliability (probability of failure); and R the system reliability. By definition, all n components must fail for a parallel system to fail. From probability theory, the system unreliability is

$$F = \Pr(\overline{E}) = \Pr(\overline{E}_1 \cdot \overline{E}_2 \cdots \overline{E}_n).$$

Because $\overline{E}_i$ ($i = 1, 2, \ldots, n$) are mutually independent, this equation can be written as

$$F = \Pr(\overline{E}_1)\Pr(\overline{E}_2)\cdots\Pr(\overline{E}_n) = \prod_{i=1}^{n}(1 - R_i). \tag{4.9}$$

The system reliability is the complement of the system unreliability: namely,

$$R = 1 - \prod_{i=1}^{n}(1 - R_i). \tag{4.10}$$

exponential

$R_0 = e^{-xt}$

If the n components are identical, (4.10) becomes

$$R = 1 - (1 - R_0)^n, \tag{4.11}$$

weibull $e^{(-t/\alpha)}$

$R_0 = e$

where R_0 is the reliability of a component. If R is specified in advance as a target, the minimum number of components required to achieve the target is

$$n = \frac{\ln(1 - R)}{\ln(1 - R_0)}. \tag{4.12}$$

If the life of the n identical components is modeled with the exponential distribution with failure rate λ, (4.11) can be written as

$$R(t) = 1 - [1 - \exp(-\lambda t)]^n. \tag{4.13}$$

$e^{-(f_1 + f_2) x x t}$

The mean time to failure of the system is given by

$$\text{MTTF} = \int_0^\infty R(t)\, dt = \frac{1}{\lambda}\sum_{i=1}^{n}\frac{1}{i}. \tag{4.14}$$

In contrast to a series system, the reliability of a parallel system increases with the number of components within the system, as indicated in (4.10). Thus, a parallel configuration is a method of increasing system reliability and is often implemented in safety-critical systems such as aircraft and spaceships. However, use of the method is often restricted by other considerations, such as the extra cost and weight due to the increased number of components. For instance, parallel design is rarely used for improving automobile reliability because of its cost.

Example 4.4   Refer to Figure 4.6. Suppose that the lighting system uses three identical bulbs and that other components within the system are 100% reliable. The times to failure of the bulbs are Weibull with parameters $\alpha = 1.35$ and $\beta = 35,800$ hours. Calculate the reliability of the system after 8760 hours of use. If the system reliability target is 99.99% at this time, how many bulbs should be connected in parallel?

SOLUTION   Since the life of the bulbs is modeled with the Weibull distribution, the reliability of a single bulb after 8760 hours of use is

$$R_0 = \exp\left[-\left(\frac{8760}{35,800}\right)^{1.35}\right] = 0.8611.$$

Substituting the value of R_0 into (4.11) gives the system reliability at 8760 hours as $R = 1 - (1 - 0.8611)^3 = 0.9973$. From (4.12), the minimum number of bulbs required to achieve 99.99% reliability is

$$n = \frac{\ln(1 - 0.9999)}{\ln(1 - 0.8611)} = 5.$$

4.5   MIXED CONFIGURATIONS

There are situations in which series and parallel configurations are mixed in a system design to achieve functional or reliability requirements. The combinations form series–parallel and parallel–series configurations. In this section we discuss the reliability of these two types of systems.

4.5.1   Series–Parallel Systems

In general, a *series–parallel system* is comprised of n subsystems in series with m_i ($i = 1, 2, \ldots, n$) components in parallel in subsystem i, as shown in Figure 4.7. The configuration is sometimes called the *low-level redundancy*

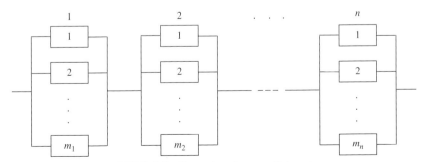

FIGURE 4.7   General series–parallel system

FIGURE 4.8   Reliability block diagram equivalent to Figure 4.7

design. To calculate the system reliability, we first reduce each parallel subsystem to an equivalent reliability block. From (4.10), the reliability R_i of block i is

$$R_i = 1 - \prod_{j=1}^{m_i}(1 - R_{ij}),\tag{4.15}$$

where R_{ij} is the reliability of component j in subsystem i; $i = 1, 2, \ldots, n$ and $j = 1, 2, \ldots, m_i$. The n blocks constitute a series system equivalent to the original system, as shown in Figure 4.8. Then the system reliability R is obtained from (4.1) and (4.15) as

$$R = \prod_{i=1}^{n}\left[1 - \prod_{j=1}^{m_i}(1 - R_{ij})\right].\tag{4.16}$$

When all components in the series–parallel system are identical and the number of components in each subsystem is equal, (4.16) simplifies to

$$R = [1 - (1 - R_0)^m]^n,\tag{4.17}$$

where R_0 is the reliability of an individual component and m is the number of components in each subsystem.

4.5.2   Parallel–Series Systems

A general *parallel–series system* consists of m subsystems in parallel with n_i ($i = 1, 2, \ldots, m$) components in subsystem i, as shown in Figure 4.9. The configuration is also known as the *high-level redundancy design*. To calculate the system reliability, we first collapse each series subsystem to an equivalent reliability block. From (4.1) the reliability R_i of block i is

$$R_i = \prod_{j=1}^{n_i} R_{ij}, \qquad i = 1, 2, \ldots, m,\tag{4.18}$$

where R_{ij} is the reliability of component j in subsystem i. The m blocks form a parallel system equivalent to the original one, as shown in Figure 4.10. Substituting (4.18) into (4.10) gives the reliability of the parallel–series system as

$$R = 1 - \prod_{i=1}^{m}\left(1 - \prod_{j=1}^{n_i} R_{ij}\right)\tag{4.19}$$

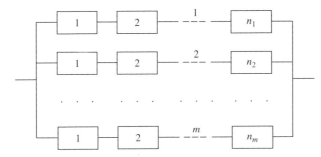

FIGURE 4.9   General parallel–series system

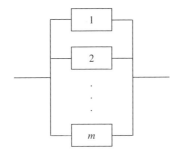

FIGURE 4.10   Reliability block diagram equivalent to Figure 4.9

If all components in the parallel–series system are identical and the number of components in each subsystem is equal, the system reliability can be written as

$$R = 1 - (1 - R_0^n)^m, \tag{4.20}$$

where R_0 is the reliability of an individual component and n is the number of components in each series subsystem.

Example 4.5   Suppose that an engineer is given four identical components, each having 90% reliability at the design life. The engineer wants to choose the system design that has a higher reliability from between the series–parallel and parallel–series configurations. The two configurations are shown in Figures 4.11 and 4.12. Which design should the engineer select from the reliability perspective?

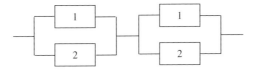

FIGURE 4.11   Series–parallel design

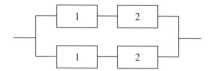

FIGURE 4.12   Parallel–series design

SOLUTION   From (4.17), the reliability of the series–parallel design is

$$R = [1 - (1 - 0.9)^2]^2 = 0.9801.$$

From (4.20), the reliability of the parallel–series design is

$$R = 1 - (1 - 0.9^2)^2 = 0.9639.$$

Obviously, the series–parallel design should be chosen.

In general, the reliability of a series–parallel system is larger than that of a parallel–series system if both use the same number of components. To illustrate this statement numerically, Figure 4.13 plots the reliabilities of the two systems versus the component reliability for different combinations of the values of m and n. In the figure, S-P stands for series–parallel and P-S for parallel–series. It is seen that the difference between the reliabilities is considerable if the component

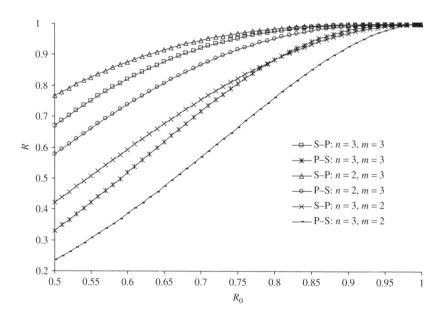

FIGURE 4.13   Reliability of series–parallel and parallel–series systems

reliability is low. However, delta decreases as the component reliability increases and becomes negligible when it is very high, say 0.99. Figure 4.13 also indicates that, given the same number of components, a system with $n > m$ has a lower reliability than one with $m > n$.

4.6 *k*-OUT-OF-*n* SYSTEMS

As presented in Section 4.4, a parallel system is operationally successful if at least one component functions. In reality, there are systems that require more than one component to succeed in order for the entire system to operate. Such systems are often encountered. A power-generating system that consists of four generators working in a derating mode may require at least two generators to operate in full mode simultaneously to deliver sufficient power. Web hosts may be installed with five servers; at least three of them must be functional so that the web service is not interrupted. In a positioning system equipped with five sensors, a minimum of three sensors operable is required to determine the location of an object. Systems of this type are usually referred to as the *k*-out-of-*n*:G systems, where n is the total number of components in the system, k is the minimum number of n components that must function for the system to operate successfully, and G stands for "good," meaning *success*. By the definition, a parallel system is a 1-out-of-*n*:G system, whereas a series system is an *n*-out-of-*n*:G system. Sometimes, we may be interested in defining a system in terms of failure. A system is known as a *k*-out-of-*n*:F system, where F stands for "failure," if and only if the failure of at least *k* components causes the *n*-component system to fail. Following this definition, a parallel system is an *n*-out-of-*n*:F system, and a series system is a 1-out-of-*n*:F system. Apparently, a *k*-out-of-*n*:G system is equivalent to an $(n - k + 1)$-out-of-*n*:F system. Because of the equivalence relationship, in this chapter we study only the *k*-out-of-*n*:G system.

Suppose that the times to failure of n components in a *k*-out-of-*n*:G system are independently and identically distributed. Let x be the number of operational components in the system. Then x is a random variable and follows the binomial distribution. The probability of having exactly k components operational is

$$\text{Pr}(x = k) = C_n^k R_0^k (1 - R_0)^{n-k}, \qquad k = 0, 1, \ldots, n, \qquad (4.21)$$

where R_0 is the reliability of a component. $\qquad C_n^k = n! / k!(n-k)!$

Since an operable *k*-out-of-*n*:G system requires at least k components to be functional, the system reliability R is

$$R = \text{Pr}(x \geq k) = \sum_{i=k}^{n} C_n^i R_0^i (1 - R_0)^{n-i}. \qquad (4.22)$$

$k = 0, 1, 2, \ldots, n)$

When $k = 1$, that is, the n components are in parallel, (4.22) becomes $R = 1 - (1 - R_0)^n$. This is the same as (4.11).

When $k = n$, that is, the n components are in series, (4.22) can be written as $R = R_0^n$. This is identical to (4.2).

If the time to failure is exponential, the system reliability is

$$R(t) = \sum_{i=k}^{n} C_n^i e^{-\lambda i t} (1 - e^{-\lambda t})^{n-i}, \tag{4.23}$$

where λ is the component failure rate. The mean time to failure of the system is

$$\text{MTTF} = \int_0^{\infty} R(t)\,dt = \frac{1}{\lambda} \sum_{i=k}^{n} \frac{1}{i}. \tag{4.24}$$

Note that (4.24) and (4.14) are the same when $k = 1$.

Example 4.6   A web host has five independent and identical servers connected in parallel. At least three of them must operate successfully for the web service not to be interrupted. The server life is modeled with the exponential distribution with $\lambda = 2.7 \times 10^{-5}$ failures per hour. Calculate the mean time between failures (MTBF) and the reliability of the web host after one year of continuous service.

SOLUTION   The web host is a 3-out-of-5:G system. If a failed server is repaired immediately to a good-as-new condition, the MTBF is the same as the MTTF and can be calculated from (4.24) as

$$\text{MTBF} = \frac{1}{2.7 \times 10^{-5}} \sum_{i=3}^{5} \frac{1}{i} = 2.9 \times 10^4 \text{ hours.}$$

Substituting the given data into (4.23) yields the reliability of the web host at 8760 hours (one year) as

$$R(8760) = \sum_{i=3}^{5} C_5^i e^{-2.7 \times 10^{-5} \times 8760 i} (1 - e^{-2.7 \times 10^{-5} \times 8760})^{5-i} = 0.9336.$$

As discussed above, a 1-out-of-n:G system is a pure parallel system. In general, a k-out-of-n:G system can be transformed to a parallel system which consists of C_n^k paths, each with k different components. To illustrate this transformation, we consider a 2-out-of-3:G system. The equivalent parallel system has $C_3^2 = 3$ parallel paths, and each path has two components. The reliability block diagram of the parallel system is shown in Figure 4.14. With the notation defined in Section 4.4, the probability of failure of the parallel system can be written as

$$F = \Pr(\overline{E_1 \cdot E_2} \cdot \overline{E_1 \cdot E_3} \cdot \overline{E_2 \cdot E_3}) = \Pr[(\overline{E}_1 + \overline{E}_2) \cdot (\overline{E}_1 + \overline{E}_3) \cdot (\overline{E}_2 + \overline{E}_3)].$$

By using the Boolean rules (Chapter 6), this equation simplifies to

$$F = \Pr(\overline{E}_1 \cdot \overline{E}_2 + \overline{E}_1 \cdot \overline{E}_3 + \overline{E}_2 \cdot \overline{E}_3). \tag{4.25}$$

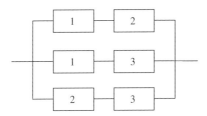

FIGURE 4.14   Reliability block diagram equivalent to a 2-out-of-3:G system

The equation indicates that the system fails if any of the three events $\overline{E}_1 \cdot \overline{E}_2$, $\overline{E}_1 \cdot \overline{E}_3$, or $\overline{E}_2 \cdot \overline{E}_3$ occurs. The event is called a *minimal cut set*. The definition and application of the minimal cut set is presented in Section 4.8 and discussed further in Chapter 6. As shown in (4.25), a 2-out-of-3:G system has three minimal cut sets, and each contains two elements. In general, a k-out-of-n:G system contains C_n^{n-k+1} minimal cut sets, and each consists of exactly k elements.

Let's continue the computation of the probability of failure. Equation (4.25) can be expanded to

$$F = \Pr(\overline{E}_1 \cdot \overline{E}_2) + \Pr(\overline{E}_1 \cdot \overline{E}_3) + \Pr(\overline{E}_2 \cdot \overline{E}_3) - 2\Pr(\overline{E}_1 \cdot \overline{E}_2 \cdot \overline{E}_3).$$

Since $\overline{E}_1$, $\overline{E}_2$, and $\overline{E}_3$ are mutually independent, the system reliability can be written as

$$
\begin{aligned}
R &= 1 - F \\
&= 1 - (1 - R_1)(1 - R_2) - (1 - R_1)(1 - R_3) - (1 - R_2)(1 - R_3) \\
&\quad + 2(1 - R_1)(1 - R_2)(1 - R_3).
\end{aligned}
\tag{4.26}
$$

If the components are identical and have a common reliability R_0, (4.26) becomes

$$R = 1 - (1 + 2R_0)(1 - R_0)^2.$$

The reliability is the same as that obtained from (4.22). Note that unlike (4.22), (4.26) does not require the components to be identical in order to calculate the system reliability. Hence, transformation of a k-out-of-n:G system to an equivalent parallel system provides a method for calculating the system reliability for cases where component reliabilities are unequal.

4.7   REDUNDANT SYSTEMS

A redundant system contains one or more standby components or subsystems in system configuration. These standby units will enable the system to continue the function when the primary unit fails. Failure of the system occurs only when some or all of standby units fail. Hence, redundancy is a system design technique that can increase system reliability. Such a technique is used widely in critical systems. A simple example is an automobile equipped with a spare tire. Whenever a tire fails, it is replaced with the spare tire so that the vehicle is still drivable. A more complicated example is described in W. Wang and Loman (2002). A power plant designed by General Electric consists of n active and one or more standby generators. Normally, each of the n generators runs at $100(n - 1)/n$ percent of its full load and together supplies 100% load to end users, where $n - 1$ generators can fully cover the load. When any one of the active generators fails, the remaining $n - 1$ generators will make up the power loss such that the output is still 100%. Meanwhile, the standby generator is activated and ramps to $100(n - 1)/n$ percent, while the other $n - 1$ generators ramp back down to $100(n - 1)/n$ percent.

If a redundant unit is fully energized when the system is in use, the redundancy is called *active* or *hot standby*. Parallel and k-out-n:G systems described in the preceding sections are typical examples of active standby systems. If a redundant unit is fully energized only when the primary unit fails, the redundancy is known as *passive standby*. When the primary unit is successfully operational, the redundant unit may be kept in reserve. Such a unit is said to be in *cold standby*. A cold standby system needs a sensing mechanism to detect failure of the primary unit and a switching actuator to activate the redundant unit when a failure occurs. In the following discussion we use the term *switching system* to include both the sensing mechanism and the switching actuator. On the other hand, if the redundant unit is partially loaded in the waiting period, the redundancy is a *warm standby*. A warm standby unit usually is subjected to a reduced level of stress and may fail before it is fully activated. According to the classification scheme above, the spare tire and redundant generators described earlier are in cold standby. In the remainder of this section we consider cold standby systems with a perfect or imperfect switching system. Figure 4.15 shows a cold standby system consisting of n components and a switching system; in this figure, component 1 is the primary component and S represents the switching system.

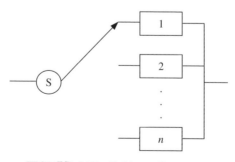

FIGURE 4.15   Cold standby system

4.7.1 Cold Standby Systems with a Perfect Switching System

If the switching system is 100% reliable, system reliability is determined by the n components. Let T_i denote the time to failure of component i $(i = 1, 2, \ldots, n)$ and T denote that of the entire system. Obviously,

$$T = \sum_{i=1}^{n} T_i. \tag{4.27}$$

If $T_1, T_2, \ldots, T_n$ are independently and exponentially distributed with failure rate λ, T follows a gamma distribution with parameters n and λ. The probability density function (pdf) is

$$f(t) = \frac{\lambda^n}{\Gamma(n)} t^{n-1} e^{-\lambda t}, \tag{4.28}$$

where $\Gamma(\cdot)$ is the gamma function, defined in Section 2.5. The system reliability is

$$R(t) = \int_t^\infty \frac{\lambda^n}{\Gamma(n)} t^{n-1} e^{-\lambda t} dt = e^{-\lambda t} \sum_{i=0}^{n-1} \frac{(\lambda t)^i}{i!}. \tag{4.29}$$

The mean time to failure of the system is given by the gamma distribution as

$$\text{MTTF} = \frac{n}{\lambda}. \tag{4.30}$$

Alternatively, (4.30) can also be derived from (4.27). Specifically,

$$\text{MTTF} = E(T) = \sum_{i=1}^{n} E(T_i) = \sum_{i=1}^{n} \frac{1}{\lambda} = \frac{n}{\lambda}.$$

If there is only one standby component, the system reliability is obtained from (4.29) by setting $n = 2$. Then we have

$$R(t) = (1 + \lambda t) e^{-\lambda t}. \tag{4.31}$$

Example 4.7 A small power plant is equipped with two identical generators, one active and the other in cold standby. Whenever the active generator fails, the redundant generator is switched to working condition without interruption. The life of the two generators can be modeled with the exponential distribution with $\lambda = 3.6 \times 10^{-5}$ failures per hour. Calculate the power plant reliability at 5000 hours and the mean time to failure.

SOLUTION   Substituting the data into (4.31) yields

$$R(5000) = (1 + 3.6 \times 10^{-5} \times 5000) e^{-3.6 \times 10^{-5} \times 5000} = 0.9856.$$

By setting $n = 2$ in (4.30), we obtain the mean time to failure as

$$\text{MTTF} = \frac{2}{3.6 \times 10^{-5}} = 5.56 \times 10^4 \text{ hours.}$$

If the n components are not identically and exponentially distributed, the computation of system reliability is rather complicated. Now let's consider a simple case where the cold standby system is comprised of two components. The system will survive time t if any of the following two events occurs:

- The primary component (whose life is T_1) does not fail in time t; that is, $T_1 \geq t$.
- If the primary component fails at time $\tau (\tau < t)$, the cold standby component (whose life is T_2) continues the function and does not fail in the remaining time $(t - \tau)$. Probabilistically, the event is described by $(T_1 < t) \cdot (T_2 \geq t - \tau)$.

Since the above two events are mutually exclusive, the system reliability is

$$R(t) = \Pr[(T_1 \geq t) + (T_1 < t) \cdot (T_2 \geq t - \tau)] = \Pr(T_1 \geq t)$$
$$+ \Pr[(T_1 < t) \cdot (T_2 \geq t - \tau)]$$
$$= R_1(t) + \int_0^t f_1(\tau) R_2(t - \tau) \, d\tau, \tag{4.32}$$

where R_i and f_i are, respectively, the reliability and pdf of component i. In most situations, evaluation of (4.32) requires a numerical method. As a special case, when the two components are identically and exponentially distributed, (4.32) can result in (4.31).

4.7.2 Cold Standby Systems with an Imperfect Switching System

A switching system consists of a failure detection mechanism and a switching actuator, and thus may be complicated in nature. In practice, it is subject to failure. Now we consider a two-component cold standby system. By modifying (4.32), we can obtain the system reliability as

$$R(t) = R_1(t) + \int_0^t R_0(\tau) f_1(\tau) R_2(t - \tau) \, d\tau, \tag{4.33}$$

where $R_0(\tau)$ is the reliability of the switching system at time τ. In the following discussion we assume that the two components are identically and exponentially distributed with parameter λ, and deal with two cases in which $R_0(\tau)$ is static or dynamic.

For some switching systems, such as human operators, the reliability may not change over time. In these situations, $R_0(\tau)$ is static or independent of time. Let $R_0(\tau) = p_0$. Then (4.33) can be written as

$$R(t) = e^{-\lambda t} + p_0 \int_0^t \lambda e^{-\lambda \tau} e^{-\lambda(t-\tau)} d\tau = (1 + p_0 \lambda t) e^{-\lambda t}. \tag{4.34}$$

Note the similarity and difference between (4.31) for a perfect switching system and (4.34) for an imperfect one. Equation (4.34) reduces to (4.31) when $p_0 = 1$. The mean time to failure of the system is

$$\text{MTTF} = \int_0^\infty R(t)\, dt = \frac{1 + p_0}{\lambda}. \tag{4.35}$$

Now we consider the situation where $R_0(\tau)$ is dynamic or dependent on time. Most modern switching systems contain both hardware and software and are complicated in nature. They can fail in different modes before the primary components break. If such failure occurs, the standby components will never be activated to undertake the function of the failed primary components. Since switching systems deteriorate over time, it is realistic to assume that the reliability of such systems is a function of time. If the life of a switching system is exponentially distributed with parameter λ_0, from (4.33) the reliability of the entire system is

$$R(t) = e^{-\lambda t} + \int_0^t e^{-\lambda_0 \tau} \lambda e^{-\lambda \tau} e^{-\lambda(t-\tau)} d\tau = e^{-\lambda t}\left[1 + \frac{\lambda}{\lambda_0}\left(1 - e^{-\lambda_0 t}\right)\right]. \tag{4.36}$$

The mean time to failure is

$$\text{MTTF} = \int_0^\infty R(t)\, dt = \frac{1}{\lambda} + \frac{1}{\lambda_0} - \frac{\lambda}{\lambda_0(\lambda + \lambda_0)}. \tag{4.37}$$

As will be shown in Example 4.8, an imperfect switching system reduces the reliability and MTTF of the entire system. To help better understand this, we first denote by r_0 the ratio of the reliability at time $1/\lambda$ with an imperfect switching system to that with a perfect system, by r_1 the ratio of the MTTF with an imperfect switching system to that with a perfect one, and by δ the ratio of λ to λ_0. Then from (4.31) and (4.36), we have

$$r_0 = \tfrac{1}{2}[1 + \delta(1 - e^{-1/\delta})]. \tag{4.38}$$

From (4.30) with $n = 2$ and (4.37), we obtain

$$r_1 = \frac{1}{2}\left[1 + \delta\left(1 - \frac{\delta}{1+\delta}\right)\right]. \tag{4.39}$$

Figure 4.16 plots r_0 and r_1 for various values of δ. It can be seen that the unreliability of the switching system has stronger effects on MTTF than on the

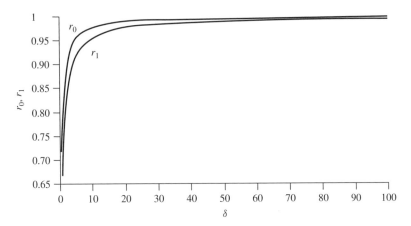

FIGURE 4.16 Plots of r_0 and r_1 for different values of δ

reliability of the entire system. Both quantities are largely reduced when λ_0 is greater than 10% of λ. The effects are alleviated by the decrease in λ_0, and become nearly negligible when λ_0 is less than 1% of λ.

Example 4.8   Refer to Example 4.7. Suppose that the switching system is subject to failure following the exponential distribution with $\lambda_0 = 2.8 \times 10^{-5}$ failures per hour. Calculate the power plant reliability at 5000 hours and the mean time to failure.

SOLUTION   Substituting the data to (4.36) yields

$$R(5000) = e^{-3.6 \times 10^{-5} \times 5000} \left[1 + \frac{3.6 \times 10^{-5}}{2.8 \times 10^{-5}} \left(1 - e^{-2.8 \times 10^{-5} \times 5000} \right) \right]$$

$$= 0.9756.$$

The mean time to failure is obtained from (4.37) as

$$\text{MTTF} = \frac{1}{3.6 \times 10^{-5}} + \frac{1}{2.8 \times 10^{-5}} - \frac{3.6 \times 10^{-5}}{2.8 \times 10^{-5}(3.6 \times 10^{-5} + 2.8 \times 10^{-5})}$$

$$= 4.34 \times 10^4 \text{ hours.}$$

Comparing these results with those in Example 4.7, we note the adverse effects of the imperfect switching system.

4.8   RELIABILITY EVALUATION OF COMPLEX SYSTEMS

So far we have studied series, parallel, series–parallel, parallel–series, k-out-of-n, and redundant systems. In reality, these configurations are frequently combined and form more complex systems in order to fulfill functional requirements. Some networks, such as power supply grids, telecommunication systems, and computing networks, are so complicated in structure that they cannot easily be decomposed into the said configurations. Reliability evaluation of complex systems requires more advanced methods. In this section we present three simple yet powerful approaches. For large-scale complex systems, manual calculation of reliability is difficult, if not prohibitive. Various commercial software packages, such as Reliasoft, Relex, and Item, are capable of calculating reliability and other measures of reliability of complex systems through simulation.

4.8.1 Reduction Method

Some systems are made up of independent series, parallel, series–parallel, parallel–series, k-out-of-n, and redundant subsystems. The system reduction method is to collapse a system sequentially into the foregoing subsystems, each represented by an equivalent reliability block. The reliability block diagram is further reduced until the entire system is represented by a single reliability block. The method is illustrated in the following example.

Example 4.9 The reliability block diagram of an engineering system is shown in Figure 4.17. The times to failures of the components are modeled with the exponential distribution with the failure rates shown by the corresponding blocks with a multiplication of 10^{-4} failures per hour. Compute the system reliability at 600 hours of mission time.

SOLUTION   The steps for calculating the system reliability are as follows:

1. Decompose the system into blocks A, B, C, and D, which represent a parallel–series, parallel, series, and cold standby subsystem, respectively, as shown in Figure 4.17.
2. Calculate the reliabilities of blocks A, B, C, and D. From (4.19), the reliability of block A is

$$R_A = 1 - (1 - R_1 R_2)(1 - R_3 R_4) = 1 - \left[1 - e^{-(1.2+2.3)\times 10^{-4}\times 600}\right]$$
$$\left[1 - e^{-(0.9+1.6)\times 10^{-4}\times 600}\right]$$
$$= 0.9736.$$

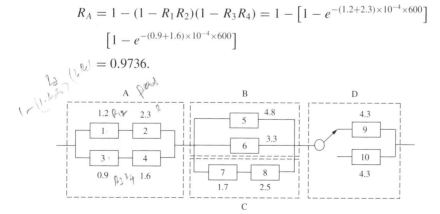

FIGURE 4.17   Engineering system of Example 4.9

From (4.10), the reliability of block B is

$$R_B = 1 - (1 - R_5)(1 - R_6) = 1 - (1 - e^{-4.8 \times 10^{-4} \times 600})(1 - e^{-3.3 \times 10^{-4} \times 600})$$
$$= 0.955.$$

From (4.1) we obtain the reliability of block C as

$$R_C = e^{-(1.7+2.5) \times 10^{-4} \times 600} = 0.7772.$$

From (4.31) the reliability of block D is

$$R_D = (1 + 4.3 \times 10^{-6} \times 600) \times e^{-4.3 \times 10^{-4} \times 600} = 0.9719.$$

The equivalent reliability block diagram is shown in Figure 4.18.

3. The equivalent system in Figure 4.18 is further reduced to blocks E and F, which are series and parallel subsystems, respectively.
4. Calculate the reliabilities of blocks E and F.

$$R_E = R_A R_D = 0.9736 \times 0.9719 = 0.9462.$$
$$R_F = 1 - (1 - R_B)(1 - R_C) = 1 - (1 - 0.955)(1 - 0.7772) = 0.99.$$

The reliability block diagram equivalent to Figure 4.18 is shown in Figure 4.19.

5. The equivalent system in Figure 4.19 consists of two units in series. It is reduced to one single block, G.
6. Calculate the reliability of block G.

$$R_G = R_E R_F = 0.9462 \times 0.99 = 0.9367.$$

Now that the original system has been reduced to a single unit, as shown in Figure 4.20, the reduction process is exhausted. Then the system reliability is $R = R_G = 0.9367.$

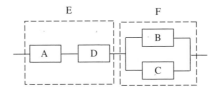

FIGURE 4.18   Reduced system equivalent to Figure 4.17

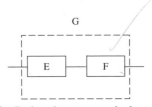

FIGURE 4.19   Reduced system equivalent to Figure 4.18

FIGURE 4.20 Reduced system equivalent to Figure 4.19

4.8.2 Decomposition Method

The system reduction method is effective when a complex system can be partitioned into a number of simple subsystems whose reliabilities are directly obtainable. In some situations we encounter more complex systems, such as the well-known bridge system shown in Figure 4.21, which cannot be solved by the reduction method. In this subsection we present a decomposition method also known as the *conditional probability approach* or *Bayes' theorem method*.

The decomposition method starts with choosing a keystone component, say A, from the system being studied. This component appears to bind the system together. In Figure 4.21, for instance, component 5 is a such keystone component. The keystone component is assumed to be 100% reliable and is replaced with a line in system structure. Then the same component is supposed to have failed and is removed from the system. The system reliability is calculated under each assumption. According to the rule of total probability, the reliability of the original system can be written as

$$R = \Pr(\text{system good} \mid A)\Pr(A) + \Pr(\text{system good} \mid \overline{A})\Pr(\overline{A}), \qquad (4.40)$$

where A is the event that keystone component A is 100% reliable, $\overline{A}$ the event that keystone component A has failed, $\Pr(\text{system good} \mid A)$ the probability that the system is functionally successful given that component A never fails, and $\Pr(\text{system good} \mid \overline{A})$ the probability that the system is functionally successful given that component A has failed. The efficiency of the method depends on the selection of the keystone component. An appropriate choice of the component leads to an efficient calculation of the conditional probabilities.

Example 4.10 Consider the bridge system in Figure 4.21. Suppose that the reliability of component i is $R_i, i = 1, 2, \ldots, 5$. Calculate the system reliability.

SOLUTION Component 5 is chosen as the keystone component, denoted A. Assume that it never fails and is replaced with a line in the system configuration.

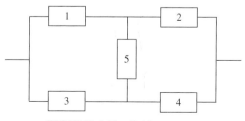

FIGURE 4.21 Bridge system

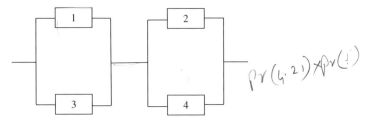

FIGURE 4.22 Bridge system when component 5 never fails

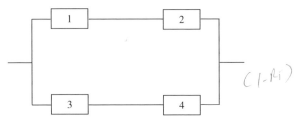

FIGURE 4.23 Bridge system when component 5 has failed

Then the system is reduced as shown in Figure 4.22. The reduced system is a series–parallel structure, and the conditional reliability is

$$\text{Pr(system good} \mid A) = [1 - (1 - R_1)(1 - R_3)][1 - (1 - R_2)(1 - R_4)].$$

The next step is to assume that component 5 has failed and is removed from the system structure. Figure 4.23 shows the new configuration, which is a parallel–series system. The conditional reliability is

$$\text{Pr(system good} \mid \overline{A}) = 1 - (1 - R_1 R_2)(1 - R_3 R_4).$$

The reliability and unreliability of component 5 are $\text{Pr}(A) = R_5$ and $\text{Pr}(\overline{A}) = 1 - R_5$, respectively. Substituting the equations above into (4.40) yields the reliability of the original system as

$$
\begin{aligned}
R &= [1 - (1 - R_1)(1 - R_3)][1 - (1 - R_2)(1 - R_4)]R_5 \\
&\quad + [1 - (1 - R_1 R_2)(1 - R_3 R_4)](1 - R_5) \\
&= R_1 R_2 + R_3 R_4 + R_1 R_4 R_5 + R_2 R_3 R_5 - R_1 R_2 R_4 R_5 \\
&\quad - R_2 R_3 R_4 R_5 - R_1 R_2 R_3 R_5 \\
&\quad - R_1 R_3 R_4 R_5 - R_1 R_2 R_3 R_4 + 2 R_1 R_2 R_3 R_4 R_5. \qquad (4.41)
\end{aligned}
$$

As illustrated in Example 4.10, calculating the reliability of the bridge system needs the selection of only one keystone component, and (4.40) is applied once.

For some complex systems, the reliabilities of decomposed systems cannot be written out directly. In these situations we may select additional keystone components and apply (4.40) successively until each term in the equation is easily obtainable. For example, if Pr(system good | A) cannot be worked out immediately, the decomposed system with A functional may be further decomposed by selecting an additional keystone component, say B. Applying (4.40) to keystone component B, we may write the reliability of the original system as

$$R = \Pr(\text{system good} \,|\, A \cdot B)\Pr(A)\Pr(B) + \Pr(\text{system good} \,|\, A \cdot \overline{B})\Pr(A)\Pr(\overline{B})$$
$$+ \Pr(\text{system good} \,|\, \overline{A})\Pr(\overline{A}). \tag{4.42}$$

The decomposition method discussed above selects one keystone component at a time. W. Wang and Jiang (2004) suggest that several such components be chosen simultaneously for some complex networks. For example, if two keystone components, say A and B, are selected, the original system will be decomposed into four subsystems with conditions $A \cdot B$, $\overline{A} \cdot B$, $A \cdot \overline{B}$, and $\overline{A} \cdot \overline{B}$, respectively, where $A \cdot B$ is the event that both A and B are functioning, $\overline{A} \cdot B$ is the event that A is not functioning and B is, $A \cdot \overline{B}$ is the event that A is functioning and B is not, and $\overline{A} \cdot \overline{B}$ is the event that both A and B are not functioning. By applying the rule of total probability, the reliability of the original system can be written as

$$R = \Pr(\text{system good} \,|\, A \cdot B)\Pr(A)\Pr(B) + \Pr(\text{system good} \,|\, \overline{A} \cdot B)\Pr(\overline{A})\Pr(B)$$
$$+ \Pr(\text{system good} \,|\, A \cdot \overline{B})\Pr(A)\Pr(\overline{B})$$
$$+ \Pr(\text{system good} \,|\, \overline{A} \cdot \overline{B})\Pr(\overline{A})\Pr(\overline{B}). \tag{4.43}$$

Equation (4.43) has four terms. In general, for binary components, if m keystone components are selected simultaneously, the reliability equation contains 2^m terms. Each term is the product of the reliability of one of the decomposed subsystems and that of the condition on which the subsystem is formed.

4.8.3 Minimal Cut Set Method

The decomposition method studied earlier is based on the rule of total probability. In this subsection we present an approach to system reliability evaluation by using a minimal cut set and the inclusion–exclusion rule. First let's discuss cut sets. A *cut set* is a set of components whose failure interrupts all connections between input and output ends and thus causes an entire system to fail. In Figure 4.21, for example, $\{1, 3, 5\}$ and $\{2, 4\}$ are cut sets. Some cut sets may contain unnecessary components. If removed, failure of the remaining components still results in system failure. In the example above, cut set $\{1, 3, 5\}$ contains component 5, which can be eliminated from the cut set without changing the failure state of the system. Such cut sets can be further reduced to form minimal cut sets. A *minimal cut set* is the smallest combination of components which if they all fail will cause the system to fail. A minimal cut set represents the smallest collection of components whose failures are necessary and sufficient to result in system

Path set = set of components required for success

failure. If any component is removed from the set, the remaining components collectively are no longer a cut set. The definitions of cut set and minimal cut set are similar to those defined in Chapter 6 for fault tree analysis.

Since every minimal cut set causes the system to fail, the event that the system breaks is the union of all minimal cut sets. Then the system reliability can be written as

$$R = 1 - \Pr(C_1 + C_2 + \cdots + C_n), \tag{4.44}$$

where C_i $(i = 1, 2, \ldots, n)$ represents the event that components in minimal cut set i are all in a failure state and n is the total number of minimal cut sets. Equation (4.44) can be evaluated by applying the inclusion–exclusion rule, which is

$$\Pr(C_1 + C_2 + \cdots + C_n) = \sum_{i=1}^{n} \Pr(C_i) - \sum_{i<j=2}^{n} \Pr(C_i \cdot C_j)$$
$$+ \sum_{i<j<k=3}^{n} \Pr(C_i \cdot C_j \cdot C_k) + \cdots$$
$$+ (-1)^{n-1} \Pr(C_1 \cdot C_2 \cdots C_n). \tag{4.45}$$

Example 4.11 Refer to Example 4.10. If the five components are identical and have a common reliability R_0, calculate the reliability of the bridge system shown in Figure 4.21 using the minimal cut set method.

SOLUTION The minimal cut sets of the bridge system are $\{1, 3\}$, $\{2, 4\}$, $\{1, 4, 5\}$, and $\{2, 3, 5\}$. Let A_i denote the event that component i has failed, $i = 1, 2, \ldots, 5$. Then the events described by the minimal cut sets can be written as $C_1 = A_1 \cdot A_3$, $C_2 = A_2 \cdot A_4$, $C_3 = A_1 \cdot A_4 \cdot A_5$, and $C_4 = A_2 \cdot A_3 \cdot A_5$. From (4.44) and (4.45) and using the rules of Boolean algebra (Chapter 6), the system reliability can be written as

$$R = 1 - \left[\sum_{i=1}^{4} \Pr(C_i) - \sum_{i<j=2}^{4} \Pr(C_i \cdot C_j) + \sum_{i<j<k=3}^{4} \Pr(C_i \cdot C_j \cdot C_k) \right.$$
$$\left. - \Pr(C_1 \cdot C_2 \cdot C_3 \cdot C_4) \right]$$

$$= 1 - [\Pr(A_1 \cdot A_3) + \Pr(A_2 \cdot A_4) + \Pr(A_1 \cdot A_4 \cdot A_5) + \Pr(A_2 \cdot A_3 \cdot A_5)$$
$$- \Pr(A_1 \cdot A_2 \cdot A_3 \cdot A_4) - \Pr(A_1 \cdot A_3 \cdot A_4 \cdot A_5) - \Pr(A_1 \cdot A_2 \cdot A_3 \cdot A_5)$$
$$- \Pr(A_1 \cdot A_2 \cdot A_4 \cdot A_5) - \Pr(A_2 \cdot A_3 \cdot A_4 \cdot A_5) + 2\Pr(A_1 \cdot A_2 \cdot A_3 \cdot A_4 \cdot A_5)]$$
$$= 1 - [2(1 - R_0)^2 + 2(1 - R_0)^3 - 5(1 - R_0)^4 + 2(1 - R_0)^5]$$
$$= 2R_0^5 - 5R_0^4 + 2R_0^3 + 2R_0^2.$$

Note that (4.41) gives the same result when all components have equal reliability R_0.

4.9 CONFIDENCE INTERVALS FOR SYSTEM RELIABILITY

In the preceding sections of this chapter we presented methods for calculating system reliability from component data. The reliabilities of components are usually unknown in practice and are estimated from test data, field failures, historical data, or other sources. The estimates inevitably contain statistical errors. In turn, system reliability calculated from such estimates deviates from the true value. Therefore, it is often desirable to estimate confidence intervals for system reliability, and the lower bound is of special interest. In this section we describe methods for estimating confidence intervals for system reliability.

In the literature, there are numerous approaches to estimation of confidence intervals, which were developed for different instances. For example, Crowder et al. (1991) present the Lindstrom and Madden approximate method for systems with binomial components. Mann (1974) develops approximately optimum confidence bounds on series and parallel system reliability for systems with binomial component data. The method applies when no component failures occur. Assuming that components are exponentially distributed and highly reliable, Gertsbakh (1982, 1989) describes methods for estimating confidence intervals for series, parallel, series–parallel, and k-out-of-n:G systems. Ushakov (1996) and Gnedenko et al. (1999) give various solution procedures for computing confidence intervals for series, parallel, series–parallel, and complex systems comprised of binomial or exponential components. In addition to these analytical methods, Monte Carlo simulation is a powerful approach to handling complicated cases where component lifetimes are Weibull, normal, lognormal, or other distributions, and systems are in complex configurations. Moore et al. (1980) provide a model for calculating confidence intervals through the maximum likelihood method. The model can be used for simple as well as complex systems comprised of Weibull or gamma components. A. Chao and Hwang (1987) present a modified Monte Carlo technique for calculating system reliability confidence intervals from pass–fail or binomial data. A more comprehensive literature review is given in, for example, Willits et al. (1997) and Tian (2002).

In this section we describe two approaches to estimating confidence intervals for system reliability. The methods make no assumptions as to time to failure distribution of components, but require known estimates of component reliabilities and variances of the estimates. Both may be obtained by using the methods described in Chapters 7 and 8 when component life is modeled with a parametric distribution. Nonparametric estimates such as the Kaplan–Meier estimates can be found in, for example, Meeker and Escobar (1998).

4.9.1 Normal-Approximation Confidence Intervals

System reliability modeling described in previous sections establishes system reliability as a function of component reliabilities. Mathematically,

$$R = h(R_1, R_2, \ldots, R_n), \tag{4.46}$$

where h implies a function, R is the system reliability, R_i $(i = 1, 2, \ldots, n)$ is the reliability of component i, and n is the number of components in the system. Note that both R and R_i may depend on time. Substituting the estimates of R_i into (4.46) gives the estimate of system reliability as

$$\hat{R} = h(\hat{R}_1, \hat{R}_2, \ldots, \hat{R}_n), \tag{4.47}$$

where $\wedge$ implies an estimate.

If the n components are in series, (4.46) reduces to (4.1). According to Coit (1997), the variance of $\hat{R}$ is

$$\text{Var}(\hat{R}) = \prod_{i=1}^{n} \left[R_i^2 + \text{Var}(\hat{R}_i) \right] - \prod_{i=1}^{n} R_i^2. \tag{4.48}$$

If the n components are in parallel, (4.46) becomes (4.10). The variance of $\hat{R}$ is

$$\text{Var}(\hat{R}) = \prod_{i=1}^{n} \left[(1 - R_i)^2 + \text{Var}(\hat{R}_i) \right] - \prod_{i=1}^{n} (1 - R_i)^2. \tag{4.49}$$

Using (4.48) and (4.49), we can easily derive the variance of $\hat{R}$ for a series–parallel or parallel–series system.

For a complex system, the variance of $\hat{R}$ can be approximated by using a Taylor series expansion of (4.46). Then we have

$$\text{Var}(\hat{R}) \approx \sum_{i=1}^{n} \left(\frac{\partial R}{\partial R_i} \right)^2 \text{Var}(\hat{R}_i). \tag{4.50}$$

Note that the covariance terms in the Taylor series expansion are zero since the n components are assumed to be mutually independent. Coefficient $\partial R / \partial R_i$ in (4.50) measures the sensitivity of the system reliability variance to the variation of individual component reliability. As we will see in the next section, the coefficient is also Birnbaum's component importance measure.

Substituting estimates of component reliabilities and component-reliability variances into (4.48), (4.49), or (4.50), we can obtain an estimate of $\text{Var}(\hat{R})$, denoted $\hat{\text{V}}\text{ar}(\hat{R})$. We often approximate the distribution of $\hat{R}$ with a normal distribution. Then the two-sided $100(1 - \alpha)\%$ confidence interval for the system reliability is

$$\hat{R} \pm z_{1-\alpha/2} \sqrt{\hat{\text{V}}\text{ar}(\hat{R})}, \tag{4.51}$$

where $z_{1-\alpha/2}$ is the $100(1 - \alpha/2)$ percentile of the standard normal distribution. The one-sided lower $100(1 - \alpha)\%$ confidence bound is

$$\hat{R} - z_{1-\alpha} \sqrt{\hat{\text{V}}\text{ar}(\hat{R})}. \tag{4.52}$$

Note that (4.51) and (4.52) can yield a negative lower confidence bound. To ensure that the lower confidence bound is always nonnegative, we use the transformation

$$p = \ln \frac{R}{1 - R}.$$ (4.53)

$\hat{p}$ is obtained by replacing R in (4.53) with $\hat{R}$. According to Meeker and Escobar (1998), the random variable

$$Z_{\hat{p}} = \frac{\hat{p} - p}{\sqrt{\hat{V}ar(\hat{p})}}$$

can be approximated by the standard normal distribution, where

$$\sqrt{\hat{V}ar(\hat{p})} = \frac{\sqrt{\hat{V}ar(\hat{R})}}{\hat{R}(1 - \hat{R})}.$$

The distribution of $Z_{\hat{p}}$ leads to the two-sided $100(1 - \alpha)\%$ confidence interval as

$$\left[\frac{\hat{R}}{\hat{R} + (1 - \hat{R}) \times w}, \frac{\hat{R}}{\hat{R} + (1 - \hat{R})/w} \right],$$ (4.54)

where

$$w = \exp \left[\frac{z_{1-\alpha/2}\sqrt{\hat{V}ar(\hat{R})}}{\hat{R}(1 - \hat{R})} \right].$$

The one-sided lower $100(1 - \alpha)\%$ confidence bound is obtained by replacing $z_{1-\alpha/2}$ with $z_{1-\alpha}$ and using the lower endpoint of (4.54).

Example 4.12   Refer to Example 4.10. Suppose that the estimates of component reliabilities and component-reliability variances at the mission time of 1000 hours have been calculated from life test data, as shown in Table 4.1. Estimate the two- and one-sided lower 95% confidence bound(s) on the system reliability.

TABLE 4.1   Estimates of Component Reliabilities and Variances

	Component				
	1	2	3	4	5
$\hat{R}_i$	0.9677	0.9358	0.9762	0.8765	0.9126
$\sqrt{\hat{V}ar(\hat{R}_i)}$	0.0245	0.0173	0.0412	0.0332	0.0141

SOLUTION Substituting estimates of the component reliabilities into (4.41), we obtain the estimate of the system reliability at 1000 hours as $\hat{R} = 0.9909$. To use (4.50) to calculate the variance of the estimate of the system reliability, we first work out the partial derivatives of (4.41) with respective to R_i. They are

$$\frac{\partial R}{\partial R_1} = R_2 + R_4 R_5 - R_2 R_4 R_5 - R_2 R_3 R_5 - R_3 R_4 R_5 - R_2 R_3 R_4 + 2 R_2 R_3 R_4 R_5,$$

$$\frac{\partial R}{\partial R_2} = R_2 + R_3 R_5 - R_1 R_4 R_5 - R_3 R_4 R_5 - R_1 R_3 R_5 - R_1 R_3 R_4 + 2 R_1 R_3 R_4 R_5,$$

$$\frac{\partial R}{\partial R_3} = R_4 + R_2 R_5 - R_2 R_4 R_5 - R_1 R_2 R_5 - R_1 R_4 R_5 - R_1 R_2 R_4 + 2 R_1 R_2 R_4 R_5,$$

$$\frac{\partial R}{\partial R_4} = R_3 + R_1 R_5 - R_1 R_2 R_5 - R_2 R_3 R_5 - R_1 R_3 R_5 - R_1 R_2 R_3 + 2 R_1 R_2 R_3 R_5,$$

$$\frac{\partial R}{\partial R_5} = R_1 R_4 + R_2 R_3 - R_1 R_2 R_4 - R_2 R_3 R_4 - R_1 R_2 R_3 - R_1 R_3 R_4 + 2 R_1 R_2 R_3 R_4.$$

Evaluating the partial derivatives at the estimates of component reliability gives

$$\left.\frac{\partial R}{\partial R_1}\right|_{R_i=\hat{R}_i} = 0.0334, \qquad \left.\frac{\partial R}{\partial R_2}\right|_{R_i=\hat{R}_i} = 0.1248, \qquad \left.\frac{\partial R}{\partial R_3}\right|_{R_i=\hat{R}_i} = 0.0365,$$

$$\left.\frac{\partial R}{\partial R_4}\right|_{R_i=\hat{R}_i} = 0.0666, \qquad \left.\frac{\partial R}{\partial R_5}\right|_{R_i=\hat{R}_i} = 0.0049.$$

Substituting into (4.50) the values of the partial derivatives and the estimates of the component-reliability variances, we obtain the estimate of the system reliability variance as

$$\hat{V}ar(\hat{R}) \approx \sum_{i=1}^{5} \left(\left.\frac{\partial R}{\partial R_i}\right|_{R_i=\hat{R}_i}\right)^2 \hat{V}ar(\hat{R}_i)$$

$$= 0.0334^2 \times 0.0245^2 + 0.1248^2 \times 0.0173^2 + 0.0365^2 \times 0.0412^2$$

$$+ 0.0666^2 \times 0.0332^2 + 0.0049^2 \times 0.0141^2 = 1.25 \times 10^{-5}.$$

Here (4.54) is used to calculate the confidence intervals. First we calculate the value of w for the two-sided 95% confidence intervals. It is

$$w = \exp\left[\frac{1.96 \times \sqrt{1.25 \times 10^{-5}}}{0.9909 \times (1 - 0.9909)}\right] = 2.157.$$

Substituting the values of w and $\hat{R}$ into (4.54), we get the two-sided 95% confidence interval for the system reliability as [0.9806, 0.9957]. Now we calculate the value of w for the one-sided 95% confidence bound. It is

$$w = \exp\left[\frac{1.6448 \times \sqrt{1.25 \times 10^{-5}}}{0.9909 \times (1 - 0.9909)}\right] - 1.9058.$$

Substituting the value of w and $\hat{R}$ into the lower endpoint of (4.54) gives 0.9828, which is the one-sided lower 95% confidence bound on the system reliability.

4.9.2 Lognormal-Approximation Confidence Intervals

The confidence intervals described in Section 4.9.1 are based on the normal approximation to the distribution of system reliability estimate. Now we present a method, due largely to Coit (1997), for the situation in which very few component failures are available. The method is similar to the normal approximation except that the estimate of system reliability (unreliability) is assumed to have the lognormal distribution. This assumption is legitimate for a large-scale system, as we show later.

Like the normal approximation, the method also requires calculation of the variance of system reliability estimate. For a pure series system, the variance is computed by using (4.48); for a pure parallel system, it is calculated from (4.49). If the system is more complex, the variance is determined by using the system reduction method discussed earlier and applying (4.48) and (4.49) sequentially. The process for determination of the variance consists of the following four steps:

1. Partition the system into blocks where each block is comprised of components in pure series or parallel configurations.
2. Calculate the reliability estimates and variances for series blocks using (4.48) and for parallel blocks using (4.49).
3. Collapse each block by replacing it in the system reliability block diagram with an equivalent hypothetic component with the reliability and variance estimates that were calculated for it.
4. Repeat steps 1 to 3 until the system reliability block diagram is represented by a single component. The variance for this component approximates the variance of the original system reliability estimate.

Once the variance is calculated, we estimate the confidence intervals for system reliability. The estimation is based on the assumption that the system reliability (unreliability) estimate has a lognormal distribution. This assumption is reasonable for a relatively large-scale system. For a series configuration of independent subsystems, the system reliability is the product of the subsystem reliability values, as formulated in (4.1). Then the logarithm of the system reliability is the sum of the logarithms of the subsystem reliabilities. According to the central limit theorem, the logarithm of system reliability approximately follows a normal distribution if there are enough subsystems regardless of their time-to-failure distributions. Therefore, the system reliability is lognormal. An analogous argument can be made for a parallel system, where the system unreliability is approximately lognormal. Experimental results from simulation reported in Coit (1997) indicate that the approximation is accurate for any system that can be partitioned into at least eight subsystems in series or parallel.

For a series system, the estimate of system reliability has a lognormal distribution with parameters μ and σ. The mean and variance are, respectively,

$$E(\hat{R}) = \exp\left(\mu + \frac{1}{2}\sigma^2\right),$$

$$\text{Var}(\hat{R}) = \exp(2\mu + \sigma^2)[\exp(\sigma^2) - 1] = [E(\hat{R})]^2[\exp(\sigma^2) - 1].$$

The mean value of the system reliability estimate is the true system reliability: namely, $E(\hat{R}) = R$. Hence, the variance of the log estimate of system reliability can be written as

$$\sigma^2 = \ln\left[1 + \frac{\text{Var}(\hat{R})}{R^2}\right]. \tag{4.55}$$

The estimate of σ, denoted $\hat{\sigma}$, can be obtained by substituting into (4.55) the estimates of system reliability and variance calculated earlier.

Because $\hat{R}$ is lognormal, the random variable $Z_{\ln(\hat{R})} = [\ln(\hat{R}) - \mu]/\sigma$ has the standard normal distribution. This yields the two-sided $100(1 - \alpha)\%$ confidence interval for the system reliability as

$$\left[\hat{R}\exp\left(\frac{1}{2}\hat{\sigma}^2 - z_{1-\alpha/2}\hat{\sigma}\right), \quad \hat{R}\exp\left(\frac{1}{2}\hat{\sigma}^2 + z_{1-\alpha/2}\hat{\sigma}\right)\right]. \tag{4.56}$$

The one-sided lower $100(1 - \alpha)\%$ confidence bound is obtained by replacing $z_{1-\alpha/2}$ with $z_{1-\alpha}$ and using the lower endpoint of (4.56).

Similarly, for a parallel system, the two-sided $100(1 - \alpha)\%$ confidence interval for the system unreliability is

$$\left[\hat{F}\exp\left(\frac{1}{2}\hat{\sigma}^2 - z_{1-\alpha/2}\hat{\sigma}\right), \hat{F}\exp\left(\frac{1}{2}\hat{\sigma}^2 + z_{1-\alpha/2}\hat{\sigma}\right)\right], \tag{4.57}$$

where

$$\hat{F} = 1 - \hat{R} \text{ and } \hat{\sigma}^2 = \ln\left[1 + \frac{\hat{\text{Var}}(\hat{R})}{\hat{F}^2}\right].$$

Note that $\text{Var}(\hat{F}) = \text{Var}(\hat{R})$. The lower and upper bounds on system reliability equal 1 minus the upper and lower bounds on system unreliability from (4.57), respectively. The one-sided lower $100(1 - \alpha)\%$ confidence bound on system reliability is

$$1 - \hat{F}\exp\left(\frac{1}{2}\hat{\sigma}^2 + z_{1-\alpha}\hat{\sigma}\right). \tag{4.58}$$

Coit (1997) restricts the confidence intervals above to systems that can be partitioned into series or parallel blocks in order to calculate the variances of block reliability estimates from (4.48) or (4.49). If the variances for complex blocks are computed from (4.50), the restriction may be relaxed and the confidence intervals are applicable to any large-scale systems that consist of subsystems (containing blocks in various configurations) in series or parallel.

Example 4.13   The reliability block diagram of an engineering system is shown in Figure 4.24. There are nine different types of components in the system. All components are independent, although three component types are used more than once. Suppose that the component reliability and variance at the mission time of 500 hours have been estimated from life tests. The data are given in Table 4.2. Calculate the one-sided lower 95% confidence bound on the system reliability.

SOLUTION   The system can be decomposed into eight subsystems in series. Thus, the lognormal approximation may apply to estimate the confidence bound. First the system is partitioned into parallel and series blocks, as shown in Figure 4.25. The estimates of block reliabilities are

$$\hat{R}_A = \hat{R}_1 \hat{R}_2 = 0.9856 \times 0.9687 = 0.9548,$$

$$\hat{R}_B = 1 - (1 - \hat{R}_3)^2 = 1 - (1 - 0.9355)^2 = 0.9958,$$

$$\hat{R}_C = \hat{R}_4 \hat{R}_5 \hat{R}_6 = 0.9566 \times 0.9651 \times 0.9862 = 0.9105,$$

$$\hat{R}_D = 1 - (1 - \hat{R}_7)^2 = 1 - (1 - 0.9421)^2 = 0.9966,$$

$$\hat{R}_E = \hat{R}_8 \hat{R}_9 = 0.9622 \times 0.9935 = 0.9559,$$

The variance estimates for series blocks, A, C, and E, are calculated from (4.48). Then we have

$$\hat{V}ar(\hat{R}_A) = [\hat{R}_1^2 + \hat{V}ar(\hat{R}_1)][\hat{R}_2^2 + \hat{V}ar(\hat{R}_2)] - \hat{R}_1^2 \hat{R}_2^2$$

$$= (0.9856^2 + 0.0372^2)(0.9687^2 + 0.0213^2) - 0.9856^2 \times 0.9687^2$$

$$= 0.00174.$$

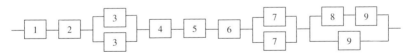

FIGURE 4.24   Reliability block diagram of Example 4.13

TABLE 4.2   Estimates of Component Reliabilities and Variances

	Component								
	1	2	3	4	5	6	7	8	9
$\hat{R}_i$	0.9856	0.9687	0.9355	0.9566	0.9651	0.9862	0.9421	0.9622	0.9935
$\sqrt{\hat{V}ar(\hat{R}_i)}$	0.0372	0.0213	0.0135	0.046	0.0185	0.0378	0.0411	0.0123	0.0158

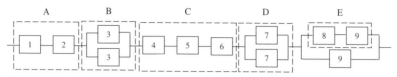

FIGURE 4.25   Reduced system equivalent to Figure 4.24

Similarly,

$$\hat{V}ar(\hat{R}_C) = 0.00344 \quad \text{and} \quad \hat{V}ar(\hat{R}_E) = 0.00038.$$

The variance estimates for parallel blocks, B and D, are calculated from (4.49). Then we have

$$\hat{V}ar(\hat{R}_B) = [(1 - \hat{R}_3)^2 + \hat{V}ar(\hat{R}_3)]^2 - (1 - \hat{R}_3)^4$$
$$= [(1 - 0.9355)^2 + 0.0135^2]^2 - (1 - 0.9355)^4 = 1.55 \times 10^{-6},$$
$$\hat{V}ar(\hat{R}_D) = 1.418 \times 10^{-5}.$$

After estimating the block reliabilities and variances, we replace each block with a hypothetical component in the system reliability block diagram. The diagram is then further partitioned into a series block G and a parallel block H, as shown in Figure 4.26. The reliability estimates for the blocks are

$$\hat{R}_G = \hat{R}_A \hat{R}_B \hat{R}_C \hat{R}_D = 0.8628 \quad \text{and} \quad \hat{R}_H = 1 - (1 - \hat{R}_9)(1 - \hat{R}_E) = 0.9997.$$

The variance estimates for blocks G and H are

$$\hat{V}ar(\hat{R}_G) = 0.0045 \quad \text{and} \quad \hat{V}ar(\hat{R}_H) = 5.9556 \times 10^{-7}.$$

Again, blocks G and H are replaced with two pseudocomponents, as shown in Figure 4.27. The two components are in series and can be represented by one

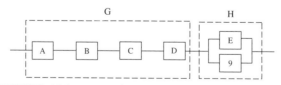

FIGURE 4.26   Reduced system equivalent to Figure 4.25

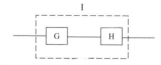

FIGURE 4.27   Reduced system equivalent to Figure 4.26

FIGURE 4.28   Reduced system equivalent to Figure 4.27

block, denoted I, as shown in Figure 4.28. The estimates of the system reliability and variance at 500 hours equal these of block I: namely,

$$\hat{R} = \hat{R}_I = 0.8625 \quad \text{and} \quad \hat{V}ar(\hat{R}) = \hat{V}ar(\hat{R}_I) = 0.0045.$$

Substituting the values of $\hat{R}$ and $\hat{V}ar(\hat{R})$ into (4.55) gives

$$\hat{\sigma}^2 = \ln\left(1 + \frac{0.0045}{0.8625}\right) = 0.0052.$$

The one-sided lower 95% confidence bound on the system reliability is

$$\hat{R}\exp(0.5\hat{\sigma}^2 - z_{1-\alpha}\hat{\sigma}) = 0.8625 \times \exp(0.5 \times 0.0052 - 1.645 \times \sqrt{0.0052})$$
$$= 0.768.$$

4.10   MEASURES OF COMPONENT IMPORTANCE

In the preceding sections of this chapter we described methods for estimating system reliability and confidence intervals for various system configurations. The evaluation may conclude that the reliability achieved for the current design does not meet a specified reliability target. In these situations, corrective actions must be taken to improve reliability. Such actions may include upgrading components, modifying system configuration, or both at the same time. No matter what actions are to be taken, the first step is to identify the weakest components or subsystems, which pose most potential for improvement. The identification can be done by ranking components or subsystems by their importance to system reliability, then priority should be given to the components or subsystems of high importance. An importance measure assigns a numerical value between 0 and 1 to each component or subsystem; 1 signifies the highest level of importance, and thus the system is most susceptible to the failure of corresponding component or subsystem, whereas 0 indicates the least level of importance and the greatest robustness of the system to the failure of relevant component or subsystem.

There are numerous measures of importance. In this section we present three major measures, including Birnbaum's measure, criticality importance, and Fussell–Vesely's importance, which are applicable to both components and subsystems. Other importance measures can be found in, for example, Barlow and Proschan (1974), Lambert (1975), Natvig (1979), Henley and Kumamoto (1992), Carot and Sanz (2000), and Hwang (2001). Boland and El-Neweihi (1995) survey the literature concerning the topic of importance measures and make critical comparisons.

4.10.1 Birnbaum's Measure of Importance

Birnbaum (1969) defines *component importance* as the probability of the component being critical to system failure, where *being critical* means that the component failure coincides with the system failure. Mathematically, it can be expressed as

$$I_B(i|t) = \frac{\partial R(t)}{\partial R_i(t)}, \tag{4.59}$$

where $I_B(i|t)$ is Birnbaum's importance measure of component i at time t, $R(t)$ the system reliability, and $R_i(t)$ the reliability of component i. The measure of importance may change with time. As a result, the components being weakest at a time may not remain weakest at another point of time. Thus, the measure should be evaluated at the times of particular interest, such as the warranty period and design life.

As indicated in (4.59) and pointed out in Section 4.9.1, $I_B(i|t)$ is actually the measure of the sensitivity of the system reliability to the reliability of component i. A large value of $I_B(i|t)$ signifies that a small variation in component reliability will result in a large change in system reliability. Naturally, components of this type should receive and deserve more resources for improvement.

Since $R(t) = 1 - F(t)$ and $R_i(t) = 1 - F_i(t)$, where $F(t)$ and $F_i(t)$ are the probabilities of failure of the system and component i, respectively, (4.59) can be written as

$$I_B(i|t) = \frac{\partial F(t)}{\partial F_i(t)}. \tag{4.60}$$

Example 4.14   A computing system consists of four computers configured according to Figure 4.29. The times to failure of individual computers are distributed exponentially with parameters $\lambda_1 = 5.5 \times 10^{-6}$, $\lambda_2 = 6.5 \times 10^{-5}$, $\lambda_3 = 4.3 \times 10^{-5}$, and $\lambda_4 = 7.3 \times 10^{-6}$ failures per hour. Calculate Birnbaum's importance measures of each computer in the system at $t = 4000$ and 8000 hours, respectively.

SOLUTION   Let $R_i(t)$ denote the reliability of computer i at time t, where $i = 1, 2, 3, 4$, and t will be omitted for notational convenience when appropriate. We first express the system reliability in terms of the reliabilities of individual

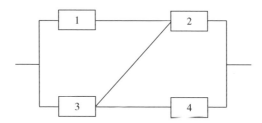

FIGURE 4.29   Reliability block diagram of the computing system

computers by using the decomposition method. Computer 2 is selected as the key-stone component, denoted A. Given that it never fails, the conditional probability of the system being good is

$$\Pr(\text{system good} \,|\, A) = 1 - (1 - R_1)(1 - R_3).$$

Similarly, provided that computer 2 is failed, the conditional probability that the system is functional is

$$\Pr(\text{system good} \,|\, \overline{A}) = R_3 R_4.$$

From (4.40), the system reliability at time t is

$$R(t) = [1 - (1 - R_1)(1 - R_3)]R_2 + R_3 R_4(1 - R_2) = R_1 R_2 + R_2 R_3 + R_3 R_4$$
$$- R_1 R_2 R_3 - R_2 R_3 R_4.$$

From (4.59), Birnbaum's importance measures for computers 1 through 4 are

$$I_B(1|t) = R_2(1 - R_3), \qquad I_B(2|t) = R_1 + R_3 - R_1 R_3 - R_3 R_4,$$
$$I_B(3|t) = R_2 + R_4 - R_1 R_2 - R_2 R_4, \qquad I_B(4|t) = R_3(1 - R_2).$$

Since the times to failure of the computers are exponential, we have $R_i(t) = e^{-\lambda_i t}$, $i = 1, 2, 3, 4$.

The reliabilities of individual computers at 4000 hours are

$$R_1(4000) = 0.9782, \qquad R_2(4000) = 0.7711,$$
$$R_3(4000) = 0.8420, \qquad R_4(4000) = 0.9712.$$

Then the values of the importance measures are

$$I_B(1|4000) = 0.1218, \qquad I_B(2|4000) = 0.1788,$$
$$I_B(3|4000) = 0.2391, \qquad I_B(4|4000) = 0.1928.$$

According to the importance measures, the priority of the computers is, in descending order, computers 3, 4, 2, and 1.

Similarly, the reliabilities of individual computers at 8000 hours are

$$R_1(8000) = 0.957, \qquad R_2(8000) = 0.5945,$$
$$R_3(8000) = 0.7089, \qquad R_4(8000) = 0.9433.$$

The importance measures at 8000 hours are

$$I_B(1|8000) = 0.173, \qquad I_B(2|8000) = 0.3188,$$
$$I_B(3|8000) = 0.4081, \qquad I_B(4|8000) = 0.2975.$$

Thus, computers 3, 2, 4, and 1 have a descending priority.

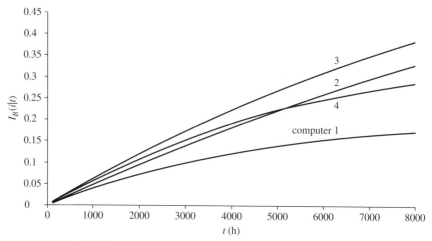

FIGURE 4.30 Birnbaum's importance measures of individual computers at different times

Comparison of the priorities at 4000 and 8000 hours shows that computer 3 is most important and computer 1 is least important, at both points of time. Computer 4 is more important than 2 at 4000 hours; however, the order is reversed at 8000 hours. The importance measures at different times (in hours) are plotted in Figure 4.30. It indicates that the short-term system reliability is more sensitive to computer 4, and computer 2 contributes more to the long-term reliability. Therefore, the importance measures should be evaluated, and the priority be made, at the time of interest (e.g., the design life).

4.10.2 Criticality Importance

Birnbaum's importance measure equals the probability that a component is critical to the system. In contrast, *criticality importance* is defined as the probability that a component is critical to the system and has failed, given that the system has failed at the same time. In other words, it is the probability that given that the system has failed, the failure is caused by the component. Mathematically, it can be expressed as

$$I_C(i|t) = \frac{\partial R(t)}{\partial R_i(t)} \frac{F_i(t)}{F(t)} = I_B(i|t) \frac{F_i(t)}{F(t)}, \qquad (4.61)$$

where $I_C(i|t)$ is the criticality importance and the other notation is the same as those for Birnbaum's importance. Equation (4.61) indicates that the criticality importance is Birnbaum's importance weighed by the component unreliability. Thus, a less reliable component will result in a higher importance.

Example 4.15   Refer to Example 4.14. Determine the criticality importance measures for the individual computers at 4000 and 8000 hours.

SOLUTION By using (4.61) and the results of $I_B(i|t)$ from Example 4.14, we obtain the criticality importance measures for the four computers as

$$I_C(1|t) = \frac{R_2(1 - R_3)(1 - R_1)}{1 - R}, \quad I_C(2|t) = \frac{(R_1 + R_3 - R_1 R_3 - R_3 R_4)(1 - R_2)}{1 - R},$$

$$I_C(3|t) = \frac{(R_2 + R_4 - R_1 R_2 - R_2 R_4)(1 - R_3)}{1 - R}, \quad I_C(4|t) = \frac{R_3(1 - R_2)(1 - R_4)}{1 - R}.$$

The reliability values of the four computers at 4000 and 8000 hours have been worked out in Example 4.14. The system reliabilities at the specified times are $R(4000) = 0.9556$ and $R(8000) = 0.8582$. Then the criticality importance measures at 4000 hours are

$$I_C(1|4000) = 0.0597, \quad I_C(2|4000) = 0.9225,$$

$$I_C(3|4000) = 0.8515, \quad I_C(4|4000) = 0.125.$$

Computers 2, 3, 4, and 1 have a descending priority order.
   Similarly, the criticality importance measures at 8000 hours are

$$I_C(1|8000) = 0.0525, \quad I_C(2|8000) = 0.9116,$$

$$I_C(3|8000) = 0.8378, \quad I_C(4|8000) = 0.115.$$

The priority order at 8000 hours is the same as that at 4000 hours. Figure 4.31 plots the criticality importance measures at different times (in hours). It is seen that the priority order of the computers is consistent over time. In addition, computers 2 and 3 are far more important than computers 1 and 4, because they are considerably less reliable.

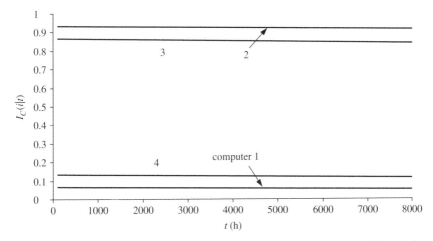

FIGURE 4.31 Criticality importance measures of individual computers at different times

4.10.3 Fussell–Vesely's Measure of Importance

Considering the fact that a component may contribute to system failure without being critical, Vesely (1970) and Fussell (1975) define the importance of a component as the probability that at least one minimal cut set containing the component is failed given that the system has failed. Mathematically, it can be expressed as

$$I_{\mathrm{FV}}(i|t) = \frac{\Pr(C_1 + C_2 + \cdots + C_{n_i})}{F(t)}, \tag{4.62}$$

where C_j is the event that the components in the minimal cut set containing component i are all failed; $j = 1, 2, \ldots, n_i$, and n_i is the total number of the minimal cut sets containing component i; $F(t)$ is the probability of failure of the system at time t. In (4.62), the probability, $\Pr(C_1 + C_2 + \cdots + C_{n_i})$, can be calculated by using the inclusion–exclusion rule expressed in (4.45). If component reliabilities are high, terms with second and higher order in (4.45) may be omitted. As a result, (4.62) can be approximated by

$$I_{\mathrm{FV}}(i|t) = \frac{1}{F(t)} \sum_{j=1}^{n_i} \Pr(C_j). \tag{4.63}$$

Example 4.16 Refer to Example 4.14. Calculate Fussell–Vesely's importance measures for the individual computers at 4000 and 8000 hours.

SOLUTION The minimal cut sets of the computing system are $\{1, 3\}$, $\{2, 4\}$, and $\{2, 3\}$. Let A_i denote the failure of computer i, where $i = 1, 2, 3, 4$. Then we have $C_1 = A_1 \cdot A_3$, $C_2 = A_2 \cdot A_4$, and $C_3 = A_2 \cdot A_3$. Since the reliabilities of individual computers are not high, (4.63) is not applicable. The importance measures are calculated from (4.62) as

$$I_{\mathrm{FV}}(1|t) = \frac{\Pr(C_1)}{F(t)} = \frac{\Pr(A_1 \cdot A_3)}{F(t)} = \frac{F_1 F_3}{F},$$

$$I_{\mathrm{FV}}(2|t) = \frac{\Pr(C_2 + C_3)}{F(t)} = \frac{\Pr(A_2 \cdot A_4) + \Pr(A_2 \cdot A_3) - \Pr(A_2 \cdot A_3 \cdot A_4)}{F(t)}$$

$$= \frac{F_2(F_4 + F_3 - F_3 F_4)}{F},$$

$$I_{\mathrm{FV}}(3|t) = \frac{\Pr(C_1 + C_3)}{F(t)} = \frac{F_3(F_1 + F_2 - F_1 F_2)}{F},$$

$$I_{\mathrm{FV}}(4|t) = \frac{\Pr(C_2)}{F(t)} = \frac{F_2 F_4}{F},$$

where $F_i = 1 - R_i$ and $F = 1 - R$. Substituting into the equations above the values of R_i and R at 4000 hours, which have been worked out in Examples 4.14

and 4.15, we obtain the importance measures as

$$I_{FV}(1|4000) = 0.0775, \qquad I_{FV}(2|4000) = 0.9403,$$

$$I_{FV}(3|4000) = 0.875, \qquad I_{FV}(4|4000) = 0.1485.$$

Ranked by the importance measures, computers 2, 3, 4, and 1 have a descending priority order.

Similarly, the importance measures at 8000 hours are

$$I_{FV}(1|8000) = 0.0884, \qquad I_{FV}(2|8000) = 0.9475,$$

$$I_{FV}(3|8000) = 0.885, \qquad I_{FV}(4|8000) = 0.1622.$$

The priority order at 8000 hours is the same as that at 4000 hours. Figure 4.32 plots the importance measures of the four individual computers at different times (in hours). It is seen that computers 2 and 3 are considerably more important than the other two at different points of time, and the relative importance order docs not change with time.

Examples 4.14 through 4.16 illustrate application of the three importance measures to the same problem. We have seen that the measures of criticality importance and Fessell–Vesely's importance yield the same priority order, which does not vary with time. The two measures are similar and should be used if we are concerned with the probability of the components being the cause of system failure. The magnitude of these measures increases with the unreliability of component (Meng, 1996), and thus a component of low reliability receives a

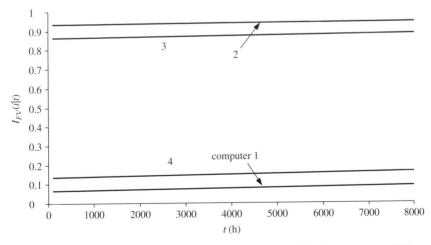

FIGURE 4.32 Fessel-Vesely's importance measures of individual computers at different times

high importance rating. The two measures are especially appropriate for systems that have a wide range of component reliabilities. In contrast, Birnbaum's measure of importance yields a different priority order in the example, which varies with time. The inconsistency at different times imposes difficulty in selecting the weakest components for improvement if more than one point of time is of interest. In addition, the measure does not depend on the unreliability of the component in question (Meng, 2000). Unlike the other two, Birnbaum's importance does not put more weight on less reliable components. Nevertheless, it is a valuable measure for identifying the fastest path to improving system reliability. When using this measure, keep in mind that the candidate components may not be economically or technically feasible if the component reliabilities are already high. To maximize the benefits, it is suggested that Birnbaum's importance be used at the same time with one of the other two. In the examples, if resources allow only two computers for improvement, concurrent use of the measures would identify computers 3 and 2 as the candidates, because both have large effects on system reliability and at the same time have a high likelihood of causing the system to fail. Although Birnbaum's measure suggests that computer 4 is the second-most important at 4000 hours, it is not selected because the other two measures indicate that it is far less important than computer 2. Clearly, the resulting priority order from the three measures is computers 3, 2, 4, and 1.

4.11 RELIABILITY ALLOCATION

In earlier sections we presented methods for estimating system reliability from component data. The methods are basically a bottom-up process; that is, we begin with estimating component reliability and end with determination of system reliability. In this section we describe a top-down process by which the system reliability target is allocated to individual components within the system in the manner that when each component achieves the allocated reliability, the overall system reliability target is attained. This process is called the *reliability allocation* or *apportionment*. For wording convenience, we use *component* to refer to a component, module, or subsystem. A reliability allocation problem can be formulated as

$$h(R_1^*, R_2^*, \ldots, R_n^*) \geq R^*, \tag{4.64}$$

where R^* is the system reliability target, R_i^* $(i = 1, 2, \ldots, n)$ is the reliability target of component i, and h denotes a functional relationship between the system reliability and component reliabilities. The functional relationship is obtained from system reliability analysis described in previous sections. Now the task of reliability allocation becomes solving inequality (4.64) for R_i^*.

Reliability allocation is an important task in a comprehensive reliability program, especially when the products under development are complicated. The distinct benefits of reliability allocation are as follows:

1. A complicated product contains a number of components, which are often planned, designed, tested, and manufactured by different external suppliers and

contractors and various internal departments. It is important that all parties involved are partnered and share the objective of delivering the end product to customers with the required reliability. From the project management point of view, to accomplish this, each and every partner should be assigned and committed to the reliability target. Reliability allocation defines a legitimate reliability goal for each component.

2. Quantitative reliability targets for components motivate responsible parties to improve reliability through use of reliability techniques in the first place, better engineering design, robust manufacturing processes, and rigorous testing methods.

3. Mandatory reliability requirements force reliability tasks, most of which were described in Chapter 3, to be considered equally with engineering activities aimed at meeting other customer expectations, such as weight, cost, and performance, in the process of product realization.

4. Reliability allocation drives a deep understanding of product hierarchical structure (i.e., the functional relationships among components, subsystems, and the end product). The process may lead to identification of design weakness and subsequent improvement.

5. Outputs of reliability allocation process can serve as inputs to other reliability tasks. For example, reliability assigned to a component will be used to design a reliability verification test for the component (Chapter 9).

Reliability allocation is essentially a repetitive process. It is conducted in the early design stage to support concept design when available information is extremely limited. As the design process proceeds and more details of the design and materials are determined, the overall reliability target should be reallocated to reduce the cost and risk of achieving the reliability goal. The allocation process may be invoked by the failure of one or more components to attain the assigned reliability due to technological limitations. The process is also repeated whenever a major design change takes place.

There are numerous methods for reliability allocation. In this section we delineate the simple, yet most commonly used ones, including the equal allocation technique, ARINC method, AGREE technique, and optimal allocation methodology. Also presented is the customer-driven approach, which was developed by the author and never published elsewhere. Prior to describing these methods, let's discuss the criteria for reliability allocation.

4.11.1 Criteria for Reliability Allocation

The task of reliability allocation is to select component reliability targets R_1^*, $R_2^*, \ldots, R_n^*$ which satisfy the inequality (4.64). Mathematically, there are an infinite number of such sets. Clearly, these sets are not equally good, and even some of them are unfeasible. For instance, some sets assign extremely high reliability goals to certain components and thus may be economically or technologically unachievable. Some sets allocate low-reliability targets to critical components whose failure would cause safety, environmental, or legal consequences.

It is important to establish some criteria that should be considered in reliability allocation. The common criteria are described below. But the appropriate ones depend on specific products. For example, Y. Wang et al. (2001) propose some for use in computerized numerical control (CNC) lathes.

1. *Likelihood of failure.* A component that has already demonstrated a high probability of failure in previous applications should be given a low-reliability target because of the arduous effort required to improve the reliability. Conversely, it is reasonable to assign a high-reliability goal to a reliable component.

2. *Complexity.* The number of constituent parts (modules or components) within a subsystem reflects the complexity of the subsystem. AGREE (1954) defines complexity as the number of modules and their associated circuitry, where a module is, for example, a transistor or a magnetic amplifier. Y. Wang et al. (2001) define the complexity of a CNC lathe subsystem as the ratio of the number of essential parts within the subsystem (whose failure causes the subsystem to fail) to the total number of such essential parts in the entire CNC machine. In general, complexity should be defined in a way to reflect the fact that a higher complexity leads to a lower reliability. It has an objective similar to that of the likelihood of failure. Thus, in reliability allocation, a more complex subsystem will be assigned a lower-reliability target.

3. *Criticality.* The failure of some components may cause severe effects, including, for example, loss of life and permanent environmental damage. The situation will be exacerbated when such components have a high likelihood of failure. Apparently, criticality is a synthesis of severity and failure probability, as defined in the FMEA technique described in Chapter 6. If a design cannot eliminate severe failure modes, it is imperative that the components attain the minimum probability of failure. As such, high-reliability goals should be assigned to them.

4. *Cost.* Cost is a major criterion that concerns the commercial industry and often is an objective subject to minimization. The costs required to achieve the same reliability increment vary with components. Some components incur a high cost to improve reliability a little because of the difficulty in design, verification, and production. It may be economically beneficial to allocate a higher-reliability target to the components that require less cost to improve reliability.

4.11.2 Equal Allocation Technique

The equal allocation technique treats equally all criteria including those described in Section 4.11.1 for all components within a system and assigns a common reliability target to all components to achieve the overall system reliability target. Although naive, this method is the simplest one and is especially useful in the early design phase when no detail information is available. For a series system, the system reliability is the product of the reliabilities of individual components.

Thus, (4.64) can be written as

$$\prod_{i=1}^{n} R_i^* \geq R^*. \tag{4.65}$$

The minimum reliability requirement of a component is given by

$$R_i^* = (R^*)^{1/n}, \qquad i = 1, 2, \ldots, n. \tag{4.66}$$

If all components are exponential, (4.65) becomes

$$\sum_{i=1}^{n} \lambda_i^* \leq \lambda^*, \tag{4.67}$$

where λ_i^* and λ^* are the maximum allowable failure rates of component i and system, respectively. Then the maximum allowable failure rate of a component is

$$\lambda_i^* = \frac{\lambda^*}{n}, \qquad i = 1, 2, \ldots, n. \tag{4.68}$$

Example 4.17 An automobile consists of a body, a powertrain, an electrical subsystem, and a chassis connected in series, as shown in Figure 4.2. The lifetimes of all subsystems are exponentially distributed and equally important. If the vehicle reliability target at 36 months in service is 0.98, determine the reliability requirement at this time and the maximum allowable failure rate of each subsystem.

SOLUTION   From (4.66), the reliability of each individual subsystem is

$$R_i^*(36) = (0.98)^{1/4} = 0.995, \qquad i = 1, 2, 3, 4.$$

The maximum allowable failure rate of the vehicle in accordance with the overall reliability target is

$$\lambda^* = -\frac{\ln[R^*(36)]}{36} = -\frac{\ln(0.98)}{36} = 5.612 \times 10^{-4} \text{ failures per month.}$$

From (4.68), the maximum allowable failure rate of each individual subsystem is

$$\lambda_i^* = \frac{5.612 \times 10^{-4}}{4} = 1.403 \times 10^{-4} \text{ failures per month,} \qquad i = 1, 2, 3, 4.$$

4.11.3   ARINC Approach

The ARINC approach, proposed by the Aeronautical Research Inc., assumes that all components are (1) connected in series, (2) independent of each other, (3) exponentially distributed, and (4) have a common mission time. Then reliability allocation becomes the task of choosing the failure rates of individual components λ_i^* such that (4.67) is satisfied. The determination of λ_i^* takes into account the likelihood of component failure (one of the criteria described earlier) by using the following weighting factors:

$$w_i = \frac{\lambda_i}{\sum_{i=1}^n \lambda_i}, \qquad i = 1, 2, \ldots, n, \tag{4.69}$$

where λ_i is the failure rate of component i obtained from historical data or prediction. The factors reflect the relative likelihood of failure. The larger the value of w_i, the more likely the component is to fail. Thus, the failure rate target allocated to a component should be proportional to the value of the weight: namely,

$$\lambda_i^* = w_i \lambda_0, \qquad i = 1, 2, \ldots, n, \tag{4.70}$$

where λ_0 is a constant. Because $\sum_{i=1}^n w_i = 1$ and if the equality holds in (4.67), inserting (4.70) into (4.67) yields $\lambda_0 = \lambda^*$. Therefore, (4.70) can be written as

$$\lambda_i^* = w_i \lambda^*, \qquad i = 1, 2, \ldots, n. \tag{4.71}$$

This gives the maximum allowable failure rate of a component. The corresponding reliability target is readily calculated as

$$R_i^*(t) = \exp(-w_i \lambda^* t), \qquad i = 1, 2, \ldots, n.$$

Example 4.18   Refer to Example 4.17. The warranty data for similar subsystems of an earlier model year have generated the failure rate estimates of the body, powertrain, electrical subsystem, and chassis as $\lambda_1 = 1.5 \times 10^{-5}$, $\lambda_2 = 1.8 \times 10^{-4}$, $\lambda_3 = 2.3 \times 10^{-5}$, and $\lambda_4 = 5.6 \times 10^{-5}$ failures per month, respectively. Determine the reliability requirement at 36 months in service and the maximum allowable failure rate of each subsystem in order to achieve the overall reliability target of 0.98.

SOLUTION   As worked out in Example 4.17, the maximum allowable failure rate of the vehicle in accordance with the reliability target of 0.98 at 36 months is $\lambda^* = 5.612 \times 10^{-4}$ failures per month. Next we calculate the weighting factors by using (4.69) and obtain

$$w_1 = \frac{1.5 \times 10^{-5}}{1.5 \times 10^{-5} + 1.8 \times 10^{-4} + 2.3 \times 10^{-5} + 5.6 \times 10^{-5}}$$

$$= \frac{1.5 \times 10^{-5}}{27.4 \times 10^{-5}} = 0.0547,$$

$$w_2 = \frac{1.8 \times 10^{-4}}{27.4 \times 10^{-5}} = 0.6569, \qquad w_3 = \frac{2.3 \times 10^{-5}}{27.4 \times 10^{-5}} = 0.0839,$$

$$w_4 = \frac{5.6 \times 10^{-5}}{27.4 \times 10^{-5}} = 0.2044.$$

Substituting the values of λ^* and the weighting factors into (4.71) gives the maximum allowable failure rates of the four subsystems. Then we have

$$\lambda_1^* = 0.0547 \times 5.612 \times 10^{-4} = 3.0698 \times 10^{-5},$$

$$\lambda_2^* = 0.6569 \times 5.612 \times 10^{-4} = 3.6865 \times 10^{-4},$$

$$\lambda_3^* = 0.0839 \times 5.612 \times 10^{-4} = 4.7085 \times 10^{-5},$$

$$\lambda_4^* = 0.2044 \times 5.612 \times 10^{-4} = 1.1471 \times 10^{-4}.$$

The minimum reliability requirements corresponding to the maximum allowable failure rates are

$$R_1^*(36) = \exp(-\lambda_1^* \times 36) = \exp(-3.0698 \times 10^{-5} \times 36) = 0.9989,$$

$$R_2^*(36) = \exp(-3.6865 \times 10^{-4} \times 36) = 0.9868,$$

$$R_3^*(36) = \exp(-4.7085 \times 10^{-5} \times 36) = 0.9983,$$

$$R_4^*(36) = \exp(-1.1471 \times 10^{-4} \times 36) = 0.9959.$$

As a check, the resulting vehicle reliability at 36 months is

$$R_1^*(36) \times R_2^*(36) \times R_3^*(36) \times R_4^*(36) = 0.9989 \times 0.9868 \times 0.9983 \times 0.9959$$

$$= 0.98.$$

This equals the reliability target specified.

4.11.4 AGREE Allocation Method

The AGREE allocation method, developed by the Advisory Group of Reliability of Electronic Equipment (AGREE), determines the minimum allowable mean time to failure for each individual subsystem to satisfy the system reliability target. This allocation approach explicitly takes into account the complexity of subsystems. Complexity is defined in terms of modules and their associated circuitry, where each module is assumed to have an equal failure rate. This assumption should be kept in mind when defining the boundary of modules. In general, module counts for highly reliable subsystems such as computers should be reduced because the failure rates are far lower than those of less reliable subsystems such as actuators.

The AGREE allocation method also considers the importance of individual subsystems, where *importance* is defined as the probability of system failure when a subsystem fails. The importance reflects the essentiality of the subsystem to the success of system. The importance of 1 means that the subsystem must function successfully for the system to operate. The importance of 0 indicates that the failure of the subsystem has no effect on system operation.

Assume that the subsystems are independently and exponentially distributed, and operate in series with respect to their effect on system success. Then (4.64) can be written as

$$\prod_{i=1}^{n} \left\{ 1 - w_i \left[1 - R_i^*(t_i) \right] \right\} = R^*(t),$$ (4.72)

where $R^*(t)$ is the system reliability target at time t, $R_i^*(t_i)$ the reliability target allocated for subsystem i at time t_i ($t_i \leq t$), w_i the importance of subsystem i, and n the number of subsystems. It can be seen that the allocation method allows the mission time of a subsystem to be less than that of the system.

Since the times to failure of subsystems are distributed exponentially and we have the approximation $\exp(-x) \approx 1 - x$ for a very small x, (4.72) can be written as

$$\sum_{i=1}^{n} \lambda_i^* w_i t_i = -\ln \left[R^*(t) \right],$$

where λ_i^* is the failure rate allocated to subsystem i. Taking the complexity into account, λ_i^* can be written as

$$\lambda_i^* = -\frac{m_i \ln[R^*(t)]}{m w_i t_i}, \qquad i = 1, 2, \ldots, n,$$ (4.73)

where m_i is the number of modules in subsystem i, m is the total number of modules in the system and equals $\sum_1^n m_i$, and w_i is the importance of subsystem i.

Considering the approximations $\exp(-x) \approx 1 - x$ for small x and $\ln(y) \approx y-1$ for y close to 1, the reliability target allocated to subsystem i can be written as

$$R_i^*(t_i) = 1 - \frac{1 - [R^*(t)]^{m_i/m}}{w_i}.$$ (4.74)

If w_i is equal or close to 1, (4.74) simplifies to

$$R_i^*(t_i) = [R^*(t)]^{m_i/m w_i}.$$ (4.75)

It can be seen that (4.73) and (4.74) would result in a very low reliability target for a subsystem of little importance. A very small value of w_i distortedly outweighs the effect of complexity and leads to an unreasonable allocation. The method works well only when the importance of each subsystem is close to 1.

Example 4.19 An on-board diagnostic system is installed in an automobile to detect the failure of emission-related components. When a failure occurs, the system generates diagnostic trouble codes corresponding to the failure type, saves the codes to a computer to facilitate subsequent repair, and illuminates the failure indicator on the panel cluster to alert the driver to the need for repair. The system consists of sensing, diagnosis, and indication subsystems, where the sensing and

TABLE 4.3 Data for AGREE Reliability Allocation

Subsystem Number	Subsystem	Number of Modules	Importance	Operating Time (h)
1	Sensing	12	1	12
2	Diagnosis	38	1	12
3	Indication	6	0.85	6

diagnosis subsystems are essential for the system to fulfill the intended functions. Failure of the indication subsystem causes the system to fail at an estimated probability of 0.85. In the case of indicator failure, it is possible that a component failure is detected by the driver due to poor drivability. Determine the reliability targets for the subsystems in order to satisfy the system reliability target of 0.99 in a driving cycle of 12 hours. Table 4.3 gives the data necessary to solve the problem.

SOLUTION From Table 4.3, the total number of modules in the system is $m = 12 + 38 + 6 = 56$. Substituting the given data into (4.73) yields the maximum allowable failure rates (in failures per hour) of the three subsystems as

$$\lambda_1^* = -\frac{12 \times \ln(0.99)}{56 \times 1 \times 12} = 1.795 \times 10^{-4}, \qquad \lambda_2^* = -\frac{38 \times \ln(0.99)}{56 \times 1 \times 12}$$

$$= 5.683 \times 10^{-4},$$

$$\lambda_3^* = -\frac{6 \times \ln(0.99)}{56 \times 0.85 \times 6} = 2.111 \times 10^{-4}.$$

From (4.74), the corresponding reliability targets are

$$R_1^*(12) = 1 - \frac{1 - [0.99]^{12/56}}{1} = 0.9978, \qquad R_2^*(12) = 1 - \frac{1 - [0.99]^{38/56}}{1}$$

$$= 0.9932,$$

$$R_3^*(6) = 1 - \frac{1 - [0.99]^{6/56}}{0.85} = 0.9987.$$

Now we substitute the allocated reliabilities into (4.72) to check the system reliability, which is

$$[1 - 1 \times (1 - 0.9978)] \times [1 - 1 \times (1 - 0.9932)] \times [1 - 0.85 \times (1 - 0.9987)]$$

$$= 0.9899.$$

This approximately equals the system reliability target of 0.99.

4.11.5 Customer-Driven Allocation Approach

In Chapter 3 we presented methods for setting the reliability required for a product to meet customer expectations. Now we describe an approach to allocating the reliability target to each subsystem. As in Chapter 3, suppose that QFD (quality function deployment) analysis has identified k important customer expectations, denoted $E_1, E_2, \ldots, E_k$. The customer expectations are linked to m independent and monotone-critical performance characteristics, denoted $Y_1, Y_2, \ldots, Y_m$, whose thresholds are $D_1, D_2, \ldots, D_m$, respectively. Let $S_1^*, S_2^*, \ldots, S_k^*$ be the minimum customer satisfactions on $E_1, E_2, \ldots, E_k$, respectively. Here S_i^* ($i = 1, 2, \ldots, k$) are specified. The probabilities of the performance characteristics not exceeding the respective thresholds are obtained by solving (3.2). Then we have

$$\Pr(Y_i \leq D_i) = p_i, \qquad i = 1, 2, \ldots, m. \tag{4.76}$$

The system (product) is supposed to consist of n subsystems, each of which has n_j ($j = 1, 2, \ldots, n$) performance characteristics strongly correlated to the system performance characteristics as indicated by the second house of quality. Let $(x_1, x_2, \ldots, x_{n_1})$, $(x_{n_1+1}, x_{n_1+2}, \ldots, x_{n_1+n_2})$, $\ldots$, $(x_{n_1+n_2+\cdots+n_{n-1}+1}, x_{n_1+n_2+\cdots+n_{n-1}+2}, \ldots, x_{n_1+n_2+\cdots+n_{n-1}+n_n})$ denote the performance characteristic vectors of subsystems $1, 2, \ldots$, and n, respectively. Also let d_i be the threshold of x_i, where $i = 1, 2, \ldots, \sum_1^n n_j$. Through QFD analysis we can identify the subsystem performance characteristics which are strongly interrelated to each system performance characteristic. The assumption regarding independence of $Y_1, Y_2, \ldots, Y_m$ prescribes that one subsystem characteristic cannot affect more than one system characteristic and that the characteristics of the same subsystem are independent. We further assume that each subsystem characteristic is correlated to a system characteristic and all are independent. Suppose that Y_i is determined by m_i subsystem characteristics and Y_i exceeds D_i if any of the m_i subsystem characteristics crosses a threshold. Thus, the probability of Y_i not exceeding D_i is

$$p_i = \Pr(Y_i \leq D_i) = \prod_{j}^{m_i} \Pr(x_j \leq d_j) = \prod_{j}^{m_i} p(x_j), \qquad i = 1, 2, \ldots, m, \tag{4.77}$$

where $p(x_j) = \Pr(x_j \leq d_j)$, x_j denotes a subsystem characteristic strongly related to Y_i, and index j may not be numerically consecutive. If the m_i subsystem characteristics are equally important and the associated probabilities are set equal, we have

$$p(x_j) = p_i^{1/m_i}, \qquad i = 1, 2, \ldots, m. \tag{4.78}$$

Using (4.78), we can specify the reliability target for each subsystem performance characteristic. Because subsystem i is measured by n_i performance characteristics, the reliability target R_i^* of subsystem i can be written as

$$R_i^* = \prod_{j=J_0+1}^{J_1} p(x_j), \qquad i = 1, 2, \ldots, n, \tag{4.79}$$

where $J_1 = \sum_1^i n_j$, $J_0 = \sum_1^{i-1} n_j$ for $i \geq 2$, and $J_0 = 0$ for $i = 1$.

Equation (4.79) gives the minimum reliability requirement that is correlated to the minimum customer satisfaction. It is worth noting that the reliability target is a function of time because the performance characteristics change with time. Thus, we should specify the time of particular interest at which minimum reliability must be achieved.

Example 4.20 A product consists of three subsystems. Through QFD analysis, we have identified that customers have three important expectations for the product, say E_1, E_2, and E_3, and that there exist strong correlations of E_1 to Y_1 and Y_2, E_2 to Y_1 and Y_3, and E_3 to Y_2, where Y_i ($i = 1, 2, 3$) are the independent product performance characteristics. The QFD study further indicates that Y_1 is strongly affected by x_1 and x_3, Y_2 by x_2, x_5 and x_6, Y_3 by x_4, where x_1 and x_2 are the performance characteristics of subsystem 1, x_3 and x_4 of subsystem 2, and x_5 and x_6 of subsystem 3. Determine the reliability target of each subsystem to achieve 88%, 90%, and 95% customer satisfactions on E_1, E_2, and E_3, respectively, at the design life.

SOLUTION   From (3.2) we have $p_1 p_2 = 0.88$, $p_1 p_3 = 0.9$, and $p_2 = 0.95$. Solving this equation system gives $p_1 = 0.93$, $p_2 = 0.95$, and $p_3 = 0.97$. Then the product reliability target at the design life is $R^* = p_1 p_2 p_3 = 0.93 \times 0.95 \times 0.97 = 0.857$. Since Y_1 is affected by x_1 and x_3, from (4.78), we obtain $p(x_1) = p(x_3) = 0.93^{1/2} = 0.9644$. Similarly, we have $p(x_2) = p(x_5) = p(x_6) = 0.95^{1/3} = 0.983$ and $p(x_4) = 0.97$.

The reliability target at the design life for each subsystem is determined from (4.79) as

$$R_1^* = \prod_{i=1}^{2} p(x_i) = 0.9644 \times 0.983 = 0.948, \qquad R_2^* = \prod_{i=3}^{4} p(x_i)$$

$$= 0.9644 \times 0.97 = 0.9355,$$

$$R_3^* = \prod_{i=5}^{6} p(x_i) = 0.983 \times 0.983 = 0.9663.$$

As a check, the minimum system reliability is $R_1^* \times R_2^* \times R_3^* = 0.948 \times 0.9355 \times 0.9663 = 0.857$. This is equal to the product reliability target R^*. It should be pointed out that meeting the subsystem reliability targets does not guarantee all customer satisfactions. To ensure all customer satisfactions, $p(x_j)$ for each subsystem characteristic must not be less than the assigned values.

4.11.6 Optimal Allocation Methods

Cost is an important factor that often dominates a reliability allocation. As described in Chapter 3 and illustrated in Figure 3.7, reliability investment cost increases with required reliability level. However, the investment is returned with savings in engineering design, verification, and production costs. The savings subtracted from the investment cost yields the net cost, which concerns a reliability allocation. In general, the cost is a nondecreasing function of required reliability. The more stringent the reliability target, the higher the cost. As the reliability required approaches 1, the cost incurred by meeting the target increases rapidly. Cost behaviors vary from subsystem to subsystem. In other words, the cost required to attain the same increment in reliability is dependent on subsystem. As such, it is economically beneficial to assign higher-reliability goals to the subsystems that demand lower costs to meet the targets. The discussion above indicates that reliability allocation heavily influences cost. A good allocation method should achieve the overall reliability requirement and low cost simultaneously.

Let $C_i(R_i)$ denote the cost of subsystem i with reliability R_i. The cost of the entire system is the total of all subsystem costs: namely,

$$C = \sum_{i=1}^{n} C_i(R_i), \qquad (4.80)$$

where C is the cost of the entire system and n is the number of subsystems. In the literature, various models for $C_i(R_i)$ have been proposed. Examples include Misra (1992), Aggarwal (1993), Mettas (2000), Kuo et al. (2001), and Kuo and Zuo (2002). In practice, it is important to develop or select the cost functions that are suitable for the specific subsystems. Unfortunately, modeling a cost function is an arduous task because it is difficult, if not impossible, to estimate the costs associated with attaining different reliability levels of a subsystem. The modeling process is further complicated by the fact that subsystems within a system often have different cost models. Given the constraints, we often employ a reasonable approximation to a cost function.

If the cost function $C_i(R_i)$ for subsystem i ($i = 1, 2, \ldots, n$) is available, the task of reliability allocation may be transformed into an optimization problem. In some applications, cost is a critical criterion in reliability allocation. Then the reliability targets for subsystems should be optimized by minimizing the cost, while the constraint on overall system reliability is satisfied. This optimization problem can be formulated as

$$\text{Min} \sum_{i=1}^{n} C_i(R_i^*), \qquad (4.81)$$

subject to $h(R_1^*, R_2^*, \ldots, R_n^*) \geq R^*$, where R_i^* ($i = 1, 2, \ldots, n$) are the decision variables, R^* is the overall system reliability target, and h has the same meaning as in (4.1). Solving (4.81) gives the optimal values of R_i^* or the optimal reliability targets for subsystems.

As a variant, the cost model $C_i(R_i^*)$ in (4.81) may be substituted by an effort function, and the allocation method is known as the *effort minimization approach*.

Let $E_i(R_i, R_i^*)$ denote the *effort function*, which describes the dollar amount of effort required to increase the reliability of subsystem i from the current reliability level R_i to a higher reliability level R_i^*. The larger the difference between R_i and R_i^*, the more the effort, and vice versa. Thus, the effort function is nonincreasing in R_i for a fixed value of R_i^* and nondecreasing in R_i^* for a fixed value of R_i. Using the effort function, (4.81) can be written as

$$\text{Min} \sum_{i=1}^{n} E_i(R_i, R_i^*), \tag{4.82}$$

subject to $h(R_1^*, R_2^*, \ldots, R_n^*) \geq R^*$, $R_i^* \geq R_i$; $i = 1, 2, \ldots, n$.

In some situations where a system failure can cause severe consequences, such as the loss of human life or permanent damage to environment, increasing reliability may be more critical than cutting cost. Then the objective of reliability allocation is to maximize the system reliability while meeting the given cost constraint. The problem can be formulated as

$$\text{Max}[h(R_1^*, R_2^*, \ldots, R_n^*)], \tag{4.83}$$

subject to $\sum_{i=1}^{n} C_i(R_i) \leq C^*$, where C^* is the maximum allowable cost for the system. Solving (4.83) yields the optimal values of reliability targets for individual subsystems.

The optimization problems above consider only two factors: system reliability and cost. In practice, we may want to include some other important dimensions, such as weight and size, which have effects on reliability and cost. Then the optimization models should be modified accordingly to take these factors into account. For example, a constraint on weight may be added to (4.81) if the objective is still to minimize the cost.

The optimization models above are known as *nonlinear programming problems*. To solve any of the models, numerical methods are required. Efficient algorithms may be found in, for example, Bazaraa et al. (1993) and Kuo et al. (2001). Nowadays, implementation of the algorithms is no longer a tough challenge, due to the availability of commercial software such as Matlab; even Microsoft Excel is able to solve small optimization problems.

Example 4.21 A system consists of four subsystems configured according to Figure 4.33. The cost (in dollars) of each subsystem is given by

$$C_i(R_i) = a_i \exp\left(\frac{b_i}{1 - R_i}\right), \qquad i = 1, 2, 3, 4,$$

where R_i is the reliability of subsystem i and a_i and b_i are cost-function parameters whose values are summarized in Table 4.4 for each subsystem. The system is required to achieve a design life reliability of 0.92. Determine the subsystem reliability targets that minimize the total cost.

FIGURE 4.33   System configuration of Example 4.21

TABLE 4.4   Parameters of the Cost Function

	Subsystem			
	1	2	3	4
a_i	3.5	3	4.5	1.2
b_i	0.07	0.11	0.13	0.22

SOLUTION   Based on the system configuration, the system reliability is $R = R_1 R_4 (R_2 + R_3 - R_2 R_3)$. Let R_i^* denote a target of R_i ($i = 1, 2, 3, 4$). Then the optimization model is

$$
\text{Min} \left[3.5 \exp \left(\frac{0.07}{1 - R_1^*} \right) + 3 \exp \left(\frac{0.11}{1 - R_2^*} \right) + 4.5 \exp \left(\frac{0.13}{1 - R_3^*} \right) \right.
$$
$$
\left. + 1.2 \exp \left(\frac{0.22}{1 - R_4^*} \right) \right],
$$

subject to $R_1^* R_4^* (R_2^* + R_3^* - R_2^* R_3^*) \geq 0.92$. The optimization model can be solved easily using a numerical algorithm. Here we employ Newton's method and the Solver in Microsoft Excel, and obtain $R_1^* = 0.9752$, $R_2^* = 0.9392$, $R_3^* = 0.9167$, and $R_4^* = 0.9482$. The costs (in dollars) associated with the subsystems are $C_1(R_1^*) = 58.81$, $C_2(R_2^*) = 18.31$, $C_3(R_3^*) = 21.42$, and $C_4(R_4^*) = 83.94$. The minimum total cost of the system is $C = \sum_{i=1}^{4} C_i(R_i^*) = \$182.48$.

PROBLEMS

4.1 An automotive V6 engine consists of six identical cylinders. For the engine to perform its intended functions, all six cylinders must be operationally successful. Suppose that the mileage to failure of a cylinder can be modeled with the Weibull distribution with shape parameter 1.5 and characteristic life 3.5×10^6 miles. Calculate the reliability and failure rate of the engine at 36,000 miles.

4.2 A special sprinkler system is comprised of three identical humidity sensors, a digital controller, and a pump, of which the reliabilities are 0.916, 0.965, and 0.983, respectively. The system configuration is shown in Figure 4.34. Calculate the reliability of the sprinkler system.

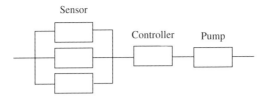

FIGURE 4.34   Reliability block diagram of a sprinkler system

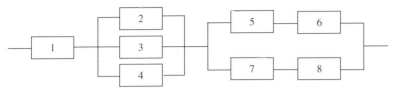

FIGURE 4.35   Reliability block diagram of Problem 4.3

4.3 Calculate the reliability of the system in Figure 4.35, where component i has reliability R_i $(i = 1, 2, \ldots, 8)$.

4.4 A power-generating plant is installed with five identical generators running simultaneously. For the plant to generate sufficient power for the end users, at least three of the five generators must operate successfully. If the time to failure of a generator can be modeled with the exponential distribution with $\lambda = 3.7 \times 10^{-5}$ failures per hour, calculate the reliability of the plant at 8760 hours.

4.5 A critical building has three power sources from separate stations. Normally, one source provides the power and the other two are in standby. Whenever the active source fails, a power supply grid switches to a standby source immediately. Suppose that the three sources are distributed identically and exponentially with $\lambda = 1.8 \times 10^{-5}$ failures per hour. Calculate the reliability of the power system at 3500 hours for the following cases:

(a) The switching system is perfect and thus never fails.

(b) The switching system is subject to failure according to the exponential distribution with a failure rate of 8.6×10^{-6} failures per hour.

4.6 A computing system consists of five individual computers, as shown in Figure 4.36. R_i $(i = 1, 2, \ldots, 5)$ is the reliability of computer i at a given time. Compute the system reliability at the time.

4.7 Refer to Problem 4.6. Suppose that the computer manufacturers have provided estimates of reliability and variance of each computer at 10,000 hours, as shown in Table 4.5. Calculate the one-sided lower 95% confidence bound on the system reliability at 10,000 hours.

4.8 Calculate the Birnbaum, criticality, and Fessel–Vesely measures of importance for the computing system in Problem 4.7. What observation can you make from the values of the three importance measures?

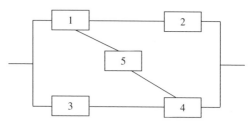

FIGURE 4.36   Reliability block diagram of a computing system

TABLE 4.5   Estimates of the Reliability and Variance of Individual Computers

	Computer				
	1	2	3	4	5
$\hat{R}_i$	0.995	0.983	0.988	0.979	0.953
$\sqrt{\hat{V}ar(\hat{R}_i)}$	0.0172	0.0225	0.0378	0.0432	0.0161

4.9 An automotive powertrain system consists of engine, transmission, and axle subsystems connected in logic series. The reliability target of the powertrain system is 0.98 at 36 months. Allocate the reliability target equally to each subsystem.

4.10 Refer to Problem 4.9. The lives of the engine, transmission, and axle subsystems are assumed to be exponentially distributed with 6.3×10^{-5}, 3.1×10^{-5}, and 2.3×10^{-5} failures per month, respectively. Determine the reliability target of each subsystem using the ARINC method.

4.11 A system is comprised of four subsystems connected in logic series. The system reliability target is 0.975 at 500 hours of continuous operation. Compute the reliability target for each subsystem using the AGREE approach with the data given in Table 4.6.

4.12 A product consists of four subsystems. QFD analysis indicates that the product has four important customer expectations (E_i, $i = 1, 2, 3, 4$), four system performance characteristics (Y_j, $j = 1, 2, 3, 4$) highly correlated to E_i, and six subsystem performance characteristics (x_k, $k = 1, 2, \ldots, 6$)

TABLE 4.6   Data for Problem 4.11

Subsystem	Number of Modules	Importance	Operating Time (h)
1	33	1	500
2	18	1	500
3	26	0.93	405
4	52	1	500

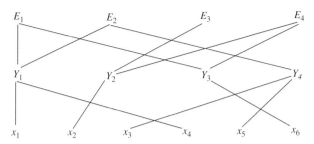

FIGURE 4.37 Strong correlations between E_i and Y_j and Y_j and x_k

cascaded from Y_j. Here x_1 and x_2 belong to subsystem 1, x_3 to subsystem 2, x_4 and x_5 to subsystem 3, and x_6 to subsystem 4. The correlations between E_i and Y_j, and Y_j and x_k are shown in Figure 4.37. The customer satisfactions on E_1, E_2, E_3, and E_4 are specified to be 90%, 92%, 95%, and 87%, respectively. Determine the reliability target required for each subsystem to achieve the customer satisfaction.

4.13 Refer to Example 4.21. If the maximum allowable cost for the system is $250, determine the subsystem reliability targets that maximize the system reliability.

5

RELIABILITY IMPROVEMENT THROUGH ROBUST DESIGN

5.1 INTRODUCTION

In Chapter 3 we presented methods for setting the reliability target of a product. The target quantifies the reliability requirement, which is a part of the design specifications that must be satisfied. In Chapter 4 we described approaches to allocating the reliability target to subsystems and components within the product and predicting product reliability. In many situations the predicted reliability is lower than the target or the allocated reliability. The difference is usually substantial for a product containing new technologies. To achieve the specified reliability, we often require rigorous reliability improvement in the product design phase. There exist numerous methods for increasing reliability. Redundancy is a classical reliability design approach; however, it adds cost and weight and finds rare applications in most commercial products (e.g., automobiles). Using high-reliability materials and components is also a common practice. But this method may be inefficient in today's competitive business climate because of the increased cost.

Robust design is a powerful technique for improving reliability at low cost in a short time. Robust design is a statistical engineering methodology for optimizing product or process conditions so that product performance is minimally sensitive to various sources of variation. The methodology was first developed and advocated by Taguchi (1987). Since the 1980s, it has been applied extensively to boost the quality of countless products and processes. A large body of the literature describes success stories on this subject. Ryoichi (2003) reports that the robust design method has been successful in airplane-engine engineering

Life Cycle Reliability Engineering, by Guangbin Yang
Copyright © 2007 John Wiley & Sons, Inc.

development and presents three wining cases in particular. Menon et al. (2002) delineate a case study on the robust design of spindle motor. Tu et al. (2006) document the robust design of a manufacturing process. Chen (2001) describes robust design of very large-scale integration (VLSI) process and device. Taguchi (2000) and Taguchi et al. (2005) present a large number of successful projects conducted in a wide spectrum of companies.

Numerous publications, most of which come with case studies, demonstrate that robust design is also an effective methodology for improving reliability. K. Yang and Yang (1998) propose a design and test method for achieving robust reliability by making products and processes insensitive to the environmental stresses. The approach is illustrated with a case study on the reliability improvement of the integrated-circuit interconnections. Chiao and Hamada (2001) describe a method for analyzing degradation data from robust design experiments. A case study on reliability enhancement of light-emitting diodes is presented. Tseng et al. (1995) report on increasing the reliability of fluorescent lamp with degradation data. C. F. Wu and Hamada (2000) present the design of experiments and dedicate a chapter to methods of reliability improvement through robust parameter design. Condra (2001) includes Taguchi's method and basic reliability knowledge in one book and discusses several case studies. Phadke and Smith (2004) apply robust design method to increase the reliability of engine control software.

In this chapter we describe the concepts of reliability and robustness and discuss their relationships. The robust design methods and processes for improving reliability are presented and illustrated with several industrial examples. Some advanced topics on robust design are described at the end of the chapter; these materials are intended for readers who want to pursue further study.

5.2 RELIABILITY AND ROBUSTNESS

In contrast to the standard reliability definition given in Chapter 2, the IEEE Reliability Society (2006) defines *reliability* as follows: "Reliability is a design engineering discipline which applies scientific knowledge to assure a product will perform its intended function for the required duration within a given environment. This includes designing in the ability to maintain, test, and support the product throughout its total life cycle. Reliability is best described as product performance over time. This is accomplished concurrently with other design disciplines by contributing to the selection of the system architecture, materials, processes, and components—both software and hardware; followed by verifying the selections made by thorough analysis and test."

Compared to the standard definition, the IEEE Reliability Society's definition is more oriented to engineering. It emphasizes that reliability is a design engineering discipline in view of the fact that the reliability, maintainability, testability, and supportability of a product depend largely on the quality of design. As described earlier, a powerful design technique is robust design, which aims at building robustness into products in the design phase.

Robustness is defined as the ability of a product to perform its intended function consistently at the presence of noise factors. Here, the noise factors are the variables that have adverse effects on the intended function and are impossible or impractical to control. Environmental stresses are the typical noise factors. This definition is widely applied in the field of quality engineering to address initial robustness when the product service time is zero. If customer satisfaction over time is concerned, the effect of time should be taken into account.

Reliability and robustness are correlated. On the one hand, reliability can be perceived as robustness over time. A robust product has high reliability during its early time in service under varions use conditions. To be reliable, the product must maintain its robustness over time. It is possible that a robust product is not reliable as time increases. Let's consider a scenario. A product is said to have failed if its performance characteristic crosses a threshold. The product is robust against the use conditions and unit-to-unit variation in the early service time. As age increases, the performance characteristic degrades rapidly, resulting in low reliability. This scenario is illustrated in Figure 5.1, where S_1 and S_2 represent two use conditions, and G is the threshold of performance characteristic y. On the other hand, robustness can be thought of as reliability at different use conditions. A reliable product has a high robustness value under the use conditions specified. To be robust, the product must maintain its reliability under different use conditions. It is possible that a reliable product is not robust. For example, a product that is reliable under S_1 may not be reliable under S_2. In this case the product is not robust against the use conditions, as illustrated in Figure 5.2.

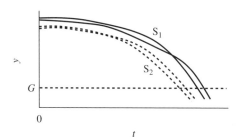

FIGURE 5.1   Robust product sensitive to aging

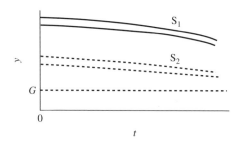

FIGURE 5.2   Reliable product sensitive to use conditions

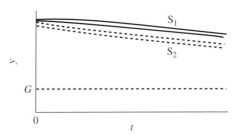

FIGURE 5.3 Product insensitive to aging and use conditions

A product should be designed to achieve both high robustness and reliability. Such a product is robust against noise factors and reliable over time, as illustrated in Figure 5.3, and thus is said to have high robust reliability. From the discussions above, robust reliability can be defined as the ability of a product to perform its intended function consistently over time at the presence of noise factors.

To achieve high robust reliability, a robust design should integrate the time variable into the design scheme. Such a design technique, called *robust reliability design*, is an extension of Taguchi's robust design. In this chapter we describe mainly methods for the improvement of robust reliability.

5.3 RELIABILITY DEGRADATION AND QUALITY LOSS

5.3.1 Quality Loss Function MSD

In engineering design, it is widely perceived that all products meeting design tolerance specifications are equally good regardless of how far a quality characteristic (i.e., a performance characteristic) deviates from its target. However, from the customer's point of view, these products have different quality levels; the closer the characteristic to the target, the better. Any deviation from the target value will cause a loss which can be measured in dollars. The perception is described meaningfully by the quadratic *quality loss function*, which can be written as

$$L(y) = K(y - m_y)^2, \tag{5.1}$$

where $L(y)$ is the quality loss, y a quality characteristic, m_y the target value of y, and K the quality loss coefficient. Quality characteristics can be categorized into three types: (1) nominal-the-best, (2) smaller-the-better, and (3) larger-the-better.

Nominal-the-Best Characteristics   In engineering design we frequently encounter the nominal-the-best characteristics. The nominal value is the target value. Due to production process variation, the characteristics are allowed to vary within a range, say $\pm \Delta_0$, where Δ_0 is called the *tolerance*. For example, the voltage output of a battery can be written as 12 ± 0.1 V. Equation (5.1) describes the quality

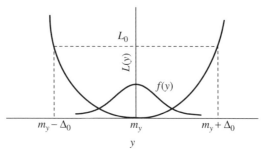

FIGURE 5.4   Quadratic quality loss for a nominal-the-best characteristic

loss of this type of characteristics. The quality loss is illustrated in Figure 5.4. If the quality loss is L_0 when the tolerance is just breached, (5.1) can be written as

L_0 = average loss to Cust ($) @ tolerance

$$L(y) = \frac{L_0}{\Delta_0^2}(y - m_y)^2. \tag{5.2}$$

The quality characteristic y is a random variable due to the unit-to-unit variation and can be modeled with a probabilistic distribution. The probability density function (pdf) of y, denoted $f(y)$, is depicted in Figure 5.4. If y has a mean μ_y and standard deviation σ_y, the expected quality loss is

$$E[L(y)] = E[K(y - m_y)^2] = K E[(y - \mu_y) + (\mu_y - m_y)]^2$$
$$= K[(\mu_y - m_y)^2 + \sigma_y^2]. \tag{5.3}$$

Equation (5.3) indicates that to achieve the minimum expected quality loss, we have to minimize the variance of y and set the mean μ_y to the target m_y.

Smaller-the-Better Characteristics   If y is a smaller-the-better characteristic, its range can be written as $[0, \Delta_0]$, where 0 is the target value and Δ_0 is the upper limit. The quality loss function is obtained by substituting $m_y = 0$ into (5.1) and can be written as

$$L(y) = Ky^2. \tag{5.4}$$

The quality loss function is depicted is Figure 5.5. If the quality loss is L_0 when y just breaches the upper limit, the quality loss at y can be written as

$K = L_0/\Delta_0^2$

$$L(y) = \frac{L_0}{\Delta_0^2}y^2. \tag{5.5}$$

The expected quality loss is

$$E[L(y)] = K(\mu_y^2 + \sigma_y^2). \tag{5.6}$$

Larger-the-Better Characteristics   If y is a larger-the-better characteristic, its range is $[\Delta_0, \infty]$, where Δ_0 is the lower limit. Because the reciprocal of a

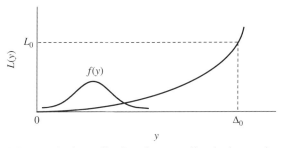

FIGURE 5.5 Quadratic quality loss for a smaller-the-better characteristic

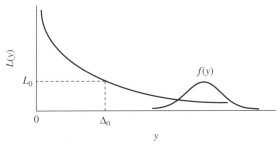

FIGURE 5.6 Quadratic quality loss for a larger-the-better characteristic

larger-the-better characteristic has the same quantitative behavior as a smaller-the-better one, the quality loss function can be obtained by substituting $1/y$ for y in (5.4). Then we have

$$L(y) = K \left(\frac{1}{y} \right)^2. \tag{5.7}$$

The quality loss function is depicted in Figure 5.6. If the quality loss is L_0 when y is at the lower limit Δ_0, the quality loss at y is

$$L(y) = \frac{L_0 \Delta_0^2}{y^2}. \tag{5.8}$$

The expected loss is obtained by using the Taylor series expansion and can be written as

$$E[L(y)] \approx \frac{K}{\mu_y^2} \left(1 + \frac{3\sigma_y^2}{\mu_y^2} \right). \tag{5.9}$$

Equation (5.9) indicates that increasing the mean or decreasing the variance reduces the quality loss.

Check Notes for MSD

5.3.2 Reliability Degradation

As described in Chapter 2, failure modes can be classified into two groups: hard failure and soft failure. *Hard failure* is catastrophic failure, and *soft failure* is degradation of product performance to an unacceptable level. The quality loss functions presented in Section 5.3.1 are applicable to these two types of failure modes.

Hard Failure   For a hard-failure product, there is generally no indication of performance degradation before the failure occurs. Customers perceive the product life as the key quality characteristic. Obviously, life is a larger-the-better characteristic. The quality loss function for life can be modeled by (5.7). Then (5.8) is used to calculate the quality loss, where Δ_0 is often the design life and L_0 is the loss due to the product failure at the design life. Here the design life is deemed as the required life span because customers usually expect a product to work without failure during its design life. L_0 may be determined by the life cycle cost. Dhillon (1999) describes methods for calculation of the cost.

The expected quality loss for life is described by (5.9). To minimize or reduce the loss due to failure, we have to increase the mean life and reduce the life variation. In robust reliability design, this can be accomplished by selecting the optimal levels of the design parameters.

Soft Failure   For a soft-failure product, failure is defined in terms of a performance characteristic crossing a prespecified threshold. Such a performance characteristic usually belongs to the smaller-the-better or larger-the-better type. Few are nominal-the-best type. Regardless of the type, the performance characteristic that defines failure is the quality characteristic that incurs the quality loss. The characteristic is often the one that most concerns customers.

The quality loss functions presented in Section 5.3.1 describe the initial loss due to the spreading performance characteristic caused by material and process variations. After the product is placed in service, the performance characteristic degrades over time. As a result, the reliability decreases and the quality loss increases with time. It is clear that the quality loss is nondecreasing in time. Taking the time effect into account, the expected loss for a smaller-the-better characteristic can be written as

$$E\{L[y(t)]\} = K[\mu_y^2(t) + \sigma_y^2(t)]. \tag{5.10}$$

Similarly, the expected loss for a larger-the-better characteristic is

$$E\{L[y(t)]\} \approx \frac{K}{\mu_y^2(t)} \left[1 + \frac{3\sigma_y^2(t)}{\mu_y^2(t)} \right]. \tag{5.11}$$

The relationships among performance degradation, reliability, and quality loss are illustrated in Figures 5.7 and 5.8. Figure 5.7 shows a smaller-the-better characteristic degrading over time and an increasing probability of failure defined by the characteristic. In Figure 5.7, Δ_0 is the threshold G, $f[y(t_i)]$ ($i = 1, 2, 3$) is the pdf of y at t_i, and the shaded area represents the probability of failure at the corresponding time. In Chapter 8 we describe the degradation reliability in detail. Figure 5.8 plots the quality loss function for the characteristic at different times. It is clear that the probability of failure increases with the quality loss. Therefore, minimizing the quality loss maximizes the reliability. As (5.10) and

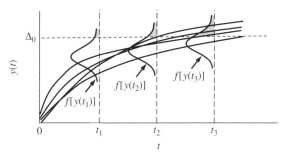

FIGURE 5.7 Relation of performance degradation and probability of failure

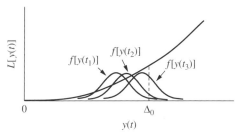

FIGURE 5.8 Quadratic quality loss as a function of time

(5.11) indicate, the quality loss may be minimized by alleviating the performance degradation, which can be accomplished by robust design (to be discussed later).

5.4 ROBUST DESIGN PROCESS

In Section 5.3 we related reliability degradation to quality loss and concluded that reliability can be improved through robust design. In this section we describe the process of robust design.

5.4.1 Three Stages of Robust Design

Robust design is a statistical engineering methodology for minimizing the performance variation of a product or process by choosing the optimal conditions of the product or process to make the performance insensitive to noise factors. According to Taguchi (1987), a robust design consists of three stages: system design, parameter design, and tolerance design.

System design involves selection of technology and components for use, design of system architecture, development of a prototype that meets customer requirements, and determination of manufacturing process. System design has significant impacts on cost, yield, reliability, maintainability, and many other performances of a product. It also plays a critical role in reducing product sensitivity to noise factors. If a system design is defective, the subsequent parameter design and tolerance design aimed at robustness improvement are fruitless. In recent years, some system design methodologies have emerged and shown effective, such as the axiomatic design and TRIZ (an innovative problem-solving method). Dieter (2000), Suh (2001), Rantanen and Domb (2002), and K. Yang and El-Haik (2003) describe the system design in detail. Nevertheless, there are few systematic approaches in the literature, largely because system design is skill intensive.

Parameter design aims at minimizing the sensitivity of the performance of a product or process to noise factors by setting its design parameters at the optimal levels. In this step, designed experiments are usually conducted to investigate the relationships between the design parameters and performance characteristics of the product or process. Using such relationships, one can determine the optimal setting of the design parameters. In this book, parameter design is the synonym of robust design in a narrow sense. In a broad sense, the former is a subset of the latter.

Tolerance design is to choose the tolerance of important components to reduce the performance sensitivity to noise factors under cost constraints. Tolerance design may be conducted after the parameter design is completed. If the parameter design cannot achieve sufficient robustness, tolerance design is necessary. In this step, the important components whose variability has the largest effects on the product sensitivity are identified through experimentation. Then the tolerance of these components is tightened by using higher-grade components based on the trade-off between the increased cost and the reduction in performance variability. Jeang (1995), P. Ross (1996), Creveling (1997), C. C. Wu and Tang (1998), C. Lee (2000), and Vlahinos (2002) describe the theory and application of the tolerance design.

5.4.2 Steps of Robust Design

As stated earlier, robust design implies the parameter design in this book. The steps of a robust design are structured to save time and cost and to improve the robust reliability in an efficient manner. The steps are as follows:

1. *Define the boundary.* Robust design is usually performed on the subsystems or components of a complex product. This step is to determine the subsystem or component within the product for robust design and to identify the impact of neighboring subsystems and components on the subsystem or component under study in terms of functional interactions and noise disturbance. In the remainder of this chapter, the subsystem or component under study is referred to as a *system* unless stated otherwise. Section 5.5 delineates the system boundary definition.

2. *Develop a P-diagram (a parameter diagram).* It shows pictorially the (a) system for robust design, (b) design parameters (control factors), (c) noise factors, (d) inputs (signals), (e) outputs (functions, responses), and (f) failure modes. A P-diagram contains all information necessary for subsequent robust design. In Section 5.6 we discuss the P-diagram in detail.

3. *Determine the key quality characteristic that characterizes the functions of the system to the greatest extent.* The characteristic is to be monitored

and measured in experimentation. For a binary-state system, life is usually the key quality characteristic and is used as the experimental response. For a degradation system, the critical performance characteristic is the key quality characteristic. In Chapter 8 we discuss selection of the critical performance characteristic.

4. *Identify the key noise factors and determine their levels.* In general, numerous noise factors apply to the system under study. It is impossible or impractical to include all noise factors in a robust design; only the key factors can be studied. The key noise factors are those that have the largest adverse impact on the functions of the system. The range of noise levels should be as broad as possible to represent real-world use conditions. The number of noise levels is constrained by the test time, cost, and available capacity of test equipment.

5. *Determine the main control factors and their levels.* The main control factors are those to which the functions of the system are most sensitive. The range of control factor levels should be as wide as possible while maintaining the intended functions. The number of levels is restricted by the availability of time, budget, and test resource. In this step it is important to identify potential interactions between control factors. An interaction between two factors exists if the effect of a factor on the system function depends on whether the other factor is present.

6. *Design the experiment.* In this step, the orthogonal arrays are employed to design the experiment. An inner array is selected to accommodate the control factors and their potential interactions. An outer array is used to lay out the noise factors. In Section 5.8 we describe the design of experiment in detail. In this step we should also decide an appropriate number of replicates at each experimental condition to obtain sufficient statistical accuracy with available resources. The order in which the experiments are conducted is randomized to avoid biased effects. The test equipment, measurement tools, and measurement frequency on the key quality characteristic are selected in this step. If necessary, a study of gauge repeatability and reproducibility, commonly known as *gauge R&R*, should be performed. Montgomery (2001a), for example, presents gauge R&R methods.

7. *Conduct the experiment.* This step is to generate and collect measurement data on the key quality characteristic of test units at each experimental condition. In experimentation it is essential to comply with the operational standards of the test facility and reduce human errors. In some applications, a computer simulation may replace physical testing to save time and cost. The simulation does not need experimental replicates, because every run of the simulation gives the same results. The most critical factor for a successful computer simulation is to create a model that represents the system adequately.

8. *Analyze the experimental data.* In this step we (a) identify the control factors that have statistically significant effects on the experimental

response, (b) determine the optimal setting of the significant control factors, and (c) predict the response under the optimal setting. Graphical response analysis or analysis of variance (ANOVA) is usually performed in this step.

9. *Run a confirmation test.* The optimality of the control factor levels is confirmed by running a test on the samples at the optimal setting.

10. *Recommend actions.* The optimal setting should be implemented in design and production. To sustain the improvement, follow-up actions such as executing a statistical process control are recommended. Montgomery (2001a) and Stamatis (2004), for example, describe quality control approaches, including statistical process control.

5.5 BOUNDARY DEFINITION AND INTERACTION ANALYSIS

A complex product is usually expensive. Robust design through experiment on the final product is usually unaffordable. Rather, it is often conducted on subsystems or components (both referred to as *systems*, as stated earlier) within the product. To perform robust design on a system, we first have to define the boundary of the system. The system selected is a part of the product, and hence has interactions with other subsystems, components, and software within the product. On the other hand, the system may also interact directly with the environment, customers, and the production process. The interaction can be physical contact, information exchange, or energy flow. The last two types of interactions can be one-way or two-way. Some interactions are integral parts of the system functionality, and some can cause noise effects that jeopardize the functionality.

A boundary diagram is often used to illustrate the boundary of a system. Figure 5.9 shows a generic example of a boundary diagram in which the one-directional arrow represents a one-way interaction, and the two-directional arrow, a two-way interaction. Defining the boundary of a system is the process of identifying the signals to the system, the outputs from the system, and the noise sources

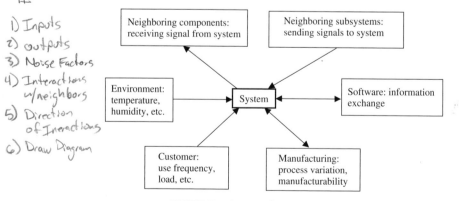

FIGURE 5.9 Generic boundary diagram

disturbing the system. Therefore, a boundary diagram provides useful information to subsequent creation of a P-diagram. In addition, a boundary diagram is a valuable input to failure mode effect and analysis (FMEA) in that it recognizes the failure effects of the system under concern. FMEA is discussed in Chapter 6.

5.6 P-DIAGRAM : No Interactions

A P-diagram illustrates the inputs (signals), outputs (intended functions or responses), control factors, noise factors, and failure modes of a system. Figure 5.10 shows a generic P-diagram, where the noise factors, signals, and intended functions may be carried over from a boundary diagram, if it has been created. The control factors and failure modes are new additions. A P-diagram contains the necessary information for robust design. The elements of a P-diagram are described below.

Signals are inputs from customers or other subsystems or components to the system under study. The system transforms the signals into functional responses and, of course, failure modes. Signals are essential to fulfilling the function of a system. For example, the force applied on an automobile braking pedal is a signal to the braking system, which converts the applied force into a braking force to stop the vehicle within a safe distance.

Noise factors are variables that have adverse effects on robustness and are impossible or impractical to control. Generally, there are three types of noise factors:

- *Internal noise*: performance degradation or deterioration as a result of product aging. For example, abrasion is an internal noise for the automobile braking system.
- *External noise*: operating conditions that disturb the functions of a system, including environmental stresses such as temperature, humidity and vibration, as well as operating load. For a braking system, road condition, driving habits, and vehicle load are external noise factors.
- *Unit-to-unit noise*: variation in performance, dimension, and geometry resulting from variability in materials and production process. This noise

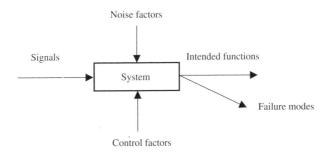

FIGURE 5.10 Generic P-diagram

factor is inevitable, but can be reduced through tolerance design and process control. In the braking system example, variation in the thickness of drums and pads is this type of noise factor.

In the automotive industry, noise factors are further detailed. For example, at Ford Motor Company, noise factors are divided into five categories: (1) unit-to-unit variation; (2) change in dimension, performance, or strength over time or mileage; (3) customer usage and duty cycle; (4) external environment, including climate and road conditions; and (5) internal environment created by stresses from neighboring components. Although the last three types belong to the external noise described above, this further itemization is instrumental in brainstorming all relevant noise factors. Strutt and Hall (2003) describe the five types of noise factors in greater detail.

Control factors are the design parameters whose levels are specified by designers. The purpose of a robust design is to choose optimal levels of the parameters. In practice, a system may have a large number of design parameters, which are not of the same importance in terms of the contribution to robustness. Often, only the key parameters are included in a robust design. These factors are identified by using engineering judgment, analytical study, a preliminary test, or historical data analysis.

Intended functions are the functionalities that a system is intended to perform. The functions depend on signals, noise factors, and control factors. Noise factors and control factors influence both the average value and variability of functional responses, whereas the signals determine the average value and do not affect the variability.

Failure modes represent the manner in which a system fails to perform its intended functions. As explained earlier, failure modes can be classified into two types: hard failure and soft failure. In the braking system example, excessive stopping distance is a soft failure, whereas complete loss of hydraulic power in the braking system is a hard failure.

Example 5.1   An on-board diagnostic (OBD) system is installed in automobiles to monitor the failure of exhaust gas recirculation (EGR) components. Such an OBD system is also referred to as an EGR monitor. When a component fails, the EGR monitor detects the failure and illuminates a malfunction light on the instrument panel to alert the driver to the need for repair. Figure 5.11 shows a P-diagram for an EGR monitor. The example would only be typical for this monitor and is not intended to be exhaustive. A full P-diagram of the monitor contains more noise factors and a large number (about 70) of calibration parameters, as well as several strategies and algorithms of calibration.

5.7   NOISE EFFECTS MANAGEMENT

The creation of a P-diagram leads to the identification of all noise factors that disturb the functional responses and generate failure modes. The noise factors

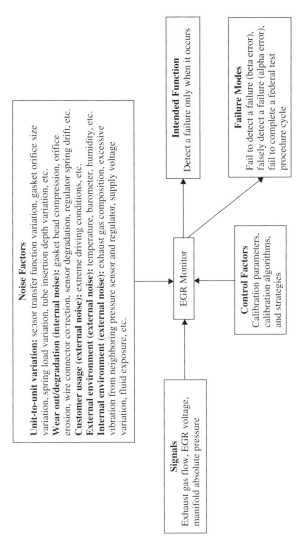

FIGURE 5.11 P-diagram of an EGR monitor

and failure modes have cause-and-effect relations, which usually are complicated. One noise factor may result in more than one failure mode; one failure mode may be the consequence of several noise factors. For example, in Figure 5.11, the failure mode "fail to detect a failure (beta error)" is caused by multiple noise factors, including sensor transfer function variation, wire connector corrosion, supply voltage variation, and others. On the other hand, the variation in sensor transfer function can cause both alpha and beta errors. Since a product usually involves numerous noise factors, it is important to identify the critical factors that cause the most troublesome failure modes, those with high risk priority number values which will surface in the FMEA process (Chapter 6).

The effects of critical noise factors must be addressed in robust design. System design, parameter design, and tolerance design are the fundamental methods for reducing noise effects. In system design, the following common techniques are often employed to eliminate or mitigate the adverse effects:

1. *Change the technology and system architecture.* This method is proactive and expensive. It is effective in alleviating the effects of internal and external noise factors.

2. *Reduce or remove noises through design.* This technique needs special design aimed at particular noises. For example, in electronic circuit design, several capacitors are paralleled to reduce the influence of electrostatic discharge.

3. *Adopt a compensation device.* Although passive, this approach is useful in many applications. For example, cooling systems are installed in automobiles to lower the engine temperature, which is a critical external noise factor for an engine system. The success of this method depends on the reliability of the compensation device. Once the device fails, the noise effects will be active.

4. *Disguise or divert noise through special design.* This technique bypasses noise to unimportant systems or environment. For example, heat sinks may be installed in electronic products to dissipate heat to the surrounding so as to bring down the temperature of heat-sensitive components.

5.8   DESIGN OF EXPERIMENTS

In earlier sections we defined the scope of robust design, identified the critical control factors and noise factors and their levels, and determined the key quality characteristic. The next step in robust design is to design the experiment.

Design of experiment is a statistical technique for studying the effects of multiple factors on the experimental response simultaneously and economically. The factors are laid out in a structured array in which each row represents a combination of levels of factors. Then experiments with each combination are conducted and response data are collected. Through experimental data analysis, we can choose the optimal levels of control factors that minimize the sensitivity of the response to noise factors.

Various types of structured arrays or experimental designs, such as full factorial designs and a variety of fractional factorial designs are described in the literature (e.g., C. F. Wu and Hamada, 2000; Montgomery, 2001b). In a full factorial design, the number of runs equals the number of levels to the power of the number of factors. For example, a two-level full factorial design with eight factors requires $2^8 = 256$ runs. If the number of factors is large, the experiments will be unaffordable in terms of time and cost. In these situations, a fractional factorial design is often employed. A fractional factorial design is a subset of a full factorial design, chosen according to certain criteria. The commonly used classical fractional factorials are 2^{k-p} and 3^{k-p}, where 2 (3) is the number of levels, k

the number of factors, and 2^{-p} (3^{-p}) the fraction. For example, a two-level half-fractional factorial with eight factors needs only $2^{8-1} = 128$ runs.

The classical fractional factorial designs require that all factors have an equal number of levels. For example, a 2^{k-p} design can accommodate only two-level factors. In practice, however, some factors are frequently required to take a different number of levels. In such situations, the classical fractional factorial designs are unable to meet the demand. A more flexible design is that of orthogonal arrays, which have been used widely in robust design. As will be shown later, the classical fractional factorial designs are special cases of orthogonal arrays. In this section we present experimental design using orthogonal arrays.

5.8.1 Structure of Orthogonal Arrays

An *orthogonal array* is a balanced fractional factorial matrix in which each row represents the levels of factors of each run and each column represents the levels of a specific factor that can be changed from each run. In a balanced matrix:

- All possible combinations of any two columns of the matrix occur an equal number of times within the two columns. The two columns are also said to be *orthogonal*.
- Each level of a specific factor within a column has an equal number of occurrences within the column.

For example, Table 5.1 shows the orthogonal array $L_8(2^7)$. The orthogonal array has seven columns. Each column may accommodate one factor with two levels, where the low and high levels are denoted by 0 and 1, respectively. From Table 5.1 we see that any two columns, for example, columns 1 and 2, have level combinations (0,0), (0,1), (1,0), and (1,1). Each combination occurs twice within the two columns. Therefore, any two columns are said to be orthogonal. In addition, levels 0 and 1 in any column repeat four times. The array contains eight rows, each representing a run. A full factorial design with seven factors and

$$L_N (I^P \times J^Q)$$

TABLE 5.1 $L_8(2^7)$ Orthogonal Array

N = No. Exps
P = No of I levels
Q = No of J levels

Run	Column						
	1	2	3	4	5	6	7
1	0	0	0	0	0	0	0
2	0	0	0	1	1	1	1
3	0	1	1	0	0	1	1
4	0	1	1	1	1	0	0
5	1	0	1	0	1	0	1
6	1	0	1	1	0	1	0
7	1	1	0	0	1	1	0
8	1	1	0	1	0	0	1

two levels of each would require $2^7 = 128$ runs. Thus, this orthogonal array is a $\frac{1}{16}$ fractional factorial design. In general, because of the reduction in run size, an orthogonal array usually saves a considerable amount of test resource. As opposed to the improved test efficiency, an orthogonal array may confound the main effects (factors) with interactions. To avoid or minimize such confounding, we should identify any interactions before design of experiment and lay out the experiment appropriately. This is discussed further in subsequent sections.

In general, an orthogonal array is indicated by $L_N(I^P \times J^Q)$, where N denotes the number of experimental runs, P is the number of I-level columns, and Q is the number of J-level columns. For example, $L_{18}(2^1 \times 3^7)$ identifies the array as having 18 runs, one two-level column, and seven three-level columns. The most commonly used orthogonal arrays have the same number of levels in all columns, and then $L_N(I^P \times J^Q)$ simplifies to $L_N(I^P)$. For instance, $L_8(2^7)$ indicates that the orthogonal array has seven columns, each with two levels. The array requires eight runs, as shown in Table 5.1.

Because of the orthogonality, some columns in $L_N(I^P)$ are fundamental (independent) columns, and all other columns are generated from two or more of the fundamental columns. The generation formula is as follows, with few exceptions.

(number in the column generated from i fundamental columns)

$$= \sum_{j=1}^{i} (\text{number in fundamental column } j)(\text{mode I}), \qquad (5.12)$$

where $2 \leq i \leq$ total number of fundamental columns. The modulus I calculation gives the remainder after the sum is divided by I.

Example 5.2 In $L_8(2^7)$ as shown in Table 5.1, columns 1, 2, and 4 are the fundamental columns, all other columns being generated from these three columns. For instance, column 3 is generated from columns 1 and 2 as follows:

column 1		column 2		column 3
$\begin{bmatrix} 0 \\ 0 \\ 0 \\ 0 \\ 1 \\ 1 \\ 1 \\ 1 \end{bmatrix}$	$+$	$\begin{bmatrix} 0 \\ 0 \\ 1 \\ 1 \\ 0 \\ 0 \\ 1 \\ 1 \end{bmatrix}$	(mode 2) $=$	$\begin{bmatrix} 0 \\ 0 \\ 1 \\ 1 \\ 1 \\ 1 \\ 0 \\ 0 \end{bmatrix}$

5.8.2 Linear Graphs

As explained in Section 5.8.1, an orthogonal array is comprised of fundamental columns and generated columns. *Fundamental columns* are the independent columns, and *generated columns* are the interaction columns. For example, in Table 5.1, the interaction between columns 1 and 2 goes to column 3. In an experimental layout, if factor A is assigned to column 1 and factor B to column 2, column 3 should be allocated to the interaction A × B if it exists. Assigning an independent factor to column 3 can lead to an incorrect data analysis and faulty conclusion because the effect of the independent factor is confounded with the interaction effect. Such experimental design errors can be prevented by using a linear graph.

A *linear graph*, a pictorial representation of the interaction information, is made up of dots and lines. Each dot indicates a column to which a factor (main effect) can be assigned. The line connecting two dots represents the interaction between the two factors represented by the dots at each end of the line segment. The number assigned to a dot or a line segment indicates the column within the array. In experimental design, a factor is assigned to a dot, and an interaction is assigned to a line. If the interaction represented by a line is negligible, a factor may be assigned to the line.

Figure 5.12 shows two linear graphs of $L_8(2^7)$. Figure 5.12*a* indicates that columns 1, 2, 4, and 7 can be used to accommodate factors. The interaction between columns 1 and 2 goes into column 3, the interaction between 2 and 4 goes into column 6, and the interaction between columns 1 and 4 goes into column 5. From (5.12) we can see that column 7 represents a three-way interaction among columns 1, 2, and 4. The linear graph assumes that three-way or higher-order interactions are negligible. Therefore, column 7 is assignable to a factor. It should be noted that all linear graphs are based on this assumption, although it may be questionable in some applications. $1 + (L-1) \times F$

Example 5.3 To assign an experiment with five two-level factors, A, B, C, D, and E, and interactions A × B and B × C to $L_8(2^7)$, using Figure 5.12*a* we can allocate factor A to column 1, factor B to column 2, factor C to column 4, factor D to column 7, factor E to column 5, interaction A × B to column 3, and interaction B × C to column 6.

Most orthogonal arrays have two or more linear graphs. The number and complexity of linear graphs increase with the size of orthogonal array. A multitude of linear graphs provide great flexibility for assigning factors and interactions.

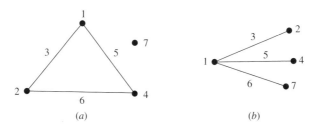

FIGURE 5.12   Linear graphs for $L_8(2^7)$

The most commonly used orthogonal arrays and their linear graphs are listed in the Appendix.

5.8.3 Two-Level Orthogonal Arrays

A two-level orthogonal array is indicated as $L_N(2^P)$. The most frequently used two-level arrays are $L_4(2^3)$, $L_8(2^7)$, $L_{12}(2^{11})$, and $L_{16}(2^{15})$.

$L_4(2^3)$, shown in Table 5.2, is a half-fractional factorial array. The first two columns can be assigned to two factors. The third column accommodates the interaction between them, as indicated by the linear graph in Figure 5.13. If a negligible interaction can be justified, the third column is assignable to an additional factor.

The layout and linear graphs of $L_8(2^7)$ are shown in Table 5.1 and Figure 5.12, respectively. This array requires only eight runs and is very flexible in investigating the effects of factors and their interactions. It is often used in small-scale experimental designs.

$L_{12}(2^{11})$, given in the Appendix, is unique in that the interaction between any two columns within the array is partially spread across the remaining columns. This property minimizes the potential of heavily confounding the effects of factors and interactions. If engineering judgment considers the interactions to be weak, the array is efficient in investigating the main effects. However, the array cannot be used if the interactions must be estimated.

$L_{16}(2^{15})$ is often used in a large-scale experimental design. The array and the commonly used linear graphs can be found in the Appendix. This array provides great flexibility for examining both simple and complicated two-way interactions. For example, it can accommodate 10 factors and five interactions, or five factors and 10 interactions.

5.8.4 Three-Level Orthogonal Arrays

Three-level orthogonal arrays are needed when experimenters want to explore the quadratic relationship between a factor and the response. Although the response

TABLE 5.2 $L_4(2^3)$ Orthogonal Array

		Column	
Run	1	2	3
1	0	0	0
2	0	1	1
3	1	0	1
4	1	1	0

FIGURE 5.13   Linear graph for $L_4(2^3)$

surface method is deemed advantageous in investigating such a relationship, experimental design using orthogonal arrays is still widely used in practice because of the simplicity in data analysis. Montgomery (2001b) describes response surface analysis.

$L_N(3^P)$ refers to a three-level orthogonal array where 0, 1, and 2 denote the low, middle, and high levels, respectively. Any arrays in which the columns contain predominantly three levels are also called three-level orthogonal arrays. For example, $L_{18}(2^1 \times 3^7)$ is considered a three-level array. The most commonly used three-level arrays are $L_9(3^4)$, $L_{18}(2^1 \times 3^7)$, and $L_{27}(3^{13})$.

$L_9(3^4)$ is the simplest three-level orthogonal array. As shown in Table 5.3, this array requires nine runs and has four columns. The first two columns are the fundamental columns, and the last two accommodate the interactions between columns 1 and 2, as shown in Figure 5.14. If no interactions can be justified, the array can accommodate four factors.

$L_{18}(2^1 \times 3^7)$ is a unique orthogonal array (see the Appendix for the layout and linear graphs of this array). The first column in this array contains two levels and all others have three levels. The interaction between the first two columns is orthogonal to all columns. Therefore, the interaction can be estimated without sacrificing additional columns. However, the interactions between any pair of three-level columns are spread to all other three-level columns. If the interactions between three-level factors are strong, the array cannot be used.

$L_{27}(3^{13})$ contains three levels in each of its 13 columns (the layout and linear graphs are given in the Appendix). The array can accommodate four interactions and five factors, or three interactions and seven factors. Because the interaction between two columns spreads to the other two columns, two columns must be sacrificed in considering one interaction.

TABLE 5.3 $L_9(3^4)$ Orthogonal Array

	Column			
Run	1	2	3	4
1	0	0	0	0
2	0	1	1	1
3	0	2	2	2
4	1	0	1	1
5	1	1	2	0
6	1	2	0	1
7	2	0	2	1
8	2	1	0	2
9	2	2	1	0

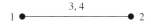

FIGURE 5.14   Linear graph for $L_9(3^4)$

5.8.5 Mixed-Level Orthogonal Arrays

The orthogonal arrays that have been discussed so far can accommodate only two-or three-level factors. Except for $L_{18}(2^1 \times 3^7)$, the arrays require that all factors have an equal number of levels. In practice, however, we frequently encounter situations in which few factors have more levels than all others. In such cases, the few factors with multiple levels will considerably increase the array size. For example, $L_8(2^7)$ is good for investigating five independent factors. However, if one of the five factors has to take four levels, $L_{16}(4^5)$ is needed simply because of the four-level factor, apparently rendering the experiment not economically efficient. To achieve the economic efficiency, in this section we describe the preparation of mixed-level orthogonal arrays using the column-merging method. The method is based on a linear graph and the concept of degrees of freedom (Section 5.8.6). In particular, in this section we describe a method of preparing a four-level column and an eight-level column in standard orthogonal arrays.

First, we study a method of creating a four-level column in a standard two-level orthogonal array. Because a four-level column has three degrees of freedom and a two-level column has one, the formation of one four-level column requires three two-level columns. The procedure of forming a four-level column has three steps:

1. Select any two independent (fundamental) columns and their interaction column. For example, to generate a four-level column in $L_8(2^7)$, columns 1, 2, and 3 may be selected, as shown in Figure 5.15.

2. Merge the numbers of the two independent (fundamental) columns selected and obtain 00, 01, 10, and 11, denoted 0, 1, 2, and 3, respectively. Then the merging forms a new column whose levels are 0, 1, 2, and 3. In the $L_8(2^7)$ example, combining the numbers of columns 1 and 2 gives a new series of numbers, as shown in Table 5.4.

3. Replace the three columns selected with the four-level column. In the $L_8(2^7)$ example, the first three columns are replaced by the new column. The new column is orthogonal to any other column, except for the original first three columns. Now, a four-level factor can be assigned to the new column, and other two-level factors go to columns 4, 5, 6, and 7.

In experimentation, an eight-level column is sometimes needed. As with a four-level column, an eight-level column can be prepared using the column-merging

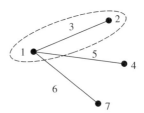

FIGURE 5.15   Selection of three columns to form a new column

TABLE 5.4   Formation of a Four-Level Column in $L_8(2^7)$

| | | | | Column | | | | | |
| | | | | Combined | New | | | | |
Run	1	2	3	Number	Column	4	5	6	7
1	0	0	0	00	0	0	0	0	0
2	0	0	0	00	0	1	1	1	1
3	0	1	1	01	1	0	0	1	1
4	0	1	1	01	1	1	1	0	0
5	1	0	1	10	2	0	1	0	1
6	1	1	0	11	3	1	1	1	0
7	1	1	0	11	3	0	1	1	0
8	1	1	0	11	3	1	0	0	1

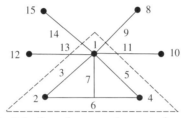

FIGURE 5.16   Selection of seven columns to form a new column

method. Because an eight-level column counts for seven degrees of freedom, it can be obtained by combining seven two-level columns. The procedure is similar to that for creating a four-level column.

1. Select any three independent (fundamental) columns and their four interaction columns with the assistance of linear graphs. For example, if an eight-level column is to be created in $L_{16}(2^{15})$, we can choose the independent columns 1, 2, and 4 and their interaction columns, 3, 5, 6 and 7, as shown in Figure 5.16. Column 7 accommodates three-column interaction (columns 1, 2, and 4) or two-column interaction (columns 1 and 6).

2. Merge the numbers of the three independent (fundamental) columns selected and obtain 000, 001, 010, 011, 100, 101, 110, and 111, denoted by 0, 1, 2, 3, 4, 5, 6, and 7, respectively. Then the merging forms a new column with these eight levels. In the $L_{16}(2^{15})$ example, combining the numbers of columns 1, 2, and 4 gives a new series of numbers, as shown in Table 5.5.

3. Replace the seven columns selected with the eight-level column. In the $L_{16}(2^{15})$ example, columns 1 to 7 are replaced by the new column. The new column is orthogonal to any other column, except for the original seven columns. Now, the new array can accommodate an eight-level factor and up to eight two-level factors.

TABLE 5.5   Formation of an Eight-Level Column in $L_{16}(2^{15})$

					Column								
Run	1	2	4	Combined Number	New Column	8	9	10	11	12	13	14	15
1	0	0	0	000	0	0	0	0	0	0	0	0	0
2	0	0	0	000	0	1	1	1	1	1	1	1	1
3	0	0	1	001	1	0	0	0	0	1	1	1	1
4	0	0	1	001	1	1	1	1	1	0	0	0	0
5	0	1	0	010	2	0	0	1	1	0	0	1	1
6	0	1	0	010	2	1	1	0	0	1	1	0	0
7	0	1	1	011	3	0	0	1	1	1	1	0	0
8	0	1	1	011	3	1	1	0	0	0	0	1	1
9	1	0	0	100	4	0	1	0	1	0	1	0	1
10	1	0	0	100	4	1	0	1	0	1	0	1	0
11	1	0	1	101	5	0	1	0	1	1	0	1	0
12	1	0	1	101	5	1	0	1	0	0	1	0	1
13	1	1	0	110	6	0	1	1	0	0	1	1	0
14	1	1	0	110	6	1	0	0	1	1	0	0	1
15	1	1	1	111	7	0	1	1	0	1	0	0	1
16	1	1	1	111	7	1	0	0	1	0	1	1	0

5.8.6   Assigning Factors to Columns

To lay out an experiment, we must select an appropriate orthogonal array and allocate the factors and interactions to the columns within the array. The selection of an orthogonal array is directed by the concept of degrees of freedom, and the assignment of factors and interactions is assisted by linear graphs.

The term *degrees of freedom* has different meanings in physics, chemistry, engineering, and statistics. In statistical analysis, it is the minimum number of comparisons that need to be made to draw a conclusion. For example, a factor of four levels, say A_0, A_1, A_2, and A_3, has three degrees of freedom because we need three comparisons between A_0 and the other three levels to derive a conclusion concerning A_0. Generally, in the context of experimental design, the number of degrees of freedom required to study a factor equals the number of factor levels minus one. For example, a two-level factor counts for 1 degree of freedom, and a three-level factor has 2 degrees of freedom.

The number of degrees of freedom of an interaction between factors equals the product of the degrees of freedom of the factors comprising the interaction. For example, the interaction between a three-level factor and a four-level factor has $(3 - 1) \times (4 - 1) = 6$ degrees of freedom.

The number of degrees of freedom in an orthogonal array equals the sum of degrees of freedom available in each column. If we continue to use $L_N(I^P)$ to denote an orthogonal array, the degrees of freedom of the array would be $(I - 1) \times P$. For example, the number of degrees of freedom available in $L_{16}(2^{15})$ is $(2 - 1) \times 15 = 15$. $L_{18}(2^1 \times 3^7)$ is a special case that deserves more attention.

As shown in Section 5.8.4, interaction between columns 1 and 2 of the array is orthogonal to all other columns. The interaction provides $(2 - 1) \times (3 - 1) = 2$ degrees of freedom. Then the total number of degrees of freedom is $2 + (2 - 1) \times 1 + (3 - 1) \times 7 = 17$. From the examples we can see that the number of degrees of freedom in an array equals the number of runs of the array minus one, that is, $N - 1$. This is generally true because N runs of an array provide $N - 1$ degrees of freedom.

Having understood the concept and calculation of degrees of freedom, we can select an appropriate orthogonal array and assign factors and interactions to the columns in the array by using its linear graphs. The procedure is as follows:

1. Calculate the total number of degrees of freedom needed to study the factors (main effects) and interactions of interest. This is the degrees of freedom required.
2. Select the smallest orthogonal array with at least as many degrees of freedom as required.
3. If necessary, modify the orthogonal array by merging columns or using other techniques to accommodate the factor levels.
4. Construct a required linear graph to represent the factors and interactions. The dots represent the factors, and the connecting lines indicate the interactions between the factors represented by the dots.
5. Choose the standard linear graph that most resembles the linear graph required.
6. Modify the required graph so that it is a subset of the standard linear graph.
7. Assign factors and interactions to the columns according to the linear graph. The unoccupied columns are error columns.

Example 5.4 The rear spade in an automobile can fracture in the ends of the structure due to fatigue under road conditions. An experiment was designed to improve the fatigue life of the structure. The fatigue life may be affected by the setting of the design parameters as well as the manufacturing process. The microfractures generated during forging grow while the spade is in use. Therefore, the control factors in this study include the design and production process parameters. The main control factors are as follows:

- *Factor A*: material; $A_0 = $ type 1, $A_1 = $ type 2
- *Factor B*: forging thickness; $B_0 = 7.5$ mm, $B_1 = 9.5$ mm
- *Factor C*: shot peening; $C_0 = $ normal, $C_1 = $ masked
- *Factor D*: bend radius; $D_0 = 5$ mm, $D_1 = 9$ mm

In addition to these main effects (factors), interactions $B \times D$ and $C \times D$ are possible and should be included in the study. Select an appropriate orthogonal array and lay out the experiment.

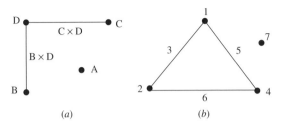

FIGURE 5.17   (*a*) Required linear graph; (*b*) standard linear graph

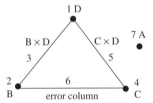

FIGURE 5.18   Assignment of factors and interactions to $L_8(2^7)$

SOLUTION   To design the experiment, we first calculate the degrees of freedom for the factors and interactions. Each factor has two levels and thus has $2 - 1 = 1$ degree of freedom. Each interaction has $(2 - 1) \times (2 - 1) = 1$ degree of freedom. The total number of degrees of freedom equals $4 \times 1 + 2 \times 1 = 6$. Then we select $L_8(2^7)$, which provides seven degrees of freedom, to lay out the experiment.

The next step is to construct a linear graph to represent the interactions and factors. The graph is shown in Figure 5.17*a*. This graph resembles the standard linear graph in Figure 5.17*b*. Then we reconstruct the linear graph by matching the graph to the standard linear graph and assign factors and interactions to the columns, as shown in Figure 5.18. Factor A is assigned to column 7 rather than column 6 to avoid factor A being confounded with any potential interaction between factors B and C. Column 6 is empty and serves as an error column.

5.8.7   Cross Arrays

So far, the design of experiment has dealt only with control factors. The purpose of robust design is to make products insensitive to noise factors by choosing the optimal levels of design parameters. To achieve this purpose, it is essential to incorporate noise factors into the experimental design. C. F. Wu and Hamada (2000), for example, suggest that a single array be used to accommodate both the control and noise factors. In the scheme of Taguchi's robust design, control factors and noise factors are allocated to separate orthogonal arrays. The array that accommodates control factors is referred to as the *inner array*, and the array that contains noise factors is called the *outer array*. The combination of an inner

TABLE 5.6 Cross Orthogonal Array

									Outer Array			
								z_1:	0	0	1	1
		Inner Array						z_2:	0	1	0	1
		Factors and Interactions										
Run	1	2	3	4	5	6	7	z_3:	0	1	1	0
1	0	0	0	0	0	0	0					
2	0	0	0	1	1	1	1					
3	0	1	1	0	0	1	1					
4	0	1	1	1	1	0	0		Experimental data			
5	1	0	1	0	1	0	1					
6	1	0	1	1	0	1	0					
7	1	1	0	0	1	1	0					
8	1	1	0	1	0	0	1					

and an outer array forms a cross array. Table 5.6 shows a cross array in which the inner array is $L_8(2^7)$ and the outer array is $L_4(2^3)$. The outer array can accommodate three noise factors (z_1, z_2, and z_3).

The run size of a cross array is $N \times l$, where N is the run size of the inner array and l is the run size of the outer array. For the cross array in Table 5.6, the total number of runs is $8 \times 4 = 32$. If more noise factors or levels are to be included in the experiment, the size of the outer array will be larger. As a result, the total run size will increase proportionally, and the experiment will become too expensive. This difficulty may be resolved by using the noise-compounding strategy.

The aim of using an outer array is to integrate into the experiment the noise conditions under which the product will operate in the field. The test units at a setting of control factors are subjected to a noise condition. The quality characteristic of a unit usually takes extreme values at extreme noise conditions. If the quality characteristic is robust against extreme conditions, it would be robust in any condition between extremes. Therefore, it is legitimate in an outer array to use only extreme noise conditions: the least and most severe conditions. The least severe condition often is a combination of the lower bounds of the ranges of the noise factors, whereas the most severe is formed by the upper bounds.

In the context of reliability testing, experiments using the levels of noise factors within the use range frequently generate few failures or little degradation. Considering this, elevated levels of carefully selected noise factors are sometimes applied to yield a shorter life or more degradation. Such noise factors must not

TABLE 5.7 Cross Array for the Rear Spade Experiment

								Outer Array	
				Inner Array					
	D	B	D $\times$ B	C	D $\times$ C		A		
Run	1	2	3	4	5	6	7	z_{11}	z_{12}
1	0	0	0	0	0	0	0		
2	0	0	0	1	1	1	1		
3	0	1	1	0	0	1	1		
4	0	1	1	1	1	0	0	Fatigue life data	
5	1	0	1	0	1	0	1		
6	1	0	1	1	0	1	0		
7	1	1	0	0	1	1	0		
8	1	1	0	1	0	0	1		

interact with the control factors. Otherwise, the accelerated test data may lead to a falsely optimal setting of the control factors (Section 5.14.4).

Example 5.5 Refer to Example 5.4. The noise factors for the rear spade are as follows:

- *Noise factor M*: stroke frequency; $M_0 = 0.5$ Hz, $M_1 = 3$ Hz
- *Noise factor S*: stroke amplitude; $S_0 = 15$ mm, $S_1 = 25$ mm

There are four combinations of noise levels, but running the experiments at the least and most severe combinations would yield sufficient information. The least severe combination is M_0 and S_0, and the most severe combination is M_1 and S_1. The two combinations are denoted by z_{11} and z_{12}. Then the outer array needs to include only these two noise conditions. Using the linear graph in Figure 5.18, we developed the cross array for the experiment of the rear spade as given in Table 5.7.

5.9 EXPERIMENTAL LIFE DATA ANALYSIS

In Section 5.8.7 we described the design of experiment with a cross array, as shown in, for example, Tables 5.6 and 5.7. Once the design is completed, the next step is to perform the experiment according to the cross array. Because reliability is the primary experimental response, experimentation is indeed a reliability test. As we know, products can fail in two distinct failure modes: hard and soft. If a product is subject to a catastrophic failure, the product is said to be *binary:* either success or failure. Life is the only meaningful quality characteristic of this type of product. In this section we describe methods for analyzing experimental life data. If a product loses its function gradually, it is possible to monitor and measure a performance characteristic during testing. The performance characteristic is the key quality characteristic used in subsequent design optimization. In Section 5.10 we discuss experimental degradation data analysis.

5.9.1 Life–Noise Factor Relationships

The life of most products can be modeled with the lognormal or Weibull distribution. As shown in Chapter 2, if life T has a lognormal distribution with shape parameter σ and scale parameter μ, $Y = \ln(T)$ follows the normal distribution with scale parameter (standard deviation) σ and location parameter (mean) μ. If T has a Weibull distribution with shape parameter β and characteristic life α, $Y = \ln(T)$ follows the smallest extreme value distribution with scale parameter $\sigma = 1/\beta$ and location parameter $\mu = \ln(\alpha)$. The four distributions above belong to the family of location-scale distributions.

If the location parameter μ at a setting of control factors is a linear function of stresses (noise factors), we have

$$\mu = \beta_0 + \beta_1 z_1 + \cdots + \beta_p z_p = \mathbf{z}^T \boldsymbol{\beta}, \tag{5.13}$$

where β_i $(i = 0, 1, \ldots, p)$ are the coefficients to be estimated from experimental data, z_i $(i = 1, 2, \ldots, p)$ the noise factors, p the number of noise factors, $\mathbf{z}^T = (1, z_1, \ldots, z_p)$, and $\boldsymbol{\beta} = (\beta_0, \beta_1, \ldots, \beta_p)^T$. In (5.13), z_i can be a transformed noise factor. For example, if temperature is the noise factor and the Arrhenius relationship is used, z_i is the reciprocal of the absolute temperature. If voltage is the noise factor and the inverse power relationship is appropriate, z_i is the logarithm of voltage. Both the Arrhenius relationship and the inverse power relationship are discussed in Chapter 7.

If second-order noise effects and noise-by-noise interaction effects are expected, the life–noise relation can be written as

$$\mu = \beta_0 + \beta_1 z_1 + \cdots + \beta_p z_p + \beta_{11} z_1^2 + \cdots + \beta_{pp} z_p^2$$
$$+ \beta_{12} z_1 z_2 + \cdots + \beta_{(p-1)p} z_{(p-1)} z_p$$
$$= \beta_0 + \mathbf{Z}^T \mathbf{b} + \mathbf{Z}^T \mathbf{B} \mathbf{Z}, \tag{5.14}$$

where

$$\mathbf{Z} = \begin{bmatrix} z_1 \\ z_2 \\ \cdots \\ z_p \end{bmatrix}, \qquad \mathbf{b} = \begin{bmatrix} \beta_1 \\ \beta_2 \\ \cdots \\ \beta_p \end{bmatrix}, \qquad \mathbf{B} = \begin{bmatrix} \beta_{11} & \beta_{12}/2 & \cdots & \beta_{1p}/2 \\ \beta_{21}/2 & \beta_{22} & \cdots & \beta_{2p}/2 \\ \cdots & \cdots & \cdots & \cdots \\ \beta_{p1}/2 & \beta_{p2}/2 & \cdots & \beta_{pp} \end{bmatrix}.$$

$\mathbf{B}$ is a symmetric matrix; that is, $\beta_{ij} = \beta_{ji}$. If second-order noise effects are nonexistent, $\beta_{ij} = 0$ for $i \neq j$.

As shown in Chapter 7, if a product is subjected to voltage (V_0) and temperature (T_0) simultaneously, (5.14) may be written as

$$\mu = \beta_0 + \beta_1 z_1 + \beta_2 z_2 + \beta_{12} z_1 z_2, \tag{5.15}$$

where $z_1 = 1/T_0$ and $z_2 = \ln(V_0)$.

As discussed in Section 5.8.7, the noise-compounding strategy may be used to reduce the size of an outer array. For the parameters in (5.13) and (5.14) to be estimable, the number of levels of the compounding noise factors must be greater than or equal to the number of unknowns.

5.9.2 Likelihood Functions

Sample likelihood function can be perceived as the joint probability of observed data. The probability depends on assumed models and model parameters. In Chapter 7 we discuss more about the likelihood function. For convenience, we denote

$$z(u) = \frac{u - \mu}{\sigma},$$

where μ and σ are the location and scale parameters, respectively.

Lognormal Distribution The log likelihood for a log exact failure time y is

$$\text{LE} = -\tfrac{1}{2}\ln(2\pi) - \ln(\sigma) - \tfrac{1}{2}[z(y)]^2. \tag{5.16}$$

The log likelihood for a log failure time between y and y' is

$$\text{LI} = \ln\{\Phi[z(y')] - \Phi[z(y)]\}, \tag{5.17}$$

where $\Phi[\cdot]$ is the cumulative distribution function (cdf) of the standard normal distribution.

The log likelihood for an observation left censored at log time y is

$$\text{LL} = \ln\{\Phi[z(y)]\}. \tag{5.18}$$

The log likelihood for an observation right-censored at log time y' is

$$\text{LR} = \ln\{1 - \Phi[z(y')]\}. \tag{5.19}$$

Weibull Distribution The log likelihood for a log exact failure time y is

$$\text{LE} = -\ln(\sigma) - \exp[z(y)] + z(y). \tag{5.20}$$

The log likelihood for a log failure time between y and y' is

$$\text{LI} = \ln\{\exp[-e^{z(y)}] - \exp[-e^{z(y')}]\}. \tag{5.21}$$

The log likelihood for an observation left-censored at log time y is

$$LL = \ln\{1 - \exp[-e^{z(y)}]\}. \tag{5.22}$$

The log likelihood for an observation right-censored at log time y' is

$$LR = -\exp[z(y')]. \tag{5.23}$$

Likelihood for a Run The experimental response (lifetime) can be observed to have an exact value, or be in an interval, or be right or left censored. The log likelihood function for a run is the sum of the log likelihoods of all test units in the run: namely,

$$LT = \sum_{i \in EXT} LE_i + \sum_{i \in INT} LI_i + \sum_{i \in LFT} LL_i + \sum_{i \in RHT} LR_i, \tag{5.24}$$

where EXT, INT, LFT, and RHT denote the sets of exact, interval, left-censored, and right-censored data in a run, respectively. In practice, there usually are only one or two types of data. Thus, the form of the log likelihood function is much simpler than it appears in (5.24).

5.9.3 Reliability as the Quality Characteristic

The sample log likelihood is a function of the model parameters σ, β, or **b** and **B**. The parameters can be estimated by maximizing the LT given in (5.24). Generally, a closed form for the estimation cannot be obtained, especially when the life–noise factor relationship is complicated. In these situations, numerical methods are needed.

The estimates that maximize LT in (5.24) are conditional on a run. There is a separate set of estimates for each run. Then the reliability at a specified time of interest can be computed for each run and each level of the noise condition. By exhausting the reliability calculations for each cross combination of inner and outer arrays, we can populate the reliability estimates in, for example, Table 5.8. The reliability estimates are observations of the quality characteristic and will be used for subsequent design optimization (Section 5.11).

As we know, the reliability of a product depends on the levels of both the noise and control factors. If the location parameter of the life distribution is modeled as a function of the control and noise factors, the log likelihood for the cross array can be obtained by running the summation in (5.24) over all samples in the cross array. C. F. Wu and Hamada (2000) describe this in detail. Then the reliability for each combination of inner and outer array may be estimated by following the procedure described above. This method is unfortunately very complex. We have seen that this method requires modeling of the relationship between the life and the noise and control factors. In general, numerous control factors are involved in experimentation, and the model can be unmanageably complicated. The tractability is worsened when there are interactions between the noise and control factors.

TABLE 5.8   Cross Orthogonal Array with Reliability Estimates

									Outer Array		
								z_1:	0	0	1	1
		Inner Array Factors and Interactions						z_2:	0	1	0	1
Run	1	2	3	4	5	6	7	z_3:	0	1	1	0
1	0	0	0	0	0	0	0		R_{11}	R_{12}	R_{13}	R_{14}
2	0	0	0	1	1	1	1		R_{21}	R_{22}	R_{23}	R_{24}
3	0	1	1	0	0	1	1					
4	0	1	1	1	1	0	0					
5	1	0	1	0	1	0	1		...	...	...	...
6	1	0	1	1	0	1	0					
7	1	1	0	0	1	1	0					
8	1	1	0	1	0	0	1		R_{81}	R_{82}	R_{83}	R_{84}

5.9.4   Life as the Quality Characteristic

The estimation of reliability from censored data needs to assume a life–noise relationship and requires complicated modeling. Certainly, the estimates include both model and residual errors. If the experiments do not involve censoring and yield complete life data, the life data observed, rather than the reliability estimated should be used for subsequent design optimization.

5.10   EXPERIMENTAL DEGRADATION DATA ANALYSIS

The failure of some products is defined in terms of performance characteristic crossing a specified threshold. The life of the products is the time at which the performance characteristic reaches the threshold. For these products it is possible to monitor and measure the performance characteristic during testing. The measurements contain credible information about product reliability and can be used for subsequent design optimization.

5.10.1   Performance Characteristic as the Quality Characteristic

During testing, samples are either monitored continuously or inspected periodically. If the latter, it is recommended that all samples have the same inspection times, such that measurements of different samples at the same times are available and can be compared without using a degradation model. In both cases,

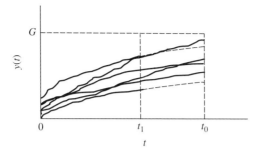

FIGURE 5.19   Projecting $y(t_1)$ to a normal censoring time

measurements at the censoring time are observations of the quality characteristic to be used in subsequent design optimization. This approach does not require modeling degradation paths and projecting the performance characteristic to a later time. Thus, it eliminates model and residual error.

In practice, samples may have different censoring times. For example, some samples have to be suspended earlier because of the failure of test equipment. In these situations, degradation models are required to model the degradation paths of samples censored earlier. In Chapter 8 we present degradation modeling in detail. By using degradation models, we project the performance characteristic to the normal censoring time. This case is illustrated in Figure 5.19, where $y(t)$ is a smaller-the-better performance characteristic at time t, t_0 a normal censoring time, t_1 an earlier censoring time, and G the threshold of y. The dashed segments of the degradation curves in Figure 5.19 represent the projection of $y(t)$. Then the projected characteristic values combined with measurements at the normal censoring time are observations of the quality characteristic for subsequent design optimization. This method requires modeling of degradation paths and projection of earlier censored samples. If such samples are few, the modeling and projection errors have little impact on selection of the optimal setting of the design parameters.

5.10.2   Life as the Quality Characteristic

Samples built with certain combinations of levels of design parameters may fail catastrophically. This situation can arise when the levels of design parameters are widely spaced. For example, electromagnetic relays may fail in different failure modes in robust design experiments. Normally, relays fail due to the excessive increase in contact resistance between two contacts. If the clearance between two contacts of a relay is too small, the relay may lose function suddenly after a period of operation because of melting contacts. Melting welds two contacts together and prevents the relay from switching. The contact resistance between the welded contacts is decreased to zero. Figure 5.20 illustrates the degradation paths of contact resistance for the test units that have these two failure modes, where $y(t)$ is the contact resistance, t the number of cycles, and G the threshold.

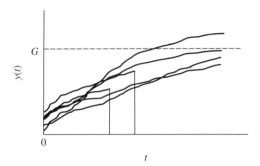

FIGURE 5.20   Degradation paths of two failure modes

If a catastrophic failure occurs before a test is completed, life should serve as the quality characteristic. For samples that have failed suddenly or gradually by censoring time, their lifetimes are observed. For samples that survive censoring, their lifetimes are calculated from degradation models. As we will see in Chapter 8, the general form of a degradation model can be written as

$$y = g(t; \beta_1, \beta_2, \ldots, \beta_p) + e, \tag{5.25}$$

where y is the observation of the performance characteristic at time t, $g(t; \beta_1, \beta_2, \ldots, \beta_p)$ is the actual degradation path, $\beta_1, \beta_2, \ldots, \beta_p$ are unknown model parameters, $e \sim N(0, \sigma_e^2)$ is the residual deviation of the observation, and σ_e is a constant. In Chapter 8 we describe various specific forms of (5.25) and the estimation of model parameters.

A product is said to have failed if y crosses threshold G. Then the life of the unit is given by

$$\hat{t} = g^{-1}(G; \beta_1, \beta_2, \ldots, \beta_p). \tag{5.26}$$

The life derived from (5.26) is actually a pseudolife; it contains model error as well as residual error. The pseudolife data and lifetimes observed are combined to serve as observations of the quality characteristic in the subsequent design optimization.

5.10.3   Reliability as the Quality Characteristic

Degradation measurement of some products requires destructive inspection; that is, the degradation of each unit can be observed only once. For example, observation on the mechanical strength of interconnection bonds or on the dielectric strength of insulators requires destruction of the unit. For these products, degradation cannot be measured over time on the same unit. As a result, modeling the degradation path of a unit is impossible.

Although the actual degradation of a single unit is unknown if it is not destructed, the degradation distribution is estimable through inspection of a few samples.

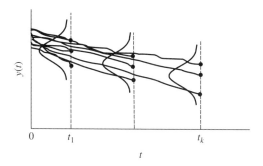

FIGURE 5.21   Destructive measurements at various times

Testing such products requires a relatively large sample size. A few samples at a time are inspected destructively. Then the statistical distribution of the degradation at the time can be estimated. As testing proceeds to the next inspection time, another group of samples is destructed for measurement. The destructive measurements are taken at each inspection time. The last inspection exhausts all remaining samples. Figure 5.21 plots the destructive measurements of a larger-the-better characteristic at various times. The bullets represent the samples destructed. The degradation paths in this plot are shown for illustration purposes and cannot be observed in reality.

Once measurements have been taken at each inspection time, a statistical distribution is fitted to these measurement data. In many applications, the measurements may be modeled with a location-scale distribution (e.g., normal, lognormal, or Weibull). For example, K. Yang and Yang (1998) report that the shear strength of copper bonds is approximately normal, and Nelson (1990, 2004) models the dielectric breakdown strength of insulators with the lognormal distribution. Let $\mu_y(t_i)$ and $\sigma_y(t_i)$ be, respectively, the location and scale parameters of the sample distribution at time t_i, where $i = 1, 2, \ldots, k$, and k is the number of inspections. The distributions are shown in Figure 5.21. The estimates $\hat{\mu}_y(t_i)$ and $\hat{\sigma}_y(t_i)$ can be obtained with graphical analysis or by the maximum likelihood method (Chapter 7). Through linear or nonlinear regression analysis on $\hat{\mu}_y(t_i)$ and $\hat{\sigma}_y(t_i)$ ($i = 1, 2, \ldots, k$), we can build the regression models $\mu_y(t; \hat{\boldsymbol{\beta}})$ and $\sigma_y(t; \hat{\boldsymbol{\theta}})$, where $\hat{\boldsymbol{\beta}}$ and $\hat{\boldsymbol{\theta}}$ are the estimated vectors of the regression model parameters. Then the reliability estimate at the time of interest, say τ, is

$$\hat{R}(\tau) = \Pr[y(\tau) \leq G] = F\left[\frac{G - \mu_y(\tau; \hat{\boldsymbol{\beta}})}{\sigma_y(\tau; \hat{\boldsymbol{\theta}})}\right] \tag{5.27}$$

for a smaller-the-better characteristic, and

$$\hat{R}(\tau) = \Pr[y(\tau) \geq G] = 1 - F\left[\frac{G - \mu_y(\tau; \hat{\boldsymbol{\beta}})}{\sigma_y(\tau; \hat{\boldsymbol{\theta}})}\right] \tag{5.28}$$

for a larger-the-better characteristic, where F denotes the cdf of a location-scale distribution. If y is modeled with a lognormal or Weibull distribution, then G is replaced with $\ln(G)$ in the equations above. In Chapter 8 we discuss the test method and data analysis of destructive inspections in greater detail.

The reliability estimate is calculated for each cross combination of the inner and outer arrays. Reliability estimates can be populated in an experimental layout similar to Table 5.8 and are observations of the quality characteristic for subsequent design optimization. A case study using destructive measurements on electronic interconnection bonds is given in Section 5.13.

5.11 DESIGN OPTIMIZATION

In earlier sections, we studied the design of experiment and preliminary analysis of experimental data. The next step in robust design is design optimization, which is aimed at finding the significant control factors and specifying the levels of these factors to maximize the robustness of products. In this section we describe statistical techniques needed for design optimization.

5.11.1 Types of Control Factors

In design optimization, control factors can be classified into four groups, depending on the influence of the factors on the quality characteristic.

1. *Dispersion factor*: a control factor that has a strong effect on the dispersion of quality characteristic (Figure 5.22a). In the figure, z is the noise factor and A is the control factor; the subscript 0 represents the low level and 1 is the high level. This figure shows that the variation of noise factor is transformed into the variability of quality characteristic. The quality characteristic at A_1 spreads more widely than at A_0. Therefore, A_0 is a better choice. Figure 5.22a also indicates that the dispersion factor interacts with the noise factor. It is this interaction that provides an opportunity for robustness improvement. In general, the level of a dispersion factor should be chosen to minimize the dispersion of the quality characteristic.

2. *Mean adjustment factor*: a control factor that has a significant influence on the mean and does not affect the dispersion of the quality characteristic (Figure 5.22b). The response line at A_0 over the noise range parallels that at A_1, indicating that the mean adjustment factor does not interact with the noise factor. In general, the level of a mean adjustment factor is selected to bring the mean of the quality characteristic on target.

3. *Dispersion and mean adjustment factor*: a control factor that influences both the dispersion and mean significantly (Figure 5.22c). This factor interacts with the noise factor and should be treated as the dispersion factor. In general, the level is set to minimize dispersion.

4. *Insignificant factor*: a control factor that affects significantly neither the dispersion nor the mean (Figure 5.22d). The response at A_0 over the noise

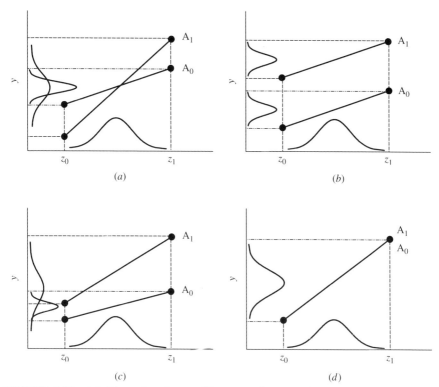

FIGURE 5.22 (*a*) Dispersion factor; (*b*) mean adjustment factor; (*c*) dispersion and mean adjustment factor; (*d*) insignificant factor

range equals that at A_1. The level is determined based on other considerations, such as the economy, manufacturability, operability, and simplicity.

5.11.2 Signal-to-Noise Ratio

Signal-to-noise ratio is a metric commonly used in the field of communication engineering. It measures the relative power of signal to noise. In the context of robust design, signals enter into a product and are transformed into the intended functions as well as failure modes, as shown in the P-diagram (Figure 5.10). The ratio of the power of the intended functions to that of the failure modes is the signal-to-noise ratio. Taguchi (1986) proposes the use of this metric to measure the robustness of products and processes. Mathematically, the signal-to-noise ratio η can be defined as

$$\eta = \frac{\mu_y^2}{\sigma_y^2}, \tag{5.29}$$

where μ_y and σ_y^2 are the mean and variance of the quality characteristic for a setting of control factors. The larger the value of η, the more robust the product.

The magnitude of η is dependent on the levels of control factors. Then a robust design becomes the task of choosing the levels of control factors that maximize η.

The quality characteristic that defines the signal-to-noise ratio may be life, reliability, or a performance characteristic, depending on the type of failure mode (soft failure or hard failure). Let y still be the quality characteristic. Without loss of generality, Table 5.9 shows the observations of y at each cross combination of the inner and outer arrays. In this table, y_{ijk} is the kth observation at the cross combination of row i and column j; $i = 1, 2, \ldots, N$, $j = 1, 2, \ldots, l$, and $k = 1, 2, \ldots, n_{ij}$; N is the number of rows in the inner array; P is the number of columns in the inner array; l is the number of columns in the outer array; n_{ij} is the number of replicates at the cross combination of row i and column j. When y is the reliability, there is no replicated observation in each test combination. That is, n_{ij} is always 1.

Nominal-the-Best Quality Characteristics   For this type of characteristic, the signal-to-noise ratio is the same as in (5.29). To simplify numerical operations, it is usually redefined as

$$\eta = 10 \log \left(\frac{\mu_y^2}{\sigma_y^2} \right), \tag{5.30}$$

where $\log(\cdot)$ denotes the common (base 10) logarithm. η is measured in decibels (dB).

In application, the mean and variance in (5.30) are unknown. They can be estimated from observations of the quality characteristic. For row i in Table 5.9, let

$$k_i = \sum_{j=1}^{l} n_{ij}, \quad \overline{y}_i = \frac{1}{k_i} \sum_{j=1}^{l} \sum_{k=1}^{n_{ij}} y_{ijk}, \quad D_i = k_i \overline{y}_i^2, \quad i = 1, 2, \ldots, N.$$

TABLE 5.9   Experimental Layout and Observations

Inner Array				Outer Array					
				1	2	$\cdots$	l		
Factors and Interactions				0	0	$\cdots$	1		
				$\cdots$	$\cdots$	$\cdots$	$\cdots$		
Run	1	2 $\cdots$	P	0	1	$\cdots$	0	$\hat{\eta}$	$\overline{y}$
1	0	0 $\cdots$	0	$y_{111}, \ldots, y_{11n_{11}}$	$y_{121}, \ldots, y_{12n_{12}}$	$\cdots$	$y_{1l1}, \ldots, y_{1ln_{1l}}$	$\hat{\eta}_1$	$\overline{y}_1$
2	0	0 $\cdots$	1	$y_{211}, \ldots, y_{21n_{21}}$	$y_{221}, \ldots, y_{22n_{22}}$	$\cdots$	$y_{2l1}, \ldots, y_{2ln_{2l}}$	$\hat{\eta}_2$	$\overline{y}_2$
$\cdots$	$\cdots$	$\cdots$ $\cdots$	$\cdots$	$\cdots$	$\cdots$	$\cdots$	$\cdots$	$\cdots$	$\cdots$
i	$\cdots$	$\cdots$ $\cdots$	$\cdots$	$y_{i11}, \ldots, y_{i1n_{i1}}$	$y_{i21}, \ldots, y_{i2n_{i2}}$	$\cdots$	$y_{il1}, \ldots, y_{iln_{il}}$	$\hat{\eta}_i$	$\overline{y}_i$
$\cdots$	$\cdots$	$\cdots$ $\cdots$	$\cdots$	$\cdots$	$\cdots$	$\cdots$	$\cdots$	$\cdots$	$\cdots$
N	1	1 $\cdots$	0	$y_{N11}, \ldots, y_{N1n_{N1}}$	$y_{N21}, \ldots, y_{N2n_{N2}}$	$\cdots$	$y_{Nl1}, \ldots, y_{Nln_{Nl}}$	$\hat{\eta}_N$	$\overline{y}_N$

Then the estimates of the mean and variance are, respectively,

$$\hat{\mu}_{yi}^2 = \frac{D_i - \hat{\sigma}_{yi}^2}{k_i},$$

$$\hat{\sigma}_{yi}^2 = \frac{1}{k_i - 1} \sum_{j=1}^{l} \sum_{k=1}^{n_{ij}} (y_{ijk} - \overline{y}_i)^2.$$

Substituting the estimates above into (5.30) gives

$$\hat{\eta}_i = 10 \log \left(\frac{D_i - \hat{\sigma}_{yi}^2}{k_i \hat{\sigma}_{yi}^2} \right) = 10 \log \left(\frac{\overline{y}_i^2 - \hat{\sigma}_{yi}^2 / k_i}{\hat{\sigma}_{yi}^2} \right). \tag{5.31}$$

If k_i is large, $\hat{\sigma}_{yi}^2 / k_i$ becomes negligible. Then the signal-to-noise ratio can be written as

$$\hat{\eta}_i \approx 10 \log \left(\frac{\overline{y}_i^2}{\hat{\sigma}_{yi}^2} \right) = 20 \log \left(\frac{\overline{y}_i}{\hat{\sigma}_{yi}} \right), \qquad i = 1, 2, \ldots, N. \tag{5.32}$$

Note that $\overline{y}_i / \hat{\sigma}_{yi}$ is the reciprocal of the coefficient of variation, which measures the dispersion of the quality characteristic. Therefore, maximizing the signal-to-noise ratio minimizes the characteristic dispersion.

Smaller-the-Better Quality Characteristics For this type of quality characteristics, the target is zero, and the estimates of the mean would be zero or negative. As a result, we cannot use the log transformation as in (5.30). Rather, the signal-to-noise ratio is defined as

$$\eta = -10 \log(\text{MSD}), \tag{5.33}$$

where MSD is the mean-squared deviation from the target value of the quality characteristic. Because the target of the smaller-the-better type is zero, the MSD for row i is given by

$$\text{MSD}_i = \frac{1}{k_i} \sum_{j=1}^{l} \sum_{k=1}^{n_{ij}} y_{ijk}^2, \qquad i = 1, 2, \ldots, N.$$

Then the signal-to-noise ratio is estimated by

$$\hat{\eta}_i = -10 \log \left(\frac{1}{k_i} \sum_{j=1}^{l} \sum_{k=1}^{n_{ij}} y_{ijk}^2 \right), \qquad i = 1, 2, \ldots, N. \tag{5.34}$$

Larger-the-Better Quality Characteristics If y is larger-the-better, then $1/y$ is smaller-the-better. Thus, the target of $1/y$ is zero. The MSD of $1/y$ for row i is given by

$$\text{MSD}_i = \frac{1}{k_i} \sum_{j=1}^{l} \sum_{k=1}^{n_{ij}} \frac{1}{y_{ijk}^2}, \qquad i = 1, 2, \ldots, N.$$

Then the signal-to-noise ratio is

$$\hat{\eta}_i = -10 \log \left(\frac{1}{k_i} \sum_{j=1}^{l} \sum_{k=1}^{n_{ij}} \frac{1}{y_{ijk}^2} \right), \qquad i = 1, 2, \ldots, N. \qquad (5.35)$$

As discussed earlier, reliability is sometimes used as the quality characteristic. Because reliability R is a larger-the-better characteristic in the range between 0 and 1, $1/R$ is a smaller-the-better type and targeted at 1. The MSD of $1/R$ is

$$\text{MSD}_i = \frac{1}{l} \sum_{j=1}^{l} \left(\frac{1}{R_{ij}} - 1 \right)^2, \qquad i = 1, 2, \ldots, N,$$

where R_{ij} is the reliability estimate at the cross combination of row i and column j. Then the signal-to-noise ratio is

$$\hat{\eta}_i = -10 \log \left[\frac{1}{l} \sum_{j=1}^{l} \left(\frac{1}{R_{ij}} - 1 \right)^2 \right], \qquad i = 1, 2, \ldots, N. \qquad (5.36)$$

Example 5.6 Refer to Table 5.8. Suppose that the reliability estimates in the first row are 0.92, 0.96, 0.8, and 0.87. Calculate the signal-to-noise ratio for this row.

SOLUTION   Substituting the reliability estimates into (5.36), we have

$$\hat{\eta}_1 = -10 \log \left\{ \frac{1}{4} \left[\left(\frac{1}{0.92} - 1 \right)^2 + \left(\frac{1}{0.96} - 1 \right)^2 \right. \right.$$

$$\left. \left. + \left(\frac{1}{0.8} - 1 \right)^2 + \left(\frac{1}{0.87} - 1 \right)^2 \right] \right\} = 16.3.$$

For each row of the inner array, a signal-to-noise ratio is calculated using (5.32), (5.34), (5.35), or (5.36), depending on the type of quality characteristic. Then a further analysis using the graphical response method or analysis of variance is performed to determine the optimal setting of control factors.

5.11.3   Steps of Design Optimization

After the experimental data y_{ijk} (as shown in Table 5.9) are available, the experimenters should analyze these data to optimize product design. Several steps are needed for design optimization. The steps differ according to the type of quality characteristic.

Smaller-the-Better and Larger-the-Better Characteristics

1. Calculate the signal-to-noise ratio for each row of the inner array, as shown in Table 5.9.

2. Identify the control factors that significantly affect the signal-to-noise ratio through graphical response analysis (Section 5.11.4) or analysis of variance (Section 5.11.5).

3. Determine the optimal setting of the significant control factors by maximizing the signal-to-noise ratio.

4. Determine the levels of insignificant control factors in light of material cost, manufacturability, operability, and simplicity.

5. Predict the signal-to-noise ratio at the optimal setting.

6. Conduct a confirmation test using the optimal setting to verify that the optimal setting chosen yields the robustness predicted.

Nominal-the-Best Characteristics

1. Calculate the signal-to-noise ratio and the mean response ($\overline{y}_i$ in Table 5.9) for each row of the inner array.

2. Identify the significant control factors and categorize them into dispersion factors, mean adjustment factors, or dispersion and mean adjustment factors (treated as dispersion factors).

3. Select the setting of the dispersion factors to maximize the signal-to-noise ratio.

4. Select the setting of the mean adjustment factors such that the estimated quality characteristic is closest to the target.

5. Determine the levels of insignificant control factors based on consideration of material cost, manufacturability, operability, and simplicity.

6. Predict the signal-to-noise ratio and the mean response at the optimal setting.

7. Conduct a confirmation test using the optimal setting to check if the optimal setting produces the signal-to-noise ratio and mean response predicted.

5.11.4   Graphical Response Analysis

The purpose of graphical response analysis is to identify the factors and interactions that significantly affect the response, and determine the combination of factor levels to achieve the most desirable response. This graphical method is intuitive, simple, and powerful, and often is a good choice for engineers. The analysis has been computerized; commercial software packages such as Minitab provide the capability for graphical analysis.

To better understand graphical analysis, let's look at an example.

Example 5.7   The attaching clips in automobiles cause audible noise while vehicles are operating. The two variables that may affect the audible noise level are length of clip and type of material. Interaction between these two variables is possible. The noise factors that influence the audible noise level include vehicle speed and temperature. The levels of the control factors and noise factors are as follows:

- *Control factor A*: length; $A_0 = 25$ cm, $A_1 = 15$ cm
- *Control factor B*: material; $B_0 = $ plastic, $B_1 = $ metal
- *Noise condition z*: driving speed and temperature; $z_1 = 40$ miles per hour and $15°C$, $z_2 = 85$ miles per hour and $30°C$

$L_4(2^3)$ is used as an inner array to accommodate the control factors. The outer array contains two columns each for a noise condition. The experimental layout is shown in Table 5.10. Then the experiments are conducted according to the cross array. The audible noise data (in dB) were collected after each vehicle accumulated 1500 miles for this test purpose. The data are summarized in Table 5.10.

The audible noise level is a smaller-the-better characteristic. The signal-to-noise ratio is calculated from (5.34) and summarized in Table 5.10. For example, the value of the ratio for the first run is

$$\hat{\eta}_1 = -10 \log \left[\tfrac{1}{2}(15^2 + 19^2) \right] = -24.7.$$

Then the average responses at levels 0 and 1 of factors A and B are computed:

Level	A	B
0	-29.1	-27.4
1	-29.5	-31.2

For example, the average response at level 0 of factor B is

$$\overline{B}_0 = \frac{-24.7 - 30.2}{2} = -27.4.$$

Next, a two-way table is constructed for the average response of the interaction between factors A and B:

TABLE 5.10   Experimental Layout for the Clip Design

Run	A	B	A × B	z_1	z_2	$\hat{\eta}$
1	0	0	0	15	19	-24.7
2	0	1	1	47	49	-33.6
3	1	0	1	28	36	-30.2
4	1	1	0	26	29	-28.8

	A_0	A_1
B_0	−24.7	−30.2
B_1	−33.6	−28.8

Having calculated the average response at each level of factors and interactions, we need to determine significant factors and interactions and then select optimal levels of the factors. The work can be accomplished using graphical response analysis.

Graphical response analysis is to plot the average response for factor and interaction levels and determine the significant factors and their optimal levels from the graphs. The average response for a level of a factor is the sum of the observations corresponding to the level divided by the total number of observations. Example 5.7 shows the calculation of average response at B_0. Plot the average responses on a chart in which the x-axis is the level of a factor and y-axis is the response. Then connect the dots on the chart. This graph is known as a *main effect plot*. Figure 5.23a shows the main effect plots for factors A and B of Example 5.7. The average response of an interaction between two factors is usually obtained using a two-way table in which a cross entry is the average response, corresponding to the combination of the levels of the two factors (see the two-way table for Example 5.7). Plot the tabulated average responses on a chart where the x-axis is the level of a factor. The chart has more than one line segment, each representing a level of the other factor, and is called an *interaction plot*. The interaction plot for Example 5.7 is shown in Figure 5.23b.

The significance of factors and interactions can be assessed by viewing the graphs. A steep line segment in the main effect plot indicates a strong effect of the factor. The factor is insignificant if the line segment is flat. In an interaction plot, parallel line segments indicate no interaction between the two factors. Otherwise an interaction is existent. Let's revisit Example 5.7. Figure 5.23a indicates that factor B has a strong effect on the response because the line segment has a steep slope. Factor A has little influence on the response because the corresponding line segment is practically horizontal. The interaction plot in Figure 5.23b indicates the lack of parallelism of the two line segments. Therefore, the interaction between factors A and B is significant.

Once the significant control factors have been identified, the optimal setting of these factors should be determined. If interaction is important, the optimal levels of the factors involved are selected on the basis of the factor-level combination that results in the most desirable response. For factors not involved in an interaction, the optimal setting is the combination of factor levels at which the most desirable average response is achieved. When the interaction between two factors is strong, the main effects of the factors involved do not have much meaning. The levels determined through interaction analysis should override the levels selected from main effect plots. In Example 5.7, the interaction plot shows that interaction between factors A and B is important. The levels of A and B should be dictated by the interaction plot. From Figure 5.23b it is seen that A_0B_0

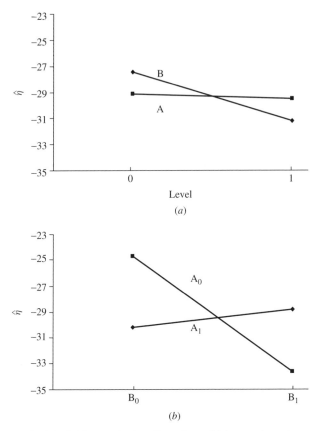

FIGURE 5.23   (*a*) Main effect plots; (*b*) interaction plot

produces the largest signal-to-noise ratio. This level combination must be used in design, although the main effect plot indicates that factor A is an insignificant variable whose level may be chosen for other considerations (e.g., using a shorter clip to save material cost).

If the experimental response is a nominal-the-best characteristic, we should generate the main effect and interaction plots for both signal-to-noise ratio and mean response. If a factor is identified to be both a dispersion factor and a mean adjustment factor, it is treated as a dispersion factor. Its level is selected to maximize the signal-to-noise ratio by using the strategy described above. To determine the optimal levels of the mean adjustment factors, we enumerate the average response at each combination of mean adjustment factor levels. The average response at a level combination is usually obtained by the prediction method described below. Then the combination is chosen to bring the average response on target.

Once the optimal levels of factors have been selected, the mean response at the optimal setting should be predicted for the following reasons. First, the prediction indicates how much improvement the robust design will potentially make. If the gain is not sufficient, additional improvement using other techniques, such as the tolerance design, may be required. Second, a subsequent confirmation test should be conducted and the result compared against the predicted value to verify the optimality of the design. The prediction is made based on an estimation of the effects of significant factors and interactions. For convenience, we denote by T and $\overline{T}$ the total of responses and the average of responses, respectively. Then we have

$$T = \sum_{i=1}^{N} y_i, \qquad \overline{T} = \frac{T}{N},$$

where y_i represents $\hat{\eta}_i$ or $\overline{y}_i$ as shown in, for example, Table 5.9.

The average response predicted at optimal levels of significant factors is

$$\hat{y} = \overline{T} + \sum_{i \in \text{MET}} (\overline{F}_i - \overline{T}) + \sum_{j > i \in \text{INT}} [(\overline{F}_{ij} - \overline{T}) - (\overline{F}_i - \overline{T}) - (\overline{F}_j - \overline{T})], \quad (5.37)$$

where MET is a set of significant main effects, INT a set of significant interactions, $\overline{F}_i$ the average response of factor F_i at the optimal level, and $\overline{F}_{ij}$ the average response of the interaction between factors F_i and F_j at the optimal levels. Because the effect of an interaction includes the main effects of the factors involved, the main effects should be subtracted from the interaction effect as shown in the second term of (5.37). If the response is a nominal-the-best characteristic, (5.37) should include all significant dispersion factors and mean adjustment factors and interactions. Then apply the equation to estimate the signal-to-noise ratio and the mean response.

In Example 5.7, B and A × B are significant and A_0B_0 is the optimal setting. The grand average response is $\overline{T} = -29.3$. The signal-to-noise ratio predicted at the optimal setting is obtained from (5.37) as

$$\hat{\eta} = \overline{T} + (\overline{B}_0 - \overline{T}) + [(\overline{A_0B_0} - \overline{T}) - (\overline{A}_0 - \overline{T}) - (\overline{B}_0 - \overline{T})] = \overline{A_0B_0} - \overline{A}_0 + \overline{T}$$

$$= -24.7 + 29.1 - 29.3 = -24.9,$$

which is close to -24.7, the signal-to-noise ratio calculated from the experimental data at A_0B_0 and shown in Table 5.10.

In general, a confirmation experiment should be conducted before implementation of the optimal setting in production. The optimality of the setting is verified if the confirmation result is close to the value predicted. A statistical hypothesis test may be needed to arrive at a statistically valid conclusion.

5.11.5 Analysis of Variance

Graphical response analysis is an intuitive method for identifying significant factors and interactions. The method is easy to understand and use when the number of factors is small. However, the analysis may become tedious if a

TABLE 5.11   One-Factor Experimental Layout

Factor Level	Observation				Total	Average
1	y_{11}	y_{12}	$\cdots$	y_{1n}	$y_{1.}$	$\bar{y}_{1.}$
2	y_{21}	y_{22}	$\cdots$	y_{2n}	$y_{2.}$	$\bar{y}_{2.}$
$\vdots$	$\vdots$	$\vdots$	$\ldots$	$\vdots$	$\vdots$	$\vdots$
p	y_{p1}	y_{p2}	$\cdots$	y_{pn}	$y_{p.}$	$\bar{y}_{p.}$

relatively large number of factors are involved. In these situations, analysis of variance (ANOVA) is more efficient.

ANOVA for One-Factor Experiments   To understand the concept and procedure for ANOVA, we first consider a one-factor experiment designed to determine the effect of factor A. Factor A has p levels, each of which contains n replicates. Let y_{ij} be the jth observation of quality characteristic y taken at level i. The experimental layout is shown in Table 5.11. Statistically, the experiment is to test the hypothesis that the mean responses at all levels are equal. Let $y_{i.}$ and $\bar{y}_{i.}$ denote the total and average of the observations at level i, respectively. Also let $y_{..}$ and $\bar{y}_{..}$ be the grand total and grand average of all observations. Then we have

$$y_{i.} = \sum_{j=1}^{n} y_{ij}, \quad \bar{y}_{i.} = \frac{y_{i.}}{n}, \qquad i = 1, 2, \ldots, p,$$

$$y_{..} = \sum_{i=1}^{p} \sum_{j=1}^{n} y_{ij}, \quad \bar{y}_{..} = \frac{y_{..}}{N},$$

where $N = pn$ is the total number of observations in the experiment.

We define the total corrected sum of squares as

$$SS_T = \sum_{i=1}^{p} \sum_{j=1}^{n} (y_{ij} - \bar{y}_{..})^2$$

to measure the overall variability in the data. SS_T can be written as

$$SS_T = \sum_{i=1}^{p} \sum_{j=1}^{n} (y_{ij} - \bar{y}_{..})^2 = \sum_{i=1}^{p} \sum_{j=1}^{n} \left[(\bar{y}_{i.} - \bar{y}_{..}) + (y_{ij} - \bar{y}_{i.}) \right]^2$$

$$= n \sum_{i=1}^{p} (\bar{y}_{i.} - \bar{y}_{..})^2 + \sum_{i=1}^{p} \sum_{j=1}^{n} (y_{ij} - \bar{y}_{i.})^2 = SS_A + SS_E, \qquad (5.38)$$

where

$$SS_A = n \sum_{i=1}^{p} (\bar{y}_{i.} - \bar{y}_{..})^2 \quad \text{and} \quad SS_E = \sum_{i=1}^{p} \sum_{j=1}^{n} (y_{ij} - \bar{y}_{i.})^2.$$

SS_A is called the *sum of squares of the factor*, and SS_E is the *sum of squares of the error*. Equation (5.38) indicates that the total corrected sum of squares can be partitioned into these two portions.

Factor A has p levels; thus, SS_A has $p - 1$ degrees of freedom. There are N observations in the experiment, so SS_T has $N - 1$ degrees of freedom. Because there are n observations in each of p levels providing $n - 1$ degrees of freedom for estimating the experimental error, SS_E has $p(n - 1) = N - p$ degrees of freedom. Note that the degrees of freedom for SS_T equals the sum of degrees of freedom for SS_A and SS_E. Dividing the sum of squares by its respective degrees of freedom gives the mean square MS: namely,

$$MS_x = \frac{SS_x}{df_x}, \tag{5.39}$$

where x denotes A or E and df_x is the number of degrees of freedom for SS_x.

The F statistic for testing the hypothesis that the mean responses for all levels are equal is

$$F_0 = \frac{MS_A}{MS_E}, \tag{5.40}$$

which has an F distribution with $p - 1$ and $N - p$ degrees of freedom. We conclude that factor A has a statistically significant effect at $100\alpha\%$ significance level if

$$F_0 > F_{\alpha, p-1, N-p}.$$

For the convenience of numerical calculation, the sums of squares may be rewritten as

$$SS_T = \sum_{i=1}^{p} \sum_{j=1}^{n} y_{ij}^2 - \frac{y_{..}^2}{N}, \tag{5.41}$$

$$SS_A = \sum_{i=1}^{p} \frac{y_{i.}^2}{n} - \frac{y_{..}^2}{N}, \tag{5.42}$$

$$SS_E = SS_T - SS_A. \tag{5.43}$$

The procedure for the analysis of variance can be tabulated as shown in Table 5.12. The table is called an *ANOVA table*.

TABLE 5.12 One-Factor ANOVA Table

Source of Variation	Sum of Squares	Degrees of Freedom	Mean Square	F_0
Factor	SS_A	$p - 1$	MS_A	MS_A/MS_E
Error	SS_E	$N - p$	MS_E	
Total	SS_T	$N - 1$		

TABLE 5.13   Temperature Data from the Engine Testing

A/F Ratio	Temperature (°C)				Total	Average
10.6	701	713	722	716	2852	713.00
11.6	745	738	751	761	2995	748.75
12.6	773	782	776	768	3099	774.75

Example 5.8   An experiment was designed to investigate the effects of the air-to-fuel (A/F) ratio on the temperature of the exhaust valve in an automobile engine. The experiment was replicated with four samples at each A/F ratio. The experimental data are summarized in Table 5.13. Determine whether A/F ratio has a strong influence at the 5% significance level.

SOLUTION   The total and average of the temperature observations are computed and summarized in Table 5.13. The grand total and grand average are

$$y_{..} = \sum_{i=1}^{3} \sum_{j=1}^{4} y_{ij} = 701 + 713 + \cdots + 768 = 8946, \qquad \overline{y}_{..} = \frac{8946}{12} = 745.5.$$

The sums of squares are

$$SS_T = \sum_{i=1}^{3} \sum_{j=1}^{4} y_{ij}^2 - \frac{y_{..}^2}{12} = 701^2 + 713^2 + \cdots + 768^2 - \frac{8946^2}{12} = 8311,$$

$$SS_A = \sum_{i=1}^{3} \frac{y_{i.}^2}{4} - \frac{y_{..}^2}{12} = \frac{2852^2 + 2995^2 + 3099^2}{4} - \frac{8946^2}{12} = 7689.5,$$

$$SS_E = SS_T - SS_A = 8311 - 7689.5 = 621.5.$$

The calculation of mean squares and F_0 is straightforward. The values are summarized in the ANOVA table as shown in Table 5.14. Because $F_0 = 55.64 > F_{0.05,2,9} = 4.26$, we conclude that the A/F ratio has a strong effect on the exhaust valve temperature at the 5% significance level.

ANOVA for Orthogonal Inner Arrays   In the design of experiment, the purpose of an outer array is to expose samples to noise factors. After the experimental data

TABLE 5.14   ANOVA Table for the Exhaust Temperature Data

Source of Variation	Sum of Squares	Degrees of Freedom	Mean Square	F_0
Factor	7689.5	2	3844.75	55.67
Error	621.5	9	69.06	
Total	8311	11		

are collected, the outer array has completed its role. The outer array usually is not involved in subsequent ANOVA for design optimization unless we are interested in understanding the effects of noise factors on the quality characteristic. The optimal levels of design parameters are determined using analysis of variance for the inner array.

A column of an inner array may be assigned to a factor, an interaction, or an error (empty column). An I-level column in $L_N(I^P)$ can be considered as an I-level factor, each level having $n = N/I$ replicates. Thus, (5.41) can be employed to calculate the total corrected sum of squares of an inner array, and (5.42) applies to a column of an inner array. Let T be the total of observations: namely,

$$T = \sum_{i=1}^{N} y_i,$$

where y_i represents $\hat{\eta}_i$ or $\bar{y}_i$ as shown in, for example, Table 5.9. Then the total corrected sum of squares can be written as

$$SS_T = \sum_{i=1}^{N} y_i^2 - \frac{T^2}{N}. \tag{5.44}$$

Also, let T_j denote the total of observations taken at level j in a column. The sum of squares of column i having I levels is

$$SS_i = \frac{I}{N} \sum_{j=0}^{I-1} T_j^2 - \frac{T^2}{N}. \tag{5.45}$$

For a two-level column, (5.45) reduces to

$$SS_i = \frac{(T_0 - T_1)^2}{N}. \tag{5.46}$$

Now let's look at a simple array, $L_9(3^4)$. From (5.45), the sum of squares of column 1 is

$$SS_1 = \frac{3}{9}[(y_1 + y_2 + y_3)^2 + (y_4 + y_5 + y_6)^2 + (y_7 + y_8 + y_9)^2] - \frac{1}{9}\left(\sum_{i=1}^{9} y_i\right)^2.$$

The sum of squares of column 2 of the array is

$$SS_2 = \frac{3}{9}[(y_1 + y_4 + y_7)^2 + (y_2 + y_5 + y_8)^2 + (y_3 + y_6 + y_9)^2] - \frac{1}{9}\left(\sum_{i=1}^{9} y_i\right)^2.$$

In an inner array, some columns may be empty and are treated as error columns. The sum of squares for an error column is computed with (5.45). Then the sums of squares for all error columns are added together. If an assigned column has a small sum of squares, it may be treated as an error column, and the sum of squares should be pooled into the error term. The total corrected sum of squares equals the total of the sums of squares for factor columns, interaction columns, and error columns. Recall that the number of degrees of freedom is $I - 1$ for an I-level column and is $N - 1$ for $L_N(I^P)$. The number of degrees of freedom for error is the sum of the degrees of freedom for error columns. The mean square and F statistic for a factor or interaction are computed from (5.39) and (5.40), respectively. It is concluded that the factor or interaction is important at the $100\alpha\%$ significance level if

$$F_0 > F_{\alpha, I-1, \mathrm{dfe}},$$

where dfe is the number of degrees of freedom for error.

The computation for ANOVA may be burdensome, especially when a large number of factors and interactions are involved. There are several commercial software packages, such as Minitab, which can perform the calculation.

If the quality characteristic is smaller-the-better or larger-the-better, the ANOVA for signal-to-noise ratio data determines the significance of the factors and interactions. The next step is to select the optimum levels of the significant factors and interactions. The selection method was discussed in Section 5.11.4.

ANOVA should be performed for both signal-to-noise ratio and mean response data if the quality characteristic belongs to the nominal-the-best type. In such cases, the levels of dispersion factors are selected to maximize the signal-to-noise ratio, while the mean adjustment factors are set at the levels that bring the response on target. The procedure for choosing the optimal setting is the same as that for the graphical analysis.

Once the optimal setting is specified, the average response at the optimal setting should be predicted by using (5.37). A confirmation test is run to verify that the predicted value is achieved.

Example 5.9 Refer to Examples 5.4 and 5.5. The design of experiment for the rear spade has four control factors and two interactions. $L_8(2^7)$ is used as an inner array to accommodate the control factors. The outer array is filled with two combinations of noise levels. Two test units of the same setting of control factors were run at each of the two noise combinations. The fatigue life data (in 1000 cycles) are shown in Table 5.15.

Fatigue life is a larger-the-better characteristic. The signal-to-noise ratio for each row of the inner array is computed using (5.35) and is shown in Table 5.15. For example, the value of the ratio for the first row is

$$\hat{\eta}_1 = -10\log\left(\frac{1}{4}\sum_{i=1}^{4}\frac{1}{y_i^2}\right)$$

$$= -10\log\left[\frac{1}{4}\left(\frac{1}{7.6^2} + \frac{1}{8.2^2} + \frac{1}{6.2^2} + \frac{1}{6.9^2}\right)\right] = 17.03.$$

TABLE 5.15 Cross Array and Fatigue Life Data for the Rear Spade

		Inner Array								
D	B	D × B	C	D × C		A	Outer Array			
Run 1	2	3	4	5	6	7	z_{11}		z_{12}	$\hat{\eta}$
1 0	0	0	0	0	0	0	7.6	8.2	6.2 6.9	17.03
2 0	0	0	1	1	1	1	7.1	6.7	4.9 4.2	14.55
3 0	1	1	0	0	1	1	4.8	6.3	5.2 3.9	13.68
4 0	1	1	1	1	0	0	6.2	5.2	4.4 5.1	14.17
5 1	0	1	0	1	0	1	3.9	4.3	3.6 4.7	12.18
6 1	0	1	1	0	1	0	5.7	5.1	4.7 3.8	13.38
7 1	1	0	0	1	1	0	6.4	5.9	5.7 6.4	15.67
8 1	1	0	1	0	0	1	6.8	6.2	4.3 5.5	14.72

The total of the values of the signal-to-noise ratio is

$$T = \sum_{i=1}^{8} \hat{\eta}_i = 17.03 + 14.55 + \cdots + 14.72 = 115.38.$$

The sum of squares for each column is calculated from (5.46). For example, the sum of squares for column 2 (factor B) is

$$SS_2 = \frac{1}{8} \times (17.03 + 14.55 + 12.18 + 13.38$$
$$- 13.68 - 14.17 - 15.67 - 14.72)^2 = 0.15.$$

The sums of squares of the factors, interactions, and error are given in Table 5.16. Because each column has two levels, the degrees of freedom for each factor, interaction, and error is 1. Note that in the table the sum of squares for factor B

TABLE 5.16 ANOVA Table for the Fatigue Life Data of the Rear Spade

Source of Variation	Sum of Squares	Degrees of Freedom	Mean Square	F_0
A	3.28	1	3.28	27.33
B	0.15	1	0.15	
C	0.38	1	0.38	3.17
D	1.51	1	1.51	12.58
D × B	9.17	1	9.17	76.42
D × C	0.63	1	0.63	5.25
e	0.09	1	0.09	
(e)	(0.24)	(2)	(0.12)	
Total	15.21	7		

is pooled into error e to give the new error term (e). Thus, the new error term has 2 degrees of freedom.

The critical value of the F statistic is $F_{0.1,1,2} = 8.53$ at the 10% significance level. By comparing the critical value with the F_0 values for the factors and interactions in the ANOVA table, we conclude that A, D, and D $\times$ B have significant effects, whereas B, C, and D $\times$ C are not statistically important.

For comparison, we use the graphical response method described in Section 5.11.4 to generate the main effect plots and interaction plots, as shown Figure 5.24. Figure 5.24a indicates that the slopes for factors A and D are steep, and thus these factors have strong effects, whereas factor B is clearly insignificant. From the figure it is difficult to judge the importance of factor C because of the marginal slope. This suggests that ANOVA should be performed. Figure 5.24b shows that the interaction between factors B and D is very strong, although factor B itself is not significant. As indicated in Figure 5.24c, there appears an interaction between factors C and D because of the lack of parallelism of the two line segments. The interaction, however, is considerably less severe than that between B and D. ANOVA shows that this interaction is statistically insignificant, but the value of F_0 is close to the critical value.

Once the significant factors and interactions are identified, the optimal levels should be selected. Because the interaction D $\times$ B has a strong effect, the levels of B and D are determined by the interaction effect. From the interaction plot, we choose $B_0 D_0$. Figure 5.24a or T_0 and T_1 of factor A calculated for ANOVA suggests that A_0 be selected. Because factor C is deemed insignificant, C_0 is chosen to maintain the current manufacturing process. In summary, the design should use material type 1, forging thickness 7.5 mm, and bend radius 5 mm with normal shot peening in manufacturing.

The value of the signal-to-noise ratio predicted at the optimal setting $A_0 B_0 C_0 D_0$ is obtained from (5.37) as

$$\hat{\eta} = \overline{T} + (\overline{A_0} - \overline{T}) + (\overline{D_0} - \overline{T}) + [(\overline{D_0 B_0} - \overline{T}) - (\overline{D_0} - \overline{T}) - (\overline{B_0} - \overline{T})]$$

$$= \overline{A_0} + \overline{D_0 B_0} - \overline{B_0} = 15.06 + 15.49 - 14.28 = 16.27.$$

A confirmation test should be run to verify that the signal-to-noise ratio predicted is achieved.

The estimated average fatigue life at the optimal setting $A_0 B_0 C_0 D_0$ is $\hat{y} = 10^{\hat{\eta}/20} \times 1000 = 6509$ cycles. This life estimate is the average of life data over the noise levels and unit-to-unit variability.

5.12   ROBUST RELIABILITY DESIGN OF DIAGNOSTIC SYSTEMS

In this section we describe the development of the robust reliability design method for diagnostic systems whose functionality is different from that in common hardware systems in that the signal and response of the systems are binary. In particular, in this section we define and measure the reliability and robustness of the systems. The noise effects are evaluated and the noise factors are prioritized.

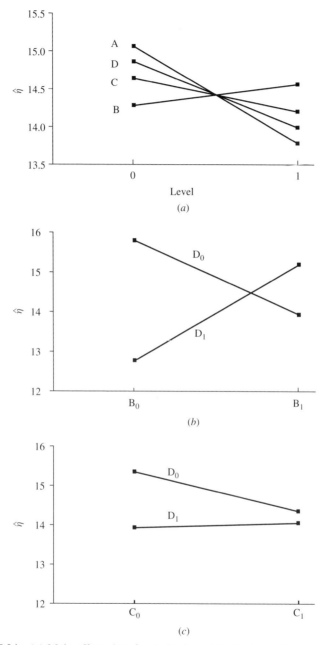

FIGURE 5.24 (*a*) Main effect plots for A, B, C, and D; (*b*) interaction plot for D × B; (*c*) interaction plot for D × C

The steps for robust reliability design are described in detail. An automotive example is given to illustrate the method.

5.12.1   Problem Statement

Diagnostic systems are software-based built-in-test systems which detect, isolate, and indicate the failures of the prime systems, where *prime systems* refers to the hardware systems monitored by the diagnostic systems. Use of diagnostic systems reduces the loss due to the failure of the prime systems and facilitates subsequent repairs. Because of the benefits, diagnostic systems have found extensive applications in industry, especially where failure of the prime systems results in critical consequences. For instance, on-board diagnostic (OBD) systems are integrated into automobiles to monitor components and systems whose failure would cause emission concerns. When failure of such components or systems occurs, the OBD system detects the failure, illuminates a light on the instrument panel cluster saying "Service Engine Soon," to alert the driver to the need for repair, and stores the diagnostic trouble codes related to the failure to aid failure isolation.

In modern software-intensive diagnostic systems, algorithms are coded to perform operations for diagnosis. Ideally, the diagnosis should indicate the true state (failure or success) of the prime systems. However, if not designed adequately, the algorithms are sensitive to noise sources and thus cause diagnostic systems to commit the following two types of errors:

- *Type I error (α error).* This error, denoted by α, is measured by the probability that the diagnostic system detected a failure given that one did not occur.

- *Type II error (β error).* This error, denoted by β, is measured by the probability that the diagnostic system failed to detect a failure given that one did occur.

Because of α error, diagnostic systems may generate failure indications on surviving prime systems. Thus, α error results in unnecessary repairs to products. Manufacturers have an intense interest in eliminating or minimizing this type of error because unnecessary repairs incur remarkable warranty expenses. On the other hand, diagnostic systems may not generate failure indications on failed prime systems because of β error. As a result, β error causes potential losses to customers, so manufacturers are also responsible for reducing β error. In the automotive industry, a large β error of an OBD system can trigger vehicle recalls, usually issued by a government agency. Therefore, it is imperative that both α and β errors be minimized over a wide range of noise factors. A powerful technique to accomplish this objective is robust reliability design.

5.12.2   Definition and Metrics of Reliability and Robustness

A prime system usually has a binary state: success or failure. The intended function of a diagnostic system is to diagnose the states correctly over time. That is, a diagnostic system should indicate a failure of the prime system when it occurs, and not indicate a failure if it does not occur. Thus, the reliability and robustness of a diagnostic system can be defined as follows.

- Reliability of a diagnostic system is defined as the probability of the system detecting the true states of the prime system under specified conditions for a specified period of time.
- Robustness of a diagnostic system is the capability of the system to detect the true states of the prime system consistently in the presence of noise sources.

Robustness can be measured by α and β. These two types of errors are correlated negatively; that is, α increases as β decreases, and vice versa. Therefore, it is frequently difficult to judge the performance of a diagnostic system using α and β only. Reliability is a more reasonable and comprehensive metric to measure performance.

G. Yang and Zaghati (2003) employ the total probability law and give the reliability of a diagnostic system as

$$R(t) = (1 - \alpha) - (\beta - \alpha)M(t), \tag{5.47}$$

where $R(t)$ is the reliability of the diagnostic system and $M(t)$ is the failure probability of the prime system. Equation (5.47) indicates that:

- If the prime system is 100% reliable [i.e., $M(t) = 0$], the reliability of the diagnostic system becomes $1 - \alpha$. This implies that the unreliability is due to false detection only.
- If the prime system fails [i.e., $M(t) = 1$], the reliability of the diagnostic system becomes $1 - \beta$. This implies that the unreliability is due only to the inability of the system to detect failures.
- If $\alpha = \beta$, the reliability becomes $1 - \alpha$ or $1 - \beta$. This implies that $M(t)$ has no influence on the reliability.
- The interval of $R(t)$ is $1 - \beta \le R(t) \le 1 - \alpha$ if $\beta > \alpha$ (which holds in most applications).

Taking the derivatives of (5.47) gives

$$\frac{\partial R(t)}{\partial \alpha} = M(t) - 1, \qquad \frac{\partial R(t)}{\partial \beta} = -M(t), \qquad \frac{\partial R(t)}{\partial M(t)} = -(\beta - \alpha). \tag{5.48}$$

Because the derivatives are negative, $R(t)$ decreases as α, β, or $M(t)$ increases. In most applications, $M(t)$ is smaller than 0.5. Hence, $|\partial R(t)/\partial \alpha| > |\partial R(t)/\partial \beta|$. This indicates that $R(t)$ is influenced more by α than by β.

Since reliability is considered as the quality characteristic, the signal-to-noise ratio for a run is computed from (5.36) as

$$\hat{\eta} = -10 \log \left[\frac{1}{l} \sum_{j=1}^{l} \left(\frac{1}{R_j} - 1 \right)^2 \right],$$

TABLE 5.17   Grouped and Prioritized Noise Factors[a]

Noise Type	Variable α	β	$M(t)$	Sensitivity	Priority
1	×	×	×	$-(1 + \beta - \alpha)$	1
2	×	×		-1	2
3	×		×	$-(1 + \beta - \alpha - M)$	3
4		×	×	$-(M + \beta - \alpha)$	5
5	×			$-(1 - M)$	4
6		×		$-M$	6
7			×	$-(\beta - \alpha)$	7

[a] ×, affected.

where l is the number of columns in an outer array and R_j is the reliability at the jth noise level.

5.12.3   Noise Effects Assessment

As discussed in Section 5.6, there are three types of noise factors: external noise, internal noise and unit-to-unit noise. Some of these noise factors disturb the diagnostic systems directly and increase α and β errors. Meanwhile, others may jeopardize the function of the prime systems and deteriorate their reliability. In general, a noise factor may affect one or more of the variables α, β and $M(t)$. Depending on what variables are disturbed, the noise factors can be categorized into seven types, as shown in Table 5.17. The noise factors in different types have unequal influences on the reliability of the diagnostic systems. The significance of a noise factor can be evaluated by the sensitivity of reliability to the noise factor. The sensitivity is obtained by using (5.48) and is summarized in Table 5.17. The table also lists the priority of the seven types of noise factors ordered by the sensitivity, assuming that $M(t) > \beta > \alpha$. Because it is impossible to include all noise factors in an experiment, only the noise factors in high-priority groups should be considered.

5.12.4   Experimental Layout

Signals from prime systems to diagnostic systems have a binary state: success or failure. Diagnostic systems should be robust against the states and noise factors. In robust design the signals and noise factors go to an outer array, with the design parameters placed in an inner array. A generic experimental layout for the robust design is shown in Table 5.18. In this table, $M_1 = 0$ indicates that the prime system is functioning and $M_2 = 1$ indicates that the prime system has failed; α_{ij} and β_{ij} ($i = 1, 2, \ldots, N; j = 1, 2, \ldots, l$) are the values of α and β at the cross combination of row i and column j.

TABLE 5.18   Generic Experimental Layout for Diagnostic Systems

	Design Parameter				$M_1 = 0$				$M_2 = 1$		
Run	A	B	C	......	z_1	z_2	$\cdots$	z_l	z_1	z_2	$\cdots$	z_l
1					α_{11}	α_{12}	$\cdots$	α_{1l}	β_{11}	β_{12}	$\cdots$	β_{1l}
2					α_{21}	α_{22}	$\cdots$	α_{2l}	β_{21}	β_{22}	$\cdots$	β_{2l}
3		Orthogonal array			α_{31}	α_{32}	$\cdots$	α_{3l}	β_{31}	β_{32}	$\cdots$	β_{3l}
$\vdots$						$\vdots$				$\vdots$		
N					α_{N1}	α_{N2}	$\cdots$	α_{Nl}	β_{N1}	β_{N2}	$\cdots$	β_{Nl}

Experiments are conducted according to the layout. The layout requires a diagnostic system with the same setting of design parameters to monitor both functioning and failed prime systems at various noise levels. For example, in the first run, a diagnostic system with the first set of parameters is built to diagnose the functioning prime system working at each of the l noise levels. Then the same diagnostic system is used to monitor the failed prime system at each noise level. During experimentation, record the number of failure occurrences detected by the diagnostic system while running at $M_1 = 0$, and the number of failure occurrences that are not detected by the diagnostic system while running at $M_2 = 1$. By definition, α_{ij} is estimated by the number of failure occurrences detected, divided by the total number of replicates given $M_1 = 0$; the estimate of β_{ij} is the number of undetected failure occurrences divided by the total number of replicates given $M_2 = 1$.

5.12.5   Experimental Data Analysis

At the time of interest τ (e.g., the warranty period or design life), the reliability of the diagnostic system at the cross combination of row i and column j is calculated from (5.47) as

$$R_{ij}(\tau) = (1 - \alpha_{ij}) - (\beta_{ij} - \alpha_{ij})M(\tau).$$

Estimates of reliability are used to compute the signal-to-noise ratio using (5.36). Table 5.19 summarizes the estimates of reliability and signal-to-noise ratio.

Once the estimates of the signal-to-noise ratio are calculated, ANOVA or graphical response analysis should be performed to identify the significant factors. Optimal levels of these factors are chosen to maximize the signal-to-noise ratio. Finally, the optimality of the setting of design parameters selected should be verified through a confirmation test.

5.12.6   Application Example

The example is to show how α, β, reliability, and signal-to-noise ratio are calculated with the automobile test data. The steps for robust design are standard and thus are not given in this example.

**TABLE 5.19   Estimates of Reliability and
Signal-to-Noise Ratio for Diagnostic Systems**

Run	z_1	z_2	$\cdots$	z_l	$\hat{\eta}$
1	$\hat{R}_{11}$	$\hat{R}_{12}$	$\cdots$	$\hat{R}_{1l}$	$\hat{\eta}_1$
2	$\hat{R}_{21}$	$\hat{R}_{22}$	$\cdots$	$\hat{R}_{2l}$	$\hat{\eta}_2$
3	$\hat{R}_{31}$	$\hat{R}_{32}$	$\cdots$	$\hat{R}_{3l}$	$\hat{\eta}_3$
$\vdots$	$\vdots$	$\vdots$	$\vdots$	$\vdots$	$\vdots$
N	$\hat{R}_{N1}$	$\hat{R}_{N2}$	$\cdots$	$\hat{R}_{Nl}$	$\hat{\eta}_N$

Test Method   A sport utility vehicle installed with an on-board diagnostic monitor with a current setting of design parameters was tested to evaluate the robustness of the monitor. Load and engine speed [revolutions per minute (RPM)] are
the key noise factors disturbing the monitor. The combinations of load and RPM
are grouped into seven noise levels; at each level both the load and RPM vary
over an interval because of the difficulty in controlling the noise factors at fixed
levels. Table 5.20 shows the noise levels. The vehicle was driven at different
combinations of load and RPM. The prime system (component) being monitored
is expected to have 10% failure probability at the end of design life ($\tau = 10$
years). Thus, failures at 10% probability were injected into the component under
monitor during the test trips. The test recorded the number of failures undetected
when failures were injected ($M_2 = 1$), and the number of failures detected when
no failures were injected ($M_1 = 0$).

Test Data   At each noise level, the number of injected failures, the number
of injected failures undetected, the number of successful operations, and the
number of failures detected from the successful operations, denoted I_1, I_2, S_1,
and S_2, respectively, are shown in Table 5.20. The test data are coded to protect
the proprietary information.

Data Analysis   The estimates of α and β equal the values of S_2/S_1 and I_2/I_1,
respectively. The reliability of the monitor at 10 years at each noise level is

TABLE 5.20   Noise Levels and Coded Test Data

Noise Level	Load	RPM ($\times 1000$)	S_2/S_1	I_2/I_1
z_1	(0.0, 0.3)	(0.0, 1.6)	1/3200	0/400
z_2	(0.0, 0.3)	[1.6, 3.2)	100/10,400	110/1200
z_3	[0.3, 0.6)	[1.6, 3.2)	30/7500	40/800
z_4	[0.6, 0.9)	[1.6, 3.2)	30/3700	100/400
z_5	(0.0, 0.3)	[3.2, 4.8)	20/600	20/80
z_6	[0.3, 0.6)	[3.2, 4.8)	30/4800	300/600
z_7	[0.6, 0.9)	[3.2, 4.8)	160/7800	800/900

TABLE 5.21 Estimates of α, β, Reliability, and Signal-to-Noise Ratio

	z_1	z_2	z_3	z_4	z_5	z_6	z_7
$\hat{\alpha}$	0.0003	0.01	0.004	0.008	0.033	0.006	0.02
$\hat{\beta}$	0	0.09	0.05	0.25	0.25	0.5	0.89
$\hat{R}(\tau)$	0.9997	0.982	0.9914	0.9678	0.9453	0.9446	0.893
$\hat{\eta}$				34.83			

calculated from (5.47) with the α and β estimates and $M(\tau) = 0.1$. Then the signal-to-noise ratio of the monitor is computed from (5.36). Table 5.21 summarizes the estimates of α, β, reliability, and signal-to-noise ratio.

5.13 CASE STUDY

In this section a case study is presented to illustrate application of the robust reliability design methods described earlier in the chapter. The case study is aimed at improving the reliability and robustness of integrated-circuit (IC) wire bonds by optimizing the wire bonding parameters. This example deals with destructive inspection, discussed in Section 5.10.3.

5.13.1 The Problem

In semiconductor device manufacturing, one of the critical processes is making the electrical interconnections between chips and packages. Interconnections by gold wire have proved robust and reliable. However, the associated cost is high and subject to reduction in the current competitive business environment. Copper wire interconnection technology, expected to replace gold wire in some applications, is being developed. Copper wire is usually bonded using thermosonic energy.

In this case study, copper wire is bonded onto a new type of substrate. In the bonding process there are four important process parameters whose optimal levels are to be determined. The parameters and their levels are shown in Table 5.22. Preliminary experiments indicate that interactions between these parameters are not important. Because thermal cycling is the major stress that causes failure of

TABLE 5.22 Wire Bonding Process Parameters and Their Levels

		Level		
Notation	Process Parameter	1	2	3
A	Stage temperature (°C)	100	125	150
B	Ultrasonic power (units)	6	7	8
C	Bonding force (gf)	60	80	100
D	Bonding time (ms)	40	50	60

TABLE 5.23   Parameters of Thermal Cycling

	Level	
Parameter	1	2
T_{max} (°C)	150	120
T_{min} (°C)	−6	−30
ΔT (°C)	215	150
dT/dt (°C/min)	15	20

wire bonds, it is desirable to make the wire bonds insensitive to thermal cycling. Table 5.23 presents the parameters of thermal cycling used in the experiment as a noise factor. In Table 5.23, T_{max} and T_{min} are the high and low temperatures of a thermal cycle, respectively, ΔT is the delta temperature between the high and low temperatures, and dT/dt is the temperature change rate.

5.13.2   Process Parameter Design

Since there is no important interaction, an $L_9(3^4)$ inner array and an outer array with two levels are used to accommodate the process parameters and noise factor, respectively. Bonds were generated using each setting of the process parameters and then underwent thermal cycle testing. During testing, to measure the shear strength (y; in grams force or gf), 20 bonds were sheared at 0, 50, 100, 200, 300, 500, and 800 cycles, respectively. The shear strength can be modeled with the normal distribution with mean μ_y and standard deviation σ_y. Figure 5.25 shows μ_y and σ_y varying with the number of thermal cycles (N_{TC}) for wire bonds tested at noise factor levels 1 and 2. Each curve on the plots represents a setting of the process parameters in the $L_9(3^4)$ array.

A wire bond is said to have failed if its shear strength is less than or equal to 18 gf. Then from (5.28) the reliability of the wire bonds can be written as

$$\hat{R}(N_{TC}) = 1 - \Phi\left[\frac{18 - \mu_y(N_{TC})}{\sigma_y(N_{TC})}\right].$$

This equation is used to calculate the reliability of wire bonds at the censoring time (800 cycles). The resulting estimates are shown in Table 5.24. The signal-to-noise ratio for each run is then computed using (5.36). Table 5.24 also summarizes estimates of the signal-to-noise ratio.

ANOVA was performed based on the values of the signal-to-noise ratio. The analysis is shown in Table 5.25. Because the sum of squares of factor D is small, the factor is treated as an error term. Values of the F statistic for factors A, B, and C are larger than the critical value $F_{0.1,2,2} = 9.0$. Therefore, the stage temperature, ultrasonic power, and bonding force are statistically important at 10% significance level. The optimal levels of these design parameters are: stage temperature = 150°C, ultrasonic power = 7 units, bonding force = 60 gf. Since the bonding time is not significant, its level is set at 40 ms to increase productivity.

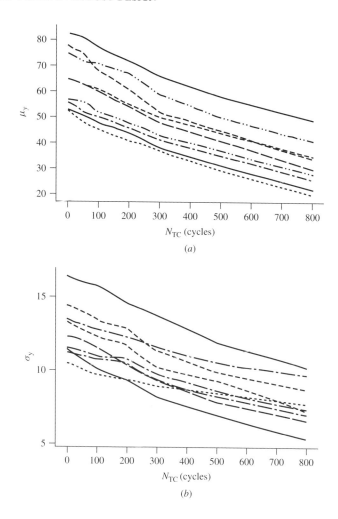

(a)

(b)

5.13.3 Signal-to-Noise Ratio Comparison

Before performing this study, the process engineer used the following design parameters: stage temperature $= 100°C$, ultrasonic power $= 7$ units, bonding force $= 80$ gf, bonding time $= 50$ ms. The combination of levels yields $\hat{\eta} = 25.10$. The optimal levels from this study improve the signal-to-noise ratio by $(46.42 - 25.10)/25.10 = 85\%$. Therefore, the robustness of reliability against thermal cycling has been increased remarkably.

5.14 ADVANCED TOPICS IN ROBUST DESIGN

In this section we introduce advanced topics, including an alternative to the signal-to-noise ratio, multiple responses, the response surface method, and accelerated testing, which are related to the subjects discussed earlier in the chapter.

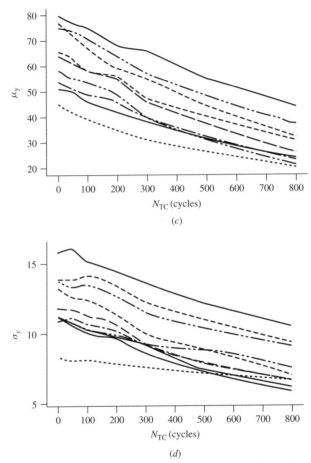

FIGURE 5.25 (a) μ_y varying with N_{TC} for wire bonds tested at noise factor level 1; (b) σ_y varying with N_{TC} for wire bonds tested at noise factor level 1; (c) μ_y varying with N_{TC} for wire bonds tested at noise factor level 2; (d) σ_y varying with N_{TC} for wire bonds tested at noise factor level 2

Recent advancements in these topics are described briefly. The materials in this section are helpful for performing a more efficient robust design.

5.14.1 Alternative to the Signal-to-Noise Ratio

Taguchi (1986, 1987) proposes using the signal-to-noise ratio to measure the robustness of a product performance. This metric has been used extensively in industry because of its simplicity. When response is a smaller-the-better or larger-the-better characteristic, as described in Section 5.11.3, the optimal setting can be found by using a one-step procedure: Select the levels of control factors that maximize the signal-to-noise ratio. This metric is a good choice for engineers.

TABLE 5.24 Estimates of Reliability and Signal-to-Noise Ratio for Wire Bonds

Run	Reliability Noise Level 1	Reliability Noise Level 2	$\hat{\eta}$
1	0.9911	0.9816	36.66
2	0.9670	0.9339	25.10
3	0.7748	0.8495	12.37
4	0.9176	0.7475	12.14
5	0.9655	0.9052	22.13
6	0.8735	0.6753	8.99
7	0.9901	0.9684	32.35
8	0.9988	0.9934	46.42
9	0.7455	0.6937	8.08

TABLE 5.25 ANOVA Table for Wire Bond Reliability

Source of Variation	Sum of Squares	Degrees of Freedom	Mean Square	F_0
A	334.85	2	167.42	81.67
B	772.65	2	386.32	188.45
C	365.02	2	182.51	89.03
D	4.10	2	2.05	
(e)	(4.10)	(2)	(2.05)	
Total	1476.62	8		

As pointed out by Nair (1992), the signal-to-noise ratio has some drawbacks, the major one being that minimizing it does not lead automatically to minimization of the quadratic quality loss (5.1) in situations where the variability of a nominal-the-best characteristic is affected by all significant control factors. The effect of this problem can, however, be mitigated by data transformation through which the variability of the transformed data is made independent of mean adjustment factors (Robinson et al., 2004). Box (1988) proposes the use of lambda plots to identify the transformation that yields the independence.

A measure alternative to the signal-to-noise ratio is the *location and dispersion model*. For each run in Table 5.9, $\bar{y}_i$ and $\ln(s_i^2)$, representing the sample mean and log sample variance over the noise replicates, are used as the measures of location and dispersion. They are

$$\bar{y}_i = \frac{1}{k_i} \sum_{j=1}^{l} \sum_{k=1}^{n_{ij}} y_{ijk}, \quad s_i^2 = \frac{1}{k_i - 1} \sum_{j=1}^{l} \sum_{k=1}^{n_{ij}} (y_{ijk} - \bar{y}_i)^2, \quad i = 1, 2, \ldots, N,$$

$$(5.49)$$

where $k_i = \sum_{j=1}^{l} n_{ij}$.

For a nominal-the-best characteristic, the procedure for obtaining the optimal setting of control factors is the same as that for optimizing the signal-to-noise ratio described in Section 5.11.3. For a smaller-the-better or larger-the-better problem, the procedure consists of two steps:

1. Select the levels of the mean adjustment factors to minimize (or maximize) the location.
2. Choose the levels of the dispersion factors that are not mean adjustment factors to minimize the dispersion.

C. F. Wu and Hamada (2000) and Nair et al. (2002) discuss in greater detail use of the location and dispersion model to achieve robustness.

5.14.2 Multiple Responses

In robust reliability design, the response or quality characteristic may be life, reliability, or performance. If life or reliability is used, the product has a single response. Multiple responses arise when a product has several performance characteristics, and some or all of them are equally important. For practical purposes, the robust reliability design described in this chapter uses one characteristic, that most closely reflecting the reliability of the product. Selection of the characteristic is based on engineering judgment, customer expectation, experience, or test data; in Chapter 8 we describe selection of the characteristic.

In some applications, using multiple responses can lead to a better design. When analyzing multiple response data, for the sake of simplicity, each response is sometimes analyzed separately to determine the optimal setting of design parameters for that response. This naive treatment may work reasonably well if there is little correlation between responses. However, when the multiple responses are highly correlated, a design setting that is optimal for one response may degrade the quality of another. In such cases, simultaneous treatment of the multiple responses is necessary. In the literature, there are two main approaches to handling multiple-response optimization problems: the desirability function method and the loss function approach.

Desirability Function Method   This method, proposed by Derringer and Such (1980) and modified by Del Casttillo et al. (1996), turns a multiple-response problem into a single-response case using a desirability function. The function is given by

$$D = [d_1(y_1) \times d_2(y_2) \times \cdots \times d_m(y_m)]^{1/m}, \qquad (5.50)$$

where d_i $(i = 1, 2, \ldots, m)$ is the desirability of response y_i, m the number of responses, and D the total desirability. Now the response of the product is D, which is a larger-the-better characteristic. In the context of robust design described in this chapter, the signal-to-noise ratio should be computed from the values of D. Then the robust design is to choose the optimal setting of control factors that maximizes the signal-to-noise ratio.

The desirability for each response depends on the type of response. For a nominal-the-best response, the individual desirability is

$$
d_i = \begin{cases}
\left(\dfrac{y_i - L_i}{m_i - L_i}\right)^{w_L}, & L_i \leq y_i \leq m_i, \\[2mm]
\left(\dfrac{y_i - H_i}{m_i - H_i}\right)^{w_H}, & m_i < y_i \leq H_i, \\[2mm]
0, & y_i < L_i \text{ or } y_i > H_i,
\end{cases}
\tag{5.51}
$$

where m_i, L_i, and H_i are the target and minimum and maximum allowable values of y, respectively, and w_L and w_H are positive constants. These two constants are equal if the value of a response smaller than the target is as undesirable as a value greater than the target.

For a smaller-the-better response, the individual desirability is

$$
d_i = \begin{cases}
1, & y_i \leq L_i, \\[2mm]
\left(\dfrac{y_i - H_i}{L_i - H_i}\right)^{w}, & L_i < y_i \leq H_i, \\[2mm]
0, & y_i > H_i,
\end{cases}
\tag{5.52}
$$

where w is a positive constant and L_i is a small enough number.

For a larger-the-better response, the individual desirability is

$$
d_i = \begin{cases}
0, & y_i \leq L_i, \\[2mm]
\left(\dfrac{y_i - L_i}{H_i - L_i}\right)^{w}, & L_i < y_i \leq H_i, \\[2mm]
1, & y_i > H_i,
\end{cases}
\tag{5.53}
$$

where w is a positive constant and H_i is a large enough number.

The desirability of each response depends on the value of the exponent. The choice of the value is arbitrary and thus subjective. In many situations it is difficult to specify meaningful minimum and maximum allowable values for a smaller-the-better or larger-the-better response. Nevertheless, the method has found many applications in industry (see, e.g., Dabbas et al., 2003; Corzo and Gomez, 2004).

Loss Function Approach Loss function for multiple responses, described by Pignatiello (1993), Ames et al. (1997), and Vining (1998), is a natural extension of the quality loss function for a single response. The loss function for a nominal-the-best response is

$$
L = (\mathbf{Y} - \mathbf{m_y})^{\mathrm{T}} \mathbf{K} (\mathbf{Y} - \mathbf{m_y}),
\tag{5.54}
$$

where $\mathbf{Y} = (y_1, y_2, \ldots, y_m)$ is the response vector, $\mathbf{m_y} = (m_{y1}, m_{y2}, \ldots, m_{ym})$ is the target vector, and $\mathbf{K}$ is a $m \times m$ matrix of which the elements are constants. The values of the constants are related to the repair and scrap cost of the product and may be determined based on the functional requirements of the product. In general, the diagonal elements of $\mathbf{K}$ measure the weights of the m responses, and the off-diagonal elements represent the correlations between these responses.

Like the single-response case, the loss function (5.54) can be extended to measure the loss for a smaller-the-better or larger-the-better response (Tsui, 1999). For a smaller-the-better response, we replace the fixed target m_{yi} with zero. For a larger-the-better response, the reciprocal of the response is substituted into (5.54) and treated as the smaller-the-better response.

If $\mathbf{Y}$ has a multivariate normal distribution with mean vector $\boldsymbol{\mu}$ and variance–covariance matrix $\boldsymbol{\Sigma}$, the expected loss can be written as

$$E(L) = (\boldsymbol{\mu} - \mathbf{m_y})^{\mathrm{T}} \mathbf{K} (\boldsymbol{\mu} - \mathbf{m_y}) + \mathrm{trace}(\mathbf{K}\boldsymbol{\Sigma}), \tag{5.55}$$

where $\boldsymbol{\mu}$ and $\boldsymbol{\Sigma}$ are the functions of control factors and noise factors and can be estimated from experimental data by multivariate analysis methods. The methods are described in, for example, Johnson (1998).

The simplest approach to obtaining the optimal setting of control factors is to directly minimize the expected loss (5.55). The direct optimization approach is used by, for example, Romano et al. (2004). Because the approach may require excessive time to find the optimal setting when the number of control factors is large, some indirect but more efficient optimization procedures have been proposed. The most common one is the two-step approach, which finds its root in Taguchi's two-step optimization approach for a single response. The approach first minimizes an appropriate variability measure and then brings the mean response on its target. Pignatiello (1993) and Tsui (1999) describe this two-step approach in detail.

5.14.3 Response Surface Methodology

The experimental analysis described in this chapter may lead to local optima because of the lack of informative relationship between response and control factors. A more effective design of experiment is the *response surface methodology* (RSM), which is known as a sequential experimental technique. The objective of RSM is to ascertain the global optimal setting of design parameters by establishing and analyzing the relationship between response and experimental variables. The variables may include both control and noise factors. The RSM experiment usually begins with a first-order experiment aimed at establishing the first-order relationship given by

$$y = \beta_0 + \sum_{i=1}^{n} \beta_i x_i + e, \tag{5.56}$$

where y is the response, x_i the experimental variable, e the residual error, β_i the coefficient representing the linear effect of x_i to be estimated from experimental data, and n the number of experimental variables. Once the relationship is built, a search must be conducted over the experimental region to determine if a curvature of the response is present. If this is the case, a second-order experiment should be conducted to build and estimate the second-order relationship given by

$$y = \beta_0 + \sum_{i=1}^{n} \beta_i x_i + \sum_{i<j}^{n} \beta_{ij} x_i x_j + \sum_{i=1}^{n} \beta_{ii} x_i^2 + e, \tag{5.57}$$

where β_i is the coefficient representing the linear effect of x_i, β_{ij} the coefficient representing the linear-by-linear interaction between x_i and x_j, and β_{ii} the coefficient representing the quadratic effect of x_i. The optimum region for experimental variables is solved by differentiating y in (5.57) with respect to x_i and setting it zero. C. F. Wu and Hamada (2000) and Myers and Montgomery (2002), for example, describe in detail the design and analysis of RSM experiments.

The principle of RSM can be applied to improve the optimality of the design setting obtained from ANOVA or graphical response analysis (K. Yang and Yang, 1998). In the context of robust design presented in this chapter, the response y in (5.57) is the signal-to-noise ratio. If there exists an interaction or quadratic effect, the relationship between signal-to-noise ratio and control factors may be modeled by (5.57), where y is replaced with η. The model contains $1 + 2n + n(n-1)/2$ parameters. To estimate the parameters, the experimental run must have the size of at least $1 + 2n + n(n-1)/2$, and each factor must involve at least three levels. The use of some orthogonal arrays, such as $L_9(3^4)$ and $L_{27}(3^{13})$, satisfies the requirements; thus, it is possible to continue response surface analysis for the experimental design. The optimal setting may be obtained through the use of standard methods for response surface analysis as described in C. F. Wu and Hamada (2000) and Myers and Montgomery (2002).

RSM assumes that all variables are continuous and derivatives exist. In practice, however, some design parameters may be discrete variables such as the type of materials. In these situations, (5.57) is still valid. But it cannot be used to determine the optima because the derivative with respect to a discrete variable is not defined. To continue the response surface analysis, we suppose that there are n_1 discrete variables and n_2 continuous variables, where $n_1 + n_2 = n$. Because the optimal levels of the n_1 discrete factors have been determined in previous analysis by using the graphical response method or ANOVA, the response surface analysis is performed for the n_2 continuous variables. Since the levels of n_1 variables have been selected, only the parameter settings that contain combinations of the selected levels of the n_1 variables can be used for response surface analysis. In general, the number of such settings is

$$w_s = N \prod_{i=1}^{n_1} \frac{1}{q_i}, \tag{5.58}$$

where N is the run size of an inner array and q_i is the number of levels of discrete variable x_i. Refer to Table 5.3, for example. $L_9(3^4)$ is used to accommodate four design parameters, one of which is assumed to be discrete and assigned to column 1. Suppose that ANOVA has concluded that level 1 is the optimal level for this variable. Then $w_s = 9/3 = 3$, because only the responses from runs 4, 5, and 6 can apply to our response surface analysis.

Excluding discrete variables from modeling, (5.57) includes only n_2 continuous variables and has $1 + 2n_2 + n_2(n_2 - 1)/2$ parameters to be estimated. Therefore, we require that $w_s \geq 1 + 2n_2 + n_2(n_2 - 1)/2$. This requirement is frequently unachievable when $n_1 \geq 2$. Thus, in most situations the response surface analysis is applicable when only one discrete design parameter is involved.

5.14.4   Accelerated Testing in Robust Design

Life or performance degradation is the primary response of an experiment designed for improving robustness and reliability. The experiment may yield few failures or little degradation at censoring time, when the levels of noise factors are within the normal use spectrum. In these situations it is difficult or impossible to perform data analysis and find the truly optimal setting of design parameters. Clearly, obtaining more life data or degradation information is necessary. This may be accomplished by carefully applying the concept of accelerated testing. Although accelerated testing has been studied and applied extensively, it is seldom discussed in the context of robust design aimed at improving reliability.

To produce a shorter life and more degradation in testing, it is natural to think of elevating noise factor levels as is usually done in a typical accelerated test. Then the accelerated test data are analyzed to draw conclusions about the optimal setting of control factors. In the context of robust design, however, the conclusions may be faulty if the accelerating noise factors interact with control factors. Without loss of generality, we assume that a robust design concerns one control factor and one noise factor. If the life has a location-scale distribution, the location parameter μ can be written as

$$\mu = \beta_0 + \beta_1 x + \beta_2 z + \beta_{12} xz, \tag{5.59}$$

where x is the control factor, z the noise factor, and β_0, β_1, β_2, and β_{12} are the coefficients to be estimated from experimental data.

The acceleration factor between the life at noise level z_1 and that at noise level z_2 is

$$A_f = \frac{\exp(\mu_1)}{\exp(\mu_2)} = \exp(\mu_1 - \mu_2), \tag{5.60}$$

where A_f is the acceleration factor and μ_i is the location parameter at noise level i. Chapter 7 presents in detail definition, explanation, and computation of the acceleration factor. For a given level of control factor, the acceleration factor between noise levels z_1 and z_2 is obtained by substituting (5.59) into (5.60). Then we have

$$A_f = \exp[(z_1 - z_2)(\beta_2 + \beta_{12} x)], \tag{5.61}$$

which indicates that the acceleration factor is a function of the control factor level. This is generally true when there are interactions between control factors and accelerating noise factors.

Accelerated test data may lead to a falsely optimal setting of design parameters if an acceleration factor depends on the level of control factor. This is illustrated by the following arguments. For the sake of convenience, we still assume that robust design involves one design parameter and one noise factor. The control factor has two levels: x_0 and x_1. The noise factor also has two levels: z_1 and z_2, where z_1 is within the normal use range and z_2 is an elevated level. Let y_{ij} ($i = 0, 1; j = 1, 2$) denote the life at x_i and z_j. Also, let A_{f0} and A_{f1} denote the acceleration factors at x_0 and x_1, respectively. Then from (5.35), the signal-to-noise ratio $\hat{\eta}'_0$ at x_0 is

$$\hat{\eta}'_0 = -10 \log\left[\frac{1}{2}\left(\frac{1}{y^2_{01}} + \frac{1}{y^2_{02}}\right)\right] = 10\log(2) - 10\log\left[\frac{1}{y^2_{01}}(1 + A^2_{f0})\right].$$

Similarly, the signal-to-noise ratio $\hat{\eta}'_1$ at x_1 is

$$\hat{\eta}'_1 = 10\log(2) - 10\log\left[\frac{1}{y^2_{11}}(1 + A^2_{f1})\right].$$

If we have

$$\frac{1}{y^2_{01}}(1 + A^2_{f0}) < \frac{1}{y^2_{11}}(1 + A^2_{f1}), \tag{5.62}$$

then $\hat{\eta}'_0 > \hat{\eta}'_1$. This indicates that x_0 is the optimal level for design. Note that this conclusion is drawn from the accelerated test data.

Now suppose that the experiment is conducted at noise factor levels z_0 and z_1, where both z_0 and z_1 are within the normal use range. The noise level z_1 remains the same and $z_0 = 2z_1 - z_2$. Then, from (5.61), we know that A_{f0} equals the acceleration factor at x_0 between the life at z_0 and the life at z_1, and A_{f1} equals the acceleration factor at x_1 between the life at z_0 and the life at z_1. Let y_{ij} ($i = 0, 1; j = 0, 1$) be the life at x_i and z_j. The signal-to-noise ratio $\hat{\eta}_0$ at x_0 is

$$\hat{\eta}_0 = -10\log\left[\frac{1}{2}\left(\frac{1}{y^2_{00}} + \frac{1}{y^2_{01}}\right)\right] = 10\log(2) - 10\log\left(\frac{1}{y^2_{01}}\frac{1 + A^2_{f0}}{A^2_{f0}}\right).$$

Similarly, the signal-to-noise ratio $\hat{\eta}_1$ at x_1 is

$$\hat{\eta}_1 = 10\log(2) - 10\log\left(\frac{1}{y^2_{11}}\frac{1 + A^2_{f1}}{A^2_{f1}}\right).$$

If (5.62) holds and $A_{f0} \geq A_{f1}$, we have

$$\frac{1}{y^2_{01}}\frac{1 + A^2_{f0}}{A^2_{f0}} < \frac{1}{y^2_{11}}\frac{1 + A^2_{f1}}{A^2_{f1}}; \tag{5.63}$$

that is, $\hat{\eta}_0 > \hat{\eta}_1$. Therefore, x_0 is the optimal level. This agrees with the conclusion drawn from the accelerated test data. However, if (5.62) holds and $A_{f0} < A_{f1}$, (5.63) may not be valid. That is, x_0 may not be the optimal level. In such case, the conclusion derived from the accelerated test data is faulty.

Let's consider a simple example that illustrates the accelerated test data leading to an erroneous conclusion. Suppose that the life y is 50 at ($x_0 = 0, z_1 = 1$), 25 at ($x_0 = 0, z_2 = 2$), 60 at ($x_1 = 1, z_1 = 1$), and 15 at ($x_1 = 1, z_2 = 2$). The value of the acceleration factor is 2 at x_0 and 4 at x_1 calculated from these life data. The value of the signal-to-noise ratio is 30 at x_0 and 26.3 at x_1. It would be concluded that x_0 is the optimal level based on the accelerated test data. Now suppose that the experiment is conducted at the noise factor levels within the normal use range: z_0 and z_1. The life y is 100 at $(x_0 = 0, z_0 = 0)$ and 240 at $(x_1 = 1, z_0 = 0)$; both are derived by applying the acceleration factors. The value of the signal-to-noise ratio is 36.0 at x_0 and 38.3 at x_1. Then we would conclude that x_1 is the optimal level, which contradicts the conclusion drawn from the accelerated test data.

In robust design, accelerating noise factors should be those that are independent of control factors, to avoid possible faulty conclusions. The high levels of accelerating noise factors should be as high as possible to maximize the number of failures or the amount of degradation, but they should not induce failure modes that are different from those in the normal use range. The low levels of the accelerating noise factors should be as low as possible to maximize the range of noise levels, but they should generate sufficient failures or degradation information. Unfortunately, such independent noise factors, if any, may be difficult to identify before experiments are conducted. In these situations, experience, engineering judgment, or similar data may be used. A preliminary accelerated test may also be performed to determine the independence.

In addition to noise factors, some control factors may also serve as accelerating variables, as described in Chapter 7 and by Joseph and Wu (2004). Such control factors should have large effects on failure or degradation and the effects are known based on physical knowledge of the product. In traditional experiments, these factors are of no direct interest to designers; they are not involved in the design of experiment and their levels are kept at normal values in experiment. In contrast, in accelerated robust testing, these factors are elevated to higher levels to increase the number of failures or the amount of degradation. The analysis of accelerated test data yields the optimal setting of design parameters. The accelerating control factors are set at normal levels in actual product design. For the conclusion to be valid at normal levels of the accelerating control factors, these factors should not interact with other control factors. This requirement restricts use of the method because of the difficulty in identifying independent accelerating control factors.

In summary, accelerated testing is needed in many experiments. However, accelerating variables should not interact with control factors. Otherwise, the optimality of design setting may be faulty.

PROBLEMS

5.1 Develop a boundary diagram, P-diagram, and strategy of noise effect management for a product of your choice. Explain the cause-and-effect relations between the noise factors and failure modes.

5.2 A robust design is to be conducted for improving the reliability and robustness of an on–off solenoid installed in automobiles. The solenoid is tested to failure, and life (in on–off cycles) is the experimental response. The design parameters selected for experiment are as follows:

- A: spring force (gf); $A_0 = 12$; $A_1 = 18$; $A_2 = 21$

- B: spring material; $B_0 = $ type 1, $B_1 = $ type 2, $B_2 = $ type 3
- C: ball lift (mm); $C_0 = 0.3$, $C_1 = 0.4$
- D: air gap (mm); $D_0 = 0.8$, $D_1 = 1.6$, $D_2 = 2.5$
- E: number of coil turns: $E_0 = 800$, $E_1 = 1000$, $E_2 = 1200$

Interaction between factors C and D is possible and needs investigation. The life of the solenoid is affected significantly by voltage and temperature, which are considered to be the key noise factors. Their levels are:

- V: voltage (V); $V_0 = 10$, $V_1 = 12$, $V_2 = 15$
- T: temperature (°C); $T_0 = 100$, $T_1 = 230$

(a) Select an inner array to assign the design parameters.

(b) Lay out the noise factors in an outer array.

(c) V_1 is the nominal value of voltage. Develop a compounding noise strategy to reduce the size of the outer array.

(d) Describe the procedure for determining the optimal levels of the design parameters.

5.3 Create a nine-level column using $L_{27}(3^{13})$. Is this nine-level column orthogonal to other columns? Write down the sum of squares for the nine-level column.

5.4 Using the $L_{16}(2^{15})$ orthogonal array, assign the following factors and interactions to find an appropriate experimental design for the following cases:

(a) Two-level factors A, B, C, D, F, G, and H and interactions A × B, B × C, and F × G.

(b) Two-level factors A, B, C, D, F, G, H, and I and interactions A × B, A × C, A × I, G × F, and G × H.

(c) One eight-level factor A, and three two-level factors B, C, and D.

5.5 Condra (2001) describes a robust design of surface-mounted capacitors. In the design, dielectric composition (factor A) and process temperature (factor B) are the control factors subject to optimization. The control factors have two levels and are accommodated in $L_4(2^2)$. The noise factors are operating voltage (factor C) and temperature (factor D). The voltage has four levels: 200, 250, 300, and 350 V. The temperature has two levels: 175 and 190°C. The usual voltage and temperature are 50 V and 50°C. Thus the experiment is an accelerated test. Table 5.26 shows the experimental layout and the response data. The response is the mean time to failure (MTTF, in hours) estimated from 10 lognormally distributed samples.

(a) Calculate the signal-to-noise ratio for each run of the inner array. Analyze the signal-to-noise ratio data by using graphical response method. Identify the significant factor(s) and interaction. What are the optimal levels of the control factors?

(b) Reanalyze the signal-to-noise ratio data calculated in part (a) by using ANOVA. Are the conclusions the same as those from the graphical response analysis?

TABLE 5.26   Experimental Layout and MTTF

	Inner Array		Outer Array								C
			0	0	1	1	2	2	3	3	
Run	A	B	0	1	0	1	0	1	0	1	D
1	0	0	430	950	560	210	310	230	250	230	
2	0	1	1080	1060	890	450	430	320	340	430	
3	1	0	890	1060	680	310	310	310	250	230	
4	1	1	1100	1080	1080	460	620	370	580	430	

(c) Plot the interaction graphs for C × A, C × B, D × A, and D × B. Do the plots suggest any interactions between the noise and control factors? Is it appropriate to use voltage and temperature as accelerating variables?

(d) Plot the interaction graph for C × D. Is it evident that the voltage interacts with temperature?

5.6 The fatigue life of an automotive timing belt and its variability were improved through robust reliability design. The design parameters are belt width (factor A), belt tension (B), belt coating (C), and tension damping (D). Each design parameter has two levels. The interaction between B and D is of interest. The noise factor is temperature, which was set at three levels: 60, 100, and 140°C. The experimental response is cycles to failure measured by a life index, which is a larger-the-better characteristic. The experiment was censored at the number of cycles translated to 2350 of the life index. The design life was equivalent to 1500 of the life index. The experimental layout and life index are given in Table 5.27. Suppose that the life index is Weibull and its relation with temperature can be modeled with the Arrhenius relationship (Chapter 7).

TABLE 5.27   Experimental Data for the Timing Belt

	Inner Array				B × D	C	Outer Array Temperature (°C)					
	A	B	D				60		100		140	
Run	1	2	3	4	5	6	7					
1	0	0	0	0	0	0	0	1635	1677	1457	1433	1172	1232
2	0	0	0	1	1	1	1	1578	1723	1354	1457	1149	1222
3	0	1	1	0	0	1	1	1757	1673	1247	1178	1080	1109
4	0	1	1	1	1	0	0	1575	1507	1103	1077	985	937
5	1	0	1	0	1	0	1	2350	2350	2173	2237	2058	1983
6	1	0	1	1	0	1	0	2035	2147	1657	1573	1338	1435
7	1	1	0	0	1	1	0	1758	1850	1713	1543	1478	1573
8	1	1	0	1	0	0	1	2350	1996	1769	1704	1503	1374

TABLE 5.28   Experimental Layout and Response Data for the Oxygen Sensor

	Inner Array				Outer Array			
	A	B	C	D				
Run	1	2	3	4	Noise Level 1		Noise Level 2	
1	0	0	0	0	11	15	22	16
2	0	1	1	1	8	12	16	14
3	0	2	2	2	15	19	26	23
4	1	0	1	2	17	23	28	35
5	1	1	2	0	7	11	18	22
6	1	2	0	1	8	5	15	18
7	2	0	2	1	23	19	20	26
8	2	1	0	2	16	20	24	26
9	2	2	1	0	8	13	17	15

(a) Estimate the reliability at the design life for each combination of design setting and temperature.

(b) Perform ANOVA to identify the significant factors. Determine optimal levels of the design parameters.

(c) Predict the signal-to-noise ratio at the optimal setting. Calculate the average reliability over the three temperatures.

5.7 To improve the reliability of an oxygen sensor, four design parameters (A to D) were chosen and assigned to $L_9(3^4)$ as shown in Table 5.28. The sensors were tested at two levels of compounding noise (temperature and humidity) and the response voltages at a given oxygen level were recorded. The sensors are said to have failed if the response voltage drifts more than 30% from the specified value. The drift percentages at termination of the test are given in Table 5.28.

(a) Perform graphical response analysis to identify the significant factors.

(b) Carry out ANOVA to determine the significant factors. What are optimal levels of the design parameters?

(c) Predict the signal-to-noise ratio at the optimal setting.

(d) The current design setting is $A_0B_0C_0D_0$. How much robustness improvement will the optimal setting achieve?

(e) Calculate the average drift percentage over the noise levels.

(f) Are the interactions between the design parameters and the compounding noise significant?

6

POTENTIAL FAILURE MODE AVOIDANCE

6.1 INTRODUCTION

In Chapter 5 we explained that a failure is the consequence of either a lack of robustness or the presence of mistakes, and presented methodologies for building robust reliability into products. In this chapter we describe techniques for detecting and eliminating mistakes.

Reliable products must be robust over time and free of mistakes. Unfortunately, to err is human, as the saying goes. Engineers design values into products as well as mistakes. Mistakes are unknowingly embedded into products in design and production. The mistakes can be errors ranging from misuse of materials to misspecification of system requirements, and may be classified into different categories according to the nature of the error. L. Chao and Ishii (2003) associate design errors with six areas: knowledge, analysis, communication, execution, change, and organization. Specific subjects, such as software engineering and civil engineering, often have their own error classification systems.

Any mistakes built into a product will compromise product features, time to market, life cycle cost, and even human safety. Although it is obvious that mistakes should be eliminated, the criticality of correcting errors at the earliest time is frequently overlooked. An error found and fixed in the early design phase usually costs considerably less than it would after the design is released to production. In general, the later the errors are removed, the higher the cost. In many situations, the cost is an exponential function of time delay. It is apparently desirable to correct any errors as soon as they emerge. The powerful

Life Cycle Reliability Engineering, by Guangbin Yang
Copyright © 2007 John Wiley & Sons, Inc.

techniques that help achieve this objective include failure mode and effects analysis (FMEA) and fault tree analysis (FTA). Computer-aided design analysis is another error detection technique, which is being widely implemented nowadays thanks to the advance in computer technology and software engineering. This technique includes mechanical stress analysis, thermal analysis, vibration analysis, and other methods. In this chapter we describe these three types of techniques.

6.2 FAILURE MODE AND EFFECTS ANALYSIS

Failure mode and effects analysis, commonly known as FMEA, is a proactive tool for discovering and correcting design deficiencies through the analysis of potential failure modes, effects, and mechanisms, followed by a recommendation of corrective action. It may be described as a systemized group of activities intended to recognize and evaluate the potential failure of a product or process and its effects, identify actions that eliminate or reduce the likelihood of the potential failure occurrence, and document the process (SAE, 2002). Essentially, FMEA is a bottom-up process consisting of a series of steps. It begins with identifying the failure mode at the lowest level (e.g., component) and works its way up to determine the effects at the highest level (e.g., end customer). The process involves an inductive approach to consider how a low-level failure can lead to one or more effects at a high level.

FMEA has numerous benefits. The results of FMEA are valuable to design engineers in determining and prioritizing the area of action. In particular, FMEA helps identify which potential failure modes result in severe effects that must be designed out of a product, which ones can be handled by corrective or mitigating actions, and which ones can be safely ignored. The outputs of an FMEA are instrumental in the development of design verification plans and production control plans. FMEA is also useful for field service engineers in diagnosing problems and determining repair strategy.

Because of these benefits, FMEA has been used extensively in various private industries. The overwhelming implementation lies in the automotive sector, where FMEA has been standardized as SAE J1739 (SAE, 2002). This standard was developed and has been advocated by major U.S. automakers. Although originating in the automotive industry, the standard is prevalent in other sectors of industry: for example, the marine industry. Another influential standard is IEC 60812 (IEC, 1985), which is used primarily in electrical engineering. In the defense industry, failure mode, effects, and criticality analysis (FMECA) is more common and performed by following MIL-STD-1629A (U.S. DoD, 1984). FMECA is similar to FMEA except that each potential failure effect is classified according to its severity. Because of the similarity, we present FMEA in this section. FMEA has become an integral step in the design process in many companies. Its implementation can be facilitated by using commercial software packages (e.g., Relex, Item), which often support various standards, such as those mentioned above.

6.2.1   FMEA Types and Benefits

Basically, FMEA may be classified into three categories according to the level of analysis: system FMEA, design FMEA, and process FMEA. There are some other types of FMEA, which can be considered as extensions of these three types. For example, software and machinery FMEA can be thought of as special cases of design FMEA.

System FMEA   *System FMEA* is sometimes called *concept FMEA* because the analysis is carried out in the concept development stage. This is the highest-level FMEA that can be performed and is used to analyze and prevent failures related to technology and system configuration. This type of FMEA should be carried out as soon as a system design (i.e., the first stage of robust design) is completed, to validate that the system design minimizes the risk of functional failure during operation. Performed properly, system FMEA is most efficient economically because any changes in the concept design stage would incur considerably less cost than in subsequent stages.

   System FMEA has numerous benefits. It helps identify potential systemic failure modes caused by the deficiency of system configurations and interactions with other systems or subsystems. This type of FMEA also aids in (1) examining system specifications that may induce subsequent design deficiencies, (2) selecting the optimal system design alternative, (3) determining if a hardware system redundancy is required, and (4) specifying system-level test requirements. It acts as a basis for the development of system-level diagnosis techniques and fault management strategy. More important, system FMEA enables actions to ensure customer satisfaction to be taken as early as in the concept design stage, and is an important input to the design FMEA to follow.

Design FMEA   *Design FMEA* is an analytical tool that is used to (1) identify potential failure modes and mechanisms, (2) assess the risk of failures, and (3) provide corrective actions before the design is released to production. To achieve the greatest value, design FMEA should start before a failure mode is unknowingly designed into a product. In this sense, FMEA serves as an error prevention tool. In reality, however, design FMEA is frequently performed as soon as the first version of the design is available, and remedial actions are developed based on the analysis to eliminate or alleviate the failure modes identified. The bottom line is that a design should avoid critical failure modes.

   The major benefit of design FMEA is in reducing the risk of failure. This is achieved by identifying and addressing the potential failure modes that may have adverse effects on environment, safety, or compliance with government regulations in the early design stage. Design FMEA also aids in objective evaluation of design in terms of functional requirements, design alternatives, manufacturability, serviceability, and environmental friendliness. It enables actions to ensure that customer satisfaction is initiated as early in the design stage as possible. In addition, this type of FMEA helps the development of design verification plans, production control plans, and field service strategy. The outputs from design FMEA are the inputs to process FMEA.

Process FMEA   *Process FMEA* is a structured logic and systematic analysis intended to (1) identify potential failure modes and mechanisms, (2) assess the risk of failure, and (3) provide corrective action before the first production run takes place. The potential failure mode of a process is defined as the manner in which the process could potentially fail to meet the process requirements and/or design intent. Because a variety of factors may contribute to the failure of a process, this type of FMEA is usually accomplished through a series of steps to consider and analyze human, machine, method, material, measurement, and environment impacts. Thus, process FMEA is often more complicated and time consuming than system and design FMEA.

Process FMEA is capable of discovering potential product and process-related failure modes and responsible failure mechanisms. It determines the key process parameters on which to focus controls for the reduction or detection of failure conditions. This type of FMEA also assesses the effects of the potential failures on end customers and enables actions to ensure that customer satisfaction is accounted for in the process development stage. Process FMEA addresses the concerns of critical failure modes discovered in design FMEA through manufacturing or assembly process design improvement.

It can be seen that there are close similarities among the three types of FMEAs. In the remainder of this section we discuss design FMEA.

6.2.2   FMEA Process

Performing an FMEA begins with defining the system for study. The interactions between this system and others should be fully understood to determine the effects and mechanisms of a potential failure mode. For this purpose, the boundary diagram described in Chapter 5 is helpful.

Once the scope for FMEA study has been defined, the lowest-level component within the system is chosen and its functions are analyzed. Each function should be technically specific, and the failure criteria of the function must be defined completely. The next step is to identify the failure modes of the component. This is followed by revealing the effects of each failure mode and evaluating the *severity* of each associated effect. For each failure mode, the responsible failure mechanisms and their *occurrences* are determined. The subsequent step is to develop control plans that help obviate or detect failure mechanisms, modes, or effects. The effectiveness of each plan is evaluated by *detection* ranking. The next step is to assess the overall risk of a failure mode. The overall risk is measured by the risk priority number (RPN), which is a product of severity, occurrence, and detection. A high RPN value indicates a high risk of failure. Appropriate corrective action should be taken to reduce the risk. Finally, the results of FMEA are documented using a standardized format.

The steps above for performing FMEA are depicted in Figure 6.1. Figure 6.2 shows a typical worksheet format for design FMEA. In application, it may be

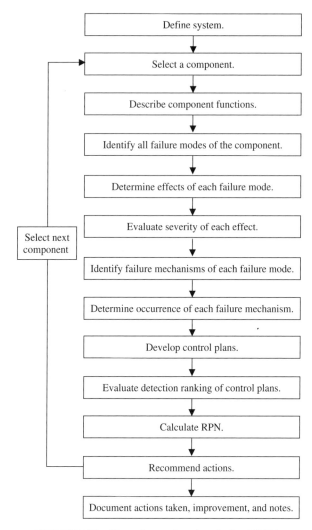

FIGURE 6.1   Steps for performing design FMEA

substituted with a different one that reflects the distinctions of a product and the needs of an organization.

6.2.3   System Definition

Similar to defining a system for robust design, described in Chapter 5, system definition for FMEA study is to determine the boundary of a system for which FMEA should be performed. For some systems that are relatively independent of others, the boundary may be obvious. However, describing the boundary of a complicated system that interacts heavily with other systems requires thorough

FAILURE MODE AND EFFECTS ANALYSIS

System: _____
Subsystem: _____
Component: _____
Core Team: _____

Design Responsibility: _____
Key Date: _____
Page: _____ of _____

FMEA Number: _____
Prepared by: _____
FMEA Date: _____

Item / Function	Potential Failure Mode	Potential Effects of Failure	S E V	Potential Failure Mechanisms	O C C	Design Controls	D E T	R P N	Recommended Actions	Responsibility and Target Completion Date	Action Results				
											Actions Taken	S E V	O C C	D E T	R P N

FIGURE 6.2 Typical worksheet for design FMEA

understanding of the target and neighboring systems and their interactions. In this step it is often required to perform an interaction analysis and create a boundary diagram using the method presented in Chapter 5. Interaction analysis also provides assistance in subsequent FMEA steps in (1) understanding the effects of failure modes on other systems and end customers, (2) evaluating the severity of the effects, and (3) discovering the failure mechanisms that may have originated from other systems.

Once defined, the system should be broken down into subsystems, modules, or components, depending on the objective of the FMEA study. Analysis at a high level (subsystem or module level) is usually intended to determine the area of high priority for further study. Analysis at the component level is technically more desirable and valuable, in that it usually leads to a determination of the causes of failure. As such, FMEA is performed mostly at the component level in practice.

6.2.4   Potential Failure Mode ← What happened

A *potential failure mode* is defined as the manner in which a component, module, subsystem, or system could potentially fail to meet design intents, performance requirements, or customer expectations. A potential failure mode can be the cause of a higher-level (module, subsystem, or system) failure. It may also be the effect of a lower-level (component, module, or subsystem) failure.

As shown in Figure 6.1, all failure modes of a component should be scoped and listed. The failure modes include these that are already known, as well as these that are unknown but that could occur under certain operating conditions, such as hot, cold, humid, and wet environments. In FMEA, two types of failure modes should be considered: hard and soft. A *hard failure* is a complete loss of function. It can be further divided into the following two categories according to the persistence of failure:

- *Irreversible failure*: complete and permanent loss of function.
- *Intermittent failure*: complete loss of function in a brief time. The function then recovers automatically.

A *soft failure* is a degradation of performance over time and causes partial loss of function. Performance degradation is usually unrecoverable and leads to complete ceasation of function. See Chapter 2 for more discussion of these two types of failure modes.

A P-diagram is a useful input to the identification of failure modes. All failure modes listed in a P-diagram created for a system should be included in the FMEA. Fault tree analysis (discussed later) is also of assistance in discovering failure modes; the top event of a fault tree is a failure mode of the system under consideration.

6.2.5   Potential Effects of Failure and Severity

The *potential effect* of a failure mode is defined as the impact of the failure mode on the function of neighboring components and higher-level systems. The potential effects are identified by asking the question: What will be the consequences if this failure happens? The consequences are evaluated with respect to the function of the item being analyzed. Because there exists a hierarchical relationship between the component, module, subsystem, and system levels, the item failure under consideration may affect the system adversely at several levels. The lowest-level effects are the *local effects*, which are the consequences of the failure on the local operation or function of the item being analyzed. The *second-level effects* are the next-level effects, which are the impacts of the failure on the next-higher-level operation or function. The failure effect at one level of system hierarchy is the item failure mode of the next-higher level of which the item is a component. The highest-level effects are the *end effects*, which are the effects of the failure on the system functions and can be noticed by the end customers. For instance, the end effects of a failure occurring to an automobile can be noise, unpleasant odor, thermal event, erratic operation, intermittent operation, inoperativeness, roughness, instability, leak, or others.

The end effect of a failure is assessed in terms of severity. In FMEA, severity is a relative ranking within the scope of a particular study. The military industry commonly employs a four-level classification system ranking from "catastrophic" to "minor," as shown in Table 6.1. In this ranking system, the lowest-ranking index measures the highest severity, and vice versa. The automotive industry generally uses a 10-level ranking system, as shown in Table 6.2; the effect of a failure mode is described as the effect on customers or conformability to government regulations. A failure mode may have multiple effects, each of which has its own severity rating. Only the most serious effect rating is entered into the severity column of the FMEA worksheet for calculating RPN.

6.2.6 Potential Failure Mechanisms and Occurrence ← Why it happened

A *potential failure mechanism* is an indication of design weakness; its consequences are failure modes. Identification of failure mechanisms is the process of

TABLE 6.1 Four-Level Military Severity Ranking System

Category	Explanation
I	*Catastrophic:* a failure that can cause death or system (e.g., aircraft, tank, missile, ship) loss
II	*Critical:* a failure that can cause severe injury, major property damage, or minor system damage which will result in mission loss
III	*Marginal:* a failure that may cause minor injury, minor property damage, or minor system damage which will result in delay or loss of availability or mission degradation
IV	*Minor:* a failure not serious enough to cause injury, property damage, or system damage, but which will result in unscheduled maintenance or repair

Source: U.S. DoD (1984).

TABLE 6.2   Ten-Level Automotive Severity Ranking System

Effect	Criteria: Severity of Effect	Ranking
Hazardous without warning	Very high severity ranking when a potential failure mode affects safe vehicle operation and/or involves noncompliance with government regulation without warning	10
Hazardous with warning	Very high severity ranking when a potential failure mode affects safe vehicle operation and/or involves noncompliance with government regulation with warning	9
Very high	Vehicle/item inoperable (loss of primary function)	8
High	Vehicle/item operable but at a reduced level of performance; customer very dissatisfied	7
Moderate	Vehicle/item operable but comfort/convenience item(s) inoperable; customer dissatisfied	6
Low	Vehicle/item operable but comfort/convenience item(s) operable at a reduced level of performance; customer somewhat dissatisfied	5
Very low	Fit and finish/squeak and rattle item does not conform; defect noticed by most customers (greater than 75%)	4
Minor	Fit and finish/squeak and rattle item does not conform; defect noticed by 50% of customers	3
Very minor	Fit and finish/squeak and rattle item does not conform; defect noticed by discriminating customers (less than 25%)	2
None	No discernible effect	1

Source: SAE (2002).

hunting design mistakes. It is this trait that enables FMEA to be a technique for detecting and correcting design errors.

Failure mechanisms are discovered by asking and addressing a number of "what" and "why" questions such as: What could cause the item to fail in this manner? Why could the item lose its function under the operating conditions? Techniques such as fault tree analysis and cause-and-effect diagrams are instrumental in determining the failure mechanisms of a particular failure mode. Examples of failure mechanisms include incorrect choice of components, misuse of materials, improper installation, overstressing, fatigue, corrosion, and others.

Identified failure mechanisms are assigned with respective occurrence values. Here, the occurrence is defined as the likelihood that a specific failure mechanism will occur during the design life. Occurrence is not the value of probability in the absolute sense; rather, it is a relative ranking within the scope of the FMEA. The automotive industry uses a 10-level ranking system, as shown in Table 6.3. The highest-ranking number indicates the most probable occurrence, whereas the lowest one is for the least probable occurrence. The probable failure rate of a failure mechanism can be estimated with the assistance of historical data, such as the previous accelerated test data or warranty data. When historical data are used, the impacts of changes in design and operating condition on the failure rate during the design life should be taken into account.

TABLE 6.3 Ten-Level Automotive Occurrence Ranking System

Probability of Failure	Likely Failure Rates During the Design Life	Ranking
Very high: persistent failures	≥ 100 per 1000 vehicles or items	10
	50 per 1000 vehicles or items	9
High: frequent failures	20 per 1000 vehicles or items	8
	10 per 1000 vehicles or items	7
Moderate: occasional failures	5 per 1000 vehicles or items	6
	2 per 1000 vehicles or items	5
	1 per 1000 vehicles or items	4
Low: relatively few failures	0.5 per 1000 vehicles or items	3
	0.1 per vehicles or items	2
Remote: failure is unlikely	≤ 0.01 per 1000 vehicles or items	1

Source: SAE (2002).

A failure mode can be the consequence of multiple failure mechanisms, which may have different occurrence values. In the FMEA worksheet, enter the occurrence value for each failure mechanism for the RPN calculation.

6.2.7 Design Controls and Detection

Design controls are methods of preventing and detecting failure modes and mechanisms. Prevention measures are aimed at obviating failure mechanisms, modes, or effects from occurrence or at reducing the rate of occurrence. Detection measures are designed to detect failure mechanisms and modes before an item is released to production. Apparently, the prevention method is proactive and preferable, and should be used whenever possible. With this method in the design process, the initial occurrence ranking of a failure mechanism under prevention control should be lowered. Since it is not always possible to use prevention control methods for a failure mechanism, detection methods should be devised. Examples of detection methods are computer-aided design analysis (described later), design review, and testing.

The effectiveness of a detection method is assessed in terms of *detection*, a ranking index measuring the relative capability of a detection method. The SAE J1739 (SAE, 2002) uses a 10-level ranking system in which 1 represents a detection method being almost certain to detect a potential failure mechanism or subsequent failure mode, and 10 represents the detection method being absolutely incapable. Table 6.4 shows an automotive detection ranking system.

6.2.8 RPN and Recommended Actions

The overall risk of a failure mode can be assessed by the risk priority number (RPN), which is the product of severity, occurrence, and detection:

$$\text{RPN} = S \times O \times D, \tag{6.1}$$

TABLE 6.4   Ten-Level Automotive Detection Ranking System

Detection	Criteria: Likelihood of Detection by Design Control	Ranking
Absolute uncertainty	Design control will not and/or cannot detect a potential cause/mechanism and subsequent failure mode; or there is no design control.	10
Very remote	Very remote chance that the design control will detect a potential cause/mechanism and subsequent failure mode.	9
Remote	Remote chance that the design control will detect a potential cause/mechanism and subsequent failure mode.	8
Very low	Very low chance that the design control will detect a potential cause/mechanism and subsequent failure mode.	7
Low	Low chance that the design control will detect a potential cause/mechanism and subsequent failure mode.	6
Moderate	Moderate chance that the design control will detect a potential cause/mechanism and subsequent failure mode.	5
Moderately high	Moderately high chance that the design control will detect a potential cause/mechanism and subsequent failure mode.	4
High	High chance that the design control will detect a potential cause/mechanism and subsequent failure mode.	3
Very high	Very high chance that the design control will detect a potential cause/mechanism and subsequent failure mode.	2
Almost certain	Design control will almost certainly detect a potential cause/mechanism and subsequent failure mode.	1

Source: SAE (2002).

where S is the severity ranking, O is the occurrence ranking, and D is the detection ranking. If the three factors are assigned rankings from 1 to 10, the RPN is in the range between 1 and 1000.

The higher the RPN value, the higher the risk of failure. Failure modes with high RPN values usually require corrective action. A Pareto chart is helpful in prioritizing the failure modes by RPN and thus determining the areas of actions. Such actions may include (but are not limited to) design change, material upgrade, and revision of test plans. Design change can reduce the severity of a failure effect and the occurrence of a failure mechanism. In contrast, the modification of test plans can only increase the capability of a detection method and lower the detection ranking. Increasing the rigorousness of test plans is a less desirable engineering action because it requires more test resources and does not address the severity and occurrence of the failure.

In general practice, when a failure mode has a high value of S, immediate remedial actions are required to prevent the serious consequence of the failure regardless of the RPN value. In the automotive industry, for example, special attention must be given to any failure mode with a 9 or 10 severity ranking, to ensure that the risk is managed through preventive or corrective action. The reduction in severity can only be achieved by elimination of the failure mode or mitigation of the effects associated with the failure mode through design change.

The recommended actions should be implemented before a design is released to production. Then the improvement will result in lower severity, occurrence, and detection rankings. The resulting RPN is calculated and recorded. The entire process of FMEA should be documented in a standardized format as shown, for example, in Figure 6.2.

6.2.9 Design FMEA Example

Example 6.1 To reduce the amount of nitrogen oxide emission and increase fuel efficiency, automobiles include an exhaust gas recirculation (EGR) system that recycles a fraction of exhaust gases to the inlet manifold, where the exhaust gas is mixed with fresh air. A typical EGR system consists of several subsystems, including the EGR valve, delta pressure feedback EGR (DFPE) sensor, EGR vacuum regulator (EVR), powertrain control module (PCM), and tubes, as shown in Figure 6.3. The exhaust gas is directed from the exhaust pipeline to the EGR valve via the EGR tube. The EGR valve regulates EGR flow to the intake manifold. The desired amount of EGR is determined by the EGR control strategy and calculated by the PCM. The PCM sends a control signal to the EVR, which regulates the vacuum directed to the EGR valve. Energized by the vacuum, the EGR valve allows the right amount of EGR to flow into the intake manifold. The EGR system is a closed-loop system in which a DPFE sensor is used to measure the EGR flow and provide a feedback to the PCM. Then the PCM redoes the calculation and adjusts the vacuum until the desired level of EGR is achieved. The EGR control process indicates that the subsystems are connected in logic series, and failure of any subsystem causes the entire system to fail. The design FMEA is performed at subsystem level by following SAE J1739. Figure 6.4 shows a part of the FMEA as an illustration; it is not intended to exhaust all elements of the analysis.

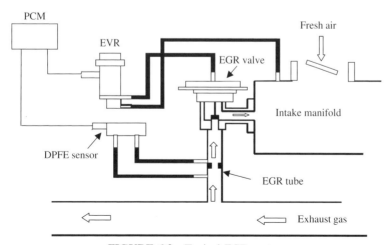

FIGURE 6.3 Typical EGR system

FAILURE MODE AND EFFECTS ANALYSIS

x System: EGR System   Design Responsibility: _____   FMEA Number: _____
_ Subsystem: _____   Key Date: _____   Prepared by: _____
_ Component: _____   Page: ___ of ___   FMEA Date: _____
_ Core Team: _____

Item / Function	Potential Failure Mode	Potential Effects of Failure	S E V	Potential Failure Mechanisms	O C C	Design Controls	D E T	R P N	Recommended Actions	Responsibility and Completion Date	Action Results				
											Actions Taken	S E V	O C C	D E T	R P N
EVR must deliver required amount of exhaust gas in target life of vehicle (10 years/150k miles)	Stuck open	Engine stall (10) Emission exceeding standard (9) Engine rough idle (7)	10	Contamination	7	Design review Design verification test Review historical data and failure cases	5	350	Develop process control plans Improve component acceptance criteria		Process control plans developed Component acceptance criteria improved	10	3	5	150
		Engine hesitation/surging when accelerating (7)		Spring breaks	3	CAE durability analysis Design verification test	3	90							

FIGURE 6.4 Design FMEA of the automobile EGR system

Stuck closed	Emission exceeding standard (9) Reduced fuel efficiency (5)	EVR applies constant vacuum	4	Design review CAE circuit analysis and accelerated testing of EVR assembly	3	120						
		EGR valve diaphragm breaks	6	Design review Design verification test Review historical data and failure cases	6	324	Change diaphragm material CAE durability analysis	Diaphragm material changed CAE durability analysis integrated to design process	9	4	3	108
		EVR fails to deliver vacuum	5	Design review CAE circuit analysis and accelerated testing of EVR assembly	3	135						

FIGURE 6.4 (*continued*)

6.3   ADVANCED TOPICS IN FMEA

6.3.1   RPN Property

RPN is an important measure in the framework of FMEA. To use it in a more effective manner, we should understand its properties, which are due largely to Bowles (2003).

The value of the RPN is in the range 1 to 1000 if the severity, occurrence, and detection take values from 1 to 10. Since the RPN is not on a continuous scale, 88% of the range is empty. Only 120 of the 1000 numbers generated from the product of S, O, and D are unique, and 67 numbers are smaller than 200. No number having a prime factor greater than 10 can be formed from (6.1). For example, the numbers 11, 22, 33, ... , 990, which are all multiples of 11, cannot be generated and are excluded from further analysis; 1000 is the largest value of RPN, 900 is the second largest, followed by 810, 800, 729, and so on. Figure 6.5 shows the distribution of possible RPN values. Table 6.5 summarizes the frequencies of RPN values in various intervals of the same length. Figure 6.5 indicates that there are many holes where the RPN cannot take values. The presence of such holes poses confusions to FMEA analysts in prioritizing and selecting failure modes for further study. For example, it would be difficult to answer the question as to whether the difference between 270 and 256 is the same as, or larger than, that between 256 and 254, because the three numbers are three consecutive RPN values. In addition, the large drop in some consecutive RPN values may lead to the omission of the failure mode with the second-largest RPN value.

As shown in Figure 6.5, there are many duplicate RPN numbers. Among 120 unique RPN numbers, only six are generated by a single unique combination of S, O and D. Most RPN numbers can be formed in several different ways. For example, an RPN value of 252 can be formed from nine different combinations of S, O, and D values, as shown in Table 6.6. Ranking failure modes by RPN values treats all these combinations equally as well as the associated failure

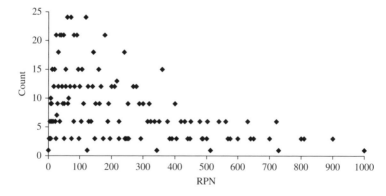

FIGURE 6.5   Distribution of unique RPN numbers

TABLE 6.5   Frequencies of RPN Values in Intervals

Interval	Count	Percent
[1,200]	67	55.8
(200,400]	26	21.7
(400,600]	17	14.2
(600,800]	7	5.8
(800,1000]	3	2.5

TABLE 6.6   Possible Combinations of S, O, and D Values Forming an RPN Value of 252

Severity	Occurrence	Detection
4	7	9
4	9	7
6	6	7
6	7	6
7	4	9
7	6	6
7	9	4
9	4	7
9	7	4

modes. From the practical point of view, however, the failure mode resulting in the combination ($S = 9, O = 7, D = 4$) may be more serious than that yielding ($S = 6, O = 6, D = 7$) even though both form the same RPN value of 252. Therefore, it is suggested that the product of S and O be used to further assess the risk when RPN values tie or are close.

Misspecifying the value of one of the three factors S, O, D has a large effect on the RPN value, especially when the other two factors have high ratings. Let S_0, O_0, and D_0 be the true values of S, O, and D, respectively. The corresponding RPN value is RPN_0. Without loss of generality, assume that D is overestimated by 1 and that S and O take the true values. Then the RPN value is

$$RPN = S_0 \times O_0 \times (D_0 + 1).$$

The increase in RPN value due to the overestimation is

$$RPN - RPN_0 = S_0 \times O_0.$$

It is seen that the increase (decrease) in RPN value due to overestimating (underestimating) the rating of one factor by 1 equals the product of the ratings of the other two factors. Such change is especially significant when the other two factors have large ratings. An example follows. If $S_0 = 9$, $O_0 = 9$, and $D_0 = 5$, $RPN_0 = 405$. The D is overestimated by 1, and the RPN value increases to 486.

In summary, the RPN technique for prioritizing failure modes has some deficiencies. Various attempts have been made to modify the prioritization technique by using fuzzy theory (see, e.g., Bowles and Pelaez, 1995; Franceschini and Galetto, 2001; Pillay and Wang, 2003). The methods proposed result in more objective and robust prioritization. But the complicated mathematical operations required by these methods restrict their application in industry.

6.3.2 Software FMEA

The complexity and sophistication of modern systems have been increasing constantly, due largely to advances in software engineering. The integration of software into systems enables systems to perform many new functions economically, but it also brings sources of failure to these systems. Indeed, software has been accountable for a large portion of system failures. Reducing the probability of software failure and minimizing the failure effects have become an essential task in system design. Software FMEA is a tool for accomplishing this task successfully.

Software FMEA is a means of determining whether any single failure in software can cause catastrophic system effects and identifying other possible consequences of unexpected software performance, where a *software failure* is defined as a software variable that is assigned an unintended value. Like hardware FMEA, software FMEA can be performed at different levels of the software hierarchical structure, which may be classified into code level, method level, class level, module level, and package level (Ozarin, 2004). For simplicity, the levels can be roughly divided into two groups: system level and detailed level.

Software FMEA at the system level is to assess the ability of a software architecture to provide protection from the effects of software and hardware failures. The analysis should be performed as early in the design process as possible in order to reduce the impact of design changes resulting from the analysis. Generally, the analysis can be initiated once the software design team has developed an initial architecture and allocated functional requirements to the software elements of the design. The FMEA at system level would treat software elements as black boxes that contain unknown software codes conforming to the requirements assigned to the elements. Typical failure modes of software elements include failure to execute, incomplete execution, erroneous output, and incorrect timing. In addition, failure modes at the system level should be identified, which may include incorrect input value, corrupted output value, blocked interruption, incorrect interrupt return, priority errors, and resource confliction (Goddard, 2000). Additional potential failure modes at both the element and system levels should be uncovered for specific software applications. Once the failure modes are determined, their effects on system outcomes are ascertained and assessed. If a failure results in hazardous outcomes, the system architecture and the adequacy of system requirements must be reviewed and improved.

Detailed software FMEA is intended to assess the ability of an as-implemented software design to achieve the specified safety requirements and to provide all needed system protection. Analysis at this level requires the existence of software design and an expression of that design, at least in pseudocode. To perform the analysis, we determine the failure modes for each variable and algorithm implemented in each software element. The possible types of failure modes depend on the type of variables. Table 6.7 lists typical failure modes of software variables (Goddard, 1993). The effects of a failure are traced through the code to the system outcomes. If a failure produces a system hazard, the system architecture, algorithms, and codes should be reviewed to assess if safety requirements have been fully implemented. If missing requirements are discovered, design changes must be recommended.

Unlike hardware FMEA, research on and application of software FMEA are very limited. Although software FMEA is recommended for evaluating critical systems in some standards, such as IEC 61508 (IEC, 1998, 2000) and SAE ARP 5580 (SAE, 2000), there are no industrial standards or generally accepted processes for performing software FMEA. The current status remains at the homemade stage; the processes, techniques, and formats of software FMEA vary from user to user.

TABLE 6.7   Software Variable Failure Modes

Variable Type	Failure Mode
Analog (real, integer)	Value exceeds allowed tolerance high.
	Value exceeds allowed tolerance low.
Enumerated (allowed values a, b, c)	Value is set to a when it should be b.
	Value is set to a when it should be c.
	Value is set to b when it should be a.
	Value is set to b when it should be c.
	Value is set to c when it should be a.
	Value is set to c when it should be b.
Enumerated with validity flag	Value is set to a when it should be b; validity flag is set to valid.
	Value is set to a when it should be c; validity flag is set to valid.
	Value is set to b when it should be a; validity flag is set to valid.
	Value is set to b when it should be c; validity flag is set to valid.
	Value is set to c when it should be a; validity flag is set to valid.
	Value is set to c when it should be b; validity flag is set to valid.
	Value is correct; validity flag is set to invalid.
Boolean (true, false)	Value is set to true when it should be false.
	Value is set to false when it should be true.

Source: Goddard (1993).

6.4   FAULT TREE ANALYSIS

A *fault tree model* is a graphical representation of logical relationships between failure events, where the top event is logically branched into contributing events through cause-and-effect analysis. In contrast to the inductive FMEA approach, *fault tree analysis* (FTA) is a deductive methodology in that it begins with the top event and proceeds through all known and possible causes that could lead to the occurrence of the top event. Because of the top-down process nature, FTA has been widely used to determine the failure causes of a specific failure mode of concern since its inception in the 1960s. In the early days, the methodology was used primarily on safety-critical systems such as military and aviation equipment. Later, the U.S. Nuclear Regulatory Commission published the notable *Fault Tree Handbook* (Vesely et al., 1981), which to a certain extent, standardized, facilitated, and promoted the use of FTA for systems of this type. Today, the development of many commercial products has adopted FTA for cause-and-effect analysis. Often it is applied in conjunction with FMEA to enhance the identification of failure mechanisms for failure modes that cause severe failure effects. FTA can now be facilitated by using commercial software packages such as these by Relex, Isograph, and Item.

6.4.1   FTA Benefits and Process

FTA is a top-down process by which an undesirable event, referred to as the *top event*, is logically decomposed into possible causes in increasing detail to determine the causes or combinations of causes of the top event. A complete FTA can yield both qualitative and quantitative information about the system under study. Qualitative information may include failure paths, root causes, and weak areas of the system. The construction of a fault tree in itself provides the analyst with a better understanding of system functional relationships and potential sources of failure, and thereby serves as a means to review the design to eliminate or reduce potential failures. The results from FTA are useful inputs to the development of design verification plans, operation maintenance policies, and diagnosis and repair strategies. Quantitative analysis of a fault tree gives a probabilistic estimation of the top event and can lead to a conclusion as to whether the design is adequate in terms of reliability and safety. The analysis also yields important failure paths through which the causes are easily propagated to the top event and result in the occurrence of failure, and thus indicates critical areas in which corrective actions are required.

Like FMEA, FTA begins with defining the system and failure, and concludes with understanding the causes of failure. Although steps for performing FTA are essentially dependent on the purpose of the analysis and the system in question, we can summarize the typical steps as follows:

1. Define the system, assumptions, and failure criteria. The interactions between the system and neighbors, including the human interface, should be

fully analyzed to take account of all potential failure causes in the FTA. For this purpose, the boundary diagram described in Chapter 5 is helpful.

2. Understand the hierarchical structure of the system and functional relationships between subsystems and components. A block diagram representing the system function may be instrumental for this purpose.

3. Identify and prioritize the top-level fault events of the system. When FTA is performed in conjunction with FMEA, the top events should include failure modes that have high severity values. A separate fault tree is needed for a selected top event.

4. Construct a fault tree for the selected top event using the symbols and logic described in Section 6.4.2. Identify all possible causes leading to the occurrence of the top event. These causes can be considered as the intermediate effects.

5. List all possible causes that can result in the intermediate effects and expand the fault tree accordingly. Continue the identification of all possible causes at a lower level until all possible root causes are determined.

6. Once the fault tree is completed, analyze it to understand the cause-and-effect logic and interrelationships among the fault paths.

7. Identify all single failures and prioritize cut sets (Section 6.4.3) by the likelihood of occurrence.

8. If quantitative information is needed, calculate the probability of the top event to occur.

9. Determine whether corrective actions are required. If necessary, develop measures to eradicate fault paths or to minimize the probability of fault occurrence.

10. Document the analysis and then follow up to ensure that the corrective actions proposed have been implemented. Update the analysis whenever a design change takes place.

6.4.2 Event and Logic Symbols

As defined earlier, a fault tree is a graphical representation of logical relationships between failure events. Thus, a fault tree may be viewed as a system of event and logic symbols. *Event symbols* indicate whether events are normal, independent, or insignificant. Table 6.8 shows the most commonly used event symbols and their meanings. The symbols can be described as follows:

- *Circle*: a basic event that requires no further development. It represents the type of events at the lowest level and thus indicates the termination of tree ramification. Reliability information of the events should be available for quantitative analysis of a fault tree.

- *Rectangle*: an intermediate event that can be developed further. It denotes an event that results from a combination of more basic events through logic gates.

TABLE 6.8   Fault Tree Event Symbols

Name	Event Symbol	Description
Circle		Basic event with sufficient data
Rectangle		Event represented by a gate
Diamond		Undeveloped event
Oval		Conditional event
House		House event
Triangle in		Transfer-in symbol
Triangle out		Transfer-out symbol

- *Diamond*: an undeveloped event that is not developed further either because it is of insufficient consequence or because information necessary for further development is unavailable.
- *Oval*: a conditional event that is used in conjunction with other logic gates, such as INHIBIT gates (described below).
- *House*: an external event that is not itself a fault that is expected to cause the output event to occur. The house event is artificially turned on or off to examine various special cases of fault trees.
- *Triangle in*: a symbol indicating that the tree is developed further elsewhere (e.g., on another page). It is used in a pair with a triangle-out symbol.
- *Triangle out*: a symbol indicating that this portion of the tree must be connected to the corresponding triangle in.

Logic symbols graphically represent the gates used to interconnect the low-level events that contribute to the top-level event according to their causal relations. A gate may have one or more input events, but only one output event. Table 6.9 lists frequently used logic symbols. The meanings of the symbols are described below.

- *AND gate.* An output event is produced if all the input events occur simultaneously.

TABLE 6.9   Fault Tree Logic Symbols

Name	Event Symbol	Description
AND gate		Output event occurs if all input events occur simultaneously.
OR gate		Output event occurs if any one of the input events occurs.
INHIBIT gate		Input produces output when a conditional event occurs.
EXCLUSIVE OR gate		Output event occurs if only one of the input events occurs.
VOTING gate	k/n	Output event occurs if at least k of n input events occur.

- *OR gate.* An output event is produced if any one or more of the input events occurs.
- *INHIBIT gate.* Input produces output only when a certain condition is satisfied. It is used in a pair with the conditional event symbol. An INHIBIT gate is a special type of AND gate.
- *EXCLUSIVE OR gate.* Input events cause an output event if only one of the input events occurs. The output event will not occur if more than one input event occurs. This gate can be replaced with the combination of AND gates and OR gates.
- *VOTING gate.* Input events produce an output event if at least k of n input events occur.

Example 6.2   Refer to Example 6.1. Develop a fault tree for the top event that no EGR flows into the intake manifold.

SOLUTION   The fault tree for the top event is shown in Figure 6.6, where the gates are alphabetized and the basic events are numbered for the convenience

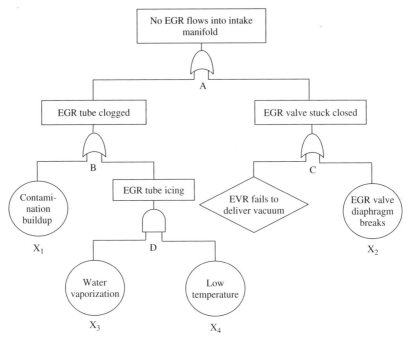

FIGURE 6.6   Fault tree for the top event of no EGR flow to the intake manifold

of future reference. As shown in Figure 6.6, the intermediate event "EGR valve stuck closed" is caused either by the "EGR valve diaphragm breaks" or by the "EVR fails to deliver vacuum." It is worth noting that these two causes have been identified in the FMEA of the EGR system, as given in Figure 6.4. In general, FMEA results can facilitate the development of fault trees.

Example 6.3   As described in Example 5.1, an on-board diagnostic (OBD) system is installed in automobiles to diagnose the failure of EGR components. If a component fails, the OBD system should detect the failure and illuminate the malfunction indicator light (MIL), designated as, say, "Service Engine Soon" on the instrument panel to alert the driver to the need for repair. The OBD system should not turn the light on if no relevant failure occurs. However, a false MIL is sometimes illuminated in certain conditions; the error is type I or alpha error. Develop a fault tree to determine the causes leading to the false MIL.

SOLUTION   The false MIL is the top event. The fault tree for this top event is shown in Figure 6.7. An INHIBIT gate is used here to describe the logic that a false MIL occurs only on vehicles without an EGR component failure, where the logic agrees with the definition of type I error. In the fault tree, only the event "MIL criteria falsely satisfied" is fully expanded to the lowest level, because the causes of this event are of most interest. The fault tree indicates that the false MIL is caused by software algorithm problems coupled with sensor error.

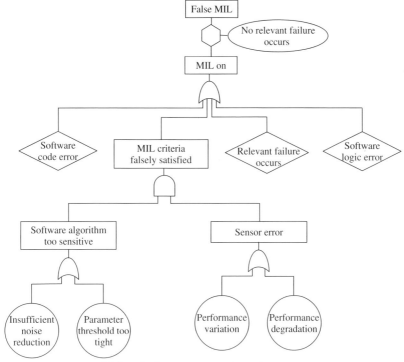

FIGURE 6.7   Fault tree for a false MIL

6.4.3   Qualitative Analysis by Cut Sets

One of the primary purposes for performing FTA is to discover what causes the top event to occur and how it occurs. A *cut set*, defined as a collection of basic events whose occurrence will cause the top event to take place, is a common tool for this purpose. For example, in Figure 6.6, if events X_3 and X_4 occur simultaneously, the top event, "No EGR flows into intake manifold," will happen. Thus, $\{X_3, X_4\}$ is a cut set. Similarly, $\{X_1, X_3\}$ and $\{X_2, X_4\}$ are cut sets. Generally, a fault tree has several cut sets. Some cut sets, such as $\{X_1, X_3\}$ and $\{X_2, X_4\}$ in the example, may not represent the simplest configuration of the basic events. If this is the case, the cut sets can be reduced to form minimal cut sets.

A *minimal cut set* is the smallest combination of basic events which if they all occur will cause the top event to occur. A minimal cut set represents the smallest combination of basic events whose failures are necessary and sufficient to cause the occurrence of the top event. If any event is removed from the set, the remaining events collectively are no longer a cut set. For example, the fault tree in Figure 6.6 has three minimal cut sets: $\{X_1\}$, $\{X_2\}$, and $\{X_3, X_4\}$. The cut sets $\{X_1, X_3\}$ and $\{X_2, X_4\}$ mentioned above are not minimal and can be reduced to $\{X_1\}$ and $\{X_2\}$, respectively. Minimal cut sets can be found using Boolean algebra (Section 6.4.5).

The minimal cut sets of a fault tree can provide insightful information about the potential weak points of a complex system, even when it is not possible to compute the probability of either the cut sets or the top event. The failure probabilities of different basic components in the same system are usually in the same order of magnitude. Thus, the failure probability of a minimal cut set decreases in the order of magnitude as the size of the minimal cut set increases. With this observation, we can analyze the importance of the minimal cut sets by prioritizing them according to their size. Loosely, the smaller the size, the more important the minimal cut set. A single-event minimal cut set always has the highest importance because the single-point failure will result in the occurrence of the top event. The importance is followed by that of double-event cut sets, then triple-event cut sets, and so on. The prioritization of minimal cut sets directs the area of design improvement and provides clues to developing corrective actions.

Another application of minimal cut sets is *common cause analysis*. A *common cause* is a condition or event that causes multiple basic events to occur. For example, fire is a common cause of equipment failures in a plant. In a qualitative analysis, all potential common causes are listed, and the susceptibility of each basic event is assessed to each common cause. The number of vulnerable basic events in a minimal cut set determines the relative importance of the cut set. If a minimal cut set contains two or more basic events that are susceptible to the same common cause failure, these basic events are treated as one event, and the size of the minimal cut set should be reduced accordingly. Then the importance of minimal cut sets should be reevaluated according to the reduced size of the cut sets. Furthermore, the analysis should result in recommended actions that minimize the occurrence of common causes and protect basic events from common cause failures.

6.4.4 Quantitative Analysis by Reliability Block Diagrams

A fault tree can be converted to a reliability block diagram on which a probability evaluation can be performed using the methods described in Chapter 4. When a fault tree contains only AND and OR gates, the analysis is especially simple and pragmatic.

An AND gate in a fault tree is logically equivalent to a parallel reliability block diagram, both describing the same logic that the top event occurs only when all contributing causes occur. For example, Figure 6.8a shows an AND gate fault tree containing two basic events, and Figure 6.8b illustrates the corresponding parallel reliability block diagram. Suppose that the failure probabilities of basic events 1 and 2 are $p_1 = 0.05$ and $p_2 = 0.1$, respectively. Then the reliability of the top event is

$$R = 1 - p_1 p_2 = 1 - 0.05 \times 0.1 = 0.995.$$

An OR gate in a fault tree logically corresponds to a series reliability block diagram because both graphically represent the same logic that the top event

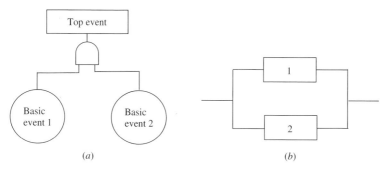

FIGURE 6.8 (*a*) AND gate fault tree; (*b*) parallel reliability block diagram equivalent to part (*a*)

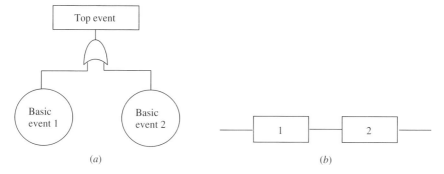

FIGURE 6.9 (*a*) OR gate fault tree; (*b*) series reliability block diagram equivalent to part (*a*)

occurs when one of the basic events occurs. For example, Figure 6.9*a* shows an OR gate fault tree containing two basic events, and Figure 6.9*b* illustrates the corresponding series reliability block diagram. Suppose that the failure probabilities of basic events 1 and 2 are $p_1 = 0.05$ and $p_2 = 0.1$, respectively. Then the reliability of the top event is

$$R = (1 - p_1)(1 - p_2) = (1 - 0.05)(1 - 0.1) = 0.855.$$

The conversion of a fault tree to a reliability block diagram usually starts from the bottom of the tree. The basic events under the same gate at the lowest level of the tree form a block depending on the type of the gate. The block is treated as a single event under the next high-level gate. The block, along with other basic events, generates an expanded block. This expanded block is again considered as a single event, and conversion continues until an intermediate event under a gate is seen. Then the intermediate event is converted to a block by the same process. The block and existing blocks, as well as the basic events, are put together according to the type of the gate. The process is repeated until the top gate is converted.

Example 6.4   Convert the fault tree in Figure 6.6 to a reliability block diagram. Suppose that the EVR never fails to deliver vacuum and that the probabilities of basic events X_1 to X_4 are $p_1 = 0.02$, $p_2 = 0.05$, $p_3 = 0.01$, and $p_4 = 0.1$, respectively. Calculate the probability that no EGR flows into the intake manifold.

SOLUTION   Conversion of the fault tree starts from gate D, which is an AND gate. Basic events X_3 and X_4 form a parallel block. This block, thought of as a single component, connects with basic event X_1 in series because the next high-level gate (B) is an OR gate. When the conversion moves up to gate A, an intermediate event "EGR valve stuck closed" is encountered. The intermediate event requires a separate conversion. In this particular case, the EVR is assumed to be 100% reliable and basic event X_2 is fully responsible for the intermediate event. Because gate A is an AND gate, X_2 connects in series with the block converted from gate B. The complete block diagram is shown in Figure 6.10.

The probability that no EGR flows into the intake manifold is

$$R = (1 - p_1)(1 - p_2)(1 - p_3 p_4) = (1 - 0.02)(1 - 0.05)(1 - 0.01 \times 0.1)$$
$$= 0.93.$$

6.4.5   Determination of Minimal Cut Sets by Boolean Algebra

Determination of minimal cut sets is an important step for both qualitative and quantitative analysis of a fault tree. A powerful tool for accomplishing this task is *Boolean algebra*, the algebra of sets and binary logic with only sentential connectives. The theory and application of Boolean algebra are described in, for example, Whitesitt (1995). Basically, a fault tree can be deemed to be the graphical representation of Boolean relationships among fault tree events that cause the top event to occur. It is possible to translate a fault tree into an entirely equivalent set of Boolean expressions by employing the rules of Boolean algebra.

The following rules of Boolean algebra are commonly used in FTA, where the symbol · represents an intersection, the symbol + stands for a union, and X, Y, and Z are sets. In the context of FTA, intersection is equivalent to AND logic and union corresponds to OR logic. More Boolean rules may be found in Vesely et al. (1981) and Henley and Kumamoto (1992).

- *Commutative law*: $X \cdot Y = Y \cdot X$; $X + Y = Y + X$.
- *Idempotent law*: $X \cdot X = X$; $X + X = X$.
- *Associative law*: $X \cdot (Y \cdot Z) = (X \cdot Y) \cdot Z$; $X + (Y + Z) = (X + Y) + Z$.

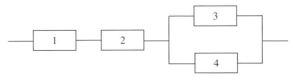

FIGURE 6.10   Reliability block diagram equivalent to the fault tree in Figure 6.6

- *Law of absorption*: $X \cdot (X + Y) = X$; $X + X \cdot Y = X$.
- *Distributive   law*: $X \cdot (Y + Z) = X \cdot Y + X \cdot Z$; $X + Y \cdot Z = (X + Y) \cdot (X + Z)$.

Given a fault tree, the corresponding Boolean expression can be worked out through a top-down or bottom-up process. In the *top-down process*, we begin at the top event and work our way down through the levels of the tree, translating the gates into Boolean equations. The *bottom-up process* starts at the bottom level and proceeds upward to the top event, replacing the gates with Boolean expressions. By either process, the equations for all gates are combined and reduced to a single equation. The equation is further simplified using Boolean algebra rules and thus is written as a union of minimal cut sets. Here we illustrate the bottom-up process with the following example.

Example 6.5   Determine the minimal cut sets of the fault tree in Figure 6.11 using Boolean algebra.

SOLUTION   The fault tree in Figure 6.11 is converted to a Boolean expression through the bottom-up process. To do so, we first write the expressions for the gates at the bottom of the tree. Then we have

$$E_5 = X_4 \cdot X_5. \tag{6.2}$$

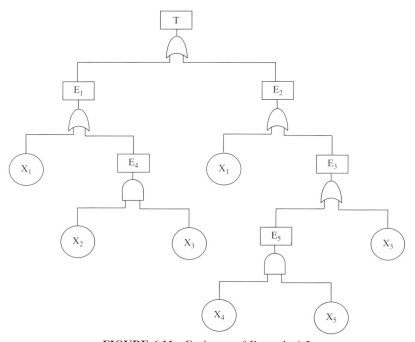

FIGURE 6.11   Fault tree of Example 6.5

Working upward to the intermediate events, we obtain

$$E_3 = X_3 + E_5, \tag{6.3}$$

$$E_4 = X_2 \cdot X_3. \tag{6.4}$$

Substituting (6.2) into (6.3) gives

$$E_3 = X_3 + X_4 \cdot X_5. \tag{6.5}$$

Moving up to the higher-level intermediate events, we obtain

$$E_1 = X_1 + E_4, \tag{6.6}$$

$$E_2 = X_1 + E_3. \tag{6.7}$$

Substituting (6.4) into (6.6) and (6.5) into (6.7), we have

$$E_1 = X_1 + X_2 \cdot X_3, \tag{6.8}$$

$$E_2 = X_1 + X_3 + X_4 \cdot X_5. \tag{6.9}$$

We proceed to the top event of the tree and translate the OR gate into

$$T = E_1 + E_2. \tag{6.10}$$

Substituting (6.8) and (6.9) into (6.10) yields

$$T = X_1 + X_2 \cdot X_3 + X_1 + X_3 + X_4 \cdot X_5. \tag{6.11}$$

Equation (6.11) is logically equivalent to the fault tree in Figure 6.11. Each term in (6.11) is a cut set that will lead to occurrence of the top event. However, the cut sets are not minimal. Reducing the expression by the rules of Boolean algebra results in

$$T = X_1 + X_3 + X_4 \cdot X_5. \tag{6.12}$$

Now X_1, X_3, and $X_4 \cdot X_5$ are the minimal cut sets of the fault tree. Equation (6.12) indicates that the top event can be expressed as the union of three minimal cut sets.

In general, a top event can be written as the union of a finite number of minimal cut sets. Mathematically, we have

$$T = C_1 + C_2 + \cdots + C_n, \tag{6.13}$$

where T is the top event, C_i ($i = 1, 2, \ldots, n$) is a minimal cut set, and n is the number of minimal cut sets. C_i consists of the intersection of the minimum number of basic events required to cause the top event to occur.

For a small fault tree, translation of a top event into the union of minimal cut sets can be done by hand, as illustrated above. However, the problem will become intractable when a fault tree consists of perhaps 20 or more gates, because the complexity of mathematical operation increases significantly with the number of gates. In these situations, FTA software is useful or essential. Nowadays, there are various commercial software packages, such as these by Relex, Isograph, and Item, which are capable of determining minimal cut sets and calculating the probability of a top event.

6.4.6 Quantitative Analysis by Minimal Cut Sets

Quantatitive analysis of a fault tree is to evaluate the probability of a top event. A top event, as formulated in (6.13), can be expressed as the union of a finite number of minimal cut sets. Thus, the probability of a top event is given by

$$\Pr(T) = \Pr(C_1 + C_2 + \cdots + C_n). \tag{6.14}$$

By using the inclusion–exclusion rule, (6.14) can be expanded to

$$\Pr(T) = \sum_{i=1}^{n} \Pr(C_i) - \sum_{i<j=2}^{n} \Pr(C_i \cdot C_j) + \sum_{i<j<k=3}^{n} \Pr(C_i \cdot C_j \cdot C_k)$$

$$+ \cdots + (-1)^{n-1} \Pr(C_1 \cdot C_2 \cdots C_n). \tag{6.15}$$

Evaluation of (6.15) involves the following three steps:

1. Determine the probabilities of basic events. This step usually requires the use of different data sources, such as accelerated test data, field and warranty data, historical data, and benchmarking analysis.
2. Compute the probabilities of all minimal cut sets contributing to the top event. Essentially, this step is to compute the probability of the intersection of basic events.
3. Calculate the probability of the top event by evaluating (6.15).

The first step is discussed in great detail in other chapters of the book. Now we focus on the second and third steps.

If a minimal cut set, say C, consists of an intersection of m basic evelts, say $X_1, X_2, \ldots, X_m$, the probability of the minimal cut set is

$$\Pr(C) = \Pr(X_1 \cdot X_2 \cdots X_m). \tag{6.16}$$

If the m basic events are independent, (6.16) simplifies to

$$\Pr(C) = \Pr(X_1) \cdot \Pr(X_2) \cdots \Pr(X_m), \tag{6.17}$$

where $\Pr(X_i)$ is the probability of basic event X_i. In many situations, the assumption of independence is valid because the failure of one component usually does not depend on the failure of other components in the system unless the components are subject to a common failure. If they are dependent, other methods, such as the Markov model, may be used (see, e.g., Henley and Kumamoto, 1992).

Before the probability of the top event can be calculated, the probabilities of the intersections of minimal cut sets in (6.15) should be determined. As a special case, if the n minimal cut sets are mutually exclusive, (6.15) reduces to

$$\Pr(T) = \sum_{i=1}^{n} \Pr(C_i). \tag{6.18}$$

If the minimal cut sets are not mutually exclusive but independent, the probability of the intersection of the minimal cut sets can be written as the product of the probabilities of individual minimal cut sets. For example, the probability of the intersection of two minimal cut sets C_1 and C_2 is

$$\Pr(C_1 \cdot C_2) = \Pr(C_1) \cdot \Pr(C_2). \tag{6.19}$$

In many situations, the minimal cut sets of a system are dependent because the sets may contain one or more common basic events. Nevertheless, the probability of the intersection of minimal cut sets may still be expressed in a simplified form if the basic events are independent. For example, if $X_1, X_2, \ldots, X_k$ are the independent basic events that appear in minimal cut sets C_1, C_2, or both, the probability of the intersection of C_1 and C_2 can be written as

$$\Pr(C_1 \cdot C_2) = \Pr(X_1) \cdot \Pr(X_2) \cdots \Pr(X_k). \tag{6.20}$$

The validity of (6.20) may be illustrated with the following example. Suppose that $C_1 = X_1 \cdot X_2 \cdots X_i$ and $C_2 = X_i \cdot X_{i+1} \cdots X_k$. C_1 and C_2 are dependent because both contain a common basic event X_i. The intersection of C_1 and C_2 can be written as

$$C_1 \cdot C_2 = X_1 \cdot X_2 \cdots X_i \cdot X_i \cdot X_{i+1} \cdots X_k.$$

According to the idempotent law, $X_i \cdot X_i = X_i$. Thus,

$$\Pr(C_1 \cdot C_2) = \Pr(X_1 \cdot X_2 \cdots X_i \cdot X_{i+1} \cdots X_k) = \Pr(X_1) \Pr(X_2) \cdots \Pr(X_k).$$

Once the probabilities of the intersections of minimal cut sets are calculated, (6.15) is ready for evaluating the probability of the top event. The evaluation process appears simple under the assumption of independent basic events, but it is tedious when a top event consists of a large number of minimal cut sets. Because a high-order intersection usually has a low probability, the third and higher terms in (6.15) may be omitted in practice.

Example 6.6 Refer to Example 6.5. Suppose that basic events $X_1, X_2, \ldots, X_5$ are independent of each other and that their probabilities are $p_1 = 0.01$, $p_2 = 0.005$, $p_3 = 0.005$, $p_4 = 0.003$, and $p_5 = 0.008$. Calculate the probability of the top event.

SOLUTION   The top event of the fault tree has been formulated in (6.12) as the union of the minimal cut sets. Then the probability of the top event is

$$Pr(T) = Pr(X_1 + X_3 + X_4 \cdot X_5) = Pr(X_1) + Pr(X_3) + Pr(X_4 \cdot X_5) - Pr(X_1 \cdot X_3)$$
$$- Pr(X_1 \cdot X_4 \cdot X_5) - Pr(X_3 \cdot X_4 \cdot X_5) + Pr(X_1 \cdot X_3 \cdot X_4 \cdot X_5)$$
$$= p_1 + p_3 + p_4 p_5 - p_1 p_3 - p_1 p_4 p_5 - p_3 p_4 p_5 + p_1 p_3 p_4 p_5$$
$$= 0.01 + 0.005 + 0.003 \times 0.008 - 0.01 \times 0.005 - 0.01 \times 0.003 \times 0.008$$
$$- 0.005 \times 0.003 \times 0.008$$
$$+ 0.01 \times 0.005 \times 0.003 \times 0.008 = 0.015.$$

In a quantitative analysis, we often are interested in measuring the relative importance of a minimal cut set or basic event. The information is helpful in determining the area of design improvement and in developing effective corrective actions. The simplest, yet useful measures are given below. More sophisticated ones can be found in, for example, Henley and Kumamoto (1992).

The relative importance of a minimal cut set can be defined as the ratio of the probability of the minimal cut set to that of the top event. Mathematically,

$$I_C = \frac{Pr(C)}{Pr(T)}, \tag{6.21}$$

where I_C is the relative importance of minimal cut set C and T is the top event.

A basic event may contribute to multiple minimal cut sets. The more minimal cut sets in which the basic event is involved and the higher probability of the sets, the more important the basic event. Thus, the relative importance of a basic event can be defined as

$$I_X = \frac{\sum_{i=1}^{k} Pr(C_i)}{Pr(T)}, \tag{6.22}$$

where I_X is the relative importance of basic event X, k the number of minimal cut sets that contain X, and T the top event.

It is worth noting that the probability of a top event and the value of an importance measure are time dependent, because the probability of basic events usually increases with time. Therefore, it is vital to evaluate these quantities at the design life and other times of interest (e.g., the warranty period).

6.5   ADVANCED TOPICS IN FTA

6.5.1   FTA by a Binary Decision Diagram

The Boolean algebra method described in this chapter is the most common approach to quantitative and qualitative analysis of fault trees. This method requires the determination of minimal cut sets, which may be difficult to obtain when a fault tree is large. More recently, a method based on the binary decision diagram (BDD) has been developed and used (see, e.g., Rauzy, 1993; Bouissou, 1996; Sinnamon and Andrews, 1996, 1997a,b; and Dugan, 2003). The BDD method does not require minimal cut sets for quantitative analysis and can be more efficient and accurate in probability computation.

A BDD is a directed acyclic graph representing a Boolean function. All paths through a BDD terminate in one of two states: a 1 state or a 0 state, with 1 representing system failure (occurrence of the top event) and 0 corresponding to system success (nonoccurrence of the top event). All paths terminating in a 1 state form a cut set of the fault tree. A BDD consists of a root vertex, internal vertices and terminal vertices, which are connected by *branches*. Sometimes branches are called *edges*. Terminal vertices end with the value 0 or 1, while internal vertices represent the corresponding basic events. The *root vertex*, the top internal vertex, always has two branches. Branches (edges) are assigned a value 0 or 1, where 0 corresponds to the basic event nonoccurrence and 1 indicates occurrence of a basic event. All left-hand branches leaving each vertex are assigned the value 1 and called *1 branches*; all right-hand branches are given the value 0 and called *0 branches*. Figure 6.12 shows an example BDD in which X_i is the basic event.

The cut sets can be found from a BDD. First we select a terminal 1 vertex and proceed upward through the internal vertices to the root vertex. All alternative paths that start from the same terminal 1 vertex and lead to the root vertex should be identified. A cut set is formed by the 1 branches of each path. The process is repeated for other terminal 1 vertices, and the corresponding cut sets are determined in the same way. In the example shown in Figure 6.12, starting from terminal 1 vertex of X_4 produces two cut sets: $X_4 \cdot X_3$ and $X_4 \cdot X_3 \cdot X_1$. Originating from terminal 1 vertex of X_3 yields only one cut set: $X_3 \cdot X_2 \cdot X_1$. Thus, the BDD of Figure 6.12 has three cut sets.

A BDD is converted from a fault tree through the use of an if–then–else structure. The structure is denoted **ite**(X, f_1, f_2), which means: If X fails, consider

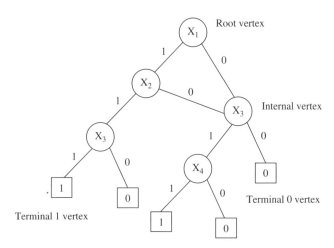

FIGURE 6.12   BDD concepts

function f_1, else consider function f_2. Function f_1 lies on the 1 branch of X and f_2 on the 0 branch. We define the following operation procedures for the **ite** structure.

Let $J = \textbf{ite}(X_1, J_1, J_2)$ and $H = \textbf{ite}(X_2, H_1, H_2)$; then

$$J * H = \begin{cases} \textbf{ite}(X_1, J_1 * H, J_2 * H) & \text{if } X_1 < X_2, \\ \textbf{ite}(X_1, J_1 * H_1, J_2 * H_2) & \text{if } X_1 = X_2, \end{cases} \quad (6.23)$$

where $*$ represents AND $(\cdot)$ or OR $(+)$. Generally, (6.23) can be simplified by employing the following rules:

$$1 \cdot \textbf{ite}(X, f_1, f_2) = \textbf{ite}(X, f_1, f_2),$$

$$0 \cdot \textbf{ite}(X, f_1, f_2) = 0,$$

$$1 + \textbf{ite}(X, f_1, f_2) = 1,$$

$$0 + \textbf{ite}(X, f_1, f_2) = \textbf{ite}(X, f_1, f_2).$$

With the foregoing definitions and notation, a fault tree can be converted to a BDD. The conversion procedure is as follows:

1. Give the basic events an ordering, such as $X_1 < X_2$. Normally, a top-down ordering can be used; that is, the basic events placed higher up the fault tree are listed first and considered as being "less than" those farther down the tree. An inefficient ordering may largely increase the size of the resulting BDD.
2. Convert each intermediate event of the fault tree to the **ite** structure in a bottom-up manner. Reduce the structure to its simplest form.
3. Express the top event in the **ite** structure, and simplify the expression.

The conversion procedure is illustrated with the following example.

Example 6.7   Figure 6.13 shows a fault tree. Construct a BDD for this fault tree.

SOLUTION   First we give the basic events an arbitrary ordering: $X_1 < X_2 < X_3$. The intermediate events E_1 and E_2 are written in terms of **ite** structure as

$$E_1 = X_1 + X_3 = \textbf{ite}(X_1, 1, 0) + \textbf{ite}(X_3, 1, 0) = \textbf{ite}(X_1, 1, \textbf{ite}(X_3, 1, 0)),$$

$$E_2 = X_3 + X_2 = \textbf{ite}(X_3, 1, 0) + \textbf{ite}(X_2, 1, 0) = \textbf{ite}(X_3, 1, \textbf{ite}(X_2, 1, 0)).$$

The top event can be expressed as

$$T = E_1 \cdot E_2 = \textbf{ite}(X_1, 1, \textbf{ite}(X_3, 1, 0)) \cdot \textbf{ite}(X_3, 1, \textbf{ite}(X_2, 1, 0))$$

$$= \textbf{ite}(X_1, \textbf{ite}(X_3, 1, \textbf{ite}(X_2, 1, 0)), \textbf{ite}(X_3, 1, 0) \cdot \textbf{ite}(X_3, 1, \textbf{ite}(X_2, 1, 0)))$$

$$= \textbf{ite}(X_1, \textbf{ite}(X_3, 1, \textbf{ite}(X_2, 1, 0)), \textbf{ite}(X_3, 1, 0)). \quad (6.24)$$

Based on (6.24), a BDD can be constructed as shown in Figure 6.14. The cut sets of the BDD are $X_3 \cdot X_1$, $X_2 \cdot X_1$, and X_3.

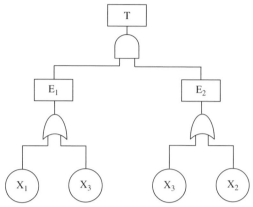

FIGURE 6.13 Example fault tree with repeated event X_3 (From Sinnamon and Andrews, 1997b.)

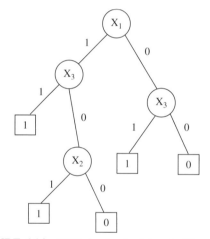

FIGURE 6.14 BDD for the fault tree in Figure 6.13

The probability of a top event can be evaluated from the corresponding BDD. First we find the disjoint paths through the BDD. This is done using the method for determining cut sets and including in a path the basic events that lie on a 0 branch. Such basic events are denoted by $\overline{X}_i$, meaning X_i not occurring. The probability of the top event is equal to the probability of the sum of the disjoint paths through the BDD. In Example 6.7, the disjoint paths are $X_3 \cdot X_1$, $X_2 \cdot \overline{X}_3 \cdot X_1$, and $X_3 \cdot \overline{X}_1$. The probability of the top event can be written as

$$\Pr(T) = \Pr(X_3 \cdot X_1 + X_2 \cdot \overline{X}_3 \cdot X_1 + X_3 \cdot \overline{X}_1)$$

$$= \Pr(X_3 \cdot X_1) + \Pr(X_2 \cdot \overline{X}_3 \cdot X_1) + \Pr(X_3 \cdot \overline{X}_1)$$

$$= p_1 p_3 + p_1 p_2 (1 - p_3) + p_3 (1 - p_1) = p_1 p_2 (1 - p_3) + p_3, \quad (6.25)$$

where p_i is the probability of X_i, and the independence of basic events is assumed.

As illustrated above, calculation of the top event probability from BDD does not require minimal cut sets. Therefore, it is more efficient when a fault tree is large. However, the size of a BDD is largely affected by basic events ordering, and an inefficient BDD may be generated with a poor ordering. Although a BDD may improve the efficiency of probability computation, it loses the logic information contained in the corresponding fault tree. Therefore, a BDD cannot be used for a cause-and-effect analysis.

6.5.2 Dynamic FTA

The gates described so far in this chapter capture the logic of failure events at the same time frame; the fault trees are frequently called the *static fault trees*. In some systems the occurrences of failure events are sequentially dependent. For example, consider a system with one active component and one standby spare connected with a switch controller. If the switch controller fails after the standby is in operation, failure of the active component does not interrupt the function of the system. However, if the switch controller breaks down before the active component fails, the system ceases to function as soon as the active component fails because the standby unit cannot be activated. Therefore, whether the system fails is determined not only by the combination of the events, but also by the sequence in which the events occur. To model the dynamic behavior of the failure event occurrence, some dynamic gates have been defined (Dugan, 2003) as follows:

- *COLD SPARE (CSP) gate.* An output event occurs when the primary component and all cold spare units have failed, where the primary component is the one that is initially powered on, and the cold spare units are those used as replacements for the primary component. Cold spare units may have zero failure rates before being switched into active operation.
- *WARM SPARE (WSP) gate.* An output event occurs when the primary component and all warm spare units have failed. Warm spare units may have reduced failure rates before being switched into active operation.
- *HOT SPARE (HSP) gate.* An output event occurs when the primary component and all hot spare units have failed. Hot spare units may have the same failure rates before and after being switched into active operation.
- *FUNCTIONAL DEPENDENCE (FDEP) gate.* The gate has a single trigger event (either a basic event or the output of another gate of the tree) and one or more dependent basic events. The dependent basic events are forced to occur when the trigger event occurs. The output reflects the status of the trigger event.
- *SEQUENCE ENFORCING (SENF) gate.* The input events are forced to occur in the left-to-right order in which they appear under the gate.

- *PRIORITY AND (PAND) gate.* An output event occurs if all input events occur in order. It is logically equivalent to an AND gate, with the added condition that the events must take place in a specific order.

A fault tree may be comprised of both static and dynamic subtrees. Static subtrees are solved using the methods described earlier in this chapter, while dynamic parts are usually converted to Markov models and worked out by using Markov methods (Manian et al., 1999).

Example 6.8 Consider a hypothetical sprinkler system (Meshkat et al., 2000). The system consists of three temperature sensors, one digital controller, and two pumps. Each pump has a support stream composed of valves and filters. The sensors send signals to the digital controller, which activates one of the pumps when temperature readings at two of the three sensors exceed a preset threshold. The other pump is in standby. The system fails if both pumps fail. The minimum requirements for the sprinkler system to operate are the success of two sensors and one pump. Develop a fault tree for the top event that the system fails to sprinkle water.

SOLUTION   Modeling the system failure needs dynamic gates, which in particular describe the cold standby spare and functional dependency of the pumps on the valves and filters. The fault tree is shown in Figure 6.15.

6.6   COMPUTER-AIDED DESIGN CONTROLS

A conventional design cycle is typically a design–test–fix process. In this process, design is completed and subjected to limited design reviews. Then prototypes are built and tested in the presence of various noise factors. The test usually reveals numerous design deficiencies and results in design changes. Then new prototypes have to be built for the next wave of tests, which may expose the same and/or other design inadequacies. Failure to pass a design verification test will again bring about design changes. Typically, the design–test–fix process has to be repeated several times before the design is ready for release. Obviously, the design process is inefficient in both time and cost. In the current competitive business environment, as discussed in Chapter 3, the design process is being changed by integrating reliability techniques into the process in the early design phase before prototypes are built. Design control is one of the reliability approaches and particularly aimed at identification of potential failure modes of a design. There are a wide variety of design control methods, including mechanical stress analysis, thermal analysis, vibration analysis, tolerance analysis, EMC (electromagnetic compatibility) analysis, and others. Methods suitable for a specific design depend on the type of product, expected operating environments and loads, potential failure modes as shown in FMEA, and other factors.

In most applications, design analysis employs analytical methods because of the fact that the analysis has to be performed before prototypes are

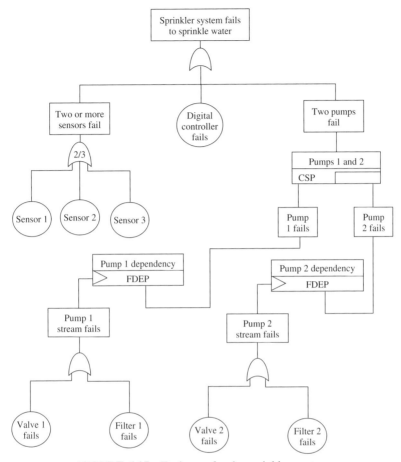

FIGURE 6.15 Fault tree for the sprinkler system

built. Analytical methods usually require complicated mathematical models and intensive computations even for a relatively simple design. Thus, computer simulation is helpful and often essential. In this section we describe briefly mechanical stress analysis, thermal analysis, and vibration analysis, which are based on the finite element method. Finite element analysis (FEA) approximates geometrically complicated structures with many small, simply shaped elements. Then a complex calculation for the original structure can be replaced with many easy calculations for simply shaped elements. The calculations are done using dedicated computer software.

6.6.1 Mechanical Stress Analysis

Mechanical structures, including the electronic packages, often fail, due to excessive stress, in various modes, such as cracking, exorbitant deformation, fatigue, and creep. Ireson et al. (1996) describe more mechanical failure modes. As an integral part of the design process, design analysis should be performed to determine stress distribution and reveal overstressing conditions that will result in premature failures or other concerns.

As mentioned above, mechanical stress analysis requires use of the finite element method. In FEA modeling, the boundary conditions, forces/loads, material properties, and perhaps manufacturing process are integrated into the finite element model for the calculation of stress and deformation. Detailed description of theory and application of the mechanical stress analysis using FEA can be found in, for example, Adams and Askenazi (1998). The analysis is usually performed using a commercial software package such as Hypermesh, Ansys, or Cosmos. As a result of the analysis, potential unacceptable stress conditions and deformations can be discovered. Then design changes should be recommended to address these concerns. For example, an engine component was analyzed using FEA to identify excessive deformation and overstressing problems. The FEA model and stress distribution for the component are shown in Figure 6.16. The analysis indicates an overstress area on the top of the component. This finding resulted in corrective actions that were taken before the design was prototyped.

6.6.2 Thermal Analysis

Thermal stress is the cause of many failures. For mechanical structures that include electronic packages and interconnections, thermal stress due to thermal expansion and contraction can result in cracking, fatigue, creep, and excessive deformation. High temperature is also a notorious stress that contributes to the failure of electronic components in various modes, such as excessive leakage current and degraded output. The increase in temperature significantly reduces the life of a product according to the Arrhenius relationship (Chapter 7), so it is important to determine the temperature distribution in a temperature-sensitive product, and to evaluate its effects on the safety and reliability of the product. This can be done by performing an FEA-based thermal analysis.

One of the major applications of the thermal analysis lies in the design of a printed circuit board (PCB). The primary purpose of the analysis is to determine the temperature distribution on the board and locate hot regions. To accomplish this task, a software package such as BetaSoft or CARMA is generally needed to build and calculate FEA models. Unlike FEA software for mechanical stress

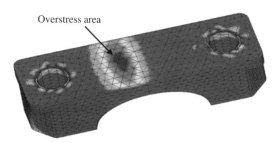

Overstress area

FIGURE 6.16   FEA model and stress distribution of an engine component

analysis, thermal analysis software can generate FEA models automatically based on the dimensions, geometry, and layers of the board. Using the models, the software then calculates the temperature distribution with the defined boundary conditions, electrical loads, board and component thermal properties, packaging method, and the ambient temperature. The resulting temperature distribution indicates hot regions. The components within these hot regions should be checked for functionality and reliability. If the temperature of a hot region raises concerns, design changes are required, including, for example, the use of heat sink, repopulation of components, and modification of circuitry. Sergent and Krum (1998) describe in detail the thermal analysis of electronic assemblies, including PCB.

Let's look at an example. Right after the schematic design and PCB layout of an automobile body control board were completed, a thermal analysis was performed to examine the design for potential deficiencies. The temperature distribution on the top layer of the board is shown in Figure 6.17, where the rectangles and ovals on the board represent electronic components. The hot region on the board was found to coincide with the area where two resistors were populated. Even though the temperature of the hot region generated no major concerns on the current-carrying capability of copper trace and solder joint integrity, the high case temperatures of these two resistors would reduce long-term reliability in the field. Therefore, design changes were enforced to lower the temperature.

6.6.3   Vibration Analysis

Products such as airplanes and automobiles are subjected to severe vibrating conditions during operation. Products such as personal computers work in stationary environments but undergo vibration during transportation from manufacturers to customers. It is safe to say that almost all products have to encounter vibration during their lifetime. For most products, vibration has adverse effects on product functionality and reliability. Failure modes due to this stress

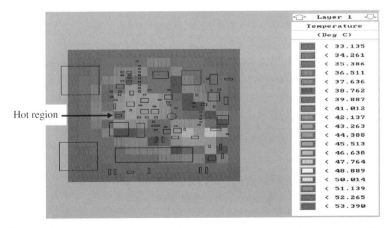

FIGURE 6.17   Temperature distribution of an automobile body control PCB

may include cracking, fatigue, loose connection, and others. See Chapter 7 for more discussions on vibration. It is vital to determine the product behavior at the presence of vibration. This task can be accomplished by performing a vibration analysis.

Vibration analysis is commonly based on FEA and employs a commercial software package such as MATLAB, MathCAD, or CARMA. The analysis calculates natural frequencies and displacements with inputs of boundary conditions and vibration environment. The method of mounting defines the boundary conditions, and the type of vibration (sinusoidal or random vibration) and its severity specify the vibration environment.

Once the natural frequencies and displacements are computed, further analyses can be performed. For example, maximum displacement should be checked against minimum clearance to prevent any potential mechanical interference. The first natural frequency is often used to calculate the stress caused by the vibration and to determine the fatigue life. A low natural frequency indicates high stress and displacement. If problems are detected, corrective actions should be taken to increase the natural frequency. Such measures may include use of a rib, modification of the mounting method, and others.

Vibration analysis is frequently performed on PCB design to uncover potential problems such as PCB and solder joint cracking and low fatigue life. For example, for the automobile body control board discussed in Section 6.6.2, Figure 6.18 shows the FEA-based vibration analysis results, including the first three natural frequencies and the first fundamental mode shape of vibration. The board was simply restrained at four edges and subjected to random vibration. Further calculations on bending stress and fatigue life indicated no concerns due to the vibration condition specified. Detailed vibration analysis for electronic equipment is described in, for example, Steinberg (2000).

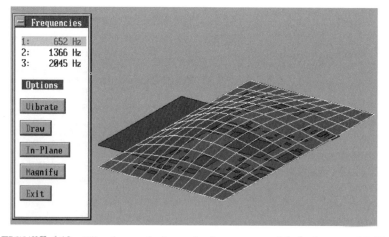

FIGURE 6.18   Vibration analysis results for an automobile body control PCB

PROBLEMS

6.1 Explain the processes by which design FMEA and FTA detect design mistakes. Can human errors be discovered through the use of FMEA or FTA?

6.2 Explain the correlations and differences between the following terms used in design FMEA and FTA:

(a) *Failure mode* and *top event*.

(b) *Failure mechanism/cause* and *basic event*.

(c) *Failure effect* and *intermediate event*.

(d) *Occurrence* and *failure probability*.

6.3 Perform a design FMEA on a product of your choice using the worksheet of Figure 6.2 and answer the following questions:

(a) What are the top three concerns by RPN?

(b) What are the top three concerns by $S \times O$? Is the result the same as that by RPN? Is $S \times O$ a more meaningful index than RPN in your case?

(c) Construct a fault tree for the failure mode with the highest severity. Does the fault tree provide more insights about how the failure mode occurs?

6.4 What are the impacts of the following actions on severity, occurrence, and detection rankings?

(a) Add a new test method.

(b) Implement a design control before prototypes are built.

(c) Take a failure prevention measure in design.

6.5 Describe the purposes of qualitative and quantitative analyses in FTA. What are the roles of minimal cut sets?

6.6 Construct a fault tree for the circuit shown in Figure 6.19, where the top event is "blackout." Convert the fault tree to a reliability block diagram.

6.7 Figure 6.20 depicts a simplified four-cylinder automobile engine system. The throttle controls the amount of air flowing to the intake manifold. While the engine is at idle, the throttle is closed and a small amount of air is bypassed through the inlet air control solenoid to the manifold to prevent engine

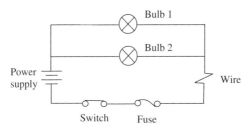

FIGURE 6.19 Two-bulb lighting circuit

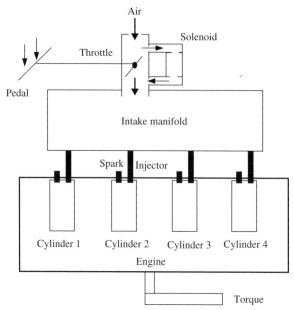

FIGURE 6.20   Simplified automotive engine system

stalling. Fuel is mixed with air, injected into each cylinder, and ignited by spark. Suppose that the failure probabilities of the throttle, solenoid, sparks, and fuel injectors are 0.001, 0.003, 0.01, and 0.008, respectively. For the top event "Engine stalls while vehicle is at idle," complete the following tasks:

(a) Construct a fault tree for the top event.

(b) Determine the minimal cut sets.

(c) Evaluate the probability of the top event.

(d) Convert the fault tree to a reliability block diagram and calculate the top event probability.

6.8 Refer to Problem 6.7. If the top event is "Engine stalls while vehicle is moving," complete the following tasks:

(a) Construct a fault tree for the top event.

(b) Determine the minimal cut sets.

(c) Evaluate the probability of the top event.

(d) Convert the fault tree to a BDD and evaluate the top event probability.

7

ACCELERATED LIFE TESTS

7.1 INTRODUCTION

Increasing global competition has placed great pressure on manufacturers to deliver products with more features and higher reliability at a lower cost and in less time. The unprecedented challenges have motivated manufacturers to develop and deploy effective reliability programs, which include accelerated life tests (ALTs), as described in Chapter 3. An ALT subjects test units to higher-than-use stress levels to shorten their times to failure. The life data so obtained are then extrapolated using a life–stress relationship to estimate the life distribution at a use condition. Because they yield failure information in a short time, ALTs are used extensively in various phases of a product life cycle. Early in the product design phase, the reliability of materials and components can be assessed and qualified by testing them at higher stress levels. As the design moves forward, robust reliability design is often performed to improve the reliability by choosing the optimal settings of design parameters. As we have seen in Chapter 5, robust reliability design requires extensive testing, which may be conducted at elevated levels of noise factors. As soon as the design is completed, prototypes are subjected to design verification (DV) testing. If successful, the next step is process validation (PV) testing. Both DV and PV were discussed in Chapter 3. These two types of testing often include ALTs, which are intended to demonstrate the achievement of a specified reliability target. ALTs are sometimes needed even after the design is released to full production. For example, we may rely on such tests to determine the causes of excessive process variation and to duplicate the

Life Cycle Reliability Engineering, by Guangbin Yang
Copyright © 2007 John Wiley & Sons, Inc.

critical failure modes observed in the field. In short, ALTs are essential in all effective reliability programs, attributing to their irreplaceable role in improving and estimating reliability. The author has been present at five consecutive Annual Reliability and Maintainability Symposia and noticed that the sessions on ALT topics were far more heavily attended than any other concurrent sessions.

An ALT can be (1) qualitative or (2) quantitative, depending on the purpose of the test. A qualitative test is usually designed and conducted to generate failures as quickly as possible in the design and development phase. Subsequent failure analyses and corrective actions lead to the improvement of reliability. This type of test, also known as *highly accelerated life testing* (HALT), is discussed in Section 7.9. Other sections of the chapter are dedicated to quantitative tests, which are aimed at estimating product life distribution: in particular, percentiles and the probability of failure (i.e., the population fraction failing).

7.2 DEVELOPMENT OF TEST PLANS

Planning an ALT in advance is a critical step toward success in obtaining valid and accurate information. A feasible and reasonable test plan should include managerial, logistical, and technical considerations. Managerial considerations deal with formation of a team, definition of the roles and responsibilities of each team member, coordination of the team, and other personnel management tasks. Logistical tasks are to secure the availability of test facilities such as chambers, functionality inspection systems, and measuring devices. Technical considerations include determining the test's purpose, sample units and size, failure definition, time scale, acceleration methodology, and data collection and analysis methods, which are discussed in this section.

7.2.1 Test Purposes

From a product life cycle perspective, ALTs can be classified into three categories: design ALT, qualification ALT, and production ALT. The purposes of these tests are usually different.

In the design and development phase, the common test purposes are as follows:

1. *Compare and assess the reliability of materials and components.* Such tests take place in the early stage of the design and development phase to select the appropriate vendors of the materials and components.

2. *Determine optimal design alternatives.* Design engineers often develop multiple design alternatives at a low level of the product hierarchical structure. Prototypes at this level are functionally operable and inexpensive. ALTs are conducted to evaluate the reliability performance of each design alternative, on which the selection of the best candidate may be based. ALTs performed in robust reliability design have such purpose.

3. *Confirm the effectiveness of a design change.* In designing a new product, design changes are nearly inevitable during the design and development phase.

Even for a carryover design, some fixes are often necessary. The changes must be verified as early as possible. ALTs are needed for this purpose.

4. *Evaluate the relationship between reliability and stress.* Sometimes ALTs are performed to assess the sensitivity of reliability to certain stresses. The resulting information is used to improve the robustness of the design and/or to specify the limit of use condition.

5. *Discover potential failure modes.* A test serving this purpose is important for a new product. Critical failure modes, which can cause severe effects such as safety hazards, must be eradicated or mitigated in the design and development phase.

After the design is completed, prototypes are built using a manufacturing process similar to that for full production, and then subjected to DV testing. In this stage, the common purposes of ALTs are as follows:

1. *Demonstrate that the design achieves a specified reliability target.* An ALT conducted by a supplier for this purpose must use the sample size, test time, and stress levels agreed upon by original equipment manufacturers. Indeed, the test for this purpose is a reliability verification test (Chapter 9).

2. *Estimate the reliability of the design.* Often, an ALT is needed to measure the reliability of the design at a use condition to assess its competitiveness and to estimate warranty cost.

Once a design passes DV testing, the established manufacturing process builds units for PV testing, which usually includes ALTs serving the following purposes:

1. *Demonstrate that the manufacturing process is capable of producing products that meet a specified reliability target.* As in the DV phase, if performed by a supplier, an ALT must have the sample size, test time, and stress levels endorsed by the original equipment manufacturers. The test is actually a reliability verification test.

2. *Estimate the product reliability.* Since the life test data contain the information about design reliability as well as process variation, the estimated reliability is relatively close to the reliability level that customers will experience.

Production at full capacity may begin after the design passes PV testing. ALTs may be required in the production phase for the following purposes:

1. *Identify the special causes for a statistically significant process shift.* Statistical process control tools detect such a shift and trigger a series of investigations, which can include ALTs to find causes of a change in failure mode or life distribution due to process variation.

2. *Duplicate the critical failure modes observed in the field for determination of the failure mechanisms.*

3. *Acceptance sampling.* ALTs may be performed to decide if a particular lot should be stopped from shipping to customers.

7.2.2 Sample Representation and Sample Size

In the early design phase, ALTs are aimed at evaluating the reliability of components and materials proposed for the design. Thus, the tests should employ representative sample units drawn randomly from a large population made up of various lots. Such a random sampling also applies to the assessment of end-product reliability in the full production phase. For development purposes, such as the determination of optimal levels of design parameters and the assessment of design change, ALTs are usually conducted on prototypes especially created for the tests. These test units are built using specified levels of control factors (design configuration or design parameters), while the levels of noise factors such as the unit-to-unit variation are minimized such that the effects of the control factors are not disguised. Control and noise factors are defined and classified in Chapter 5.

In the DV testing stage, test units resemble the final product to a great extent, but essentially they are still prototypes, because a manufacturing process set up at this stage is subject to change. In addition, the process variations in full production usually do not occur in a pilot production. Thus, it is important to realize that the reliability estimate in this phase does not fully characterize the reliability that customers will see. The estimate is primarily optimistic.

Samples for ALTs in the PV phase are regarded representative of the final product because both use the same materials, components, design, production process, and process monitoring techniques. However, the test units are built within a short period of time and thus do not vary from lot to lot. Strictly speaking, such samples cannot fully characterize a full-production population. Nevertheless, a reliability estimate obtained from an ALT in this stage is reasonably realistic and useful for decision making.

As discussed earlier, ALTs are sometimes conducted to investigate special causes or to determine the acceptance of a lot. Samples for such tests should be drawn from the lots under concern, not from the entire population. For example, ALTs dedicated to identifying the special causes for process shift must test samples from a batch made during the process shift. Otherwise, the test will not produce useful information.

Sample size is an important number that must be determined before testing. It largely affects the test cost, required capacity of test equipment, test time, and estimate accuracy. If an ALT is part of DV or PV testing, the sample size is determined by the reliability target and consumer and producer risks. In Chapter 9 we describe methods of calculating a suitable sample size. An ALT aimed at evaluating reliability should have a large sample size whenever possible. The size can be calculated by specifying the statistical accuracy of the reliability estimates, as described in, for example, Nelson (1990, 2004) and Meeker and Escobar (1998). The statistical sample size, however, is frequently too large to be affordable. In practice, the sample size used is considerably smaller than the statistical one, but should not be less than the total number of potential failure modes of the test units. This minimum sample size allows each failure mode to have an opportunity to appear during testing. This requirement must be met when the test purpose is to duplicate the critical failure modes observed in the field.

7.2.3 Acceleration Methods

The purpose of acceleration is to yield reliability information more quickly. Any means that serves this purpose is an acceleration method. Basically, there are four types of acceleration methods: (1) overstressing, (2) increasing usage rate, (3) changing level of a control factor, and (4) tightening the failure threshold. The methods may be classified further as shown in Figure 7.1. The appropriate method to use for a specific product depends on the purpose of the test and the product itself. In practice, an ALT often utilizes one or two of the four types of method.

Overstressing   Overstressing, the most common acceleration method, consists of running test units at higher-than-use levels of stresses. The stresses applied in a test should be those that stimulate the product to fail in the field. Such stresses include environmental, electrical, mechanical, and chemical stimuli, such as temperature, humidity, thermal cycling, radiation, voltage, electrical current, vibration, and mechanical load. Stress may be applied in different patterns over time, which include constant stress, step stress, progressive stress, cyclic stress, and random stress. They are described briefly below.

In *constant-stress testing,* the stress level of a unit is held constant over time. An ALT may have several groups of units, each subjected to a different constant level of stress. Figure 7.2 shows life distributions of two groups of units at high and low levels of stress, respectively. The units at high stress yield shorter lives;

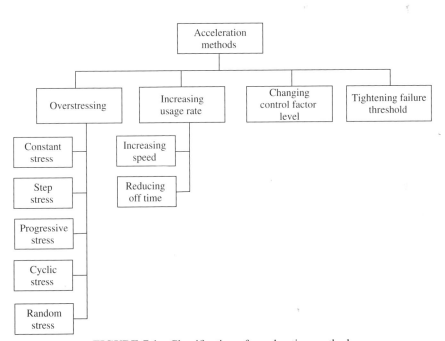

FIGURE 7.1   Classification of acceleration methods

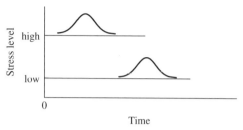

FIGURE 7.2   Life distributions at two levels of constant stress

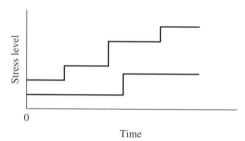

FIGURE 7.3   Two step-stress loading patterns

the other group survives longer. This test method is most common in practice because of the simplicity of stress application and data analysis.

In *step-stress testing,* units are subjected to a stress level held constant for a specified period of time, at the end of which, if some units survive, the stress level is increased and held constant for another specified period. This process is continued until a predetermined number of units fail or until a predetermined test time is reached. When a test uses only two steps, the test is called a simple step-stress test. Figure 7.3 shows two- and multiple-step loading patterns. A step-stress test yields failures in a shorter time than does a constant-stress test. Thus, it is an effective test method for discovering failure modes of highly reliable products. However, models for the effect of step stressing are not well developed and may result in inaccurate conclusions. Nelson (1990, 2004), and Pham (2003) describe test method, data analysis, and examples.

In *progressive stress testing,* the stress level is increased constantly (usually, linearly) until a predetermined number of test units fail or until a predetermined test time is reached. The stress loading method is shown in Figure 7.4. The slopes of the straight lines are the rates at which the stress levels are increased and represent the severity of the stress. The higher the rate, the shorter the times to failure. Like step-stress testing, the test method is effective in yielding failures and imposes difficulties for modeling the data. Nelson (1990, 2004) presents test method, data analysis, and examples.

In *cyclic stress loading,* the stress level is changed according to a fixed cyclic pattern. Common examples of such stress are thermal cycling and sinusoidal vibration. In contrast to the fixed amplitude of a cyclic stress, the level of a

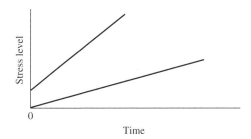

FIGURE 7.4 Progressive stress loading patterns

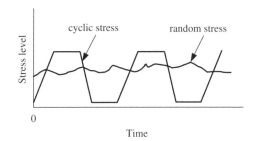

FIGURE 7.5 Cyclic and random stress loading

random stress changes at random and has a probabilistic distribution. Random vibration is a typical example of random stress. Figure 7.5 illustrates the two types of stress loading methods. The loading patterns are used primarily for the purpose of simulating the actual stress that a product will encounter in the field.

Increasing the Usage Rate *Usage* is the amount of use of a product. It may be expressed in miles, cycles, revolutions, pages, or other measures. *Usage rate* is the frequency of a product being operated, and may be measured by hertz (Hz), cycles per hour, revolutions per minute, miles per month, pages per minute, or others. Many commercial products are operated intermittently. In contrast, test units are run continuously or more frequently, to reduce the test time. For example, most automobiles are operated less than two hours a day and may accumulate 100 miles. In a proving ground, test vehicles may be driven eight or more hours a day and accumulate 500 or more miles. On the other hand, some products run at a low speed in normal use. Such products include bearings, motors, relays, switches, and many others. In testing they are operated at a higher speed to shorten the test time. For the two types of products, the life is usually measured by the usage to failure, such as cycles to failure and miles to failure.

Special care should be exercised when applying the acceleration method. It is usually assumed that usage to failure at a higher usage rate is equal to that at the usual usage rate. The assumption does not hold in situations where an increased usage rate results in an additional environmental, mechanical, electrical, or chemical stress: for example, when raising the operational speed generates a higher temperature, or reducing the off time decreases the time for heat to dissipate. Then the equal-usage assumption may be invalid unless such effects are eliminated by using a compensation measure such as a cooling fan. In many tests the use of compensation is impractical. Then we must take into account the effect on life of increased usage rate. Considering the fact that usage to failure may be shorter or longer at a higher usage rate, G. Yang (2005) proposes an acceleration model to quantify usage rate effects (also discussed in Section 7.4.5).

Changing the Level of a Control Factor   A control factor is a design parameter whose level can be specified by designers. We have seen in Chapter 5 that the level of a control factor can affect the life of a product. Therefore, we can intentionally choose the level of one or more control factors to shorten the life of test units. This acceleration method requires the known effects on life of the control factors. The known relationship between life and the level of control factors may be developed in robust reliability design (Chapter 5). Change of dimension is a common application of this test method. For example, a smaller-diameter shaft is tested to determine the fatigue life of a larger shaft because the former yields a shorter life. Large capacitors are subjected to electrical voltage stress to estimate the life of small capacitors with the same design, on the assumption that the large ones will fail sooner because of a larger dielectric area. Nelson (1990, 2004) describes a size-effect model relating the failure rate of one size to that of another size. Bai and Yun (1996) generalize that model. A change of geometry in favor of failure is another use of the acceleration method. For example, reducing the fillet radius of a mechanical component increases the stress concentration and thus shortens life. In practice, other design parameters may be used as accelerating variables. The control factors must not, however, interact with other accelerating stresses. Otherwise, the test results may be invalid, as described in Chapter 5.

Tightening the Failure Threshold   For some products, failure is said to have occurred when one of its performance characteristics crosses a specified threshold. Clearly, the life of the products is determined by the threshold. The tighter the threshold, the shorter the life, and vice versa. Thus, we can accelerate the life by specifying a tighter threshold. For example, a light-emitting diode at a normal threshold of 30% degradation in luminous flux may survive 5000 hours. If the threshold is reduced to 20%, the life may be shortened to 3000 hours. This acceleration method requires a model that relates life to threshold and is discussed in Chapter 8.

Acceleration Factor   An important concept in ALTs is the *acceleration factor*, defined as the ratio of a life percentile at stress level S to that at stress level S'. Mathematically,

$$A_f = \frac{t_p}{t'_p},$$
(7.1)

where A_f is the acceleration factor, p the specified probability of failure (i.e., the population fraction failing), and t_p (t'_p) the $100p$th percentile of the life distribution at S (S'). Often, p is chosen to be 0.5 or 0.632.

The primary use of the acceleration factor is to calculate a life percentile at a low stress level from the percentile observed at a high stress level. It also expresses the equivalent number of test hours at a low stress level to one hour at a high stress level. The use of an acceleration factor implicitly assumes that life distributions have the same shape at the two stress levels. For example, for a (transformed) location-scale distribution, the scale parameter values at different stress levels are assumed to be equal. If a Weibull distribution is involved, the shape parameter should not depend on the stress level if the acceleration factor is to remain simple.

The acceleration factor can be expressed in terms of life distribution parameters. For a lognormal distribution with constant σ, from (2.43) and (7.1), the acceleration factor can be written as

$$A_f = \exp(\mu - \mu'),\tag{7.2}$$

where μ and μ' are the scale parameter values at S and S', respectively.

Similarly, from (2.25) and (7.1), the acceleration factor for a Weibull distribution with constant β can be expressed as

$$A_f = \frac{\alpha}{\alpha'},\tag{7.3}$$

where α and α' are the characteristic lives of a Weibull distribution at S and S', respectively.

7.2.4 Data Collection Methods

Whenever possible, test units should be monitored continuously during testing to get exact failure times. For a binary-state product, the monitoring system detects a failure exactly when a unit ceases to function. The life of the unit is the time to catastrophic failure. To monitor a degrading unit, the system should measure the performance characteristics continuously or very frequently. Doing so enables generating nearly exact failure times, which yield more accurate estimates. The life of the unit is the time at which a performance characteristic crosses its failure threshold. This data collection method usually needs an automated data acquisition system and advanced software.

In many situations, inspection or measurement in intervals is more technically and economically practical than continuously monitoring. If this method is used, we should take efforts to avoid the circumstance where most failures fall into a few intervals. This can be accomplished to a great extent by shortening times between inspections. If the product is exponential or Weibull with a shape parameter less than 1, more frequent inspections should be taken early in the test because more failures are expected. If we use a lognormal or Weibull distribution with a shape parameter greater than 1, the intervals can be longer in the early time, then shorter, and then longer in the late time of the test. To make a better inspection schedule, we may preestimate the life distribution using reliability handbooks, historical data, or preliminary tests.

7.2.5 Optimum Test Plans

ALTs are often conducted to estimate life distribution at use conditions. The statistical error of the estimate depends on the test plan. Certainly, it is desirable to devise the optimal test plans that minimize error. For a constant-stress test, a test plan consists of stress levels, the number of test units allocated to each stress level, and values of other variables. A step-stress test plan usually consists of the times at which a stress level is increased or the number of failures to trigger the increase in stress level. Constant-stress tests are more common and their plans are studied in detail later in this chapter.

7.3 COMMON STRESSES AND THEIR EFFECTS

Used either separately or with other acceleration methods, overstressing is usually the first (or even the best) choice for accelerating a test. The stresses applied in a test should be the ones that accelerate the failure modes that the product will experience in the field. Laboratory tests must not use a stress that causes failure mechanisms never seen in the field; otherwise, the reliability estimates are meaningless. If multiple stresses are expected in the field and few can be applied in testing, the ones selected should be effective in producing relevant failure modes. To help select the appropriate stresses, in this section we introduce the most frequently used stresses and discuss briefly common failure mechanisms and modes caused by such stresses.

7.3.1 Constant Temperature

Elevated constant temperature is perhaps the most common stress in accelerated testing. This is largely because high temperature accelerates many failure mechanisms of most products. Some products, such as personal computers, appear to work at room temperature. But many components within the products, such as the central processing units of computers, may be at an elevated temperature. High temperature can produce various failure modes. Some common ones are discussed below, but there are many more. As we will see, these failure modes are fundamental and can cause high-level failure modes.

1. *Oxidation.* In a narrow sense, oxidation is the reaction of a substance with oxygen. When some materials are in contact with oxygen, an oxide compound is formed as a result of chemical reaction between the oxygen and the materials. Metal is most susceptible to oxidation. High temperature provides energy to the chemical reaction and thus accelerates the oxidation process. Oxidation is the cause of many failure modes that are directly observable. For example, oxidation causes corrosion of metals and may result in fracture of a structure. Oxidation of electronic components increases contact resistance and causes deterioration of electrical performance. As a countermeasure, most electronic products are hermetically sealed.

2. *Electromigration.* When an electrical current flows through a metal, electrons exchange momentum with metal atoms. This results in a mass transport

along the direction of electron movement. On the other hand, when metal atoms are activated by the momentum exchange, they are subjected to an applied electrical field opposite to the electron movement, and move against that movement. The two movements are accelerated by high temperature and interact to determine the direction of net mass transfer. As a result of the mass transfer, vacancies and interstitials are created on the metal. *Vacancies* develop voids and microcracks, which may cause, for example, an increased contact resistance or open circuit. *Interstitials* are the exotic mass on the surface of the metal and may result in a short circuit. In addition to temperature and electrical current density, the susceptibility to electromigration also depends on the material. Silver is the metal most subject to this failure.

   3. *Creep.* Creep is a gradual plastic deformation of a component exposed to high temperature and mechanical stress, resulting in elongation of the component. Before a component fractures, the creep process typically consists of three stages, as shown in Figure 7.6. Initially, the transient creep occurs in the first stage, where the creep rate (the slope of the strain–time curve) is high. Then the rate decreases and remains approximately constant over a long period of time called the *steady-state stage* (i.e., the second stage). As time proceeds, creep develops to the third stage, where the creep rate increases rapidly and the strain becomes so large that fracture occurs. In practice, many products fail far before creep progresses to the third stage, due to the loss of elastic strength. For example, the contact reeds of an electromagnetic relay are subjected to a cyclic load and high temperature when in operation. Creep occurs to the reeds and results in stress relaxation or loss of elastic strength, which, in turn, reduces the contact force, increases the contact resistance, and causes failure.

   4. *Interdiffusion.* When two different bulk materials are in intimate contact at a surface, molecules or atoms of one material can migrate into the other, and vice versa. Like electromigration, interdiffusion is a mass transport process which is sensitive to temperature. When a high temperature is applied, the molecules and atoms are thermally activated and their motion speeds up, increasing the interdiffusion rate. If the diffusion rates for both materials are not equal, interdiffusion can generate voids in one of the materials and cause the product's electrical, chemical, and mechanical performance to deteriorate. Interdiffusion can be the cause of various observable failure modes, such as increased electrical resistance and fracture of material.

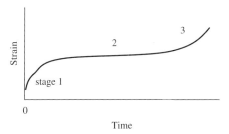

FIGURE 7.6   Successive stages of a creep process

7.3.2 Thermal Cycling

Thermal cycling involves applying high and low temperatures repeatedly over time. The variables that define the profile of a thermal cycle include high temperature (T_{max}), low temperature (T_{min}), dwell time at high temperature (t_{max}), dwell time at low temperature (t_{min}), and rate of temperature change (dT/dt), shown graphically in Figure 7.7. Thermal cycling is a widely used test method that is stipulated in various engineering standards as an integral part of the environmental stress testing. For example, MIL-STD-883F (U.S. DoD, 2004) recommends six thermal cycling profiles (from A to F) for testing microcircuits, with profile A the most lenient, with low and high temperatures at $-55°C$ and $85°C$, respectively, and profile F the most harsh, with extremes at $-65°C$ and $300°C$. The test method prevails partially because many products are subjected to thermal cycling in the field. Automotive engine components, for instance, experience this type of stress when the engine is ignited in cold weather or when the vehicle is driven through a flooding road. The engine components have to withstand rapidly increasing temperature in the former situation and a sharp temperature drop in the latter. More important, thermal cycling is effective in precipitating fatigue failures in a test, especially for connections between two different materials, such as die attachments, wire bonds, and the plated vias of electronic products.

Fatigue is the most common failure mode caused by thermal cycling. When cyclic stress is applied to a product, two different materials in mechanical connection within the product are subjected to repeated expansion and contraction. Due to mismatch of the coefficients of thermal expansion of the two materials, the repeated expansion and contraction generate a cyclic mechanical stress under which a microcrack is initiated, typically at a point of discontinuity or defect in the materials. Once a microcrack is formed, stress concentrates at the tip of the crack, where the local stress is much higher than that in the bulk of the material. The crack propagates over time under the cyclic stress. When the size of the crack develops to a threshold, the resisting strength of the material is less than the applied stress amplitude, and a fatigue failure results. For most electrical products, electrical and thermal performance degrades significantly before fatigue failure occurs. The fatigue life under thermal cycling depends on the thermal cycling profile, the coefficients of thermal expansion, and other material properties that determine crack initiation and propagation rates.

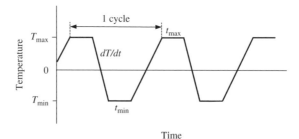

FIGURE 7.7   Thermal cycling profile

7.3.3 Humidity

There are two types of humidity measures in use: absolute humidity and relative humidity. *Absolute humidity* is the amount of water contained in a unit volume of moist air. In scientific and engineering applications, we generally employ *relative humidity*, defined as the ratio (in percent) of the amount of atmospheric moisture present relative to the amount that would be present if the air were saturated. Since the latter amount is dependent on temperature, relative humidity is a function of both moisture content and temperature. In particular, relative humidity is inversely proportional to temperature until the dew point is reached, below which moisture condenses onto surfaces.

Important failure modes due to moisture include short circuit and corrosion. Corrosion is the gradual destruction of a metal or alloy caused by chemical attack or electrochemical reaction. The primary corrosion in a humid environment is an electrochemical process in which oxidation and reduction reactions occur simultaneously. When metal atoms are exposed to a damp environment, they can yield electrons and thus become positively charged ions, provided that an electrochemical cell is complete. The electrons are then consumed in the reduction process. The reaction processes may occur locally to form pits or microcracks, which provide sites for fatigue initiation and develop further to fatigue failure. Corrosion occurring extensively on the surface of a component causes electrical performance and mechanical strength to deteriorate. The corrosion process is accelerated with high temperatures. This is the reason that humidity stress is frequently used concurrently with high temperature. For example, 85/85, which means 85°C and 85% relative humidity, is a recommended test condition in various engineering standards.

In addition to corrosion, short circuiting is sometimes a concern for electronic products working in a humid environment. Moisture condenses onto surfaces when the temperature is below the dew point. Liquid water that is deposited on a circuit may cause catastrophic failures, such as a short circuit. To minimize the detrimental effects of humidity, most electronic products are hermetically sealed.

7.3.4 Voltage

Voltage is the difference in electrical potential between two points. When voltage is applied between any two points, it is resisted by the dielectric strength of the material in between. When Ohm's law applies, the current through the material is directly proportional to the voltage. Thus, if the material is insulation, the current, which is negligible and sometimes called *leakage current*, increases with applied voltage. If the voltage is elevated to a certain level, the insulation breaks down and the current jumps. The failure usually occurs at weak spots or flaws in the material, where the dielectric strength is relatively low. In general, the higher the voltage, the shorter the insulation life. Considering this effect, voltage is often employed as an accelerating variable for testing insulators and electronic components such as capacitors.

For conductors and electronic components, high voltage means high current; thus, failure modes caused by high current (which are discussed next) may be observed at high voltage. In addition, high voltage is an important stimulus for arcing. Arcing energy and frequency are largely increased when moisture is also present. Arcing generates electromagnetic noise in neighboring components and erodes electrical contact surfaces. Failures due to arcing are common for electromechanical components. Let's consider electromagnetic relays again. Relays loaded with voltage create arcing while the contacts are being separated. Arcing wears out the noble materials on the contact surfaces and damages the contact geometry. As a result, the contact resistance is increased. In the worst case, the contacts are welded together by the high temperature generated by arcing, and thus cannot be separated for normal operation.

7.3.5 Electrical Current

Electrical current is sometimes used as an accelerating stress for electrical and electronic products such as motors, relays, conductors, and light-emitting diodes. When an electrical current flows through a conductor, heat is generated and transferred to the neighboring components, causing their temperature to rise. In this respect, the electrical current has the same effect as temperature stress applied externally to the product. In addition, current may produce the following effects:

1. *Electrical current speeds up electromigration.* When the applied current generates a current density higher than threshold, electromigration is initiated. Then the rate of electromigration increases with the current density (Young and Christou, 1994). Thus, a high electrical current may also result in electromigration-induced failure modes such as increased electrical resistance and drifting of electrical parameters.

2. *Corrosion is accelerated by electrical current.* As we saw earlier, corrosion is an electrochemical process in which the oxidation reaction generates metal ions and free electrons, and the reduction reaction consumes such electrons. When components are run with an electrical current, the reaction processes take place at a faster pace. As a result, the rate of corrosion process is increased. Indeed, electrical current is a frequently used accelerating stress in accelerated corrosion tests.

3. *Electrical current generates magnetic fields.* These fields interfere with neighboring electronic components, an effect known as *electromagnetic interference* (EMI).

7.3.6 Mechanical Vibration

In a physical sense, *mechanical vibration* is a limited reciprocating motion of an object in alternately opposite directions from its position of equilibrium. Many products are subject to vibration in normal use. For example, all automobile components experience this stress while the vehicle is operated on the road. The same happens to the components of a flying airplane. Not only is vibration ubiquitous, but it is fierce in producing failures. Consequently, engineering standards recommend vibration tests. For example, MIL-STD-810F (U.S. DoD, 2000) and MIL-STD-202G (U.S. DoD, 2002) specify various vibration test conditions.

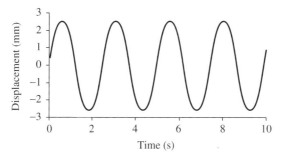

FIGURE 7.8   Sinusoidal vibration

Two types of vibration are common: sinusoidal and random vibrations. *Sinusoidal vibration* takes place at a predominant frequency, and displacement at a future time is predictable. This vibration is measured by the frequency (Hz) and displacement (mm), velocity (mm/s), or acceleration (mm/s^2 or g). Figure 7.8 shows an example of vibration at a frequency of 0.406 Hz, where the y-axis is the displacement. In reality, this type of vibration is usually caused by the cyclic operation of a product. For example, automotive engine firing is a source of sinusoidal vibration, which disturbs components under the hood. Most products are more likely to experience the second type of vibration, *random vibration*. In contrast to a sinusoidal vibration, a random vibration occurs in a wide range of frequencies, and instantaneous displacement at a future time is unpredictable. Figure 7.9 shows a 5-second random vibration where the y-axis is acceleration. Because of the random nature, the vibration is described by the power spectral density (PSD), expressed in g^2/ Hz. Since the PSD is a function of frequency, a random vibration profile should specify the PSD at various values of frequency. Figure 7.10 shows an example of such a profile, which is the vibration test condition for an automobile component. Steinberg (2000), for example, discusses mechanical vibration in detail.

In some circumstances, vibration is generated purposefully to fulfill certain functions. For example, ultrasonic vibration welds two parts in the wire bonding process. In most situations, vibration results in undesirable effects such as fatigue,

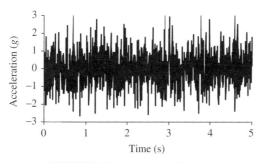

FIGURE 7.9   Random vibration

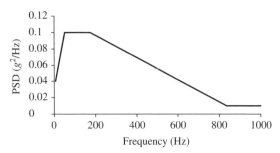

FIGURE 7.10   Random vibration test profile

wear, and loosening of connections. Due to the change in acceleration over time, vibration generates a cyclic load. As discussed earlier, cyclic stressing initiates and develops microcracks and eventually causes a fatigue failure. Vibration also induces mechanical wear, which is the attrition of materials from the surfaces between two mating components in relative movement. Mechanical wear can be adhesive, abrasive, fretting, or a combination. The wear mechanisms of each type are described in books on tribology and wear. Interested readers may consult, for example, Stachowiak and Batchelor (2000) and Bhushan (2002). Excessive wear, in turn, causes different apparent failure modes, including acoustic noise, worse vibration, local overheating, leaking, and loss of machinery precision. A loosening connection is another failure mode that is frequently observed in a vibration environment. This failure mode can result in various effects, such as leaking, deterioration of connection strength, and intermittent electrical contact.

7.4 LIFE–STRESS RELATIONSHIPS

The primary purpose of a quantitative ALT is to estimate the life distribution at a use condition. This can be accomplished by extrapolating the life data obtained at elevated stress levels. To do this, we need a model that relates life to accelerating stress, such as temperature, humidity, and voltage. Such models, usually called *acceleration models*, can be classified into the following three types:

1. *Physical models.* In a few situations we understand well how a material or component responds to applied stresses at the micro level, and how a failure process develops over time under stress. Equipped with the knowledge, we use established models that relate life to the applied stresses. The models are usually complicated, because many factors contribute simultaneously to the initiation and development of failure mechanisms. Justifiable simplification of models is acceptable, and often essential. Proper physical models provide a high level of accuracy in extrapolation. Of course, the elevated stresses must be selected appropriately so that the failure mechanisms at higher stress levels are the same as those at the use conditions. Because such a model is suitable for a specific failure mechanism, it is usually invalid for other failure mechanisms, even in the same product.

2. *Quasiphysical models.* This type of model is not based directly on specific failure mechanisms that govern the failure process of concern. However, such a model either has roots in known physical or chemical theories, or is grounded on macro-level failure mechanisms. Most commonly used acceleration models, such as the Arrhenius relationship (discussed later), belong to this type. Because the models are not derived from specific failure mechanisms, they have more applications than do physical models. Generally, these models provide better extrapolation accuracy than that of empirical models.

3. *Empirical models.* In many situations we have little knowledge of the physical or chemical reactions taking place in a material or component under the applied stresses. Without understanding the failure mechanisms, it is impossible to develop a physical model. Instead, we fit empirical models to experimental data by using linear or nonlinear regression methods. A typical example is the polynomial model. Such models may be adequate in fitting the existing data, but extrapolation to use conditions is risky.

7.4.1 Life–Temperature Models

As explained earlier, temperature is widely used in ALTs. It is effective in stimulating certain failure mechanisms and shortening times to failure. In many applications, the dependence of life on temperature can be well described by the Arrhenius and Eyring relationships.

Arrhenius Relationship The Arrhenius relationship models the effect of temperature on the rate of a first-order chemical reaction and can be written as

$$v = A_0 \exp\left(-\frac{E_a}{kT}\right), \tag{7.4}$$

where v is the chemical reaction rate in moles per second, E_a the activation energy in electron-volts (eV), k is Boltzmann's constant ($k = 8.6171 \times 10^{-5}$ eV/°C), T the absolute temperature (the Celsius temperature plus 273.15 degrees), and A_0 is a constant related to material characteristics.

The rate of chemical reaction is the amount of a reactant reacted per unit time. Assume that a failure occurs when a critical amount (in moles) of reactant reacted is reached. Then the time to reach the critical amount is the time to failure. Since the time to failure is proportional to the reciprocal of the reaction rate, (7.4) can be written as

$$L = A \exp\left(\frac{E_a}{kT}\right), \tag{7.5}$$

where L is the life and A is a constant that depends on material properties, failure criteria, product design, and other factors. Equation (7.5) is called the *Arrhenius life relationship.* Life here is the nominal life, and may represent a life percentile. For example, it can be the median life of the lognormal and normal distributions, the characteristic life of the Weibull distribution, and the mean life of the exponential distribution.

For the sake of data analysis, (7.5) is linearized. Then

$$\ln(L) = a + \frac{b}{T}, \quad (\text{taking logs}) \quad (7.6)$$

where $a = \ln(A)$ and $b = E_a/k$. Equation (7.6) indicates that the natural logarithm of nominal life is a linear function of the reciprocal temperature. The transformed linearity provides a great convenience for fitting the relationship to experimental data and visually checking the goodness of fit.

Activation energy is an important concept associated with the Arrhenius relationship. We understand that a chemical reaction is the result of the collisions between the reactant molecules. The collisions take place very frequently, but only a small fraction of the collisions convert reactants into products of the reaction. The necessary condition for the collision to cause a reaction is that the molecules must carry a minimum amount of energy to break bonds and form products. The minimum amount of energy is called the activation energy; it poses a barrier for molecules to climb over. The higher the activation energy, the lower the reaction rate and the longer the life. Activation energy is unique to a reaction, which determines the failure mechanism leading to failure. Therefore, each failure mechanism usually has a different activation energy even for the same component. For most mechanisms in electronic components or devices, the activation energy is in the range between 0.3 and 1.5 eV.

Since the Arrhenius life can represent a percentile, the acceleration factor A_f between the life L at temperature T and the life L' at temperature T' is

$$A_f = \frac{L}{L'} = \exp\left[\frac{E_a}{k}\left(\frac{1}{T} - \frac{1}{T'}\right)\right], \quad (7.7)$$

which indicates that the acceleration factor increases with the activation energy. Roughly, as the temperature increases every $10°C$, the life would reduce approximately one-half, known as the *10°C rule*. As technology progresses, some products have gained a greater immunity to temperature. Applicability of the 10°C rule to these products may be seriously questionable.

Example 7.1 In the robust reliability design of an electronic sensor, temperature was identified as the key noise factor. The levels of the noise factor used in testing were 85, 100, and 115°C. In each setting of the design parameters, two units were tested to failure at each temperature. The times to failure in hours for a setting are shown in Table 7.1. We use the Arrhenius relationship to model the dependence of life on temperature. Estimate the activation energy, the mean life at the use temperature of 35°C, and the acceleration factor between the lives at 35 and 70°C.

SOLUTION   First, we calculate the mean life at each temperature, which is shown in Table 7.1. Then (7.6) is used to fit the relationship between mean life and temperature. We perform the linear regression analysis using Minitab (a commercial statistical and reliability software) and obtain estimates of a and b as

TABLE 7.1 Life Data at Different Temperatures

	Temperature (°C)		
	85	100	115
Life (h)	2385, 2537	1655, 1738	1025, 1163
Mean Life (h)	2461	1696.5	1094

$\hat{a} = -2.648$ and $\hat{b} = 3750.636$. Since $b = E_a/k$, the estimate of the activation energy is

$$\hat{E}_a = k\hat{b} = 8.6171 \times 10^{-5} \times 3750.636 = 0.323 \text{ eV}.$$

The activation energy is relatively small, indicating that the setting of the design parameters is probably not optimal.

From (7.6), the mean life at 35°C is estimated by

$$\hat{L} = \exp\left(-2.648 + \frac{3750.636}{35 + 273.15}\right) = 13,677 \text{ hours.}$$

The estimate of the acceleration factor between the mean lives at 35 and 70°C is

$$\hat{A}_f = \exp\left[\frac{0.323}{8.6171 \times 10^{-5}}\left(\frac{1}{35 + 273.15} - \frac{1}{70 + 273.15}\right)\right] = 3.5.$$

The acceleration factor can be roughly interpreted as meaning that testing a sensor at 70°C for 1 hour is equivalent to testing the sensor at 35°C for 3.5 hours. In other words, if a sensor failed in 1 hour at 70°C, the life of the sensor would have been 3.5 hours at 35°C. Similarly, if a sensor ran 1 hour without failure at 70°C, the sensor would have survived 3.5 hours at 35°C.

The Arrhenius relationship has been widely used for decades. Some recent applications are in, for example, medical devices (Jiang et al., 2003), lithium ion cells (Broussely et al., 2001), petroleum-based ferrofluid (Segal et al., 1999), and motor insulation systems (Oraee, 2000). But note that the Arrhenius relationship is not universally applicable to all cases where temperature is an accelerating stress. Some examples are reported in Gillen et al. (2005) on a commercial chloroprene rubber cable jacketing material, and in Dimaria and Stathis (1999) on ultrathin silicon dioxide film. It is important to check the adequacy of the model by using the test data.

Eyring Relationship In some applications, the Arrhenius relationship does not adequately describe the dependence of life L on temperature. Instead, the *Eyring relationship*, derived from quantum mechanics, may be more appropriate. The relationship is

$$L = \frac{A}{T}\exp\left(\frac{E_a}{kT}\right), \tag{7.8}$$

where the notation is the same as in (7.5). Compared with the Arrhenius relationship, the Eyring model has an additional term, $1/T$. Hence, it may be more suitable when the temperature has stronger effects on the reaction rate. Despite the advantage, it has few applications in the literature.

The acceleration factor between temperatures T and T' for the Eyring relationship is

$$A_f = \frac{T'}{T} \exp\left[\frac{E_a}{k}\left(\frac{1}{T} - \frac{1}{T'}\right)\right], \tag{7.9}$$

which indicates that the acceleration factor for the Eyring relationship is T'/T times the acceleration factor for the Arrhenius relationship.

7.4.2 Life–Thermal Cycling Models

Although thermal cycling is a temperature stress, it usually stimulates failure modes different from those caused by a constant temperature, as discussed earlier. The Coffin–Manson relationship and its generalized form are often used to model the effects of thermal cycling. Nachlas (1986) proposes a different general model.

(Inverse Model).

Coffin–Manson Relationship   The life of a product subjected to thermal cycling is often measured by cycles to failure. Coffin (1954) and Manson (1966) give their relationship between the nominal number L of cycles to failure and the temperature range as

$$L = \frac{A}{(\Delta T)^B}, \tag{7.10}$$

where ΔT is the temperature range $T_{max} - T_{min}$ and A and B are constants characteristic of material properties and product design. B is usually positive. In some applications, A may be a function of cycling variables such as the frequency and maximum temperature, in which case the Norris–Landzberg relationship discussed next is more appropriate. We will see later that (7.10) is a special form of the inverse power relationship.

For the sake of data analysis, we transform (7.10) into a linear function. Taking the natural logarithm of (7.10) gives

$$\ln(L) = a + b\ln(\Delta T), \tag{7.11}$$

where $a = \ln(A)$ and $b = -B$. If A is independent of thermal cycling variables, the acceleration factor between two temperature ranges ΔT and $\Delta T'$ is

$$A_f = \left(\frac{\Delta T'}{\Delta T}\right)^B. \tag{7.12}$$

The Coffin–Manson relationship was developed to describe fatigue failure of metal subjected to thermal cycling. Since then it has been widely used for mechanical and electronic components. The model is a variant of the $S-N$ curve, which describes the relationship between the number (N) of cycles to failure and the strain (S). Recent applications of the model include Naderman and Rongen (1999), Cory (2000), Sumikawa et al. (2001), Basaran et al. (2004), R. Li (2004), and many others.

Norris–Landzberg Relationship   The Coffin–Manson relationship assumes implicitly that fatigue life depends only on the temperature range of a thermal cycle. In some applications, fatigue life is also a function of the cycling frequency and high temperature, as shown in, for example, Ghaffarian (2000), Teng and Brillhart (2002), and Shohji et al. (2004). Taking into account the effects of these thermal cycling variables, Norris and Landzberg (1969) modify the conventional Coffin–Manson relationship and propose

$$L = A(\Delta T)^{-B} f^C \exp\left(\frac{E_a}{kT_{\max}}\right), \tag{7.13}$$

where L is the nominal number of cycles to failure, A, B, and C are constants characteristic of material properties and product design and failure criteria, $T_{\max}$ is the high absolute temperature, f is the cycling frequency, and E_a, k, and ΔT have the same meanings as in (7.5) and (7.10). The unit of f may be cycles per hour, cycles per day, or another, whichever is more convenient or customary. Note that as the cycling frequency increases, the fatigue life increases when $C > 0$, decreases when $C < 0$, and does not change when $C = 0$. This provides the relationship with great flexibility for accommodating a variety of effects of frequency. Equation (7.13) having roots in (7.10) is sometimes called the *modified Coffin–Manson relationship*.

For the convenience of data analysis, we transform (7.13) into a linear function: namely,

$$\ln(L) = a + b\ln(\Delta T) + c\ln(f) + \frac{d}{T_{\max}}, \tag{7.14}$$

where $a = \ln(A)$, $b = -B$, $c = C$, and $d = E_a/k$. The unknown coefficients a, b, c, and d may be estimated by using the multiple linear regression method, which is described in, for example, Hines et al. (2002). Various commercial software packages such as Minitab can be used for the calculations.

The acceleration factor between two thermal cycling profiles is

$$A_f = \left(\frac{\Delta T'}{\Delta T}\right)^B \left(\frac{f}{f'}\right)^C \exp\left[\frac{E_a}{k}\left(\frac{1}{T_{\max}} - \frac{1}{T'_{\max}}\right)\right], \tag{7.15}$$

where a prime denotes an acceleration profile. In calculation, E_a/k may be replaced by d, whose value is estimated in linear regression analysis.

Example 7.2   Shohji et al. (2004) evaluate the reliability of chip-scale package (CSP) solder joints by subjecting them to thermal cycling, where the solder joints

TABLE 7.2 Thermal Cycling Profiles and Test Results for CSP Solder Joints

Group	$T_{\min}$ (°C)	$T_{\max}$ (°C)	ΔT (°C)	f (cycles/h)	Mean Life (cycles)
1	−40	80	120	1	208
2	−40	80	120	2	225
3	−40	80	120	3	308
4	−40	100	140	2	142
5	−40	120	160	2	108
6	−20	100	120	2	169
7	0	120	120	2	131
8	30	80	50	2	1300
9	30	100	70	2	650
10	30	120	90	2	258
11	−20	30	50	2	6231
12	−40	30	70	2	1450

are the alloy Sn−37Pb. In the experiment, 12 thermal cycling profiles were used as shown in Table 7.2. Under each test condition, five CSPs were tested, each with multiple solder joints. A CSP is said to have failed when one of its solder joints disconnects. The tests were run until all units failed. (Note that running all units to failure is generally a poor practice when we are interested in estimating the lower tail of the life distribution.) The mean life for a test profile is the average of the numbers of cycles to failure of the five units that underwent the same condition. The mean life data are also shown in Table 7.2. By using the Norris–Landzberg relationship, estimate the activation energy and the mean life under the use profile, where we assume that $T_{\min} = -10°C$, $T_{\max} = 25°C$, and $f = 1$ cycle per hour. Also calculate the acceleration factor between the use profile and the accelerating profile, where $T'_{\min} = -30°C$, $T'_{\max} = 105°C$, and $f' = 2$ cycles per hour.

SOLUTION   Equation (7.14) is fitted to the data. The multiple linear regression analysis was performed with Minitab. The analysis results are summarized in Table 7.3.

The large F value in the analysis of variance summarized in Table 7.3 indicates that there exists a transformed linear relationship between the mean life and at least some of the cycling variables. In general, (7.14) needs to be checked for lack of fit. Doing so usually requires repeated observations at the same test conditions (Hines et al., 2002). Such observations were not given in this paper (Shohji et al., 2004). The analysis in Table 7.3 also shows that the cycling frequency is not statistically significant due to its small T value, and may be excluded from the model. In this example, we keep this term and have

$$\ln(\hat{L}) = 9.517 - 2.064 \ln(\Delta T) + 0.345 \ln(f) + \frac{2006.4}{T_{\max}}. \qquad (7.16)$$

TABLE 7.3 Multiple Linear Regression Analysis Results from Minitab

```
The regression equation is
ln(Life) = 9.52 - 2.06 ln(DT) + 0.345 ln(f) + 2006 1/Tmax


Predictor        Coef      SE Coef          T         P
Constant        9.517        1.918       4.96     0.001
ln(DT)        -2.0635       0.2388      -8.64     0.000
ln(f)          0.3452       0.3091       1.12     0.296
1/Tmax(K)      2006.4        361.5       5.55     0.001


S = 0.2459        R-Sq = 97.1%      R-Sq(adj) = 96.0%


Analysis of Variance


Source            DF          SS         MS         F        P
Regression         3     16.1083     5.3694     88.81    0.000
Residual Error     8      0.4837     0.0605
Total             11     16.5920
```

Since $d = E_a/k$, the activation energy is $\hat{E}_a = 8.6171 \times 10^{-5} \times 2006.4 = 0.17$ eV. Substituting the use profile into (7.16) gives

$$\ln(\hat{L}) = 9.517 - 2.064 \ln(25 + 10) + 0.345 \ln(1) + \frac{2006.4}{25 + 273.15} = 8.908.$$

The mean life under the use profile is $\hat{L} = \exp(8.908) = 7391$ cycles. The estimates of parameters B and C are $\hat{B} = -\hat{b} = 2.064$ and $\hat{C} = \hat{c} = 0.345$. To estimate the acceleration factor between the use and accelerating profiles, we substitute the estimates of B, C, and E_a and the values of the profile variables into (7.15) and obtain

$$\hat{A}_f = \left(\frac{135}{35}\right)^{2.064} \left(\frac{1}{2}\right)^{0.345} \exp\left[2006.4 \times \left(\frac{1}{25+273.15} - \frac{1}{105+273.15}\right)\right] = 53.$$

7.4.3 Life–Voltage Relationship

Voltage is effective in accelerating various failure mechanisms discussed earlier. Thus, it is frequently used as an accelerating stress for products such as capacitors, transformers, and insulators. The effect of voltage on life is often modeled with an inverse power relationship. Some applications of the relationship include Montanari et al. (1988), Kalkanis and Rosso (1989), and Feilat et al. (2000). In a few situations, the dependence of life on voltage may be better explained by the exponential model (see, e.g., Yassine et al., 2000; Vollertsen and Wu, 2004). Here we discuss the inverse power relationship.

Inverse Power Relationship   The *inverse power relationship* can be written as

$$L = \frac{A}{V^B}, \tag{7.17}$$

where L is the nominal life, V the stress, and A and B are constants dependent on material properties, product design, failure criteria, and other factors. It is often used for the life of dielectrics subjected to voltage V. It is worth noting that the inverse power relationship may apply to a stress other than voltage, including mechanical load, pressure, electrical current, and some others. For example, Harris (2001) applies the relationship to the life of a bearing as a function of mechanical load, and Black (1969) expresses the median life to electromigration failure of microcircuit conductors as an inverse power function of the current density at a given temperature. In addition, the Coffin–Manson relationship (7.10) and the usage rate model (discussed later) are special cases of the inverse power relationship.

For the convenience of data analysis, we transform (7.17) into a linear relationship as

$$\ln(L) = a + b \ln(V), \tag{7.18}$$

where $a = \ln(A)$ and $b = -B$. Both a and b are estimated from test data.

The acceleration factor between two stress levels is

$$A_f = \left(\frac{V'}{V}\right)^B, \tag{7.19}$$

where the prime denotes higher stress.

Example 7.3   To evaluate the reliability of a type of surface-mounted electrolytic capacitor, three tests, each with eight units, were conducted at elevated voltage levels of 80, 100, and 120 V, respectively. All units were run to failure, where a failure is said to have occurred when the capacitance drifts more than 25%. The failure times in hours are shown in Table 7.4. Estimate the mean life at the rated voltage of 50 V. If a capacitor ran 1500 hours without failure at 120 V, calculate the equivalent time the capacitor would have survived at the rated voltage.

TABLE 7.4   Life Data at Elevated Voltages

	Voltage (V)		
	80	100	120
Life (h)	1770	1090	630
	2448	1907	848
	3230	2147	1121
	3445	2645	1307
	3538	2903	1321
	5809	3357	1357
	6590	4135	1984
	6744	4381	2331
Mean Life (h)	4197	2821	1362

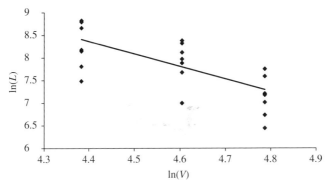

FIGURE 7.11 Scatter plot and regression line fitted to the mean life of the capacitors

SOLUTION The mean life at an elevated voltage is the average of the lifetimes at that voltage. The resulting mean lives are shown in Table 7.4. Then (7.18) is used to fit the mean life data at each voltage level. Simple linear regression analysis gives $\hat{a} = 20.407$ and $\hat{b} = -2.738$. The regression line and raw life data are plotted in Figure 7.11. The estimates of A and B are $\hat{A} = \exp(20.407) = 7.289 \times 10^8$ and $\hat{B} = 2.738$. The mean life at 50 V is $\hat{L}_{50} = 7.289 \times 10^8/50^{2.738} = 16,251$ hours.

The acceleration factor between 50 and 120 V is $\hat{A}_f = (120/50)^{2.738} = 10.99$. Then 1500 hours at 120 V is equivalent to $1500 \times 10.99 = 16,485$ hours at 50 V. That is, if a capacitor ran 1500 hours at 120 V without failure, the capacitor would have survived 16,485 hours at 50 V.

7.4.4 Life–Vibration Relationship

Vibration is sometimes used as an accelerating variable to accelerate fatigue failure for electronic and mechanical products. Often, the fatigue life L can be modeled with the inverse power relationship and can be written as

$$L = \frac{A}{G^B},\qquad(7.20)$$

where A and B are constants, and G represents G_{rms} (g), known as *root-mean-square acceleration*. G_{rms} equals the peak acceleration times 0.707 for a sinusoidal vibration and the square root of the area under the power spectral density (PSD, $g^2/$ Hz) for a random vibration. This relation is used in MIL-STD-810F (U.S. DoD, 2000), which gives the values of B for different types of products. For example, B takes a value of 4 for Air Force avionics under random vibration, and 6 under sinusoidal vibration. In general, B is estimated from test data.

7.4.5 Life–Usage Rate Relationship

Increasing usage rate is an acceleration method for some products that are operated at a low rate in the field, as discussed earlier. Increased usage rate may affect the usage to failure, where the usage is in cycles, revolutions, miles, or other measures. In other words, the usage to failure at different usage rates may not be the same. Some experimental results and theoretical explanations are shown in, for example, Popinceanu et al. (1977), Tamai et al. (1997), Harris (2001), and Tanner et al. (2002). G. Yang (2005) models nominal life as a power function of the usage rate. The model is written as

$$L = Af^B,$$ (7.21)

where L is the nominal usage to failure, f is the usage rate, and A and B are constants dependent on material properties, product design, failure criteria, and other factors. A may be a function of other stresses if applied simultaneously. For example, if test units are also subjected to a temperature stress, A may be a function of temperature, say, the Arrhenius relationship. Then (7.21) is extended to a combination model containing both the usage rate and temperature.

Increase in usage rate may prolong, shorten, or not change the usage to failure. Equation (7.21) is flexible in accommodating these different effects. In particular, the usage to failure decreases as the usage rate increases when $B < 0$, increases with usage rate when $B > 0$, and is not affected by usage rate when $B = 0$.

In testing a group of units, the usage rate is usually held constant over time. Then the nominal clock time τ to failure is given by

$$\tau = \frac{Af^B}{f} = Af^{B-1},$$ (7.22)

which indicates that (1) increasing usage rate in a test shortens the clock lifetime and test length when $B < 1$, (2) does not affect the clock lifetime and test length when $B = 1$, and (3) prolongs the clock lifetime and test length when $B > 1$. Clearly, the effectiveness of the acceleration method depends on the value of B. Acceleration is achieved only when $B < 1$. In reality, the value of B is unknown before testing. It can be preestimated using historical data, preliminary tests, engineering experience, or reliability handbooks such as MIL-HDBK-217F (U.S. DoD, 1995).

Note that (7.21) is a variant of the inverse power relationship. The linear transformation and the acceleration factor for the usage rate model are similar to those for the inverse power relationship. The linearized relationship is

$$\ln(L) = a + b\ln(f),$$ (7.23)

where $a = \ln(A)$ and $b = B$. When $B < 1$, the acceleration factor between two usage rates is

$$A_f = \frac{L}{L'} = \left(\frac{f}{f'}\right)^B,$$ (7.24)

where a prime denotes the increased usage rate. It is worth noting that $A_f < 1$ when $0 < B < 1$. This also indicates that usage to failure increases with the usage rate. Nevertheless, the clock time to failure is accelerated, and the test time is shortened.

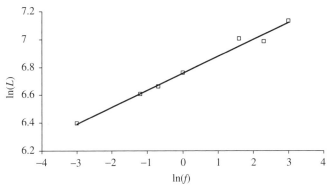

FIGURE 7.12   Regression line fitted to the mean life data of the micro relays

Example 7.4 Tamai et al. (1997) study the effects of switching rate on the contact resistance and life of micro relays. They report that the number of cycles to failure increases with the switching rate before a monolayer is formed. A sample of the relays was exposed to an environment containing silicon vapor at the concentration of 1300 ppm and loaded with 10 V and 0.5 A dc. The mean cycles to failure of the relays at switching rates of 0.05, 0.3, 0.5, 1, 5, 10, and 20 Hz are approximately 600, 740, 780, 860, 1100, 1080, and 1250 cycles, respectively, which were read from the charts in the paper. Estimate both the mean cycles to failure at a switching rate of 0.01 Hz, and the usage rate acceleration factor between 0.01 and 5 Hz for the given environment.

SOLUTION   Equation (7.23) is fitted to the mean cycles to failure and the switching rate. Simple linear regression analysis gives $\ln(\hat{L}) = 6.756 + 0.121 \ln(f)$. Hence, $\hat{L} = 859.45 f^{0.121}$. This regression line is shown in Figure 7.12, which suggests that (7.21) models the relationship adequately. The estimate of the mean cycles to failure at 0.01 Hz is $\hat{L} = 859.45 \times 0.01^{0.121} = 491$ cycles. The usage rate acceleration factor between 0.01 and 5 Hz is $\hat{A}_f = (0.01/5)^{0.121} = 0.47$. Note that the acceleration factor is less than 1. This indicates that the number of cycles to failure at 5 Hz is larger than the number at 0.01 Hz. However, the use of 5 Hz reduces the test clock time because $\hat{B} = 0.121 < 1$.

7.4.6   Life–Size Model

To meet a variety of customer demands, products are often fabricated in a number of sizes, and product size may affect the lifetime. For example, Brooks (1974) states that the life of a short piece of test cable is likely to be different from that of a long cable. Nelson (1990, 2004) gives more examples, including capacitor dielectric, conductors in microelectronics, and others, which appear to have failure rates proportional to product size. In practice, because of the size effect, we may purposefully test specimens larger (smaller) than the actual product to accelerate the test. Similar acceleration methods may be used by changing the level of other design control factors.

Motivated by Nelson (1990, 2004), Bai and Yun (1996) propose a relationship between failure rate and product size as

$$\lambda'(t) = \left(\frac{s'}{s}\right)^B \lambda(t), \tag{7.25}$$

where $\lambda(t)$ and $\lambda'(t)$ are the product failure rates at sizes s and s', respectively, and B is the size effect, a constant dependent on material properties, failure criteria, product design, and other factors. When $B = 1$, the model reduces to the one in Nelson (1990, 2004).

The size effect relationship is a special form of the proportional hazards model, which is due to Cox (1972) and discussed in, for example, Meeker and Escobar (1998) and Blischke and Murthy (2000). Since the model describes the effect of test condition (or size in this context) on failure rate rather than on lifetime, the life at the use condition is not simply the one at the test condition multiplied by an acceleration factor. Due to the complexity, the application of (7.25) to accelerated life tests is currently limited to a few situations. When the life of a product is modeled with the Weibull distribution, (7.25) can be written as

$$\frac{\alpha}{\alpha'} = \left(\frac{s'}{s}\right)^{B/\beta}, \tag{7.26}$$

where α and α' are the characteristic lives of the Weibull distribution at sizes s and s', respectively, and β is the common shape parameter. From (7.3), we see that (7.26) is the acceleration factor between the two sizes.

7.4.7 Life–Temperature and Nonthermal Stress Relationships

In an ALT, temperature is frequently applied simultaneously with a nonthermal stress such as humidity, voltage, electrical current, pressure, vibration, or mechanical load. The life relationship often is modeled as

$$L = \frac{A}{T} \exp\left(\frac{E_a}{kT}\right) \exp(BS) \exp\left(\frac{CS}{kT}\right), \tag{7.27}$$

where S is the nonthermal stress; A, B, and C are constants depending on material properties, failure criteria, product design, and other factors; and other notation is that of the Arrhenius relationship. Equation (7.27) is called the *generalized Eyring relationship*, where the last term with constant C models the interaction between T and S. The interaction term implies that the acceleration effect of temperature depends on the level of S, and vice versa. If the interaction is nonexistent, the last term is dropped off by setting $C = 0$. Readers may consult Chapter 5 for detecting interaction effects through analysis of variance or interaction plotting. In many applications, the first term $(1/T)$ is omitted and S is a transformation of the nonthermal stress V: for example, $S = \ln(V)$. The generalized Eyring relationship has variants, including life relationships with temperature and humidity, voltage, or current.

Life–Temperature and Humidity Relationship In many applications, humidity is applied along with high temperature in an accelerated life test of, for example, plastic packaging of electronic devices. Peck (1986) reviews a wide range of published test data available at that time for aluminum corrosion failures and proposes the relationship expressed as

$$L = \frac{A}{(\text{RH})^B} \exp\left(\frac{E_a}{kT}\right), \tag{7.28}$$

where RH is the relative humidity, A and B are constants, and other notation is that of the Arrhenius relationship. By analyzing the published data, Peck (1986) found values of B between -2.5 and -3.0, and values of E_a between 0.77 and 0.81 eV. Then Hallberg and Peck (1991) updated the values to $B = -3.0$ and $E_a = 0.9$ eV. Although the relationship is regressed from a limited number of products, it may be applicable to others and certainly has different parameter values. For example, the model fits test data adequately on gallium arsenide pseudomorphic high-electron-mobility transistor (GaAs pHEMT) switches and yields an estimate of $B = -10.7$ (Ersland et al., 2004).

Note that (7.28) can be derived from the generalized Eyring model by omitting the first and last terms and setting $S = \ln(RH)$. The logarithm of (7.28) gives

$$\ln(L) = a + b \ln(\text{RH}) + \frac{c}{T}, \tag{7.29}$$

where $a = \ln(A)$, $b = -B$, and $c = E_a/k$. The acceleration factor between the life at T and RH and the life at T' and $(\text{RH})'$ is

$$A_f = \left[\frac{(\text{RH})'}{\text{RH}}\right]^B \exp\left[\frac{E_a}{k}\left(\frac{1}{T} - \frac{1}{T'}\right)\right]. \tag{7.30}$$

Life–Temperature and Voltage Relationship In testing electronic and electrical products such as capacitors, resistors, diodes, microelectronic circuits, and dielectric insulators, temperature and voltage are frequently applied at the same time to increase the acceleration effect. The relationship between the life and the stresses is often modeled by

$$L = \frac{A}{V^B} \exp\left(\frac{E_a}{kT}\right) \exp\left[\frac{C \ln(V)}{kT}\right], \tag{7.31}$$

where V is the voltage, A, B, and C are constants, and other notation is that of the Arrhenius relationship. In practice, the last term is often assumed nonexistent if the interaction between temperature and voltage is not strongly evident (see, e.g., Mogilevsky and Shirn, 1988; Al-Shareef and Dimos, 1996). Then the resulting simplified relationship accounts for only the main effects of the stresses, which are described individually by the Arrhenius relationship and the inverse power relationship.

Life–Temperature and Current Relationship   Electrical current and temperature are sometimes combined to accelerate failure modes such as electromigration and corrosion. The relationship between the life and the combined stresses is often modeled by

$$L = \frac{A}{I^B} \exp\left(\frac{E_a}{kT}\right),$$ (7.32)

where I is the electrical current in amperes, A and B are constants, and other notation is that of the Arrhenius relationship. In the context of electromigration, I represents the current density in amperes per square unit length. Then (7.32) is called *Black's* (1969) *equation*, and it has been used extensively.

7.5 GRAPHICAL RELIABILITY ESTIMATION AT INDIVIDUAL TEST CONDITIONS

A constant-stress ALT consists of testing two or more groups of units under different conditions. Life data from each test condition are first analyzed individually to determine a suitable life distribution, to identify outliers in each data set, and to estimate the distribution characteristics of interest at the condition. Then life estimates at all test conditions are combined to estimate life at a use condition. In this section we focus on life data analysis at individual test conditions; reliability estimation at a use condition are discussed later. A simple, yet powerful life data analysis is the graphical method. Today, most commercial reliability software packages perform graphical analysis. They greatly reduce the time to generate graphs and estimates; however, it does not mean that we can simply treat the software as a black box. In fact, understanding the theoretical background is necessary for correct interpretation and use of software output. In this section we discuss the theory and application of graphical analysis for different types of data.

7.5.1 Censoring Mechanisms and Types of Data

Censoring mechanism and type of data are two important concepts in life data analysis. They are described here before the graphical and analytical methods are introduced.

Censoring Mechanisms   Often, tests must be terminated before all units fail. Such situations cause censoring. Censoring results in few data observations and increases statistical errors. When test resources such as time, equipment capacity, and personnel are restricted, we must use this method, albeit reluctantly, to shorten test time. In practice, there are three types of censoring: type I, type II, and random.

In *type I censoring*, also known as *time censoring*, a test is suspended when a predetermined time is reached on all unfailed units. That time is called the *censoring time*. In situations where product life is characterized by both time and usage, the censoring mechanism specifies the censoring time and usage. The test is terminated at the prespecified time or usage, whichever comes first. For example, automobiles tested in a proving ground are subject to time and mileage censoring, and a vehicle is removed from a test as soon as its accumulated time or mileage reaches the predetermined value. Type I censoring yields a random number of failures, which sometimes may be zero. It is important to ensure that the censoring time is long enough to fail some units; otherwise, data analysis is difficult or impossible. This type of censoring is common in practice, due to convenient time management.

Type II censoring, also called the *failure censoring*, results when a test is terminated when a prespecified number of failures is reached. This censoring method yields a fixed number of failures, which is appealing to the statistical data analysis. On the other hand, the censoring time is a random variable, which imposes a difficulty with time constraints. Because of this disadvantage, type II censoring is less common in practice.

Random censoring is the termination of a test at random. This type of censoring is often the result of an accident occurring during testing. For example, the failure of test equipment or damage to the sample causes suspension of a test. Random censoring also occurs when a unit fails from a mode that is not of interest. This type of censoring results in both random test time and a random number of failures.

Types of Data   ALTs may yield various types of data, depending on data collection methods and censoring methods. If the test units are monitored continuously during testing, the test yields the *exact life* when a unit fails. In contrast, test units are often inspected periodically during testing, and failures are not detected until inspection. Then the failure times are known to be between the times of the last and current inspections, and they are *interval life data*. As a special case, if a unit has failed before the first inspection time, the life of the unit is said to be *left censored*. In contrast, if a unit survives the censoring time, the life of the unit is *right censored*. If all surviving units have a common running time at test termination, their data are called *singly right censored*. For this type of data to occur, one needs to plan and conduct a test carefully. In practice, the censored units often have different running times. The data of such units are said to be *multiply right censored*. This situation arises when some units have to be removed from the test earlier or when the units are started on the test at different times and censored at the same time. If the censoring is long enough to allow all units to fail, the resulting failure times are *complete life data*. But this is usually poor practice for life tests, especially when only the lower tail of the life distribution is of interest.

7.5.2 Probability Plots

Graphical analysis of life data employs probability plots, which graphically display the relationship between time and the cumulative distribution function (cdf). As discussed in Chapter 2, such relationships are nonlinear for exponential, Weibull, normal, and lognormal distributions. For ease of plotting and visualization, data are plotted on probability paper which has special scales that linearize a cdf. If a life data set plots close to a straight line on Weibull probability paper, the Weibull distribution describes the population adequately. In general, a linearized cdf can be written as

$$y = a + bx, \tag{7.33}$$

where x and y are the transformed time and cdf, and a and b are related to the distribution parameters. Now let's work out the specific forms of a, b, x, and y for the commonly used distributions.

Exponential Distribution   The exponential cdf is

$$F(t) = 1 - \exp(-\lambda t),$$

where λ is the failure rate. This cdf is linearized and takes the form of (7.33), where $y = \ln[1/(1 - F)]$, $x = t$, $a = 0$, and $b = \lambda$. Exponential probability paper can be constructed with the transformed scale $\ln[1/(1 - F)]$ on the vertical axis and the linear scale t on the horizontal axis. Any exponential cdf is a straight line on such paper. If a data set plots near a straight line on this paper, the exponential distribution is a reasonable model.

The value of λ is the slope of the cdf line. Since $\lambda t = 1$ when $1 - F = e^{-1}$ or $F = 0.632$, the estimate of λ is equal to the reciprocal of the time at which $F = 0.632$.

Weibull Distribution   The Weibull cdf is

$$F(t) = 1 - \exp\left[-\left(\frac{t}{\alpha}\right)^{\beta}\right],$$

where α and β are the characteristic life and the shape parameter, respectively. This cdf is linearized and takes the form of (7.33), where $y = \ln\ln[1/(1 - F)]$, $x = \ln(t)$, $a = -\beta \ln(\alpha)$, and $b = \beta$. A Weibull probability paper has the transformed scale $\ln\ln[1/(1 - F)]$ on the vertical axis and $\ln(t)$ (a log scale) on the horizontal axis. The Weibull distribution adequately models a data set if the data points are near a straight line on the paper.

The parameters α and β can be estimated directly from the plot. Note that when $1 - F = e^{-1}$ or $F = 0.632$, $-\beta \ln(\alpha) + \beta \ln(t) = 0$ or $\alpha = t_{0.632}$. Namely, the value of the characteristic life is the time at which $F = 0.632$. The shape parameter is the slope of the straight line on the transformed scales. Some Weibull papers have a special scale for estimating β.

Normal Distribution   The cdf of the normal distribution is

$$F(t) = \Phi\left(\frac{t - \mu}{\sigma}\right),$$

where μ and σ are, respectively, the location and scale parameters or mean and standard deviation, and $\Phi(\cdot)$ is the cdf of the standard normal distribution. This cdf is linearized and takes the form of (7.33), where $y = \Phi^{-1}(F)$ and $\Phi^{-1}(\cdot)$ is the inverse of $\Phi(\cdot)$, $x = t$, $a = -\mu/\sigma$, $b = 1/\sigma$. Normal probability paper has a $\Phi^{-1}(F)$ scale on the vertical axis and the linear data scale t on the horizontal axis. On such paper any normal cdf is a straight line. If data plotted on such paper are near a straight line, the normal distribution is a plausible model.

The parameters μ and σ can be estimated from the plot. When $F = 0.5$, $t_{0.5} = \mu$. Thus, the value of the mean is the time at which $F = 0.5$. Similarly, when $F = 0.841$, $t_{0.841} = \mu + \sigma$. Then $\sigma = t_{0.841} - \mu$. Alternatively, σ can be estimated by the reciprocal of the slope of the straight line.

Lognormal Distribution   The cdf of the lognormal distribution is

$$F(t) = \Phi\left[\frac{\ln(t) - \mu}{\sigma}\right],$$

where μ and σ are the scale and shape parameters, respectively. This cdf is linearized and takes the form of (7.33), where $y = \Phi^{-1}(F)$, $x = \ln(t)$, $a = -\mu/\sigma$, and $b = 1/\sigma$. A lognormal probability plot has a $\Phi^{-1}(F)$ scale on the vertical axis and an $\ln(t)$ scale on the horizontal axis. The plot is similar to the plot for the normal distribution except that the horizontal axis here is the log scale. If the life data are lognormally distributed, the plot exhibits a straight line.

The median $t_{0.5}$ can be read from the time scale at the point where $F = 0.5$. Then the scale parameter is $\mu = \ln(t_{0.5})$. Similar to the normal distribution, when $F = 0.841$, $t_{0.841} = \exp(\mu + \sigma)$. Thus, $\sigma = \ln(t_{0.841}) - \mu$, where $t_{0.841}$ is read from the time scale at the point where $F = 0.841$. Alternatively, σ can be estimated by the reciprocal of the slope of the straight line. Here base e (natural) logarithms are used; base 10 logarithms are used in some applications.

7.5.3   Application of Probability Plots

In this subsection we present applications of probability plots to different types of data to assess a life distribution and estimate its parameters, percentiles, and probability of failure (population fraction failing). The data are complete exact, singly right-censored exact, multiply right-censored exact, or interval lifetimes.

Complete or Singly Right-Censored Exact Data   The probability of failure (population fraction failing) F at a failure time is unknown. To construct a probability plot, we usually approximate F with a plotting position. To obtain plotting positions, we first order the failure times from smallest to largest, say $t_{(1)} \leq t_{(2)} \leq \cdots \leq t_{(r)}$, where r is the number of failures. If r equals the sample size n, the data set is complete; otherwise, it is censored. The plotting position F_i for $t_{(i)}$ is

$$F_i = \frac{i - 0.5}{n}, \qquad (7.34)$$

where $i = 1, 2, \ldots, r$. The literature gives other plotting positions. Most statistical and reliability software packages provide multiple alternative positions. The plotting positions for censored units, if any, are not calculated and their times are not plotted. After calculating F_i for each failure, plot F_i versus $t_{(i)}$ on appropriate probability paper. A data set should be plotted on various probability papers if the type of distribution is not known from experience. The paper that gives the straightest plot is likely to be the best distribution. More important, the selection should be justified by the physics of failure. Today, most probability plotting is performed with reliability or statistical software packages, which generate probability plots and estimates of model parameters and of other quantities of interest, such as percentiles and the population fraction failing by a specified age (e.g., warranty or design life). The plotting process is illustrated below with a singly censored data set.

Example 7.5 To estimate the reliability of a type of small electronic module at a use temperature of 35°C, three groups of modules were tested: at 100, 120, and 150°C. The sample sizes at the three temperatures were 12, 8, and 10, respectively. The units allocated to 150°C were run to failure, whereas the tests at 100 and 120°C were time-censored at 5500 and 4500 hours, respectively. The life data are given in Table 7.5. Estimate the life distribution at each temperature. We will revisit this example later to estimate reliability at the use temperature.

SOLUTION   The data are singly censored on the right at 100 and 120°C, and complete at 150°C. We first analyze the data at 100°C. The failure times are ordered from smallest to largest, as shown in Table 7.5. The F_i for each $t_{(i)}$ is

TABLE 7.5   Failure Times of Electronic Modules

	Group		
	1	2	3
Temperature (°C)	100	120	150
Life (h)	1138	1121	420
	1944	1572	650
	2764	2329	703
	2846	2573	838
	3246	2702	1086
	3803	3702	1125
	5046	4277	1387
	5139	4500[a]	1673
	5500[a]		1896
	5500[a]		2037
	5500[a]		
	5500[a]		

[a] Censored age.

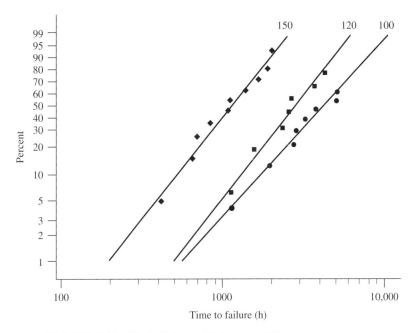

FIGURE 7.13 Weibull fits to lifetimes at different temperatures

calculated from (7.34), where $n = 12$, and $i = 1, 2, \ldots, 8$. Then F_i versus $t_{(i)}$ is plotted on lognormal and Weibull probability papers, respectively. The plot here was done using Minitab, which plots the percentage $100F_i$. The Weibull paper (Figure 7.13) gives the straightest plot. The software provided the estimates of the characteristic life and shape parameter: $\hat{\alpha}_1 = 5394$ hours and $\hat{\beta}_1 = 2.02$. Similarly, the failure times at 120 and 150°C yield straight Weibull plots in Figure 7.13. The estimates of the model parameters are $\hat{\alpha}_2 = 3285$ hours and $\hat{\beta}_2 = 2.43$ at 120°C, and $\hat{\alpha}_3 = 1330$ hours and $\hat{\beta}_3 = 2.41$ at 150°C. The software calculated these estimates using the least squares method. As we will see later, maximum likelihood method yields more accurate estimates.

Multiply Right-Censored Exact Data For multiply censored data, the plotting position calculation is more complicated than that for complete or singly censored data. Kaplan and Meier (1958) suggest a product-limit estimate given by

$$F_i = 1 - \prod_{j=1}^{i} \left(\frac{n-j}{n-j+1} \right)^{\delta_j}, \tag{7.35}$$

where n is the number of observations, i the rank of the ith ordered observation, and δ_j the indicator. If observation j is censored, $\delta_j = 0$; if observation j is uncensored, $\delta_j = 1$. Other plotting positions may be used; some software packages (e.g., Minitab) provide multiple options, including this Kaplan–Meier position. The plotting procedures are the same as those for complete or singly censored data, and are illustrated in Example 7.6.

Interval Data   Often, test units are not monitored continuously during testing, due to technological or economic limitations; rather, they are inspected periodically. Then failures are not detected until inspection. Thus, the failure times do not have exact values; they are interval data. Let t_i ($i = 1, 2, \ldots, m$) denote the ith inspection time, where m is the total number of inspections. Then the intervals preceding the m inspections are $(t_0, t_1], (t_1, t_2], \ldots, (t_{m-1}, t_m]$. Suppose that inspection at t_i yields r_i failures, where $0 \leq r_i \leq n$ and n is the sample size. The exact failure times are unknown; we spread them uniformly over the interval. Thus, the failure times in $(t_{i-1}, t_i]$ are approximated by

$$t_{ij} = t_{i-1} + j\frac{t_i - t_{i-1}}{r_i + 1}, \qquad i = 1, 2, \ldots, m; j = 1, 2, \ldots, r_i, \qquad (7.36)$$

where t_{ij} is the failure time of unit j failed in interval i. Intuitively, when only one failure occurs in an interval, the failure time is estimated by the midpoint of the interval. After each failure is assigned a failure time, we perform the probability plotting by using the approximate exact life data, where the plotting position is determined by (7.34) or (7.35), depending on the type of censoring. We illustrate the plotting procedures in the following example.

Example 7.6   A sample of 10 automobile transmission parts was tested at a high mechanical load representing the 90th percentile of the customer usage profile. The test yielded a critical failure mode in low cycles and led to a design change. To evaluate the effectiveness of the fix, 12 redesigned parts underwent a test at the same load. In both tests the parts were inspected every 20,000 cycles; any failed parts were removed from test, and the test continued until the predetermined number of cycles was accumulated. The life intervals are summarized in Table 7.6. Estimate the life distributions of the critical failure mode before and after the design change, and draw a conclusion about the effectiveness of the fix.

SOLUTION   As the data indicate, test units 4 and 7 of the "before" group have a failure mode different from the critical one of concern. This is so for units 1 and 2 of the "after" group. They are considered as censored units in the subsequent data analysis, because the critical modes observed would have occurred later. In addition, the data of both groups are censored on the right.

To estimate the two life distributions, we first approximate the life of each failed unit by using (7.36), as shown in Table 7.6. The approximate lives are treated as if they were exact. The corresponding plotting positions are calculated using (7.35) and presented in Table 7.6. Since the Weibull distribution is suggested by historical data, the data are plotted on Weibull probability paper. The plot in Figure 7.14 was produced with Minitab. Figure 7.14 suggests that the distribution is adequate for the test data. The least squares estimates of the Weibull

TABLE 7.6   Transmission Part Life Data (10^5 Cycles)

	Before			After		
Unit	Life Interval	Approximate Life	Plotting Position	Life Interval	Approximate Life	Plotting Position
1	(0.4, 0.6)	0.50	0.10	(2.4, 2.6]	2.50^a	
2	(1.2, 1.4]	1.27	0.20	(3.2, 3.4]	3.30^a	
3	(1.2, 1.4]	1.33	0.30	(3.8, 4.0]	3.90	0.10
4	(1.8, 2.0]	1.90^a		(4.2, 4.4]	4.30	0.20
5	(2.4, 2.6]	2.47	0.42	(5.0, 5.2]	5.07	0.30
6	(2.4, 2.6]	2.53	0.53	(5.0, 5.2]	5.13	0.40
7	(3.2, 3.4]	3.30^a		(6.2, 6.4]	6.30	0.50
8	(3.8, 4.0]	3.90	0.69	(7.6, 7.8]	7.70	0.60
9	(4.6, 4.8]	4.70	0.84	(8.4, 8.6]	8.47	0.70
10	(4.8, ∞)	4.80^a		(8.4, 8.6]	8.53	0.80
11				(8.8, ∞)	8.80^a	
12				(8.8, ∞)	8.80^a	

^{a}Censoring due to either termination of the test or the occurrence of a different failure mode.

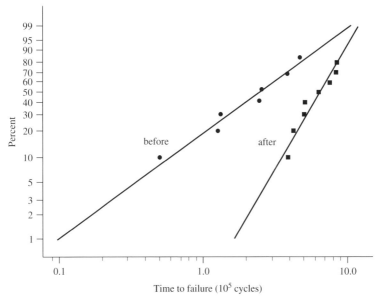

FIGURE 7.14   Weibull plots for the transmission part life data

parameters before design change are $\hat{\alpha}_B = 3.29 \times 10^5$ cycles and $\hat{\beta}_B = 1.30$. The design engineers were interested in the B_{10} life, which is estimated from (2.25) as

$$\hat{B}_{10,B} = 3.29 \times 10^5 [-\ln(1 - 0.10)]^{1/1.30} = 0.58 \times 10^5 \text{ cycles.}$$

For the after group, $\hat{\alpha}_A = 7.33 \times 10^5$ cycles and $\hat{\beta}_A = 3.08$. The B_{10} estimate is

$$\hat{B}_{10,A} = 7.33 \times 10^5 [-\ln(1 - 0.10)]^{1/3.08} = 3.53 \times 10^5 \text{ cycles}.$$

The design change greatly increased the B_{10} life. Figure 7.14 shows further that the life of the after group is considerably longer than that of the before group, especially at the lower tail. Therefore, it can be concluded that the fix is effective in delaying the occurrence of the critical failure mode.

7.6 ANALYTICAL RELIABILITY ESTIMATION AT INDIVIDUAL TEST CONDITIONS

Although the graphical approaches presented earlier are simple, analytical methods are frequently needed to obtain more accurate estimates. In this section we discuss the maximum likelihood (ML) method and its application to estimating the parameters of different life distributions, including the exponential, Weibull, normal, and lognormal, with various types of data. The ML calculations are complicated in most situations and require the use of numerical algorithms. Fortunately, a number of commercial software packages are now available for performing that laborious work. With estimated distribution parameters, we can estimate the percentiles, probability of failure (population fraction failing), and other quantities of interest.

7.6.1 Likelihood Functions for Different Types of Data

The sample likelihood function can be perceived as the joint probability of the data observed. Suppose that a sample of size n is drawn from a population with the probability density function $f(t; \theta)$, where θ is the model parameter (θ may be a vector of parameters). The sample yields n independent observations (exact failure times), denoted $t_1, t_2, \ldots, t_n$. Since the failure time is a continuous random variable, the probability of it taking an exact value is zero. The probability that an observation t_i occurs in a small time interval Δt equals $f(t_i; \theta)\Delta t$. Then the joint probability of observing $t_1, t_2, \ldots, t_n$ is

$$l(\theta) = \prod_{i=1}^{n} f(t_i; \theta)\Delta t, \qquad (7.37)$$

where $l(\theta)$ is called the *likelihood function*. Since Δt does not depend on θ, the term can be omitted in subsequent estimation of the model parameter(s). Then (7.37) simplifies to

$$l(\theta) = \prod_{i=1}^{n} f(t_i; \theta). \qquad (7.38)$$

For the sake of numerical calculation, the log likelihood is used in applications. Then (7.38) is rewritten as

$$L(\theta) = \sum_{i=1}^{n} \ln[f(t_i; \theta)], \qquad (7.39)$$

where $L(\theta) = \ln[l(\theta)]$ is the log likelihood and depends on the model parameter(s) θ. The ML estimate of θ is the value of θ that maximizes $L(\theta)$. Sometimes, the estimate of θ is obtained by solving

$$\frac{\partial L(\theta)}{\partial \theta} = 0. \tag{7.40}$$

Other times, it is found by iteratively finding the value of θ that maximizes $L(\theta)$. The resulting estimate $\hat{\theta}$, which is a function of $t_1, t_2, \ldots, t_n$, is called the *maximum likelihood estimator* (MLE). If θ is a vector of k parameters, their estimators are determined by solving k equations each like (7.40) or by iteratively maximizing $L(\theta)$ directly. In most situations, the calculation requires numerical iteration and is done using commercial software. It is easily seen that the form of the log likelihood function varies with the assumed life distribution. Furthermore, it also depends on the type of data, because the censoring mechanism and data collection method (continuous or periodical inspection) affect the joint probability shown in (7.37). The log likelihood functions for various types of data are given below.

Complete Exact Data   As discussed earlier, such data occur when all units are run to failure and subjected to continuous inspection. The log likelihood function for such data is given by (7.39). Complete exact data yield the most accurate estimates.

Right-Censored Exact Data   When test units are time censored on the right and inspected continuously during testing, the observations are right-censored exact failure times. Suppose that a sample of size n yields r failures and $n - r$ censoring times. Let $t_1, t_2, \ldots, t_r$ denote the r failure times, and $t_{r+1}, t_{r+2}, \ldots, t_n$ denote the $n - r$ censoring times. The probability that censored unit i would fail above its censoring time t_i is $[1 - F(t_i; \theta)]$, where $F(t; \theta)$ is the cdf of $f(t; \theta)$. Then the sample log likelihood function is

$$L(\theta) = \sum_{i=1}^{r} \ln[f(t_i; \theta)] + \sum_{i=r+1}^{n} \ln[1 - F(t_i; \theta)]. \tag{7.41}$$

When the censoring times $t_{r+1}, t_{r+2}, \ldots, t_n$ are all equal, the data are singly censored data. If at least two of them are unequal, the data are multiply censored data.

Complete Interval Data   Sometimes all test units are run to failure and inspected periodically during testing, usually all with the same inspection schedule. The situation results in complete interval data. Let t_i ($i = 1, 2, \ldots, m$) be the ith inspection time, where m is the total number of inspections. Then the m inspection intervals are $(t_0, t_1], (t_1, t_2], \ldots, (t_{m-1}, t_m]$. Suppose that inspection at t_i detects r_i failures, where $0 \le r_i \le n$, $n = \sum_1^m r_i$, and n is the sample size. Since a failure is known to have occurred within an interval i with probability $[F(t_i; \theta) - F(t_{i-1}; \theta)]$, the sample log likelihood function is

$$L(\theta) = \sum_{i=1}^{m} r_i \ln[F(t_i; \theta) - F(t_{i-1}; \theta)]. \tag{7.42}$$

Right-Censored Interval Data   When test units are inspected at times $t_1, t_2, \ldots,$ t_m and some surviving units are removed from test at inspection (type I censoring), the data are right-censored interval data. As above, we denote by $(t_0, t_1], (t_1, t_2], \ldots, (t_{m-1}, t_m]$ the m inspection intervals, in which $r_1, r_2, \ldots, r_m$ failures occur, respectively. Suppose that d_i units are suspended immediately after inspection at time t_i. Then t_i is the censoring time of the d_i units. The total number of censored units is $\sum_1^m d_i = n - \sum_1^m r_i$, where n is the sample size. The sample log likelihood function is

$$L(\theta) = \sum_{i=1}^{m} r_i \ln[F(t_i; \theta) - F(t_{i-1}; \theta)] + \sum_{i=1}^{m} d_i \ln[1 - F(t_i; \theta)]. \tag{7.43}$$

If only the last inspection results in suspensions, that is, $d_1 = d_2 = \cdots = d_{m-1} = 0$ and $d_m \geq 1$, the test yields singly censored data. If units are censored at two or more different inspection times, the data are multiply censored.

7.6.2   Approximate Confidence Intervals

In Section 7.6.1 we presented various likelihood functions that will be used to obtain the ML point estimators of model parameters, which often have large statistical uncertainty. The estimators may or may not be close to the true values of the population parameters being estimated. Thus, often we are also interested in confidence intervals for the parameters. In general, confidence intervals may be constructed using approximate, analytical or bootstrap, approaches. In reliability analysis involving censored data, *analytical methods* are difficult. The *bootstrap approaches* are based on computer simulation and require intensive computation. But the two types of methods provide accurate or good approximate confidence intervals. Here, we describe the normal approximation method, which is relatively simple and works well when the number of failures is moderate to large (say, 15 or more). Most commercial software packages for reliability and statistical data analysis use this method.

In Section 7.6.1 we remarked that the model parameter θ may be a vector. Now suppose that θ denotes k parameters $\theta_1, \theta_2, \ldots, \theta_k$. The MLE of the parameters are sometimes obtained by solving k equations each like (7.40) and are denoted $\hat{\theta}_1, \hat{\theta}_2, \ldots, \hat{\theta}_k$. More often they are found by maximizing $L(\theta)$ directly. The steps for constructing a confidence interval for each parameter follow:

    1. Calculate all second partial derivatives of the sample log likelihood function with respect to the model parameters.

2. Form a symmetric matrix of the negative second partial derivatives: namely,

$$
\mathbf{I} = \begin{bmatrix}
-\dfrac{\partial^2 L}{\partial \theta_1^2} & -\dfrac{\partial^2 L}{\partial \theta_1 \partial \theta_2} & \cdots & -\dfrac{\partial^2 L}{\partial \theta_1 \partial \theta_k} \\
-\dfrac{\partial^2 L}{\partial \theta_2 \partial \theta_1} & -\dfrac{\partial^2 L}{\partial \theta_2^2} & \cdots & -\dfrac{\partial^2 L}{\partial \theta_2 \partial \theta_k} \\
\cdots & \cdots & \cdots & \cdots \\
-\dfrac{\partial^2 L}{\partial \theta_k \partial \theta_1} & -\dfrac{\partial^2 L}{\partial \theta_k \partial \theta_2} & \cdots & -\dfrac{\partial^2 L}{\partial \theta_k^2}
\end{bmatrix}.
\tag{7.44}
$$

The expectation of $\mathbf{I}$ is the well-known *Fisher information matrix*.

3. Evaluating (7.44) at $\theta_i = \hat{\theta}_i$ $(i = 1, 2, \ldots, k)$ gives the local estimate of the Fisher information matrix, denoted $\hat{\mathbf{I}}$.

4. Calculate the inverse of the local Fisher information matrix, denoted $\hat{\mathbf{I}}^{-1}$.

5. Determine the estimate of the variance for $\hat{\theta}_i$ $(i = 1, 2, \ldots, k)$ from the relationship given by

$$
\hat{\boldsymbol{\Sigma}} = \begin{bmatrix}
\hat{V}ar(\hat{\theta}_1) & \hat{C}ov(\hat{\theta}_1, \hat{\theta}_2) & \cdots & \hat{C}ov(\hat{\theta}_1, \hat{\theta}_k) \\
\hat{C}ov(\hat{\theta}_2, \hat{\theta}_1) & \hat{V}ar(\hat{\theta}_2) & \cdots & \hat{C}ov(\hat{\theta}_2, \hat{\theta}_k) \\
\cdots & \cdots & \cdots & \cdots \\
\hat{C}ov(\hat{\theta}_k, \hat{\theta}_1) & \hat{C}ov(\hat{\theta}_k, \hat{\theta}_2) & \cdots & \hat{V}ar(\hat{\theta}_k)
\end{bmatrix} = \hat{\mathbf{I}}^{-1},
\tag{7.45}
$$

where $\hat{\boldsymbol{\Sigma}}$ denotes the estimate of the asymptotic variance–covariance matrix $\boldsymbol{\Sigma}$.

6. The two-sided $100(1 - \alpha)\%$ confidence interval for model parameter θ_i is

$$
[\theta_{i,L}, \theta_{i,U}] = \hat{\theta}_i \pm z_{1-\alpha/2}\sqrt{\hat{V}ar(\hat{\theta}_i)}, \qquad i = 1, 2, \ldots, k,
\tag{7.46}
$$

where $\theta_{i,L}$ and $\theta_{i,U}$ are the lower and upper bounds, and $z_{1-\alpha/2}$ is the $100(1 - \alpha/2)$th standard normal percentile. Note that (7.46) assumes that $\hat{\theta}_i$ has a normal distribution with mean θ_i and standard deviation $\sqrt{\hat{V}ar(\hat{\theta}_i)}$. The normality may be adequate when the number of failures is moderate to large (say, 15 or more). The one-sided $100(1 - \alpha)\%$ confidence bound is easily obtained by replacing $z_{1-\alpha/2}$ with $z_{1-\alpha}$ and using the appropriate sign in (7.46). When θ_i is a positive parameter, $\ln(\hat{\theta}_i)$ may be better approximated using the normal distribution. The resulting positive confidence interval is

$$
[\theta_{i,L}, \theta_{i,U}] = \hat{\theta}_i \exp\left(\pm \frac{z_{1-\alpha/2}\sqrt{\hat{V}ar(\hat{\theta}_i)}}{\hat{\theta}_i}\right).
\tag{7.47}
$$

Often, we want to calculate the confidence interval for a reliability, failure probability, percentile, or other quantities as a function of $\theta_1, \theta_2, \ldots, \theta_k$. Let $g = g(\theta_1, \theta_2, \ldots, \theta_k)$ denote the quantity and $\hat{g} = g(\hat{\theta}_1, \hat{\theta}_2, \ldots, \hat{\theta}_k)$ be the estimate. After calculating $\hat{\boldsymbol{\Sigma}}$, we estimate the variance for the $\hat{g}$ as

$$\hat{V}ar(\hat{g}) \approx \sum_{i=1}^{k} \left(\frac{\partial g}{\partial \theta_i}\right)^2 \hat{V}ar(\hat{\theta}_i) + \sum_{i=1}^{k} \sum_{\substack{j=1 \\ i \neq j}}^{k} \left(\frac{\partial g}{\partial \theta_i}\right) \left(\frac{\partial g}{\partial \theta_j}\right) \hat{C}ov(\hat{\theta}_i, \hat{\theta}_j), \quad (7.48)$$

where the $\partial g / \partial \theta_i$ are evaluated at $\hat{\theta}_1, \hat{\theta}_2, \ldots, \hat{\theta}_k$. If the correlation between the parameter estimates is weak, the second term in (7.48) may be omitted.

The two-sided approximate $100(1 - \alpha)\%$ confidence interval for g is

$$[g_L, g_U] = \hat{g} \pm z_{1-\alpha/2} \sqrt{\hat{V}ar(\hat{g})}. \quad (7.49)$$

If g must be positive, we may construct the confidence interval for g based on the log transformation and obtain bounds similar to those in (7.47). Later in this section we illustrate the calculation of the confidence intervals with examples.

7.6.3 Exponential Distribution

In this subsection we describe the maximum likelihood methods for the simple exponential distribution and provide the MLEs and confidence intervals for quantities of interest. As presented earlier, the exponential pdf is

$$f(t) = \frac{1}{\theta} \exp\left(-\frac{t}{\theta}\right), \quad (7.50)$$

where θ is the mean life. The failure rate is $\lambda = 1/\theta$.

Complete Exact Data   The sample log likelihood for such data is obtained by substituting (7.50) into (7.39). Then we have

$$L(\theta) = -n \ln(\theta) - \frac{1}{\theta} \sum_{i=1}^{n} t_i. \quad (7.51)$$

The MLE of θ is

$$\hat{\theta} = \frac{1}{n} \sum_{i=1}^{n} t_i. \quad (7.52)$$

The MLE of the failure rate is $\hat{\lambda} = 1/\hat{\theta}$. The estimate of the $100p$th percentile, reliability, probability of failure (population fraction failing), or other quantities are obtained by substituting $\hat{\theta}$ or $\hat{\lambda}$ into the appropriate formula in Chapter 2.

The estimate of the variance of $\hat{\theta}$ is

$$\hat{V}ar(\hat{\theta}) = \frac{\hat{\theta}^2}{n}. \quad (7.53)$$

From (7.46), the two-sided approximate $100(1 - \alpha)\%$ confidence interval for θ is

$$[\theta_L, \theta_U] = \hat{\theta} \pm \frac{z_{1-\alpha/2}\hat{\theta}}{\sqrt{n}}. \tag{7.54}$$

The one-sided $100(1 - \alpha)\%$ confidence bound is easily obtained by replacing $z_{1-\alpha/2}$ with $z_{1-\alpha}$ and using the appropriate sign in (7.54).

The exact confidence interval for θ is

$$[\theta_L, \theta_U] = \left[\frac{2n\hat{\theta}}{\chi^2_{(1-\alpha/2);2n}}, \frac{2n\hat{\theta}}{\chi^2_{\alpha/2;2n}} \right], \tag{7.55}$$

where $\chi^2_{p;2n}$ is the $100p$th percentile of the χ^2 (chi-square) distribution with $2n$ degrees of freedom.

The confidence interval for failure rate λ is

$$[\lambda_L, \lambda_U] = \left[\frac{1}{\theta_U}, \frac{1}{\theta_L} \right].$$

The confidence interval for failure probability at a particular time is

$$[F_L, F_U] = \left[1 - \exp\left(-\frac{t}{\theta_U} \right), 1 - \exp\left(-\frac{t}{\theta_L} \right) \right].$$

Note that the confidence interval for F depends on t.

The confidence interval for the $100p$th percentile is

$$[t_{p,L}, t_{p,U}] = [-\theta_L \ln(1 - p), -\theta_U \ln(1 - p)].$$

Example 7.7 An electromechanical module is required to achieve an MTTF of $\theta = 15,000$ hours at $40°C$. In a design verification testing, 15 units were sampled and tested at $125°C$ to shorten the test time. It is known that the life distribution is exponential, and the acceleration factor between the two temperatures is 22.7. The failure times are 88, 105, 141, 344, 430, 516, 937, 1057, 1222, 1230, 1513, 1774, 2408, 2920, and 2952 hours. Determine if the design meets the reliability requirement at the 90% confidence level.

SOLUTION   From (7.52), the MLE of the MTTF at $125°C$ is

$$\hat{\theta}' = \frac{1}{15} \times (88 + 105 + \cdots + 2952) = 1175.8 \text{ hours,}$$

where the prime implies an accelerating condition.

The approximate lower 90% confidence bound on the MTTF at $125°C$ is

$$\theta'_L = \hat{\theta}' - \frac{z_{1-\alpha}\hat{\theta}'}{\sqrt{n}} = 1175.8 - \frac{1.282 \times 1175.8}{\sqrt{15}} = 786.6 \text{ hours.}$$

The lower 90% confidence bound at 40°C is $\theta_L = 22.7 \times 786.6 = 17{,}856$ hours. Since $\theta_L = 17{,}856 > 15{,}000$, we conclude that the design surpasses the MTTF requirement at the 90% confidence level.

Right-Censored Exact Data Suppose that r out of n units fail in a test and the remainder are censored on the right (type II censoring). The failure times are $t_1, t_2, \ldots, t_r$, and the censoring times are $t_{r+1}, t_{r+2}, \ldots, t_n$. Then the sum $\sum_1^n t_i$ is the total test time. Formulas (7.52) to (7.55) can be used for the right-censored exact data by replacing the sample size n with the number of failures r. The resulting formulas may apply to type I censoring, but the confidence interval derived from (7.55) is no longer exact.

Example 7.8 Refer to Example 7.7. Suppose that the design verification test has to be censored at 1100 hours. Determine if the design meets the MTTF requirement at the 90% confidence level by using the censored data.

SOLUTION In this example, $n = 15$ and $r = 8$. The failure times $t_1, t_2, \ldots, t_8$ are known, and $t_9 = t_{10} = \cdots = t_{15} = 1100$ hours. The MLE of the MTTF at 125°C is

$$\hat{\theta}' = \frac{1}{r} \sum_{i=1}^{n} t_i = \frac{1}{8} \times (88 + 105 + \cdots + 1057 + 7 \times 1100) = 1414.8 \text{ hours,}$$

where the prime denotes an accelerating condition. Since the number of failures is small, the normal-approximation confidence interval may not be accurate. We calculate the confidence interval from the chi-square distribution. The one-sided lower 90% confidence bound is

$$\theta_L' = \frac{2r\hat{\theta}'}{\chi_{(1-\alpha);2r}^2} = 2 \times 8 \times \frac{1414.8}{23.54} = 961.6 \text{ hours.}$$

The lower 90% confidence bound at 40°C is $\theta_L = 22.7 \times 961.6 = 21{,}828$ hours. The lower 90% confidence bound is greater than 15,000 hours. So we conclude that the design surpasses the MTTF requirement at the 90% confidence level. But note that the early censoring yields an optimistic estimate of the mean life as well as a lower confidence bound.

Interval Data Following the notation in (7.42), the sample log likelihood function for the complete interval data is

$$L(\theta) = \sum_{i=1}^{m} r_i \ln\left[\exp\left(-\frac{t_{i-1}}{\theta}\right) - \exp\left(-\frac{t_i}{\theta}\right)\right]. \tag{7.56}$$

Equating to zero the derivative of (7.56) with respect to θ does not yield a closed-form expression for θ. The estimate of θ is obtained by maximizing $L(\theta)$ through a numerical algorithm: for example, the Newton–Raphson method. The Solver of Microsoft Excel provides a convenient means for solving a small optimization problem like this. Most statistical and reliability software packages calculate this estimate. Confidence intervals for the mean life and other quantities may be computed as described earlier.

Using the notation in (7.43), we obtain the sample log likelihood function for right-censored interval data as

$$L(\theta) = \sum_{i=1}^{m} r_i \ln\left[\exp\left(-\frac{t_{i-1}}{\theta}\right) - \exp\left(-\frac{t_i}{\theta}\right)\right] - \frac{1}{\theta}\sum_{i=1}^{m} d_i t_i. \qquad (7.57)$$

Like (7.56), the estimate of θ is calculated by maximizing $L(\theta)$. The approximate normal confidence interval for the mean life does not have an explicit form but can be computed by following the procedures discussed in Section 7.6.2.

7.6.4 Weibull Distribution

In this subsection we discuss ML estimation of Weibull distribution parameters with complete and type I censored data. The confidence intervals for the parameters and other quantities of interest are also presented. The methods apply to type II censored data in an obvious way.

Presented earlier, the Weibull pdf is

$$f(t) = \frac{\beta}{\alpha^{\beta}} t^{\beta-1} \exp\left[-\left(\frac{t}{\alpha}\right)^{\beta}\right], \qquad (7.58)$$

where β is the shape parameter and α is the scale parameter or characteristic life.

Complete Exact Data   When the data are complete and exact, the sample log likelihood function is obtained by substituting (7.58) into (7.39). Then we have

$$L(\alpha, \beta) = \sum_{i=1}^{n}\left[\ln(\beta) - \beta \ln(\alpha) + (\beta - 1)\ln(t_i) - \left(\frac{t_i}{\alpha}\right)^{\beta}\right]. \qquad (7.59)$$

The estimates $\hat{\alpha}$ and $\hat{\beta}$ may be got by maximizing (7.59); the numerical calculation frequently uses the Newton–Raphson method, of which the efficiency and convergence depend on the initial values. Qiao and Tsokos (1994) propose a more efficient numerical algorithm for solving the optimization problem. Alternatively, the estimators can be obtained by solving the likelihood equations. To do this, we take the derivative of (7.59) with respect to α and β, respectively. Equating the derivatives to zero and further simplification yield

$$\frac{\sum_{i=1}^{n} t_i^{\beta} \ln(t_i)}{\sum_{i=1}^{n} t_i^{\beta}} - \frac{1}{\beta} - \frac{1}{n}\sum_{i=1}^{n} \ln(t_i) = 0, \qquad (7.60)$$

$$\alpha = \left(\frac{1}{n}\sum_{i=1}^{n} t_i^{\beta}\right)^{1/\beta}. \qquad (7.61)$$

Equation (7.60) contains only one unknown parameter β and can be solved iteratively to get $\hat{\beta}$ with a numerical algorithm. Farnum and Booth (1997) provide a good starting β value for the iteration. Once $\hat{\beta}$ is obtained, it is substituted into (7.61) to calculate $\hat{\alpha}$. The estimates may be heavily biased when the number of failures is small. Then correction methods provide better estimates. Thoman et al. (1969) tabulate bias correction coefficients for various values of the sample size and shape parameter. R. Ross (1994) formulates the correction factor for the estimate of the shape parameter as a function of the sample size. Hirose (1999) also provides a simple formula for unbiased estimates of the shape and scale parameters as well as the percentiles.

The estimate of the $100p$th percentile, reliability, failure probability (population fraction failing), or other quantities can be obtained by substituting $\hat{\alpha}$ and $\hat{\beta}$ into the corresponding formula in Chapter 2.

The two-sided $100(1 - \gamma)\%$ confidence intervals for α and β are

$$[\alpha_L, \alpha_U] = \hat{\alpha} \pm z_{1-\gamma/2}\sqrt{\hat{V}ar(\hat{\alpha})}, \tag{7.62}$$

$$[\beta_L, \beta_U] = \hat{\beta} \pm z_{1-\gamma/2}\sqrt{\hat{V}ar(\hat{\beta})}. \tag{7.63}$$

The estimates of these variances are computed from the inverse local Fisher information matrix as described in Section 7.6.2. The log transformation of $\hat{\alpha}$ and $\hat{\beta}$ may result in a better normal approximation. From (7.47), the approximate confidence intervals are

$$[\alpha_L, \alpha_U] = \hat{\alpha} \exp\left(\pm\frac{z_{1-\gamma/2}\sqrt{\hat{V}ar(\hat{\alpha})}}{\hat{\alpha}}\right), \tag{7.64}$$

$$[\beta_L, \beta_U] = \hat{\beta} \exp\left(\pm\frac{z_{1-\gamma/2}\sqrt{\hat{V}ar(\hat{\beta})}}{\hat{\beta}}\right). \tag{7.65}$$

A confidence interval for the probability of failure F at a particular time t can be developed by using (7.48) and (7.49), where $g = F(t; \alpha, \beta)$. A more accurate interval is

$$[F_L, F_U] = [G(w_L), G(w_U)], \tag{7.66}$$

where

$$[w_L, w_U] = \hat{w} \pm z_{1-\gamma/2}\sqrt{\hat{V}ar(\hat{w})}, \qquad \hat{w} = \hat{\beta} \ln\left(\frac{t}{\hat{\alpha}}\right),$$

$$\hat{V}ar(\hat{w}) = \left(\frac{\hat{\beta}}{\hat{\alpha}}\right)^2 \hat{V}ar(\hat{\alpha}) + \left(\frac{\hat{w}}{\hat{\beta}}\right)^2 \hat{V}ar(\hat{\beta}) - \frac{2\hat{w}}{\hat{\alpha}}\hat{C}ov(\hat{\alpha}, \hat{\beta}),$$

$$G(w) = 1 - \exp[-\exp(w)].$$

Here $G(w)$ is the cdf of the standard smallest extreme value distribution.

An approximate $100(1 - \gamma)\%$ confidence interval for the $100p$th percentile t_p is

$$[t_{p,L}, t_{p,U}] = \hat{t}_p \exp\left(\pm \frac{z_{1-\gamma/2}\sqrt{\hat{V}ar(\hat{t}_p)}}{\hat{t}_p}\right), \tag{7.67}$$

where

$$\hat{V}ar(\hat{t}_p) = \exp\left(\frac{2u_p}{\hat{\beta}}\right)\hat{V}ar(\hat{\alpha}) + \left(\frac{\hat{\alpha}u_p}{\hat{\beta}^2}\right)^2 \exp\left(\frac{2u_p}{\hat{\beta}}\right)\hat{V}ar(\hat{\beta})$$

$$- \left(\frac{2\hat{\alpha}u_p}{\hat{\beta}^2}\right)\exp\left(\frac{2u_p}{\hat{\beta}}\right)\hat{C}ov(\hat{\alpha}, \hat{\beta}),$$

$$u_p = \ln[-\ln(1 - p)].$$

Most commercial software packages calculate confidence intervals for α and β using (7.64) and (7.65), for F using (7.66), and for t_p using (7.67). For manual computation, we may use the following approximations due to Bain and Engelhardt (1991):

$$\hat{V}ar(\hat{\alpha}) \approx \frac{1.1087\hat{\alpha}^2}{n\hat{\beta}^2}, \qquad \hat{V}ar(\hat{\beta}) \approx \frac{0.6079\hat{\beta}^2}{n}, \qquad \hat{C}ov(\hat{\alpha}, \hat{\beta}) \approx \frac{0.2570\hat{\alpha}}{n}.$$

Right-Censored Exact Data   Suppose that r out of n test units fail and the remainder are censored on the right (type I censoring). The failure times are $t_1, t_2, \ldots, t_r$, and the censoring times are $t_{r+1}, t_{r+2}, \ldots, t_n$. The sample log likelihood is

$$L(\alpha, \beta) = \sum_{i=1}^{r}\left[\ln(\beta) - \beta\ln(\alpha) + (\beta - 1)\ln(t_i) - \left(\frac{t_i}{\alpha}\right)^{\beta}\right] - \sum_{i=r+1}^{n}\left(\frac{t_i}{\alpha}\right)^{\beta}. \tag{7.68}$$

Like (7.59) for the complete exact data, (7.68) does not yield closed-form solutions for $\hat{\alpha}$ and $\hat{\beta}$. The estimates may be obtained by directly maximizing $L(\alpha, \beta)$, or by solving the likelihood equations:

$$\frac{\sum_{i=1}^{n} t_i^{\beta}\ln(t_i)}{\sum_{i=1}^{n} t_i^{\beta}} - \frac{1}{\beta} - \frac{1}{r}\sum_{i=1}^{r}\ln(t_i) = 0, \tag{7.69}$$

$$\alpha = \left(\frac{1}{r}\sum_{i=1}^{n} t_i^{\beta}\right)^{1/\beta}. \tag{7.70}$$

When $r = n$ or the test is uncensored, (7.69) and (7.70) are equivalent to (7.60) and (7.61), respectively. Like the complete data, the censored data yield biased estimates, especially when the test is heavily censored (the number of failures is small). Bain and Engelhardt (1991), and R. Ross (1996), for example, present a bias correction.

The confidence intervals (7.64) to (7.66) for the complete exact data are equally applicable to the censored data here. In practice, the calculation is done with commercial software (see the example below). Bain and Engelhardt (1991) provide approximations to the variances and covariance of the estimates, which are useful when hand computation is necessary.

Example 7.9 Refer to Example 7.6. Use the ML method to reanalyze the approximate lifetimes. Like the graphical analysis in that example, treat the lifetimes as right-censored exact data here.

SOLUTION   The plots in Example 7.6 show that the Weibull distribution is adequate for the data sets. Now we use the ML method to estimate the model parameters and calculate the confidence intervals. The estimates may be computed by solving (7.69) and (7.70) on an Excel spreadsheet or a small computer program. Then follow the procedures in Section 7.6.2 to calculate the confidence intervals. Here the computation is performed with Minitab. For the "before" group, the ML parameter estimates are $\hat{\alpha}_B = 3.61 \times 10^5$ cycles and $\hat{\beta}_B = 1.66$. The approximate two-sided 90% confidence intervals are $[\alpha_{B,L}, \alpha_{B,U}] = [2.47 \times 10^5, 5.25 \times 10^5]$, and $[\beta_{B,L}, \beta_{B,U}] = [0.98, 2.80]$, which can be derived from (7.64) and (7.65), respectively. The corresponding B_{10} life is $\hat{B}_{10,B} = 0.93 \times 10^5$ cycles. Similarly, for the "after" group, Minitab gives $\hat{\alpha}_A = 7.78 \times 10^5$ cycles, $\hat{\beta}_A = 3.50$, $[\alpha_{A,L}, \alpha_{A,U}] = [6.58 \times 10^5, 9.18 \times 10^5]$, $[\beta_{A,L}, \beta_{A,U}] = [2.17, 5.63]$, and $\hat{B}_{10,A} = 4.08 \times 10^5$ cycles. Note that the ML estimates are moderately different from the graphical estimates. In general, the ML method provides better estimates. Despite the difference, the two estimation methods yield the same conclusion; that is, the design change is effective because of the great improvement in the lower tail performance. Figure 7.15 shows the two probability plots, each with the ML fit and the two-sided 90% confidence interval curves for percentiles. It is seen that the lower bound of the confidence interval for the after-fix group in lower tail is greater than the upper bound for the before-fix group. This confirms the effectiveness of the fix.

Interval Data   When all units are on the same inspection schedule $t_1, t_2, \ldots, t_m$, the sample log likelihood for complete interval data is

$$L(\alpha, \beta) = \sum_{i=1}^{m} r_i \ln \left\{ \exp\left[-\left(\frac{t_{i-1}}{\alpha}\right)^\beta \right] - \exp\left[-\left(\frac{t_i}{\alpha}\right)^\beta \right] \right\}, \qquad (7.71)$$

where r_i is the number of failures in the ith inspection interval $(t_{i-1}, t_i]$, and m is the number of inspections. The likelihood function is more complicated than that for the exact data; the corresponding likelihood equations do not yield closed-form estimates for the model parameters. So the estimates are obtained

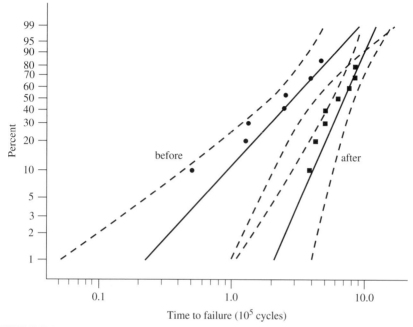

FIGURE 7.15   Weibull plots, ML fits, and percentile confidence intervals for the approximate exact life data of the transmission part

with numerical methods. This is also the case for the right-censored interval data, whose sample log likelihood function is

$$
L(\alpha, \beta) = \sum_{i=1}^{m} r_i \ln \left\{ \exp\left[-\left(\frac{t_{i-1}}{\alpha}\right)^{\beta} \right] - \exp\left[-\left(\frac{t_i}{\alpha}\right)^{\beta} \right] \right\} - \sum_{i=1}^{m} d_i \left(\frac{t_i}{\alpha}\right)^{\beta},
$$

(7.72)

where m, r_i, and t_i follow the notation in (7.71), and d_i is the number of units censored at inspection time t_i.

Confidence intervals for the interval data may be calculated from the formulas for exact data.

Example 7.10   In Examples 7.6 and 7.9, the lifetimes of the transmission part were approximated as the exact data. For the purpose of comparison, now we analyze the lifetimes as the interval data they really are. The data are rearranged according to the notation in (7.72) and are shown in Table 7.7, where only inspection intervals that result in failure, censoring, or both are listed. The inspection times are in 10^5 cycles.

For each group, the sample log likelihood function is obtained by substituting the data into (7.72). Then α and β can be estimated by maximizing $L(\alpha, \beta)$ through a numerical algorithm. Here Minitab performed the computation and gave the results in Table 7.8. The Weibull plots, ML fits, and percentile confidence intervals are depicted in Figure 7.16. There the plotted points are the

TABLE 7.7   Interval Data for the Transmission Part

	Before					After			
i	t_{i-1}	t_i	r_i	d_i	i	t_{i-1}	t_i	r_i	d_i
3	0.4	0.6	1		13	2.4	2.6		1
7	1.2	1.4	2		17	3.2	3.4		1
10	1.8	2.0		1	20	3.8	4.0	1	
13	2.4	2.6	2		22	4.2	4.4	1	
17	3.2	3.4		1	26	5.0	5.2	2	
20	3.8	4.0	1		32	6.2	6.4	1	
24	4.6	4.8	1	1	39	7.6	7.8	1	
					43	8.4	8.6	2	
					44	8.6	8.8		2

TABLE 7.8   Estimates from Different Methods and Data Types

	ML		Graphical: Approximate
Estimate	Interval Data	Approximate Exact Data	Exact Data
$\hat{\alpha}_B$	3.63×10^5	3.61×10^5	3.29×10^5
$\hat{\beta}_B$	1.65	1.66	1.30
$\hat{B}_{10,B}$	0.93×10^5	0.93×10^5	0.58×10^5
$[\alpha_{B,L}, \alpha_{B,U}]$	$[2.49, 5.29] \times 10^5$	$[2.47, 5.25] \times 10^5$	
$[\beta_{B,L}, \beta_{B,U}]$	$[0.98, 2.79]$	$[0.98, 2.80]$	
$\hat{\alpha}_A$	7.78×10^5	7.78×10^5	7.33×10^5
$\hat{\beta}_A$	3.50	3.50	3.08
$\hat{B}_{10,A}$	4.09×10^5	4.08×10^5	3.53×10^5
$[\alpha_{A,L}, \alpha_{A,U}]$	$[6.59, 9.19] \times 10^5$	$[6.58, 9.18] \times 10^5$	
$[\beta_{A,L}, \beta_{A,U}]$	$[2.17, 5.64]$	$[2.17, 5.63]$	

upper endpoints of the inspection intervals. For comparison, Table 7.8 includes the graphical estimates from Example 7.6 and those from the ML analysis of the approximate exact data in Example 7.9. Comparison indicates that:

- The ML and graphical estimates differ moderately. In general, the ML method yields more accurate results and should be used when software for ML calculations is available.

- The ML analyses based on the approximate exact data and the interval data provide close estimates. This is generally so when the inspection intervals are relatively short compared to the distribution width. The difference increases as the intervals widen.

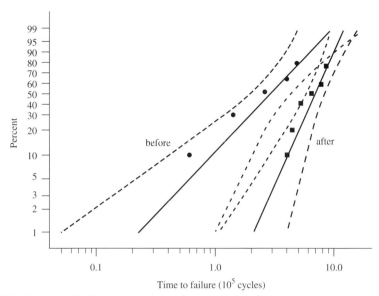

FIGURE 7.16   Weibull plots, ML fits, and percentile confidence intervals for the interval life data of the transmission part

7.6.5   Normal and Lognormal Distributions

In this subsection we describe ML estimation of the parameters of normal and lognormal distributions for complete or type I censored data. Confidence intervals for the parameters and other quantities of interest are also presented. The methods equally apply to type II censored data.

As discussed in Chapter 2, if a variable x has a lognormal distribution with scale parameter μ and shape parameter σ, $t = \ln(x)$ is normally distributed with mean μ and standard deviation σ. Thus, we fit a normal distribution to the log times $t = \ln(x)$, whose pdf is

$$f(t) = \frac{1}{\sqrt{2\pi}\sigma} \exp\left[-\frac{(t - \mu)^2}{2\sigma^2} \right]. \tag{7.73}$$

Complete Exact Data   From (7.39) the sample log likelihood of the normal distribution for complete exact data is

$$L(\mu, \sigma) = \sum_{i=1}^{n} \left[-\frac{1}{2} \ln(2\pi) - \ln(\sigma) - \frac{(t_i - \mu)^2}{2\sigma^2} \right]. \tag{7.74}$$

Equating to zero the derivatives of (7.74) with respect to μ and σ gives the likelihood equations. Solving these equations yields the MLEs

$$\hat{\mu} = \frac{1}{n} \sum_{i=1}^{n} t_i, \tag{7.75}$$

$$\hat{\sigma}^2 = \frac{1}{n} \sum_{i=1}^{n} (t_i - \hat{\mu})^2. \tag{7.76}$$

Note that (7.75) is the usual unbiased estimate of μ, whereas (7.76) is not the unbiased estimate of σ^2. The unbiased estimate is

$$s^2 = \frac{1}{n-1} \sum_{i=1}^{n} (t_i - \hat{\mu})^2. \tag{7.77}$$

When n is large, (7.76) and (7.77) are close or nearly equal.

In earlier subsections we approximated the distribution of an estimate with a normal distribution. Here the distribution of $\hat{\mu}$ is exactly normal with mean μ and standard deviation $\sigma/\sqrt{n}$. $n\hat{\sigma}^2/\sigma^2$ has a chi-square distribution with $n-1$ degrees of freedom. The two-sided exact $100(1-\alpha)\%$ confidence intervals for μ and σ are

$$[\mu_L, \mu_U] = \hat{\mu} \pm \frac{z_{1-\alpha/2}\sigma}{\sqrt{n}}, \tag{7.78}$$

$$[\sigma_L^2, \sigma_U^2] = \left[\frac{n\hat{\sigma}^2}{\chi^2_{(1-\alpha/2);(n-1)}}, \frac{n\hat{\sigma}^2}{\chi^2_{\alpha/2;(n-1)}} \right]. \tag{7.79}$$

The true value of σ is usually unknown. If σ in (7.78) is replaced by s, $\sqrt{n}(\hat{\mu} - \mu)/s$ has a t-distribution with $n-1$ degrees of freedom. The corresponding exact confidence interval is

$$[\mu_L, \mu_U] = \hat{\mu} \pm \frac{t_{\alpha/2;(n-1)}s}{\sqrt{n}}. \tag{7.80}$$

For computational convenience, most reliability and statistical software use the approximate confidence intervals:

$$[\mu_L, \mu_U] = \hat{\mu} \pm z_{1-\alpha/2}\sqrt{\hat{V}ar(\hat{\mu})}, \tag{7.81}$$

$$[\sigma_L, \sigma_U] = \hat{\sigma} \exp\left(\pm \frac{z_{1-\alpha/2}\sqrt{\hat{V}ar(\hat{\sigma})}}{\hat{\sigma}} \right), \tag{7.82}$$

where $\hat{V}ar(\hat{\mu}) \approx \hat{\sigma}^2/n$ and $\hat{V}ar(\hat{\sigma}) \approx \hat{\sigma}^2/2n$. The one-sided $100(1-\alpha)\%$ confidence bound is obtained by replacing $z_{1-\alpha/2}$ with $z_{1-\alpha}$ and using the appropriate sign.

An approximate confidence interval for the probability of failure F at a particular time t can be developed using (7.48) and (7.49), where $g(\mu, \sigma) = F(t; \mu, \sigma)$. A more accurate one is

$$[F_L, F_U] = [\Phi(w_L), \Phi(w_U)], \tag{7.83}$$

where

$$[w_L, w_U] = \hat{w} \pm z_{1-\alpha/2}\sqrt{\hat{V}ar(\hat{w})}, \qquad \hat{w} = \frac{t - \hat{\mu}}{\hat{\sigma}},$$

$$\hat{V}ar(\hat{w}) = \frac{1}{\hat{\sigma}^2}[\hat{V}ar(\hat{\mu}) + \hat{w}^2\hat{V}ar(\hat{\sigma}) + 2\hat{w}\hat{C}ov(\hat{\mu}, \hat{\sigma})],$$

and $\Phi(\cdot)$ is the standard normal cdf.

The estimate of the $100p$th percentile is

$$\hat{t}_p = \hat{\mu} + z_p\hat{\sigma}. \tag{7.84}$$

The two-sided approximate $100(1 - \alpha)\%$ confidence interval for t_p is

$$[t_{p,L}, t_{p,U}] = \hat{t}_p \pm z_{1-\alpha/2}\sqrt{\hat{V}ar(\hat{t}_p)}, \tag{7.85}$$

where $\hat{V}ar(\hat{t}_p) = \hat{V}ar(\hat{\mu}) + z_p^2\hat{V}ar(\hat{\sigma}) + 2z_p\hat{C}ov(\hat{\mu}, \hat{\sigma})$. Note that (7.85) reduces to (7.81) when $p = 0.5$.

The lognormal $100p$th percentile and confidence bounds are calculated with the antilog transformation of (7.84) and (7.85), respectively.

Example 7.11 An original equipment manufacturer wanted to choose an oxygen sensor from supplier 1 or 2. To help make the decision, the reliability of the sensors of the two suppliers was needed. This was accomplished by testing 15 sensors randomly selected from each supplier at a high temperature. It was decided that all 30 units would be run simultaneously until all units of a supplier fail, when the other group would be censored. This resulted in 15 failures for supplier 1 and 10 for supplier 2. The censoring time for supplier 2 units was 701 hours, when the last unit of supplier 1 failed. The failure times (in hours) are 170, 205, 207, 240, 275, 285, 324, 328, 334, 352, 385, 479, 500, 607, and 701 for supplier 1, and 220, 264, 269, 310, 408, 451, 489, 537, 575, and 663 for supplier 2. The supplier 1 lifetimes are complete exact data. Estimate the life distribution, the population fraction failing by 200 hours, and the median life, which interested the manufacturer. The supplier 2 data (right censored) will be analyzed in Example 7.12.

SOLUTION   The lognormal distribution adequately fits the life data of supplier 1, as indicated by a lognormal probability plot (not shown here). The next step is to calculate the ML estimates and confidence intervals. This can be done with Minitab. Here we do manual computation for illustration purposes. First the log lifetimes are calculated. Then from (7.75), the estimate of μ is

$$\hat{\mu}_1 = \frac{1}{15}[\ln(170) + \ln(205) + \cdots + \ln(701)] = 5.806.$$

In this example, the subscript 1 denotes supplier 1. The estimate of σ_1^2 is

$$\hat{\sigma}_1^2 = \frac{1}{15}\{[\ln(170) - 5.806]^2 + [\ln(205) - 5.806]^2 + \cdots + [\ln(701) - 5.806]^2\}$$
$$= 0.155,$$
$$\hat{\sigma}_1 = 0.394.$$

The estimate of the population fraction failing by 200 hours is

$$\hat{F}_1(200) = \Phi\left[\frac{\ln(200) - 5.806}{0.394}\right] = 0.097 \quad \text{or} \quad 9.7\%.$$

The lognormal estimate of the median life is

$$\hat{x}_{0.5,1} = \exp(\hat{\mu}_1) = \exp(5.806) = 332 \text{ hours.}$$

The variance estimates of $\hat{\mu}_1$ and $\hat{\sigma}_1$ are

$$\hat{V}ar(\hat{\mu}_1) \approx \frac{0.155}{15} = 0.0103, \qquad \hat{V}ar(\hat{\sigma}_1) \approx \frac{0.155}{2 \times 15} = 0.0052.$$

The two-sided approximate 90% confidence interval for μ_1 from (7.81) is

$$[\mu_{1,L}, \mu_{1,U}] = 5.806 \pm 1.6449\sqrt{0.0103} = [5.639, 5.973].$$

Similarly,

$$[\sigma_{1,L}, \sigma_{1,U}] = 0.394 \exp\left(\pm\frac{1.6449\sqrt{0.0052}}{0.394}\right) = [0.292, 0.532].$$

For complete data, $\text{Cov}(\hat{\mu}_1, \hat{\sigma}_1) = 0$. Then the estimate of the variance of $\hat{w} = (t - \hat{\mu}_1)/\hat{\sigma}_1$ is

$$\hat{V}ar(\hat{w}) = \frac{1}{0.155}[0.0103 + (-1.2987)^2 \times 0.0052] = 0.123.$$

For $t = \ln(200) = 5.298$, we have

$$[w_L, w_U] = -1.2987 \pm (-1.6449 \times \sqrt{0.123}) = [-1.8756, -0.7218].$$

Then from (7.83), the confidence interval for the population fraction failing by 200 hours is $[F_{1,L}, F_{1,U}] = [0.030, 0.235]$ or $[3.0\%, 23.5\%]$.

The lognormal confidence interval for the median life is

$$[x_{0.5,1,L}, x_{0.5,1,U}] = [\exp(\mu_{1,L}), \exp(\mu_{1,U})] = [\exp(5.639), \exp(5.973)] = [281, 393].$$

The results above will be compared with those for supplier 2 in Example 7.12.

Right-Censored Exact Data Suppose that a sample of n units has r failures. The other $n - r$ units are subjected to type I right censoring. The failure times are $t_1, t_2, \ldots, t_r$, and the censoring times are $t_{r+1}, t_{r+2}, \ldots, t_n$. The sample log likelihood is

$$L(\mu, \sigma) = \sum_{i=1}^{r} \left[-\frac{1}{2} \ln(2\pi) - \ln(\sigma) - \frac{(t_i - \mu)^2}{2\sigma^2} \right] + \sum_{i=r+1}^{n} \ln \left[1 - \Phi \left(\frac{t_i - \mu}{\sigma} \right) \right],$$

(7.86)

where $\Phi(\cdot)$ is the standard normal cdf. The estimates $\hat{\mu}$ and $\hat{\sigma}$ can be found by directly maximizing (7.86). The normal approximate confidence intervals for the mean, standard deviation, failure probability (population fraction failing), and the $100p$th percentile are calculated from (7.81), (7.82), (7.83), and (7.85), respectively, with the variance estimates derived from the censored data.

Example 7.12 Refer to Example 7.11. The life data for supplier 2 are right-censored exact data. The sample size $n = 15$, the number of failures $r = 10$, and the censoring time is 701 hours. Estimate the life distribution and the median life for supplier 2. Compare the results with those for supplier 1 in Example 7.11. Then make a recommendation as to which supplier to choose.

SOLUTION As in Example 7.11, a lognormal plot shows that the lognormal distribution adequately fits the life data of supplier 2. The distribution parameters are estimated by substituting the life data into (7.86) and maximizing the log likelihood directly. The calculation can be done by coding an Excel spreadsheet or a small computer program. Minitab provides estimates of the scale and shape parameters as $\hat{\mu}_2 = 6.287$ and $\hat{\sigma}_2 = 0.555$; the subscript 2 in this example denotes supplier 2. The population fraction failing by 200 hours is $\hat{F}_2(200) = 0.037$ (or 3.7%) and the lognormal estimate of the median life is $\hat{x}_{0.5,2} = 538$ hours. The two-sided approximate 90% confidence interval for μ_2 is $[\mu_{2,L}, \mu_{2,U}] = [6.032, 6.543]$. The confidence interval for σ_2 is $[\sigma_{2,L}, \sigma_{2,U}] = [0.373, 0.827]$. The confidence interval for the population fraction failing by 200 hours is $[F_{2,L}, F_{2,U}] = [0.006, 0.145]$ or [0.6%, 14.5%]. The confidence interval for the median life of the lognormal data is $[x_{0.5,2,L}, x_{0.5,2,U}] = [417, 694]$.

Comparing the results from Examples 7.11 and 7.12, we see that the sensor made by supplier 2 is more reliable, especially at high times in service. In particular, the median life for supplier 2 is significantly greater than that for supplier 1 (their confidence intervals do not overlap). In addition, supplier 2 has a lower probability of failure at 200 hours, although the confidence intervals for both suppliers partially intersect. Apparently, the original equipment manufacturer should choose supplier 2 from the reliability perspective.

Interval Data The sample log likelihood function for complete interval data is

$$L(\mu, \sigma) = \sum_{i=1}^{m} r_i \ln \left[\Phi \left(\frac{t_i - \mu}{\sigma} \right) - \Phi \left(\frac{t_{i-1} - \mu}{\sigma} \right) \right],$$

(7.87)

where r_i is the number of failures in the ith inspection interval $(t_{i-1}, t_i]$ and m is the number of inspection times. Unlike the likelihood function for exact data, (7.87) does not yield closed-form solutions for the parameter estimates. They must be found using a numerical method. In practice, commercial reliability software is preferred for this. If such software is not available, we may create an Excel spreadsheet and use its Solver feature to do the optimization. Excel is especially convenient for solving this problem because of its embedded standard normal distribution. Software or an Excel spreadsheet may be used to deal with the right-censored interval data, whose sample log likelihood is

$$L(\mu, \sigma) = \sum_{i=1}^{m} r_i \ln \left[\Phi \left(\frac{t_i - \mu}{\sigma} \right) - \Phi \left(\frac{t_{i-1} - \mu}{\sigma} \right) \right] + \sum_{i=1}^{m} d_i \ln \left[1 - \Phi \left(\frac{t_i - \mu}{\sigma} \right) \right],$$

(7.88)

where m, r_i, and t_i follow the notation in (7.87) and d_i is the number of units censored at inspection time t_i.

The calculation of confidence intervals for the interval data applies formulas for the exact data given earlier in this subsection.

7.7 RELIABILITY ESTIMATION AT USE CONDITION

In earlier sections we discussed the use of graphical and ML methods to estimate reliability at individual test conditions. In this section we utilize these methods and the acceleration relationships in Section 7.4 to estimate reliability at a use condition.

7.7.1 Statistical Acceleration Models

The acceleration relationships in Section 7.4 are deterministic. In other words, they describe nominal life and do not account for the scatter in life. In reliability analysis we are concerned not only with the dependence of nominal life on stress, but also with the distribution of life. The two concerns can be addressed by combining an acceleration relationship with a life distribution. Then the nominal life in the acceleration relationship is a specific percentile of the life distribution, and the resulting combination is a physical–statistical acceleration model. For example, if temperature is an accelerating variable and life is modeled with the Weibull distribution, the nominal life in the acceleration relationship (e.g., the Arrhenius relationship) is the characteristic life.

To estimate the life distribution at the use stress level from an ALT, the following assumptions are needed:

- The times to failure at the use and high stress levels can be modeled with a (transformed) location-scale distribution (e.g., exponential, Weibull, or lognormal distribution). Other distributions are less frequently employed for modeling life, but may be used.
- The scale parameter of the (transformed) location-scale distribution does not depend on the stress level. In particular, the lognormal σ and Weibull β are

constant at any stress level. As a special case of the Weibull distribution when $\beta = 1$, the exponential distribution always satisfies this assumption. The scale parameter may not be a constant in some applications, and the subsequent data analysis is much more complicated.

- The acceleration relationship is adequate over the stress range from the highest test level to the normal use level. The location parameter is a (transformed) linear function of the stresses: namely,

$$y = \gamma_0 + \gamma_1 x_1 + \cdots + \gamma_k x_k, \tag{7.89}$$

where γ_i $(i = 0, 1, 2, \ldots, k)$ is a coefficient dependent on material properties, failure criteria, product design, and other factors, x_i is the (transformed) stress, k the number of stresses, and y the location parameter of the life distribution. In particular, for the exponential distribution, $y = \ln(\theta)$; for the Weibull distribution, $y = \ln(\alpha)$; for the lognormal distribution, $y = \mu$.

In an application, the life–stress relationship and life distribution determine the specific form of (7.89). For example, if we use the Arrhenius relationship (7.6) and the Weibull distribution, (7.89) can be written as

$$\ln(\alpha) = \gamma_0 + \gamma_1 x, \tag{7.90}$$

where $x = 1/T$. If we use the inverse power relationship (7.18) and the lognormal distribution, (7.89) becomes

$$\mu = \gamma_0 + \gamma_1 x, \tag{7.91}$$

where $x = \ln(V)$. If we use the Norris–Landzberg relationship (7.14) and the exponential distribution, (7.89) simplifies to

$$\ln(\theta) = \gamma_0 + \gamma_1 x_1 + \gamma_2 x_2 + \gamma_3 x_3, \tag{7.92}$$

where $x_1 = \ln(\Delta T)$, $x_2 = \ln(f)$, and $x_3 = 1/T_{\max}$.

7.7.2 Graphical Estimation

In Section 7.5 we presented probability plotting for estimating the parameters of life distributions at individual test conditions. The following steps are used for estimating the life distribution at the use condition.

1. Plot the life data from each test condition on appropriate probability paper, and estimate the location and scale parameters, which are denoted $\hat{y}_i$ and $\hat{\sigma}_i$ $(i = 1, 2, \ldots, m)$, where m is the number of stress levels. This step has been described in detail in Section 7.5.

2. Substitute $\hat{y}_i$ and the value of x_i into the linearized relationship (7.89) and solve the equations for the coefficients using the linear regression method.

Then calculate the estimate of y at the use stress level, say $\hat{y}_0$. Alternatively, $\hat{y}_0$ may be obtained by plotting $\hat{y}_i$ versus the (linearly transformed) stress level and projecting the straight line to the use level.

3. Calculate the common scale parameter estimate $\hat{\sigma}_0$ from

$$\hat{\sigma}_0 = \frac{1}{r} \sum_{i=1}^{m} r_i \hat{\sigma}_i, \qquad (7.93)$$

where r_i is the number of failures at stress level i and $r = \sum_1^m r_i$. Equation (7.93) assumes a constant scale parameter, and is an approximate estimate of the common scale parameter. More accurate, yet complicated estimates are given in, for example, Nelson (1982).

4. Estimate the quantities of interest at the use stress level using the life distribution with location parameter $\hat{y}_0$ and scale parameter $\hat{\sigma}_0$.

It should be pointed out that the graphical method above yields approximate life estimates at the use condition. Whenever possible, the ML method (Section 7.7.3) should be used to obtain better estimates.

Example 7.13   Refer to Example 7.5. Using the Arrhenius relationship, estimate the B_{10} life and the reliability at 10,000 hours at a use temperature of $35°C$.

SOLUTION   In Example 7.5, Weibull plots of the life data at each temperature yielded the estimates $\hat{\alpha}_1 = 5394$ and $\hat{\beta}_1 = 2.02$ for the $100°C$ group, $\hat{\alpha}_2 = 3285$ and $\hat{\beta}_2 = 2.43$ for the $120°C$ group, and $\hat{\alpha}_3 = 1330$ and $\hat{\beta}_3 = 2.41$ for the $150°C$ group. As shown in (7.90), $\ln(\alpha)$ is a linear function of $1/T$. Thus, we plot $\ln(\hat{\alpha}_i)$ versus $1/T_i$ using an Excel spreadsheet and fit a regression line to the data points. Figure 7.17 shows the plot and the regression equation. The high R^2 value indicates that the Arrhenius relationship fits adequately. The estimate of

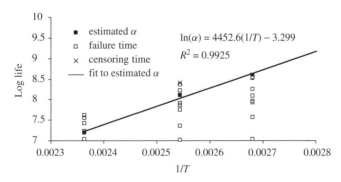

FIGURE 7.17   Plot of the fitted Arrhenius relationship for electronic modules

the Weibull scale parameter at 35°C is $\hat{\alpha}_0 = \exp(4452.6 \times 0.003245 - 3.299) = 69,542$ hours. From (7.93), the common shape parameter estimate is

$$\hat{\beta}_0 = \frac{8 \times 2.02 + 7 \times 2.43 + 10 \times 2.41}{8 + 7 + 10} = 2.29.$$

Thus, Weibull fit has shape parameter 2.29 and scale parameter 69,542 hours at use temperature 35°C. The B_{10} life is $\hat{B}_{10} = 69,542 \times [-\ln(1 - 0.1)]^{1/2.29} = 26,030$ hours. The reliability at 10,000 hours is $\hat{R}(10,000) = \exp[-(10,000/69,542)^{2.29}] = 0.9883$, which means an estimated 1.2% of the population would fail by 10,000 hours.

7.7.3 ML Estimation

In Section 7.6 we described the ML method for estimating the life distributions at individual test stress levels. The life data obtained at different stress levels were analyzed separately; each distribution is the best fit to the particular data set. The inferences from such analyses apply to these stress levels. As we know, the primary purpose of an ALT is to estimate the life distribution at a use condition. To accomplish this, in Section 7.7.1 we assumed a (transformed) location-scale distribution with a common scale parameter value and an acceleration relationship between the location parameter and the stress level. For this model we fit the model to the life data at all stress levels simultaneously. Obtain estimates at the use condition as follows:

1. Fit the life distributions at individual stress levels by probability plotting and estimate the respective scale parameters.
2. Plot all data sets and their separate cdf fits on the same probability plot. Check the assumption of a constant scale parameter. If the data plots and fitted straight lines are roughly parallel, the assumption may be reasonable and the maximum likelihood analysis may begin. Otherwise, investigation into the test method, failure modes, and others may be necessary. If no problems are found, the data should be analyzed with nonconstant scale parameters. The dependence of a scale parameter on stress level is discussed in, for example, Nelson (1990, 2004). The assumption of a constant scale parameter may be checked in an alternative way. By using (7.93) we first calculate the common scale parameter estimate. Then plot the best fits to individual data sets with this scale parameter. If all lines fit the data sets adequately, the assumption may be reasonable.
3. Write the total sample log likelihood function, which is the sum of all log likelihood functions for each test group. The sample log likelihood function for an individual group is given in Section 7.6. Now let's consider an example. An ALT consists of testing two groups at the low and high stress levels. The group at the low stress level generates right-censored exact life data, whereas the high stress group yields complete exact life data. For a

Weibull distribution, the sample log likelihood functions for the low and high stress levels are given by (7.68) and (7.59), respectively. Then the total sample log likelihood function is

$$L(\alpha_1, \alpha_2, \beta) = \sum_{i=1}^{r} \left[\ln(\beta) - \beta \ln(\alpha_1) + (\beta - 1) \ln(t_{1i}) - \left(\frac{t_{1i}}{\alpha_1}\right)^{\beta} \right] - \sum_{i=r+1}^{n_1} \left(\frac{t_{1i}}{\alpha_1}\right)^{\beta}$$

$$+ \sum_{i=1}^{n_2} \left[\ln(\beta) - \beta \ln(\alpha_2) + (\beta - 1) \ln(t_{2i}) - \left(\frac{t_{2i}}{\alpha_2}\right)^{\beta} \right], \qquad (7.94)$$

where the subscripts 1 and 2 denote the low and high stress levels, respectively.

4. Substitute an appropriate acceleration relationship into the total log likelihood function. In the example in step 3, if the Arrhenius relationship is used, substituting (7.90) into (7.94) gives

$$L(\gamma_0, \gamma_1, \beta) = \sum_{i=1}^{r} \left[\ln(\beta) - \beta(\gamma_0 + \gamma_1 x_1) + (\beta - 1) \ln(t_{1i}) - \left(\frac{t_{1i}}{e^{\gamma_0 + \gamma_1 x_1}}\right)^{\beta} \right]$$

$$- \sum_{i=r+1}^{n_1} \left(\frac{t_{1i}}{e^{\gamma_0 + \gamma_1 x_1}}\right)^{\beta}$$

$$+ \sum_{i=1}^{n_2} \left[\ln(\beta) - \beta(\gamma_0 + \gamma_1 x_2) + (\beta - 1) \ln(t_{2i}) \right.$$

$$\left. - \left(\frac{t_{2i}}{e^{\gamma_0 + \gamma_1 x_2}}\right)^{\beta} \right], \qquad (7.95)$$

where x_1 and x_2 denote the transformed low and high temperatures, respectively.

5. Estimate the model parameters [e.g., γ_0, γ_1, and β in (7.95)] by maximizing the total log likelihood function directly through a numerical method. Also, the estimates may be obtained by iteratively solving the likelihood equations; however, this approach is usually more difficult. In the example, this step yields the estimates $\hat{\gamma}_0$, $\hat{\gamma}_1$, and $\hat{\beta}$.

6. Calculate the variance–covariance matrix for the model parameters using the total log likelihood function and the local estimate of Fisher information matrix described in Section 7.6.2. In the example, this step gives

$$\hat{\Sigma} = \begin{bmatrix} \hat{V}ar(\hat{\gamma}_0) & \hat{C}ov(\hat{\gamma}_0, \hat{\gamma}_1) & \hat{C}ov(\hat{\gamma}_0, \hat{\beta}) \\ & \hat{V}ar(\hat{\gamma}_1) & \hat{C}ov(\hat{\gamma}_1, \hat{\beta}) \\ \text{symmetric} & & \hat{V}ar(\hat{\beta}) \end{bmatrix}.$$

7. Calculate the life distribution estimate at the use stress level. The location parameter estimate of the distribution is calculated from the acceleration relationship. In the example, the Weibull characteristic life at the use condition is $\hat{\alpha}_0 = \exp(\hat{\gamma}_1 + \hat{\gamma}_1 x_0)$, where $x_0 = 1/T_0$ and T_0 is the use temperature. The Weibull shape parameter estimate is $\hat{\beta}$. Having $\hat{\alpha}_0$ and $\hat{\beta}$, we can estimate the quantities of interest, such as reliability and percentiles.

8. Estimate the variance for the location parameter estimate at the use condition and the covariance for the location and scale parameter estimates. The variance is obtained from (7.48) and the acceleration relationship. For the lognormal distribution, the covariance at a given stress level is

$$\hat{C}ov(\hat{\mu}, \hat{\sigma}) = \hat{C}ov(\hat{\gamma}_0, \hat{\sigma}) + \sum_{i=1}^{k} x_i \hat{C}ov(\hat{\gamma}_i, \hat{\sigma}), \tag{7.96}$$

where k and x_i are the same as those in (7.89). Substituting the use stress levels $x_{10}, x_{20}, \ldots, x_{k0}$ into (7.96) results in the covariance of the estimate of the scale parameter and that of the location parameter at the use stress levels. Similarly, for the Weibull distribution,

$$\hat{C}ov(\hat{\alpha}, \hat{\beta}) = \hat{\alpha} \left[\hat{C}ov(\hat{\gamma}_0, \hat{\beta}) + \sum_{i=1}^{k} x_i \hat{C}ov(\hat{\gamma}_i, \hat{\beta}) \right]. \tag{7.97}$$

In the example above with a single stress, the covariance of $\hat{\alpha}_0$ and $\hat{\beta}$ from (7.97) is

$$\hat{C}ov(\hat{\alpha}_0, \hat{\beta}) = \hat{\alpha}_0 [\hat{C}ov(\hat{\gamma}_0, \hat{\beta}) + x_0 \hat{C}ov(\hat{\gamma}_1, \hat{\beta})].$$

9. Calculate the confidence intervals for the quantities estimated earlier. This is done by substituting the variance and covariance estimates of the model parameters at the use condition into the confidence intervals for an individual test condition presented in Section 7.6. In the example, the confidence interval for the probability of failure at a given time and the use condition is obtained by substituting $\hat{V}ar(\hat{\alpha}_0)$, $\hat{V}ar(\hat{\beta})$, and $\hat{C}ov(\hat{\alpha}_0, \hat{\beta})$ into (7.66).

In practice, the calculations above are performed with commercial software such as Minitab and Reliasoft ALTA. Now we illustrate the calculations with two examples below, one with a single accelerating stress and the other with two.

Example 7.14 Example 7.13 illustrates graphical reliability estimation at the use temperature for small electronic modules. The life data were presented in Example 7.5. Here we estimate the life distribution, the B_{10} life, and the reliability at 10,000 hours and at the use temperature, and calculate their confidence intervals using the ML method.

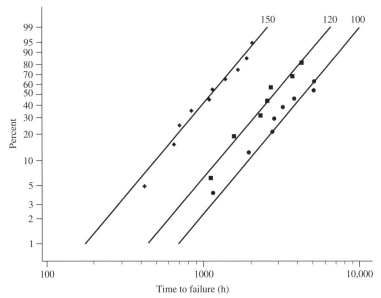

FIGURE 7.18 Weibull fits to the electronic module data with a common β

SOLUTION The graphical analysis in Example 7.5 shows that the Weibull distribution fits the life data adequately at each temperature. The Weibull fits to the three data sets were plotted in Figure 7.13, which suggests that a constant shape parameter is reasonable although the line for 100°C is not quite parallel to other two. Alternatively, in Figure 7.18, we plot the Weibull fits to the three data sets with a common shape parameter, which was calculated in Example 7.13 as $\hat{\beta}_0 = 2.29$. Figure 7.18 shows that a common shape parameter is reasonable.

The groups at 100 and 120°C have right-censored exact data, whereas that at 150°C is complete. If the life–temperature relationship is modeled with (7.90), the total sample log likelihood function is

$$L(\gamma_0, \gamma_1, \beta) = \sum_{i=1}^{8} \left[\ln(\beta) - \beta(\gamma_0 + \gamma_1 x_1) + (\beta - 1) \ln(t_{1i}) \right.$$

$$\left. - \left(\frac{t_{1i}}{e^{\gamma_0 + \gamma_1 x_1}} \right)^{\beta} \right] - 4 \times \left(\frac{5500}{e^{\gamma_0 + \gamma_1 x_1}} \right)^{\beta}$$

$$+ \sum_{i=1}^{7} \left[\ln(\beta) - \beta(\gamma_0 + \gamma_1 x_2) + (\beta - 1) \ln(t_{2i}) - \left(\frac{t_{2i}}{e^{\gamma_0 + \gamma_1 x_2}} \right)^{\beta} \right]$$

$$- \left(\frac{4500}{e^{\gamma_0 + \gamma_1 x_2}} \right)^{\beta}$$

$$+ \sum_{i=1}^{10} \left[\ln(\beta) - \beta(\gamma_0 + \gamma_1 x_3) + (\beta - 1) \ln(t_{3i}) - \left(\frac{t_{3i}}{e^{\gamma_0 + \gamma_1 x_3}} \right)^{\beta} \right],$$

where $x_1 = 1/(100 + 273.15) = 0.00268$, $x_2 = 0.00254$, $x_2 = 0.00236$, and t_{1i}, t_{2i}, and t_{3i} are the failure times observed at 100, 120, and 150°C, respectively. The estimates $\hat{\gamma}_0$, $\hat{\gamma}_1$, and $\hat{\beta}$ are readily calculated by directly maximizing $L(\gamma_0, \gamma_1, \beta)$ on the Excel spreadsheet. However, calculation of the variance–covariance matrix through Excel involves manual operation and is not recommended. Here we use Minitab to do the analysis; other software, such as Reliasoft ALTA, is also an option. Estimates of the model parameters are $\hat{\gamma}_0 = -3.156$, $\hat{\gamma}_1 = 4390$, and $\hat{\beta} = 2.27$. The variance estimates for $\hat{\gamma}_0$, $\hat{\gamma}_1$, and $\hat{\beta}$ are $\hat{V}ar(\hat{\gamma}_0) = 3.08$, $\hat{V}ar(\hat{\gamma}_1) = 484{,}819.5$, and $\hat{V}ar(\hat{\beta}) = 0.1396$. The two-sided 90% confidence intervals for the model parameters are $[\gamma_{0,L}, \gamma_{0,U}] = [-6.044, -0.269]$, $[\gamma_{1,L}, \gamma_{1,U}] = [3244.8, 5535.3]$, and $[\beta_L, \beta_U] = [1.73, 2.97]$.

The estimate of the Weibull characteristic life at 35°C is $\hat{\alpha}_0 = \exp(-3.156 + 4390 \times 0.003245) = 65{,}533$ hours. The B_{10} life at 35°C is $\hat{B}_{10} = 24{,}286$ hours. The two-sided 90% confidence interval for B_{10} life is $[B_{10,L}, B_{10,U}] = [10{,}371, 56{,}867]$. The reliability at 10,000 hours and 35°C is $\hat{R}(10{,}000) = 0.9860$. The two-sided 90% confidence interval for the reliability is $[R_L, R_U] = [0.892, 0.998]$.

Note that the $\hat{\beta}$ value here from the ML method is very close to that obtained from the graphical analysis in Example 7.13. The differences for other parameters and quantities are also small (less than 6%) in this particular case. In general, the two methods often give fairly different results, and ML estimates usually are more accurate.

Example 7.15   In this example we analyze an ALT with two accelerating stresses using the ML method. G. Yang and Zaghati (2006) present a case on reliability demonstration of a type of 18-V compact electromagnetic relays through ALT. The relays would be installed in a system and operate at 5 cycles per minute and 30°C. The system design specifications required the relays to have a lower 90% confidence bound for reliability above 99% at 200,000 cycles. A sample of 120 units was divided into four groups, each tested at a higher-than-use temperature and switching rate. In testing, the normal closed and open contacts of the relays were both loaded with 2 A of resistive load. The maximum allowable temperature and switching rate of the relays are 125°C and 30 cycles per minute, respectively. The increase in switching rate reduces the cycles to failure for this type of relay, due to the shorter time for heat dissipation and more arcing. Its effect can be described by the life–usage model (7.21). The effect of temperature on cycles to failure is modeled with the Arrhenius relationship. This ALT involved two accelerating variables. Table 7.9 shows the Yang compromise test plan, which is developed in the next section. The test plan specifies the censoring times, while the censoring cycles are the censoring times multiplied by the switching rates. The numbers of cycles to failure are summarized in Table 7.10. Estimate the life distribution at the use temperature and switching rate, and verify that the component meets the system design specification.

TABLE 7.9 Compromise Test Plan for the Compact Relays

Group	Temperature (°C)	Switching Rate (cycles/min)	Sample Size	Censoring Time (h)	Censoring Cycles ($\times 10^3$)
1	64	10	73	480	288
2	64	30	12	480	864
3	125	10	12	96	57.6
4	125	30	23	96	172.8

TABLE 7.10 Cycles to Failure of the Compact Relays

Group	Cycles to Failure[a]
1	47,154, 51,307, 86,149, 89,702, 90,044, 129,795, 218,384, 223,994, 227,383, 229,354, 244,685, 253,690, 270,150, 281,499, 288,000[+59]
2	45,663, 123,237, 192,073, 212,696, 304,669, 323,332, 346,814, 452,855, 480,915, 496,672, 557,136, 570,003
3	12,019, 18,590, 29,672, 38,586, 47,570, 56,979, 57,600[+6]
4	7,151, 11,966, 16,772, 17,691, 18,088, 18,446, 19,442, 25,952, 29,154, 30,236, 33,433, 33,492, 39,094, 51,761, 53,926, 57,124, 61,833, 67,618, 70,177, 71,534, 79,047, 91,295, 92,005

[a] A superscript $+x$ over a cycle number implies that x units would last beyond that cycle (right censored).

SOLUTION   We first graphically analyze the life data of the four groups. Groups 1 and 3 are right censored, and 2 and 4 are complete. Probability plots for individual groups indicate that the Weibull distribution is adequate for all groups, and a constant shape parameter is reasonable, as shown in Figure 7.19.

The graphical analysis should be followed by the maximum likelihood method. The total sample log likelihood function is not given here but can be worked out by summing those for the individual groups. The acceleration relationship combines the Arrhenius relationship and the life–usage model: namely,

$$\alpha(f, T) = A f^B \exp\left(\frac{E_a}{kT}\right), \tag{7.98}$$

where $\alpha(f, T)$ is the Weibull characteristic life and the other notation is the same as in (7.5) and (7.21). Linearizing (7.98) gives

$$\ln[\alpha(x_1, x_2)] = \gamma_0 + \gamma_1 x_1 + \gamma_2 x_2, \tag{7.99}$$

where $x_1 = 1/T$, $x_2 = \ln(f)$, and γ_0, γ_1, and γ_2 are constant coefficients. Note that (7.99) is a special form of (7.89). The next step is to substitute (7.99) and the life data into the total sample log likelihood function. Then estimate the model parameters γ_0, γ_1, γ_2, and β by directly maximizing the total likelihood.

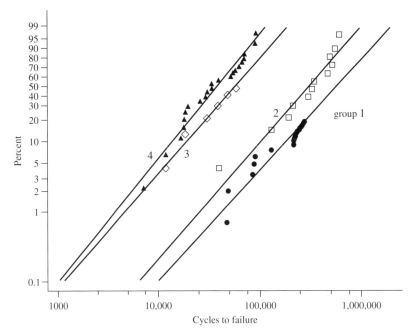

FIGURE 7.19   Weibull fits to individual groups

Instead, we use Minitab to do ML estimation and get $\hat{\gamma}_0 = 0.671$, $\hat{\gamma}_1 = 4640.1$, $\hat{\gamma}_2 = -0.445$, and $\hat{\beta} = 1.805$.

The Weibull fits to each group with the common $\hat{\beta} = 1.805$ are plotted in Figure 7.20. The Weibull characteristic life at use temperature (30°C) and the usual switching rate (5 cycles per minute) is estimated by

$$\hat{\alpha}_0 = \exp[0.671 + \frac{4640.1}{303.15} - 0.445 \times \ln(5)] = 4.244 \times 10^6 \text{ cycles.}$$

The reliability estimate at 200,000 cycles is

$$\hat{R}(200,000) = \exp\left[-\left(\frac{200,000}{4.244 \times 10^6}\right)^{1.805}\right] = 0.996.$$

The one-sided lower 90% confidence bound on the reliability is $R_L(200,000) = 0.992$. Since $R_L(200,000) = 0.992$ is marginally greater than the required reliability of 99%, we would accept that the relays meet the requirement.

Additionally, we estimate the activation energy as

$$\hat{E}_a = \hat{\gamma}_1 k = 4640.1 \times 8.6171 \times 10^{-5} = 0.4 \text{ eV.}$$

The two-sided 90% confidence interval for E_a is [0.347, 0.452]. The estimate of the switching rate effect parameter is $\hat{B} = \hat{\gamma}_2 = -0.445$. The two-sided 90%

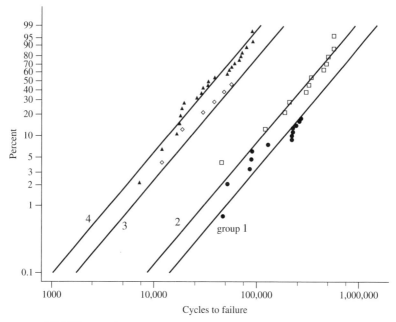

FIGURE 7.20 Weibull fits to each group with a common β

confidence interval for B is $[-0.751, -0.160]$. Because this confidence interval does not include zero, the switching rate effect on life should not be ignored. G. Yang and Zaghati (2006) discuss more about the significance of switching rate effect for the same relays.

7.8 COMPROMISE TEST PLANS

As stated and illustrated in earlier sections, the primary purpose of an ALT is to estimate the life distribution and quantities of interest at a use condition. This estimation involves extrapolation from higher stress levels by using an acceleration model, and thus includes the model error and statistical uncertainty. Sometimes, the model error outweighs the statistical one. The model error may be reduced or eliminated only by better understanding the failure mechanisms and using a more accurate model, whereas the statistical uncertainty can be reduced by carefully selecting a good test plan. A typical test plan is characterized by the stress levels, the number of test units allocated to each level, and their censoring times. From the statistical perspective, a longer censoring time often yields a more accurate estimate. However, the censoring times are constrained by the test resources and schedule and are prespecified in most applications. G. Yang (1994), G. Yang and Jin (1994), and Tang and Xu (2005) optimize the censoring times. The remaining variables (i.e., the stress levels and the number of test units allocated to each level) may be determined by various approaches, which are reviewed briefly below.

In this section we focus on compromise test plans which optimize the values of test variables to minimize the asymptotic variance of the estimate of a life percentile at the use stress level. Other test plans may be found in, for example, Nelson (1990, 2004) and Meeker and Escobar (1998). Nelson (2005) provides a nearly complete bibliography of the accelerated life (and degradation) test plans, which contains 159 publications.

7.8.1 Classification of Test Plans

There are different types of ALT plans in use, which include subjective, traditional, best traditional, statistically optimum, and compromise plans. They are discussed briefly below in order of increasing preference.

1. *Subjective plans.* Essentially, the test plans are chosen by judgment. When a test must end in a short time, engineers usually increase the low test stress level severely and reduce the space between the levels to shorten product life and meet the test schedule. On the other hand, the number of test units at each stress level is determined by the available capacity of test equipment or the cost of test units. Such test plans often yield inaccurate or even erroneous estimates of product reliability at the use condition, and should not be used.

2. *Traditional plans.* The low stress level is determined by experience, and the high one is the maximum allowable stress. The middle stress levels, if any, are equally spaced in between. The spacing is usually equal on the linearly transformed scale of the stress rather than on the original one. For example, temperature should use a $1/T$ scale. But the difference may be minor and ignorable in practice. The test units are also allocated equally to each stress level, which is a poor practice, for reasons given below. For example, a sample of 30 electronic sensors is tested at three temperatures. The low and high temperatures chosen are 85 and 150°C. Then the middle temperature is halfway between the low and high temperatures and is approximately 115°C. One-third of 30 units (i.e., 10 units) are tested at each temperature. Generally, low stress levels should have more test units to avoid no or few failures at these levels and to yield more accurate estimates of extrapolated life at a use stress level. Equal allocation of test units is against this principle and often yields poor estimates at a use condition. Traditional test plans are not recommended, although they are widely used in practice.

3. *Best traditional plans.* Like the traditional test plans, the best ones use equally spaced stress levels, each with the same number of test units. However, the low level is optimized to minimize the asymptotic variance of the estimate of a life percentile at a use condition. Nelson (1990, 2004) describes such test plans. Although better than traditional plans, the best traditional plans are less accurate than plans below because of equal allocation of test units.

4. *Statistically optimum plans.* For a single accelerating variable, the optimum plans use two stress levels. The low stress level and its sample allocation are optimized to minimize the asymptotic variance of the estimate of a specified life percentile at a use condition. The high stress level must be prespecified.

It should be as high as possible to yield more failures and reduce the variance of the estimate. However, the high stress level should not induce failure modes different from those at the use stress. Because only two stress levels are used, the test plans are sensitive to the misspecification of life distribution and preestimates of model parameters, which are required for calculating the plans. In other words, incorrect choice of the life distribution and preestimates may greatly compromise the optimality of the plans and result in a poor estimate. The use of only two stress levels does not allow checking the linearity of the assumed relationship and does not yield estimates of the relationship parameters when there are no failures at the low stress level. Hence, these plans are often not practically useful. However, they have a minimum variance, which is a benchmark for other test plans. For example, the compromise test plans below are often compared with the statistically optimum plans to evaluate the loss of accuracy for the robustness gained. The theory for optimum plans is described in Nelson and Kielpinski (1976) and Nelson and Meeker (1978). Kielpinski and Nelson (1975) and Meeker and Nelson (1975) provide the charts necessary for calculating particular plans. Nelson (1990, 2004) summarizes the theory and the charts. Meeker and Escobar (1995, 1998) describe optimum test plans with two or more accelerating variables.

5. *Compromise plans.* When a single accelerating variable is involved, such plans use three or more stress levels. The high level must be specified. The middle stress levels are often equally spaced between the low and high levels, but unequal spacing may be used. The low stress level and its number of units are optimized. The number allocated to a middle level may be specified as a fixed percentage of the total sample size or of the number of units at a low or high stress level. In the latter situation, the number is a variable. There are various optimization criteria, which include minimization of the asymptotic variance of the estimate of a life percentile at a use condition (the most commonly used criterion), the total test time, and others (Nelson, 2005). Important compromise plans with one accelerating variable are given in Meeker (1984), Meeker and Hahn (1985), G. Yang (1994), G. Yang and Jin (1994), Tang and Yang (2002), and Tang and Xu (2005). When two accelerating variables are involved, G. Yang and Yang (2002), and G. Yang (2005) give factorial compromise plans, which use four test conditions. Meeker and Escobar (1995, 1998) describe the 20% compromise plans, which employ five test conditions. In this section we focus on Yang's practical compromise test plans for Weibull and lognormal distributions with one or two accelerating variables.

7.8.2 Weibull Distribution with One Accelerating Variable

In this subsection we present *Yang's compromise test plans* for the Weibull distribution with one accelerating variable and a linear relationship. The test plans are based on the following model:

- The distribution of lifetime t is Weibull with shape parameter β and characteristic life α. Equivalently, the log life $x = \ln(t)$ has the smallest extreme value distribution with scale parameter $\sigma = 1/\beta$ and location parameter $\mu = \ln(\alpha)$.

- The scale parameter σ does not depend on the level of stress.
- The location parameter μ is a linear function of transformed stress S; namely, $\mu(S) = \gamma_0 + \gamma_1 S$, where γ_0 and γ_1 are unknown parameters to be estimated from test data.

Note that this model was stated in Section 7.7 in a more generic form.

The Yang compromise test plans use three stress levels. Here the censoring times at the three levels are predetermined by test schedule, equipment capacity, and other constraints. In general, a lower stress level needs a longer censoring time to yield enough failures. Unequal censoring times are recommended if the total test time is fixed. The high stress level must be specified. It should be as high as possible to yield more failures and decrease the variance of the estimate at the use stress; however, it should not cause failure modes that are different from those at the use stress. The low stress level and its number of test units are optimized to minimize the asymptotic variance of the estimate of the mean log life at a use stress. Here the middle stress level is midway between the low and high levels, and its number of test units is specified to be one-half that at the high level. The specifications are somewhat arbitrary, but intuitively reasonable in applications.

We use the following notation:

n = total sample size,

n_i = number of test units allocated to S_i; $i = 1, 2, 3$,

S = transformed stress,

S_i = level i of S, $i = 0, 1, 2, 3$; $i = 0$ implies use level; $i = 3$ implies high level,

$\xi_i = (S_i - S_3)/(S_0 - S_3)$ and is the transformed stress factor for S_i; $\xi_0 = 1$ for S_0,

$\quad \xi_3 = 0$ for S_3,

$\pi_i = n_i/n$ and is the proportion of total sample size n allocated to stress level i; $\quad i = 1, 2, 3$,

η_i = specified censoring time at stress level i; $i = 1, 2, 3$,

μ_i = location parameter value at stress level i; $i = 0, 1, 2, 3$,

$a_i = [\ln(\eta_i) - \mu_3]/\sigma$ and is the standardized censoring time, $i = 1, 2, 3$,

$b = (\mu_0 - \mu_3)/\sigma$.

For the smallest extreme value distribution, the mean equals the 43rd percentile. The asymptotic variance of the MLE of the mean, denoted $\hat{x}_{0.43}$, at the use stress level ($\xi_0 = 1$) is

$$\text{Var}[\hat{x}_{0.43}(1)] = \frac{\sigma^2}{n} V, \tag{7.100}$$

where V is called the *standardized variance* and is a function of a_i, b, ξ_i, and π_i ($i = 1, 2, 3$). The calculation of V is given in, for example, G. Yang and Jin (1994) and Meeker (1984). The values of a_i and b are determined by μ_0, μ_3, and σ, which are unknown in the test planning stage and must be preestimated. Methods for preestimation are described latter. The stress level ξ_1 and its sample proportion π_1 are to be optimized to minimize $\mathrm{Var}[\hat{x}_{0.43}(1)]$. Because n and σ in (7.100) are constant, the optimization model can be written as

$$\underset{\xi_1, \pi_1}{\mathrm{Min}}(V), \qquad (7.101)$$

subject to $\xi_2 = \xi_1/2$, $\xi_3 = 0$, $\pi_2 = \pi_3/2$, $\pi_3 = 1 - \pi_1 - \pi_2$, $0 \leq \xi_1 \leq 1$, $0 \leq \pi_1 \leq 1$. Here the constraint $\pi_2 = \pi_3/2$ is the same as $\pi_2 = (1 - \pi_1)/3$. Because $x = \ln(t)$, minimizing $\mathrm{Var}[\hat{x}_{0.43}(1)]$ is equivalent to minimizing the asymptotic variance of the MLE of the mean log life of the Weibull distribution at the use stress.

Given the values of a_i ($i = 1, 2, 3$) and b, we can solve (7.101) for ξ_1 and π_1 with a numerical method. Table 7.11 presents the optimum values of ξ_1, π_1, and V for various sets (a_1, a_2, a_3, b). When $a_1 = a_2 = a_3$, the test plan has a common censoring time. To find a plan from the table, one looks up the value of b first, then a_3, a_2 and a_1 in order. Linear interpolation may be needed for a combination (a_1, a_2, a_3, b) not given in the table. Extrapolation outside the table is not valid. Instead, use a numerical algorithm for solution. After obtaining these standardized values, we convert them to the transformed stress levels and actual sample allocations by using

$$S_i = S_3 + \xi_i(S_0 - S_3), \qquad n_i = \pi_i n. \qquad (7.102)$$

Then S_i are transformed back to the actual stress levels.

As stated above, the test plans depend on the values of μ_0, μ_3, and σ. They are unknown in the test planning phase and thus must be preestimated using experience, similar data, or a preliminary test. G. Yang (1994) suggests that μ_0 be approximated with the aid of a reliability prediction handbook such as MIL-HDBK-217F (U.S. DoD, 1995), which assumes constant failure rates. From the handbook we predict the failure rate, say λ_0, at the use stress. Then use $1/\lambda_0$ as the mean life. The preestimate of μ_0 is

$$\mu_0 = -\ln[\lambda_0 \Gamma(1 + \sigma)]. \qquad (7.103)$$

Example 7.16 An electronic module for pump control normally operates at 45°C. To estimate its reliability at this temperature, 50 units are to be tested at three elevated temperatures. The highest is 105°C, which is 5°C lower than the maximum allowable temperature. The specified censoring times are 1080, 600, and 380 hours for the low, middle, and high temperatures, respectively. Historical data analysis indicates that the product life can be modeled with a Weibull distribution with $\hat{\mu}_3 = 5.65$ and $\hat{\sigma} = 0.67$. Determine the Yang compromise test plan that minimizes the asymptotic variance of the MLE of the mean log life at 45°C.

TABLE 7.11 Compromise Test Plans for a Weibull Distribution with One Accelerating Variable

No.	a_1	a_2	a_3	b	π_1	ξ_1	V	No.	a_1	a_2	a_3	b	π_1	ξ_1	V
1	0	0	0	4	0.564	0.560	85.3	46	4	4	2	5	0.909	0.888	3.2
2	1	0	0	4	0.720	0.565	29.5	47	3	3	3	5	0.788	0.717	7.3
3	2	0	0	4	0.795	0.718	10.0	48	4	3	3	5	0.909	0.887	3.2
4	3	0	0	4	0.924	0.915	3.6	49	4	4	3	5	0.909	0.888	3.2
5	1	1	0	4	0.665	0.597	27.9	50	4	4	4	5	0.909	0.888	3.2
6	2	1	0	4	0.788	0.721	10.0	51	0	0	0	6	0.527	0.391	205.6
7	3	1	0	4	0.923	0.915	3.6	52	1	0	0	6	0.652	0.386	78.8
8	2	2	0	4	0.767	0.739	9.7	53	2	0	0	6	0.683	0.483	30.6
9	3	2	0	4	0.921	0.916	3.6	54	3	0	0	6	0.725	0.609	13.3
10	3	3	0	4	0.918	0.919	3.6	55	1	1	0	6	0.587	0.417	73.8
11	1	1	1	4	0.661	0.608	27.8	56	2	1	0	6	0.672	0.487	30.4
12	2	1	1	4	0.795	0.704	9.9	57	3	1	0	6	0.724	0.609	13.3
13	3	1	1	4	0.923	0.901	3.6	58	2	2	0	6	0.639	0.511	29.3
14	2	2	1	4	0.773	0.720	9.6	59	3	2	0	6	0.716	0.614	13.2
15	3	2	1	4	0.920	0.902	3.6	60	3	3	0	6	0.707	0.626	13.1
16	3	3	1	4	0.917	0.904	3.6	61	1	1	1	6	0.586	0.422	73.5
17	2	2	2	4	0.774	0.720	9.6	62	2	1	1	6	0.683	0.469	29.8
18	3	2	2	4	0.920	0.900	3.6	63	3	1	1	6	0.735	0.586	12.7
19	3	3	2	4	0.917	0.902	3.5	64	4	1	1	6	0.805	0.729	6.0
20	3	3	3	4	0.917	0.902	3.5	65	2	2	1	6	0.648	0.490	28.7
21	0	0	0	5	0.542	0.461	138.8	66	3	2	1	6	0.727	0.590	12.7
22	1	0	0	5	0.678	0.459	51.2	67	4	2	1	6	0.803	0.729	6.0
23	2	0	0	5	0.724	0.577	18.9	68	3	3	1	6	0.717	0.600	12.5
24	3	0	0	5	0.793	0.731	7.7	69	4	3	1	6	0.800	0.732	5.9
25	1	1	0	5	0.617	0.491	48.1	70	4	4	1	6	0.798	0.734	5.9
26	2	1	0	5	0.714	0.581	18.8	71	2	2	2	6	0.649	0.490	28.7
27	3	1	0	5	0.792	0.732	7.7	72	3	2	2	6	0.728	0.586	12.6
28	2	2	0	5	0.686	0.605	18.2	73	4	2	2	6	0.804	0.725	5.9
29	3	2	0	5	0.786	0.735	7.7	74	5	2	2	6	0.907	0.885	3.0
30	3	3	0	5	0.780	0.744	7.6	75	3	3	2	6	0.718	0.596	12.5
31	1	1	1	5	0.614	0.499	48.0	76	4	3	2	6	0.800	0.728	5.9
32	2	1	1	5	0.724	0.563	18.5	77	5	3	2	6	0.906	0.885	3.0
33	3	1	1	5	0.800	0.709	7.4	78	4	4	2	6	0.799	0.730	5.9
34	4	1	1	5	0.912	0.889	3.3	79	5	4	2	6	0.906	0.885	3.0
35	2	2	1	5	0.694	0.583	17.9	80	5	5	2	6	0.906	0.885	3.0
36	3	2	1	5	0.794	0.713	7.4	81	3	3	3	6	0.718	0.596	12.5
37	4	2	1	5	0.911	0.889	3.3	82	4	3	3	6	0.800	0.728	5.9
38	3	3	1	5	0.787	0.721	7.3	83	5	3	3	6	0.906	0.885	3.0
39	4	3	1	5	0.909	0.890	3.2	84	4	4	3	6	0.799	0.730	5.9
40	4	4	1	5	0.909	0.891	3.2	85	5	4	3	6	0.906	0.885	3.0
41	2	2	2	5	0.695	0.583	17.9	86	5	5	3	6	0.906	0.885	3.0
42	3	2	2	5	0.795	0.709	7.4	87	4	4	4	6	0.799	0.730	5.9
43	4	2	2	5	0.911	0.886	3.2	88	5	4	4	6	0.906	0.885	3.0
44	3	3	2	5	0.788	0.717	7.3	89	5	5	4	6	0.906	0.885	3.0
45	4	3	2	5	0.909	0.887	3.2	90	5	5	5	6	0.906	0.885	3.0

TABLE 7.11 (*continued*)

No.	a_1	a_2	a_3	b	π_1	ξ_1	V	No.	a_1	a_2	a_3	b	π_1	ξ_1	V
91	0	0	0	7	0.518	0.339	285.5	136	2	1	0	8	0.625	0.367	62.0
92	1	0	0	7	0.635	0.333	112.4	137	3	1	0	8	0.654	0.457	29.1
93	2	0	0	7	0.656	0.415	45.2	138	2	2	0	8	0.587	0.391	59.5
94	3	0	0	7	0.683	0.522	20.5	139	3	2	0	8	0.644	0.462	28.9
95	1	1	0	7	0.567	0.362	105.0	140	3	3	0	8	0.632	0.474	28.5
96	2	1	0	7	0.645	0.419	44.8	141	1	1	1	8	0.552	0.322	141.1
97	3	1	0	7	0.682	0.522	20.4	142	2	1	1	8	0.638	0.352	60.6
98	2	2	0	7	0.609	0.443	43.1	143	3	1	1	8	0.668	0.435	27.6
99	3	2	0	7	0.673	0.527	20.3	144	4	1	1	8	0.703	0.537	13.9
100	3	3	0	7	0.663	0.539	20.1	145	2	2	1	8	0.597	0.372	58.0
101	1	1	1	7	0.566	0.366	104.6	146	3	2	1	8	0.657	0.439	27.3
102	2	1	1	7	0.656	0.402	43.9	147	4	2	1	8	0.700	0.538	13.9
103	3	1	1	7	0.695	0.499	19.4	148	3	3	1	8	0.643	0.450	26.9
104	4	1	1	7	0.743	0.618	9.5	149	4	3	1	8	0.694	0.542	13.8
105	2	2	1	7	0.618	0.423	42.1	150	4	4	1	8	0.692	0.544	13.8
106	3	2	1	7	0.685	0.503	19.3	151	2	2	2	8	0.597	0.371	58.0
107	4	2	1	7	0.741	0.619	9.5	152	3	2	2	8	0.659	0.436	27.2
108	3	3	1	7	0.673	0.514	19.0	153	4	2	2	8	0.702	0.534	13.7
109	4	3	1	7	0.736	0.622	9.5	154	5	2	2	8	0.753	0.646	7.5
110	4	4	1	7	0.734	0.625	9.4	155	3	3	2	8	0.644	0.447	26.8
111	2	2	2	7	0.619	0.422	42.1	156	4	3	2	8	0.695	0.537	13.7
112	3	2	2	7	0.687	0.500	19.2	157	5	3	2	8	0.751	0.646	7.5
113	4	2	2	7	0.742	0.615	9.4	158	4	4	2	8	0.693	0.540	13.6
114	5	2	2	7	0.812	0.746	5.0	159	5	4	2	8	0.749	0.648	7.5
115	3	3	2	7	0.674	0.511	18.9	160	5	5	2	8	0.749	0.648	7.5
116	4	3	2	7	0.737	0.618	9.4	161	3	3	3	8	0.644	0.447	26.8
117	5	3	2	7	0.810	0.746	5.0	162	4	3	3	8	0.695	0.537	13.7
118	4	4	2	7	0.735	0.620	9.4	163	5	3	3	8	0.751	0.646	7.5
119	5	4	2	7	0.809	0.747	5.0	164	4	4	3	8	0.693	0.540	13.6
120	5	5	2	7	0.809	0.748	5.0	165	5	4	3	8	0.749	0.648	7.5
121	3	3	3	7	0.674	0.511	18.9	166	5	5	3	8	0.749	0.648	7.5
122	4	3	3	7	0.737	0.618	9.4	167	4	4	4	8	0.693	0.540	13.6
123	5	3	3	7	0.810	0.746	5.0	168	5	4	4	8	0.749	0.648	7.5
124	4	4	3	7	0.735	0.620	9.4	169	5	5	4	8	0.749	0.648	7.5
125	5	4	3	7	0.809	0.747	5.0	170	5	5	5	8	0.749	0.648	7.5
126	5	5	3	7	0.809	0.748	5.0	171	0	0	0	9	0.505	0.268	485.0
127	4	4	4	7	0.735	0.620	9.4	172	1	0	0	9	0.613	0.262	197.5
128	5	4	4	7	0.809	0.747	5.0	173	2	0	0	9	0.624	0.324	82.7
129	5	5	4	7	0.809	0.748	5.0	174	3	0	0	9	0.635	0.406	39.3
130	5	5	5	7	0.809	0.748	5.0	175	1	1	0	9	0.542	0.287	183.7
131	0	0	0	8	0.510	0.300	378.7	176	2	1	0	9	0.611	0.327	82.0
132	1	0	0	8	0.623	0.293	152.0	177	3	1	0	9	0.633	0.406	39.3
133	2	0	0	8	0.638	0.363	62.5	178	2	2	0	9	0.571	0.349	78.5
134	3	0	0	8	0.655	0.456	29.1	179	3	2	0	9	0.623	0.411	39.0
135	1	1	0	8	0.553	0.320	141.6	180	3	3	0	9	0.610	0.423	38.5

TABLE 7.11 (*continued*)

No.	a_1	a_2	a_3	b	π_1	ξ_1	V	No.	a_1	a_2	a_3	b	π_1	ξ_1	V
181	1	1	1	9	0.542	0.288	183.1	216	2	1	0	10	0.600	0.295	104.8
182	2	1	1	9	0.624	0.312	80.1	217	3	1	0	10	0.618	0.365	51.0
183	3	1	1	9	0.649	0.385	37.1	218	2	2	0	10	0.558	0.316	100.2
184	4	1	1	9	0.675	0.475	19.1	219	3	2	0	10	0.607	0.370	50.6
185	2	2	1	9	0.581	0.332	76.5	220	3	3	0	10	0.593	0.382	49.9
186	3	2	1	9	0.636	0.390	36.8	221	1	1	1	10	0.533	0.261	230.6
187	4	2	1	9	0.671	0.476	19.1	222	2	1	1	10	0.613	0.281	102.2
188	3	3	1	9	0.621	0.400	36.1	223	3	1	1	10	0.634	0.346	48.1
189	4	3	1	9	0.664	0.480	19.0	224	4	1	1	10	0.654	0.426	25.1
190	4	4	1	9	0.662	0.483	18.9	225	2	2	1	10	0.569	0.299	97.5
191	2	2	2	9	0.581	0.331	76.4	226	3	2	1	10	0.621	0.350	47.6
192	3	2	2	9	0.638	0.387	36.6	227	4	2	1	10	0.650	0.427	25.1
193	4	2	2	9	0.673	0.472	18.9	228	3	3	1	10	0.605	0.360	46.8
194	5	2	2	9	0.713	0.569	10.5	229	4	3	1	10	0.642	0.430	24.9
195	3	3	2	9	0.623	0.397	36.0	230	4	4	1	10	0.639	0.433	24.9
196	4	3	2	9	0.666	0.476	18.8	231	2	2	2	10	0.569	0.299	97.4
197	5	3	2	9	0.710	0.570	10.5	232	3	2	2	10	0.623	0.347	47.4
198	4	4	2	9	0.663	0.479	18.7	233	4	2	2	10	0.652	0.423	24.8
199	5	4	2	9	0.708	0.572	10.5	234	5	2	2	10	0.684	0.510	14.1
200	5	5	2	9	0.708	0.572	10.5	235	3	3	2	10	0.606	0.357	46.5
201	3	3	3	9	0.623	0.397	36.0	236	4	3	2	10	0.644	0.427	24.7
202	4	3	3	9	0.666	0.476	18.8	237	5	3	2	10	0.681	0.511	14.1
203	5	3	3	9	0.710	0.570	10.5	238	4	4	2	10	0.641	0.430	24.6
204	4	4	3	9	0.663	0.478	18.7	239	5	4	2	10	0.678	0.513	14.0
205	5	4	3	9	0.708	0.572	10.5	240	5	5	2	10	0.678	0.513	14.0
206	5	5	3	9	0.708	0.572	10.5	241	3	3	3	10	0.606	0.357	46.5
207	4	4	4	9	0.663	0.478	18.7	242	4	3	3	10	0.644	0.427	24.7
208	5	4	4	9	0.708	0.572	10.5	243	5	3	3	10	0.681	0.511	14.1
209	5	5	4	9	0.708	0.572	10.5	244	4	4	3	10	0.641	0.430	24.6
210	5	5	5	9	0.708	0.572	10.5	245	5	4	3	10	0.678	0.513	14.0
211	0	0	0	10	0.500	0.243	604.5	246	5	5	3	10	0.678	0.513	14.0
212	1	0	0	10	0.606	0.236	249.0	247	4	4	4	10	0.641	0.430	24.6
213	2	0	0	10	0.614	0.291	105.6	248	5	4	4	10	0.678	0.513	14.0
214	3	0	0	10	0.619	0.365	51.0	249	5	5	4	10	0.678	0.513	14.0
215	1	1	0	10	0.533	0.260	231.4	250	5	5	5	10	0.678	0.513	14.0

SOLUTION Using the reliability prediction handbook MIL-HDBK-217F (U.S. DoD, 1995) and each component's information about the load, quality level, operating environment, and other factors, we predict the failure rate of the module at 45°C to be $\lambda_0 = 6.83 \times 10^{-5}$ failures per hour. From (7.103),

$$\mu_0 = -\ln[6.83 \times 10^{-5} \times \Gamma(1 + 0.67)] = 9.69.$$

Then we have

$$a_1 = \frac{\ln(1080) - 5.65}{0.67} = 1.99, \quad a_2 = 1.11, \quad a_3 = 0.43, \quad b = \frac{9.69 - 5.65}{0.67} = 6.03.$$

TABLE 7.12   Actual Compromise Test Plan for the Electronic Module

Group	Temperature (°C)	Number of Test Units	Censoring Time (h)
1	74	34	1080
2	89	5	600
3	105	11	380

Table 7.11 yields $\pi_1 = 0.672$, $\xi_1 = 0.487$, and $V = 30.4$ for $(a_1, a_2, a_3, b) = (2, 1, 0, 6)$, and $\pi_1 = 0.683$, $\xi_1 = 0.469$ and $V = 29.8$ for $(a_1, a_2, a_3, b) = (2, 1, 1, 6)$. Linear interpolation to $a_3 = 0.43$ gives the optimal values $\pi_1 = 0.677$, $\xi_1 = 0.479$, and $V = 30.1$ for $(a_1, a_2, a_3, b) = (2, 1, 0.43, 6)$. Then, $\pi_2 = (1 - 0.677)/3 = 0.108$, $\xi_2 = 0.479/2 = 0.24$, $\pi_3 = 1 - 0.677 - 0.108 = 0.215$, and $\xi_3 = 0$. The standardized values are converted back to the original units. For example, the low reciprocal absolute temperature is

$$S_1 = \frac{1}{105 + 273.15} + 0.479 \times \left(\frac{1}{45 + 273.15} - \frac{1}{105 + 273.15}\right) = 0.002883.$$

The actual temperature (T_1) is $T_1 = 1/0.002883 - 273.15 = 74°C$. The number of test units at the low temperature is $n_1 = 0.677 \times 50 = 34$.

The variance of the MLE of the mean log life at 45°C can be preestimated by

$$\hat{\mathrm{Var}}[\hat{x}_{0.43}(1)] = \frac{0.67^2 \times 30.1}{50} = 0.27,$$

but the true variance should be estimated from test data when available.

The actual test plan for the electronic module is summarized in Table 7.12.

7.8.3   Lognormal Distribution with One Accelerating Variable

The Yang compromise test plans for the lognormal distribution with one accelerating variable and a linear relationship are similar to those for the Weibull distribution described in Section 7.8.2. They have the same assumptions (except for the distribution), the same notation, and the same specifications of the middle stress level and its sample allocation. Like the Weibull case, the lognormal plans also minimize the asymptotic variance of the MLE of the mean log life at a use stress.

The test plans depend on the values of μ_0, μ_3, and σ. They are unknown and should be preestimated using experience, similar data, or a preliminary test. If a reliability prediction handbook such as MIL-HDBK-217F (U.S. DoD, 1995) is used, μ_0 may be approximated by

$$\mu_0 = -\ln(\lambda_0) - 0.5\sigma^2, \tag{7.104}$$

where λ_0 is the failure rate at the use stress obtained from the reliability prediction. The test plans for various sets (a_1, a_2, a_3, b) are given in Table 7.13.

TABLE 7.13 Compromise Test Plans for a Lognormal Distribution with One Accelerating Variable

No.	a_1	a_2	a_3	b	π_1	ξ_1	V	No.	a_1	a_2	a_3	b	π_1	ξ_1	V
1	0	0	0	4	0.476	0.418	88.9	46	4	4	2	5	0.819	0.835	3.1
2	1	0	0	4	0.627	0.479	27.8	47	3	3	3	5	0.700	0.694	6.7
3	2	0	0	4	0.716	0.635	9.6	48	4	3	3	5	0.822	0.833	3.2
4	3	0	0	4	0.827	0.815	3.7	49	4	4	3	5	0.819	0.836	3.1
5	1	1	0	4	0.561	0.515	25.7	50	4	4	4	5	0.819	0.836	3.1
6	2	1	0	4	0.700	0.640	9.4	51	0	0	0	6	0.470	0.287	202.8
7	3	1	0	4	0.827	0.815	3.7	52	1	0	0	6	0.582	0.335	67.4
8	2	2	0	4	0.667	0.662	9.0	53	2	0	0	6	0.631	0.444	25.6
9	3	2	0	4	0.814	0.819	3.7	54	3	0	0	6	0.679	0.570	11.4
10	3	3	0	4	0.803	0.826	3.6	55	1	1	0	6	0.521	0.363	62.3
11	1	1	1	4	0.554	0.545	24.9	56	2	1	0	6	0.614	0.449	25.3
12	2	1	1	4	0.691	0.650	9.4	57	3	1	0	6	0.678	0.570	11.3
13	3	1	1	4	0.825	0.817	3.7	58	2	2	0	6	0.580	0.470	24.1
14	2	2	1	4	0.660	0.673	9.0	59	3	2	0	6	0.661	0.578	11.2
15	3	2	1	4	0.812	0.821	3.7	60	3	3	0	6	0.647	0.589	10.9
16	3	3	1	4	0.801	0.829	3.6	61	1	1	1	6	0.521	0.377	61.0
17	2	2	2	4	0.656	0.682	8.9	62	2	1	1	6	0.609	0.455	25.2
18	3	2	2	4	0.809	0.825	3.6	63	3	1	1	6	0.679	0.570	11.3
19	3	3	2	4	0.798	0.832	3.6	64	4	1	1	6	0.752	0.701	5.5
20	3	3	3	4	0.797	0.833	3.6	65	2	2	1	6	0.579	0.474	24.0
21	0	0	0	5	0.472	0.340	139.9	66	3	2	1	6	0.662	0.577	11.1
22	1	0	0	5	0.599	0.395	45.4	67	4	2	1	6	0.746	0.702	5.5
23	2	0	0	5	0.662	0.523	16.6	68	3	3	1	6	0.649	0.588	10.9
24	3	0	0	5	0.732	0.671	7.0	69	4	3	1	6	0.736	0.708	5.4
25	1	1	0	5	0.536	0.426	42.0	70	4	4	1	6	0.731	0.712	5.4
26	2	1	0	5	0.646	0.528	16.4	71	2	2	2	6	0.578	0.479	23.8
27	3	1	0	5	0.731	0.672	7.0	72	3	2	2	6	0.660	0.580	11.1
28	2	2	0	5	0.612	0.550	15.6	73	4	2	2	6	0.745	0.704	5.5
29	3	2	0	5	0.715	0.678	6.9	74	5	2	2	6	0.845	0.838	2.8
30	3	3	0	5	0.702	0.689	6.7	75	3	3	2	6	0.647	0.591	10.9
31	1	1	1	5	0.534	0.446	40.9	76	4	3	2	6	0.734	0.709	5.4
32	2	1	1	5	0.639	0.536	16.3	77	5	3	2	6	0.839	0.840	2.8
33	3	1	1	5	0.731	0.673	7.0	78	4	4	2	6	0.730	0.713	5.4
34	4	1	1	5	0.837	0.826	3.2	79	5	4	2	6	0.835	0.842	2.8
35	2	2	1	5	0.608	0.557	15.6	80	5	5	2	6	0.834	0.843	2.8
36	3	2	1	5	0.715	0.679	6.9	81	3	3	3	6	0.647	0.591	10.9
37	4	2	1	5	0.832	0.827	3.2	82	4	3	3	6	0.734	0.710	5.4
38	3	3	1	5	0.702	0.689	6.7	83	5	3	3	6	0.839	0.840	2.8
39	4	3	1	5	0.824	0.831	3.2	84	4	4	3	6	0.730	0.714	5.4
40	4	4	1	5	0.821	0.834	3.1	85	5	4	3	6	0.835	0.842	2.8
41	2	2	2	5	0.607	0.564	15.4	86	5	5	3	6	0.834	0.843	2.8
42	3	2	2	5	0.712	0.683	6.8	87	4	4	4	6	0.730	0.714	5.4
43	4	2	2	5	0.831	0.829	3.2	88	5	4	4	6	0.835	0.842	2.8
44	3	3	2	5	0.700	0.693	6.7	89	5	5	4	6	0.834	0.843	2.8
45	4	3	2	5	0.823	0.833	3.2	90	5	5	5	6	0.834	0.843	2.8

TABLE 7.13 (*continued*)

No.	a_1	a_2	a_3	b	π_1	ξ_1	V	No.	a_1	a_2	a_3	b	π_1	ξ_1	V
91	0	0	0	7	0.468	0.247	277.5	136	2	1	0	8	0.578	0.345	49.0
92	1	0	0	7	0.570	0.291	93.9	137	3	1	0	8	0.621	0.438	23.3
93	2	0	0	7	0.610	0.386	36.6	138	2	2	0	8	0.546	0.362	46.8
94	3	0	0	7	0.645	0.495	16.8	139	3	2	0	8	0.603	0.444	22.9
95	1	1	0	7	0.511	0.315	86.8	140	3	3	0	8	0.590	0.455	22.4
96	2	1	0	7	0.593	0.390	36.2	141	1	1	1	8	0.505	0.288	113.6
97	3	1	0	7	0.644	0.495	16.8	142	2	1	1	8	0.575	0.348	48.9
98	2	2	0	7	0.560	0.409	34.5	143	3	1	1	8	0.623	0.436	23.2
99	3	2	0	7	0.627	0.502	16.5	144	4	1	1	8	0.667	0.536	12.1
100	3	3	0	7	0.613	0.514	16.1	145	2	2	1	8	0.546	0.363	46.6
101	1	1	1	7	0.512	0.327	85.2	146	3	2	1	8	0.606	0.442	22.9
102	2	1	1	7	0.589	0.394	36.0	147	4	2	1	8	0.660	0.537	12.1
103	3	1	1	7	0.646	0.494	16.7	148	3	3	1	8	0.592	0.452	22.3
104	4	1	1	7	0.701	0.607	8.5	149	4	3	1	8	0.648	0.543	11.9
105	2	2	1	7	0.560	0.412	34.4	150	4	4	1	8	0.644	0.547	11.8
106	3	2	1	7	0.629	0.501	16.5	151	2	2	2	8	0.546	0.367	46.4
107	4	2	1	7	0.694	0.609	8.5	152	3	2	2	8	0.604	0.444	22.8
108	3	3	1	7	0.615	0.511	16.1	153	4	2	2	8	0.659	0.538	12.1
109	4	3	1	7	0.683	0.615	8.4	154	5	2	2	8	0.716	0.641	6.8
110	4	4	1	7	0.679	0.619	8.3	155	3	3	2	8	0.592	0.453	22.3
111	2	2	2	7	0.559	0.416	34.2	156	4	3	2	8	0.648	0.544	11.9
112	3	2	2	7	0.627	0.503	16.4	157	5	3	2	8	0.708	0.644	6.8
113	4	2	2	7	0.693	0.610	8.5	158	4	4	2	8	0.643	0.548	11.8
114	5	2	2	7	0.766	0.727	4.6	159	5	4	2	8	0.703	0.647	6.7
115	3	3	2	7	0.614	0.513	16.1	160	5	5	2	8	0.702	0.649	6.7
116	4	3	2	7	0.682	0.616	8.4	161	3	3	3	8	0.592	0.454	22.3
117	5	3	2	7	0.759	0.729	4.6	162	4	3	3	8	0.647	0.544	11.9
118	4	4	2	7	0.678	0.620	8.3	163	5	3	3	8	0.708	0.644	6.8
119	5	4	2	7	0.754	0.733	4.5	164	4	4	3	8	0.643	0.549	11.8
120	5	5	2	7	0.753	0.734	4.5	165	5	4	3	8	0.703	0.647	6.7
121	3	3	3	7	0.614	0.514	16.0	166	5	5	3	8	0.701	0.649	6.7
122	4	3	3	7	0.682	0.617	8.4	167	4	4	4	8	0.643	0.549	11.8
123	5	3	3	7	0.759	0.729	4.6	168	5	4	4	8	0.703	0.648	6.7
124	4	4	3	7	0.678	0.621	8.3	169	5	5	4	8	0.701	0.649	6.7
125	5	4	3	7	0.754	0.733	4.5	170	5	5	5	8	0.701	0.649	6.7
126	5	5	3	7	0.753	0.734	4.5	171	0	0	0	9	0.466	0.194	462.5
127	4	4	4	7	0.678	0.621	8.3	172	1	0	0	9	0.555	0.231	160.0
128	5	4	4	7	0.754	0.733	4.5	173	2	0	0	9	0.584	0.305	64.7
129	5	5	4	7	0.753	0.734	4.5	174	3	0	0	9	0.604	0.392	30.9
130	5	5	5	7	0.753	0.734	4.5	175	1	1	0	9	0.498	0.250	148.3
131	0	0	0	8	0.467	0.218	364.1	176	2	1	0	9	0.567	0.309	63.9
132	1	0	0	8	0.562	0.257	124.7	177	3	1	0	9	0.604	0.392	30.9
133	2	0	0	8	0.595	0.341	49.6	178	2	2	0	9	0.535	0.325	60.9
134	3	0	0	8	0.621	0.437	23.3	179	3	2	0	9	0.586	0.398	30.4
135	1	1	0	8	0.503	0.279	115.5	180	3	3	0	9	0.573	0.408	29.7

TABLE 7.13 (*continued*)

No.	a_1	a_2	a_3	b	π_1	ξ_1	V	No.	a_1	a_2	a_3	b	π_1	ξ_1	V
181	1	1	1	9	0.500	0.257	146.1	216	2	1	0	10	0.558	0.280	80.7
182	2	1	1	9	0.565	0.311	63.7	217	3	1	0	10	0.591	0.355	39.6
183	3	1	1	9	0.607	0.390	30.8	218	2	2	0	10	0.527	0.294	77.0
184	4	1	1	9	0.642	0.479	16.4	219	3	2	0	10	0.573	0.361	39.0
185	2	2	1	9	0.536	0.325	60.8	220	3	3	0	10	0.561	0.370	38.1
186	3	2	1	9	0.589	0.395	30.3	221	1	1	1	10	0.496	0.232	182.7
187	4	2	1	9	0.635	0.480	16.4	222	2	1	1	10	0.557	0.281	80.5
188	3	3	1	9	0.576	0.404	29.6	223	3	1	1	10	0.594	0.353	39.5
189	4	3	1	9	0.623	0.486	16.2	224	4	1	1	10	0.624	0.433	21.4
190	4	4	1	9	0.619	0.490	16.0	225	2	2	1	10	0.528	0.294	76.9
191	2	2	2	9	0.536	0.328	60.5	226	3	2	1	10	0.576	0.358	38.8
192	3	2	2	9	0.588	0.397	30.2	227	4	2	1	10	0.617	0.434	21.3
193	4	2	2	9	0.635	0.481	16.4	228	3	3	1	10	0.563	0.366	37.9
194	5	2	2	9	0.681	0.573	9.4	229	4	3	1	10	0.605	0.440	21.0
195	3	3	2	9	0.576	0.406	29.6	230	4	4	1	10	0.600	0.444	20.8
196	4	3	2	9	0.623	0.487	16.1	231	2	2	2	10	0.528	0.297	76.5
197	5	3	2	9	0.673	0.576	9.4	232	3	2	2	10	0.575	0.359	38.8
198	4	4	2	9	0.619	0.491	16.0	233	4	2	2	10	0.616	0.435	21.3
199	5	4	2	9	0.667	0.579	9.3	234	5	2	2	10	0.656	0.518	12.5
200	5	5	2	9	0.666	0.581	9.3	235	3	3	2	10	0.563	0.367	37.9
201	3	3	3	9	0.575	0.406	29.5	236	4	3	2	10	0.605	0.440	21.0
202	4	3	3	9	0.623	0.487	16.1	237	5	3	2	10	0.647	0.520	12.4
203	5	3	3	9	0.673	0.576	9.4	238	4	4	2	10	0.600	0.444	20.8
204	4	4	3	9	0.618	0.491	16.0	239	5	4	2	10	0.642	0.524	12.3
205	5	4	3	9	0.667	0.579	9.3	240	5	5	2	10	0.640	0.525	12.3
206	5	5	3	9	0.666	0.581	9.3	241	3	3	3	10	0.563	0.368	37.8
207	4	4	4	9	0.618	0.491	16.0	242	4	3	3	10	0.605	0.440	21.0
208	5	4	4	9	0.667	0.580	9.3	243	5	3	3	10	0.647	0.521	12.4
209	5	5	4	9	0.666	0.581	9.3	244	4	4	3	10	0.600	0.444	20.8
210	5	5	5	9	0.666	0.581	9.3	245	5	4	3	10	0.641	0.524	12.3
211	0	0	0	10	0.465	0.175	572.8	246	5	5	3	10	0.640	0.526	12.3
212	1	0	0	10	0.551	0.209	199.7	247	4	4	4	10	0.600	0.444	20.8
213	2	0	0	10	0.575	0.276	81.7	248	5	4	4	10	0.641	0.524	12.3
214	3	0	0	10	0.591	0.354	39.7	249	5	5	4	10	0.640	0.526	12.3
215	1	1	0	10	0.494	0.226	185.2	250	5	5	5	10	0.640	0.526	12.3

Example 7.17 Refer to Example 7.16. Suppose that the life distribution of the electronic module is mistakenly modeled with the lognormal, and other data are the same as those in Example 7.16. Determine the test plan. Comment on the sensitivity of the Yang compromise test plan to the incorrect choice of life distribution.

SOLUTION For the preestimates in Example 7.16, Table 7.13 and linear interpolation yield the optimal values $\pi_1 = 0.612$, $\xi_1 = 0.451$, and $V = 25.2$. Then $\pi_2 = (1 - 0.612)/3 = 0.129$, $\xi_2 = 0.451/2 = 0.23$, $\pi_3 = 1 - 0.612 - 0.129 =$

0.259, and $\xi_3 = 0$. As in Example 7.16, the standardized test plan can easily be transformed to the actual plan.

The test plan above, based on the incorrect lognormal distribution, is evaluated with the actual Weibull distribution. In other words, the standardized variance V_0 for the Weibull test plan is calculated for $\pi_1 = 0.612$ and $\xi_1 = 0.451$ obtained above. Then we have $V_0 = 31.1$. The variance increase ratio is $100 \times (31.1 - 30.1)/30.1 = 3.3\%$, where 30.1 is the standardized variance in Example 7.16. The small increase indicates that the Yang compromise test plan is not sensitive to the misspecification of life distribution for the given preestimates.

7.8.4 Weibull Distribution with Two Accelerating Variables

In this subsection we present Yang's compromise test plans for the Weibull distribution with two accelerating variables and a linear relationship. G. Yang (2005) describes the theory of the test plans. The test plans are based on the following model:

- The distribution of lifetime t is Weibull with shape parameter β and characteristic life α. Equivalently, the log life $x = \ln(t)$ has the smallest extreme value distribution with scale parameter $\sigma = 1/\beta$ and location parameter $\mu = \ln(\alpha)$.
- The scale parameter σ does not depend on the level of stresses.
- The location parameter μ is a linear function of transformed stresses (S_1 and S_2): namely,

$$\mu(S_1, S_2) = \gamma_0 + \gamma_1 S_1 + \gamma_2 S_2, \tag{7.105}$$

where γ_0, γ_1, and γ_2 are the unknown parameters to be estimated from test data.

The Yang compromise test plans use rectangular test points, as shown in Figure 7.21. The test plans are full factorial designs, where each accelerating variable is a two-level factor. Such test plans are intuitively appealing and

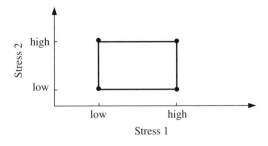

FIGURE 7.21 A 2^2 full factorial design

economically efficient in many applications. For example, if the accelerating variables are temperature and voltage, the test plans need only two chambers (one for each temperature) because each one can accommodate two groups of test units simultaneously, each loaded with a different voltage. A similar example is the test at higher temperatures and usage rates. Two groups of test units subjected to the same temperature and different usage rates may be run concurrently in one chamber.

Like test plans with one accelerating variable, the four test points here have prespecified censoring times. In general, a test point with a lower stress level needs a longer censoring time to produce enough failures. The censoring time for the southwest point should be the longest one, whereas the one for the northeast point should be shortest, if the total test time is fixed. In the situations where two groups can be tested simultaneously on the same equipment, their censoring times could be set equal. In addition to the censoring time, we also specify the high levels of the two stresses. The levels should be as high as possible to yield more failures and reduce the variance of the estimate at the use stresses; however, they should not cause failure modes that are different from those at the use levels. The low stress levels (the southwest point) and the corresponding sample allocation are optimized to minimize the asymptotic variance of the MLE of the mean log life at the use stresses. The southeast and northwest test points are each allocated 10% of the total sample size. G. Yang (2005) reports that such a choice results in less than 20% increases in variance compared with the degenerate test plans, which are statistically optimum and not useful in practice (Meeker and Escobar, 1998). Other specifications may be used, but need careful justification.

We use the following notation:

n = total sample size,

n_{ij} = number of test units allocated to test point $(S_{1i}, S_{2j}); i = 1, 2; j = 1, 2,$

S_1, S_2 = transformed stresses 1 and 2,

S_{1i} = level i of $S_1, i = 0, 1, 2; i = 0$ implies use level; $i = 2$ implies high level,

S_{2j} = level j of $S_2, j = 0, 1, 2; j = 0$ implies use level; $j = 2$ implies high level,

$\xi_{1i} = (S_{1i} - S_{12})/(S_{10} - S_{12})$ and is the transformed stress factor for S_1;

$\quad \xi_{10} = 1$ for $S_{10}, \xi_{12} = 0$ for $S_{12},$

$\xi_{2j} = (S_{2j} - S_{22})/(S_{20} - S_{22})$ and is the transformed stress factor for S_2;

$\quad \xi_{20} = 1$ for $S_{20}, \xi_{22} = 0$ for $S_{22},$

$\pi_{ij} = n_{ij}/n$ and is the proportion of total sample size n allocated to test

$\quad$ point $(S_{1i}, S_{2j}); i = 1, 2; j = 1, 2,$

η_{ij} = censoring time for test point $(S_{1i}, S_{2j}); i = 1, 2; j = 1, 2,$

μ_{ij} = location parameter value at test point $(S_{1i}, S_{2j}); i = 0, 1, 2; j = 0, 1, 2,$

$a_{ij} = [\ln(\eta_{ij}) - \mu_{22}]/\sigma$ and is the standardized censoring time; $i=1, 2; j=1, 2,$

$b = (\mu_{00} - \mu_{20})/\sigma,$

$c = (\mu_{00} - \mu_{02})/\sigma.$

Similar to the single accelerating variable case, the asymptotic variance of the MLE of the mean, denoted $\hat{x}_{0.43}$, at the use stresses ($\xi_{10} = 1, \xi_{20} = 1$) is given by

$$\text{Var}[\hat{x}_{0.43}(1, 1)] = \frac{\sigma^2}{n} V, \tag{7.106}$$

where V is called the *standardized variance*. V is a function of a_{ij}, b, c, ξ_{1i}, ξ_{2j}, and π_{ij} ($i = 1, 2; j = 1, 2$), and independent of n and σ. The calculation of V is given in, for example, Meeker and Escobar (1995), and G. Yang (2005). As specified above, $\xi_{12} = 0$, $\xi_{22} = 0$, and $\pi_{12} = \pi_{21} = 0.1$. Given the preestimates of a_{ij}, b, and c, the test plans choose the optimum values of ξ_{11}, ξ_{21}, and π_{11} by minimizing $\text{Var}[\hat{x}_{0.43}(1, 1)]$. Because n and σ in (7.106) are constant, the optimization model can be written as

$$\text{Min}(V), \tag{7.107}$$
$$\xi_{11}, \xi_{21}, \pi_{11}$$

subject to $\xi_{12} = 0$, $\xi_{22} = 0$, $\pi_{12} = \pi_{21} = 0.1$, $\pi_{22} = 1 - \pi_{11} - \pi_{12} - \pi_{21}$, $0 \leq \xi_{11} \leq 1$, $0 \leq \xi_{21} \leq 1$, and $0 \leq \pi_{11} \leq 1$.

Because $x = \ln(t)$, minimizing $\text{Var}[\hat{x}_{0.43}(1, 1)]$ is equivalent to minimizing the asymptotic variance of the MLE of the mean log life of the Weibull distribution at the use stresses.

The test plans contain six prespecified values (a_{11}, a_{12}, a_{21}, a_{22}, b, and c). To tabulate the test plans in a manageable manner, we consider only two different censoring times, one for the southwest and northwest points and the other for the southeast and northeast points. Therefore, $a_{11} = a_{12} \equiv a_1$ and $a_{21} = a_{22} \equiv a_2$. This special case is realistic and often encountered in practice, because as explained earlier, two groups may be tested concurrently on the same equipment and are subjected to censoring at the same time. When $a_1 = a_2$, all test points have a common censoring time. Table 7.14 presents the values of ξ_{11}, ξ_{21}, π_{11}, and V for various sets (a_1, a_2, b, c). To find a plan from the table, one looks up the value of c first, then b, a_2, and a_1 in order. Linear interpolation may be needed for a combination (a_1, a_2, b, c) not given in the table, but extrapolation outside the table is not valid. For a combination (a_1, a_2, b, c) outside the table, numerical calculation of the optimization model is necessary. The Excel spreadsheet for the calculation is available from the author. After obtaining the standardized values, we convert them to the transformed stress levels and sample allocations by using

$$S_{1i} = S_{12} + \xi_{1i}(S_{10} - S_{12}), \qquad S_{2j} = S_{22} + \xi_{2j}(S_{20} - S_{22}), \qquad n_{ij} = \pi_{ij}n. \tag{7.108}$$

Then S_{1i} and S_{2j} are further transformed back to the actual stress levels.

TABLE 7.14   Compromise Test Plans for a Weibull Distribution with Two Accelerating Variables

No.	a_1	a_2	b	c	π_{11}	ξ_{11}	ξ_{21}	V	No.	a_1	a_2	b	c	π_{11}	ξ_{11}	ξ_{21}	V
1	2	2	2	2	0.637	0.769	0.769	10.3	46	4	3	3	4	0.605	0.635	0.610	10.4
2	3	2	2	2	0.730	0.912	0.934	4.2	47	5	3	3	4	0.658	0.754	0.732	5.7
3	2	2	3	2	0.584	0.597	0.636	18.8	48	2	2	4	4	0.511	0.389	0.389	59.8
4	3	2	3	2	0.644	0.706	0.770	8.1	49	3	2	4	4	0.545	0.441	0.455	29.0
5	4	2	3	2	0.724	0.868	0.935	3.8	50	4	2	4	4	0.579	0.529	0.545	15.1
6	2	2	4	2	0.555	0.486	0.543	29.8	51	5	3	4	4	0.614	0.642	0.648	8.4
7	3	2	4	2	0.599	0.577	0.660	13.3	52	4	4	4	4	0.572	0.547	0.547	14.9
8	4	2	4	2	0.654	0.706	0.798	6.5	53	5	4	4	4	0.612	0.648	0.648	8.4
9	2	2	5	2	0.538	0.409	0.474	43.2	54	6	4	4	4	0.661	0.761	0.762	5.0
10	3	2	5	2	0.570	0.488	0.579	19.8	55	2	2	5	4	0.500	0.339	0.353	78.6
11	4	2	5	2	0.611	0.596	0.700	10.0	56	3	2	5	4	0.529	0.387	0.414	38.4
12	2	2	2	3	0.584	0.636	0.597	18.8	57	4	2	5	4	0.556	0.464	0.494	20.4
13	3	2	2	3	0.648	0.731	0.710	8.3	58	3	3	5	4	0.524	0.401	0.419	37.9
14	4	2	2	3	0.729	0.893	0.871	3.9	59	4	3	5	4	0.552	0.472	0.494	20.3
15	2	2	3	3	0.548	0.516	0.516	29.9	60	5	3	5	4	0.584	0.561	0.585	11.6
16	3	2	3	3	0.596	0.593	0.610	13.8	61	5	4	5	4	0.583	0.563	0.585	11.6
17	4	2	3	3	0.651	0.718	0.737	6.8	62	6	4	5	4	0.622	0.661	0.685	7.1
18	3	3	3	3	0.592	0.618	0.618	13.5	63	2	2	6	4	0.493	0.300	0.322	99.9
19	4	3	3	3	0.647	0.732	0.737	6.7	64	3	2	6	4	0.517	0.345	0.381	49.1
20	5	3	3	3	0.719	0.878	0.883	3.6	65	4	2	6	4	0.540	0.414	0.454	26.4
21	2	2	4	3	0.528	0.432	0.454	43.6	66	3	3	6	4	0.513	0.354	0.385	48.6
22	3	2	4	3	0.565	0.500	0.539	20.5	67	4	3	6	4	0.537	0.419	0.454	26.3
23	4	2	4	3	0.606	0.603	0.646	10.4	68	5	3	6	4	0.563	0.498	0.535	15.3
24	3	3	4	3	0.562	0.514	0.544	20.2	69	5	4	6	4	0.563	0.499	0.535	15.3
25	4	3	4	3	0.603	0.612	0.646	10.4	70	6	4	6	4	0.594	0.586	0.625	9.4
26	5	3	4	3	0.655	0.731	0.767	5.7	71	2	2	7	4	0.488	0.269	0.297	123.7
27	2	2	5	3	0.515	0.371	0.404	59.7	72	3	2	7	4	0.508	0.311	0.353	61.0
28	3	2	5	3	0.544	0.431	0.483	28.4	73	4	2	7	4	0.527	0.374	0.421	33.0
29	4	2	5	3	0.577	0.521	0.578	14.8	74	3	3	7	4	0.505	0.317	0.355	60.5
30	3	3	5	3	0.542	0.440	0.486	28.2	75	4	3	7	4	0.525	0.378	0.421	32.9
31	4	3	5	3	0.575	0.526	0.578	14.7	76	5	3	7	4	0.547	0.448	0.495	19.4
32	5	3	5	3	0.615	0.628	0.684	8.3	77	4	4	7	4	0.524	0.378	0.421	32.9
33	2	2	6	3	0.507	0.324	0.365	78.3	78	5	4	7	4	0.547	0.449	0.495	19.4
34	3	2	6	3	0.530	0.380	0.439	37.5	79	6	4	7	4	0.574	0.527	0.577	12.1
35	4	2	6	3	0.556	0.459	0.526	19.8	80	2	2	2	5	0.538	0.474	0.409	43.2
36	3	3	6	3	0.528	0.385	0.440	37.4	81	3	2	2	5	0.581	0.533	0.484	20.7
37	4	3	6	3	0.555	0.464	0.528	19.8	82	4	2	2	5	0.625	0.641	0.592	10.4
38	5	3	6	3	0.587	0.551	0.620	11.3	83	2	2	3	5	0.515	0.404	0.371	59.7
39	2	2	2	4	0.555	0.543	0.486	29.8	84	3	2	3	5	0.552	0.453	0.432	29.2
40	3	2	2	4	0.606	0.615	0.575	13.8	85	4	2	3	5	0.586	0.543	0.523	15.2
41	4	2	2	4	0.663	0.744	0.704	6.8	86	3	3	3	5	0.542	0.486	0.440	28.2
42	2	2	3	4	0.528	0.454	0.432	43.6	87	4	3	3	5	0.578	0.562	0.522	14.9
43	3	2	3	4	0.569	0.513	0.506	20.8	88	5	3	3	5	0.620	0.663	0.627	8.4
44	4	2	3	4	0.611	0.617	0.611	10.6	89	2	2	4	5	0.500	0.353	0.339	78.6
45	3	3	3	4	0.562	0.544	0.514	20.2	90	3	2	4	5	0.532	0.395	0.393	38.9

TABLE 7.14 (*continued*)

No.	a_1	a_2	b	c	π_{11}	ξ_{11}	ξ_{21}	V	No.	a_1	a_2	b	c	π_{11}	ξ_{11}	ξ_{21}	V
91	4	2	4	5	0.560	0.473	0.472	20.6	136	6	5	8	5	0.542	0.444	0.480	18.8
92	3	3	4	5	0.524	0.419	0.401	37.9	137	2	2	3	6	0.507	0.365	0.324	78.3
93	4	3	4	5	0.554	0.487	0.472	20.4	138	3	2	3	6	0.540	0.407	0.378	38.9
94	5	3	4	5	0.587	0.575	0.562	11.7	139	4	2	3	6	0.569	0.486	0.459	20.5
95	4	4	4	5	0.551	0.495	0.475	20.2	140	3	3	3	6	0.528	0.440	0.385	37.4
96	5	4	4	5	0.584	0.583	0.563	11.6	141	4	3	3	6	0.560	0.505	0.457	20.2
97	6	4	4	5	0.623	0.682	0.661	7.1	142	5	3	3	6	0.595	0.593	0.549	11.6
98	2	2	5	5	0.491	0.312	0.312	100.0	143	2	2	4	6	0.493	0.322	0.300	99.9
99	3	2	5	5	0.518	0.351	0.362	49.8	144	3	2	4	6	0.523	0.359	0.347	50.2
100	4	2	5	5	0.542	0.419	0.432	26.8	145	4	2	4	6	0.548	0.428	0.418	26.9
101	3	3	5	5	0.511	0.369	0.369	48.8	146	4	3	4	6	0.539	0.443	0.417	26.5
102	4	3	5	5	0.536	0.430	0.432	26.6	147	5	3	4	6	0.568	0.521	0.498	15.4
103	5	3	5	5	0.564	0.508	0.513	15.5	148	4	4	4	6	0.536	0.455	0.421	26.2
104	5	4	5	5	0.562	0.513	0.513	15.4	149	5	4	4	6	0.563	0.532	0.498	15.3
105	6	4	5	5	0.593	0.600	0.601	9.5	150	6	4	4	6	0.596	0.619	0.585	9.5
106	5	5	5	5	0.561	0.513	0.513	15.4	151	2	2	5	6	0.484	0.288	0.279	123.9
107	6	5	5	5	0.593	0.601	0.601	9.5	152	3	2	5	6	0.511	0.322	0.322	62.7
108	2	2	6	5	0.484	0.279	0.288	123.9	153	4	2	5	6	0.532	0.383	0.385	34.0
109	3	2	6	5	0.508	0.316	0.337	62.0	154	3	3	5	6	0.502	0.341	0.329	61.0
110	4	2	6	5	0.528	0.377	0.400	33.7	155	4	3	5	6	0.525	0.395	0.385	33.6
111	3	3	6	5	0.502	0.329	0.341	61.0	156	5	3	5	6	0.549	0.465	0.458	19.8
112	4	3	6	5	0.524	0.385	0.400	33.5	157	4	4	5	6	0.523	0.402	0.388	33.4
113	5	3	6	5	0.547	0.455	0.473	19.7	158	5	4	5	6	0.545	0.473	0.458	19.7
114	5	4	6	5	0.545	0.458	0.473	19.7	159	6	4	5	6	0.572	0.550	0.536	12.3
115	6	4	6	5	0.572	0.536	0.553	12.3	160	5	5	5	6	0.545	0.473	0.458	19.7
116	5	5	6	5	0.545	0.458	0.473	19.7	161	6	5	5	6	0.571	0.553	0.537	12.3
117	6	5	6	5	0.571	0.537	0.553	12.3	162	3	3	6	6	0.494	0.307	0.307	74.8
118	2	2	7	5	0.480	0.252	0.268	150.3	163	4	3	6	6	0.514	0.356	0.359	41.5
119	3	2	7	5	0.500	0.288	0.314	75.4	164	5	3	6	6	0.534	0.420	0.425	24.6
120	4	2	7	5	0.518	0.343	0.374	41.3	165	4	4	6	6	0.513	0.361	0.361	41.3
121	3	3	7	5	0.495	0.297	0.318	74.5	166	5	4	6	6	0.532	0.425	0.425	24.5
122	4	3	7	5	0.514	0.349	0.374	41.1	167	6	4	6	6	0.555	0.496	0.497	15.5
123	5	3	7	5	0.534	0.413	0.440	24.4	168	5	5	6	6	0.532	0.425	0.425	24.5
124	5	4	7	5	0.533	0.415	0.440	24.4	169	6	5	6	6	0.554	0.497	0.497	15.5
125	6	4	7	5	0.555	0.485	0.513	15.4	170	6	6	6	6	0.554	0.497	0.497	15.5
126	5	5	7	5	0.533	0.415	0.440	24.4	171	3	3	7	6	0.488	0.279	0.288	89.7
127	6	5	7	5	0.555	0.486	0.513	15.4	172	4	3	7	6	0.506	0.325	0.337	50.0
128	3	2	8	5	0.493	0.264	0.295	90.0	173	5	3	7	6	0.523	0.384	0.397	29.9
129	4	2	8	5	0.509	0.315	0.351	49.6	174	4	4	7	6	0.505	0.328	0.338	49.9
130	3	3	8	5	0.490	0.270	0.298	89.2	175	5	4	7	6	0.522	0.386	0.397	29.9
131	4	3	8	5	0.506	0.319	0.350	49.4	176	6	4	7	6	0.541	0.452	0.464	19.0
132	4	4	8	5	0.506	0.320	0.351	49.4	177	5	5	7	6	0.522	0.387	0.397	29.9
133	5	4	8	5	0.523	0.379	0.412	29.5	178	6	5	7	6	0.541	0.452	0.464	19.0
134	6	4	8	5	0.542	0.444	0.480	18.8	179	6	6	7	6	0.541	0.452	0.464	19.0
135	5	5	8	5	0.523	0.379	0.412	29.5	180	3	3	8	6	0.483	0.255	0.271	105.9

TABLE 7.14 (*continued*)

No.	a_1	a_2	b	c	π_{11}	ξ_{11}	ξ_{21}	V	No.	a_1	a_2	b	c	π_{11}	ξ_{11}	ξ_{21}	V
181	4	3	8	6	0.499	0.299	0.317	59.3	220	5	5	7	7	0.513	0.363	0.363	35.8
182	5	3	8	6	0.514	0.353	0.374	35.7	221	6	5	7	7	0.530	0.424	0.424	22.9
183	4	4	8	6	0.498	0.301	0.318	59.2	222	6	6	7	7	0.530	0.424	0.424	22.9
184	5	4	8	6	0.513	0.355	0.374	35.6	223	4	4	8	7	0.477	0.276	0.299	70.0
185	6	4	8	6	0.530	0.415	0.436	22.8	224	5	4	8	7	0.506	0.334	0.342	42.2
186	5	5	8	6	0.513	0.355	0.374	35.6	225	6	4	8	7	0.521	0.390	0.399	27.1
187	6	5	8	6	0.531	0.416	0.436	22.8	226	5	5	8	7	0.506	0.334	0.343	42.2
188	6	6	8	6	0.530	0.415	0.436	22.8	227	6	5	8	7	0.521	0.391	0.400	27.1
189	2	2	4	7	0.488	0.297	0.269	123.7	228	6	6	8	7	0.521	0.391	0.400	27.1
190	3	2	4	7	0.517	0.329	0.311	62.8	229	2	2	5	8	0.476	0.250	0.230	179.3
191	3	3	4	7	0.505	0.355	0.317	60.5	230	3	2	5	8	0.502	0.276	0.264	92.3
192	4	3	4	7	0.529	0.407	0.374	33.4	231	3	3	5	8	0.490	0.298	0.270	89.2
193	5	3	4	7	0.554	0.477	0.447	19.7	232	4	3	5	8	0.510	0.341	0.317	50.0
194	4	4	4	7	0.524	0.421	0.378	32.9	233	5	3	5	8	0.530	0.399	0.378	29.9
195	5	4	4	7	0.548	0.490	0.447	19.5	234	4	4	5	8	0.506	0.351	0.320	49.4
196	6	4	4	7	0.577	0.568	0.526	12.2	235	5	4	5	8	0.524	0.409	0.378	29.6
197	2	2	5	7	0.480	0.268	0.252	150.3	236	6	4	5	8	0.545	0.474	0.443	18.9
198	3	2	5	7	0.506	0.297	0.290	76.8	237	5	5	5	8	0.523	0.412	0.379	29.5
199	4	2	5	7	0.525	0.353	0.348	42.0	238	6	5	5	8	0.542	0.480	0.444	18.8
200	3	3	5	7	0.495	0.318	0.297	74.5	239	3	2	7	8	0.487	0.235	0.237	127.0
201	4	3	5	7	0.516	0.367	0.350	41.5	240	3	3	7	8	0.478	0.249	0.242	123.9
202	5	3	5	7	0.538	0.429	0.414	24.6	241	4	3	7	8	0.494	0.287	0.282	70.1
203	4	4	5	7	0.513	0.374	0.351	41.0	242	5	3	7	8	0.509	0.337	0.334	42.4
204	5	4	5	7	0.533	0.438	0.414	24.4	243	4	4	7	8	0.492	0.292	0.284	69.7
205	6	4	5	7	0.557	0.509	0.485	15.4	244	5	4	7	8	0.506	0.342	0.334	42.2
206	5	5	5	7	0.533	0.440	0.415	24.4	245	6	4	7	8	0.522	0.398	0.390	27.2
207	6	5	5	7	0.555	0.513	0.485	15.4	246	5	5	7	8	0.506	0.343	0.334	42.2
208	3	3	6	7	0.488	0.288	0.279	89.7	247	6	5	7	8	0.521	0.400	0.391	27.1
209	4	3	6	7	0.507	0.332	0.325	50.2	248	6	6	7	8	0.521	0.400	0.391	27.1
210	5	3	6	7	0.525	0.391	0.386	30.0	249	3	3	8	8	0.474	0.230	0.230	143.1
211	4	4	6	7	0.505	0.338	0.328	49.9	250	4	3	8	8	0.489	0.266	0.268	81.2
212	5	4	6	7	0.522	0.396	0.386	29.9	251	5	3	8	8	0.502	0.313	0.316	49.4
213	6	4	6	7	0.542	0.462	0.452	19.0	252	4	4	8	8	0.487	0.269	0.269	80.9
214	5	5	6	7	0.522	0.397	0.387	29.9	253	5	4	8	8	0.499	0.316	0.316	49.2
215	6	5	6	7	0.541	0.464	0.452	19.0	254	6	4	8	8	0.513	0.369	0.369	31.8
216	6	6	6	7	0.541	0.464	0.452	19.0	255	5	5	8	8	0.499	0.317	0.317	49.2
217	4	4	7	7	0.498	0.308	0.308	59.4	256	6	5	8	8	0.513	0.369	0.369	31.8
218	5	4	7	7	0.513	0.362	0.363	35.8	257	6	6	8	8	0.513	0.369	0.369	31.8
219	6	4	7	7	0.530	0.423	0.424	22.9									

Example 7.18 A sample of 70 air pumps is to undergo the combined constant temperature and sinusoidal vibration testing to estimate the reliability at the use condition. The use temperature and the root-mean-square (RMS) acceleration are 40°C and $1.5G_{rms}$, respectively, and the maximum allowable values are 120°C and $12G_{rms}$. It is known that the life has a Weibull distribution. Based on the Arrhenius relationship (7.5) and the life–vibration model (7.20), we model

the log characteristic life μ as a function of temperature T and G_{rms} according to (7.105), where $S_1 = 1/T$ and $S_2 = \ln(G_{rms})$. The prior-generation pump gives the preestimates $\mu_{00} = 10$, $\mu_{02} = 6.8$, $\mu_{20} = 7.5$, $\mu_{22} = 4.7$, and $\sigma = 0.53$. The censoring times are $\eta_{11} = \eta_{12} = 900$ hours and $\eta_{21} = \eta_{22} = 450$ hours. Determine the Yang compromise plan for the test.

SOLUTION   From the preestimates given, we obtain

$$a_1 = a_{11} = a_{12} = \frac{\ln(900) - 4.7}{0.53} = 3.97, \qquad a_2 = a_{21} = a_{22} = 2.66,$$

$$b = \frac{10 - 7.5}{0.53} = 4.72, \qquad c = \frac{10 - 6.8}{0.53} = 6.04.$$

Since the values of both a_2 and b are not covered in Table 7.14, repeated linear interpolations are needed. First, find the plans for $(a_1, a_2, b, c) = (4, 2, 4, 6)$ and $(4, 3, 4, 6)$, and make linear interpolation to $(4, 2.66, 4, 6)$. Next, find the plans for $(4, 2, 5, 6)$ and $(4, 3, 5, 6)$, and interpolate the plans to $(4, 2.66, 5, 6)$. Then interpolate the plans for $(4, 2.66, 4, 6)$ and $(4, 2.66, 5, 6)$ to $(4, 2.66, 4.72, 6)$, and obtain $\pi_{11} = 0.531$, $\xi_{11} = 0.399$, $\xi_{21} = 0.394$, and $V = 31.8$. For the purpose of comparison, we calculate the optimization model directly for $(3.97, 2.66, 4.72, 6.04)$ and get $\pi_{11} = 0.530$, $\xi_{11} = 0.399$, $\xi_{21} = 0.389$, and $V = 32.6$. In this case, the linear interpolation results in a good approximation.

TABLE 7.15   Actual Compromise Test Plan for the Air Pump

Group	Temperature ($^\circ$C)	RMS Acceleration (G_{rms})	Number of Test Units	Censoring Time (h)
1	84	5.3	37	900
2	84	12	7	900
3	120	5.3	7	450
4	120	12	19	450

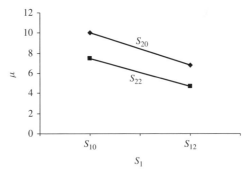

FIGURE 7.22   Interaction plot for temperature and vibration

The standardized plan derived from the linear interpolations is converted to the actual test plan using (7.108). The actual plan is summarized in Table 7.15.

When the test data are available, the transformed linear acceleration model should be checked for adequacy. As a preliminary check using the preestimates, we generate the interaction plot (Chapter 5) for the two stresses, as shown in Figure 7.22. The two lines roughly parallel, indicating that the interaction between the temperature and vibration is insignificant, and the linear model may be adequate.

7.8.5   Lognormal Distribution with Two Accelerating Variables

The Yang compromise test plans for the lognormal distribution with two accelerating variables and a linear relationship are similar to those for the Weibull distribution described in Section 7.8.4. They have the same assumptions (except for the distribution), the same notation, the rectangular test points, and the same sample allocations to the southeast and northwest points ($\pi_{12} = \pi_{21} = 0.1$). As in the Weibull case, the lognormal plans also minimize the asymptotic variance of the MLE of the mean log life at the use stresses. The test plans for various sets (a_1, a_2, b, c) are given in Table 7.16.

Example 7.19   Refer to Example 7.18. If the life distribution of the air pump is mistakenly modeled as lognormal and other data are the same as those in Example 7.18, calculate the test plan. Comment on the sensitivity of the Yang compromise test plan to the misspecification of life distribution.

SOLUTION   For the preestimates in Example 7.18, Table 7.16 and linear interpolation yield the test plan $\pi_{11} = 0.509$, $\xi_{11} = 0.422$, $\xi_{21} = 0.412$, and $V = 26.4$. These values are close to $\pi_{11} = 0.507$, $\xi_{11} = 0.418$, $\xi_{21} = 0.407$, and $V = 27.0$, which are obtained from the direct calculation of the optimization model for the set (3.97, 2.66, 4.72, 6.04). With the correct Weibull distribution, the approximate test plan yields the standardized variance of 32. The variance increase is $100 \times (32 - 31.8)/31.8 = 0.6\%$, where 31.8 is the standardized variance derived in Example 7.18. The small increase in variance indicates that the Yang compromise test plan is robust against the incorrect choice of life distribution for the preestimates given.

7.8.6   Tests with Higher Usage Rates

As discussed earlier, the life of some products may be measured by usage (e.g., mileage and cycles). In testing such products we often use an elevated stress along with an increased usage rate to reduce the test time. The increase in usage rate may prolong or shorten the usage to failure, and the effect can be modeled by (7.21). Thus, such a test involves two accelerating variables. Suppose that the relationship between the location parameter and the transformed stress S_1 and usage rate S_2 can be modeled with (7.105), where $S_2 = \ln(f)$ and f is

TABLE 7.16   Compromise Test Plans for a Lognormal Distribution with Two Accelerating Variables

No.	a_1	a_2	b	c	π_{11}	ξ_{11}	ξ_{21}	V	No.	a_1	a_2	b	c	π_{11}	ξ_{11}	ξ_{21}	V
1	2	2	2	2	0.565	0.708	0.708	9.3	46	4	3	3	4	0.574	0.640	0.617	9.0
2	3	2	2	2	0.666	0.847	0.861	4.0	47	5	3	3	4	0.630	0.747	0.730	5.1
3	2	2	3	2	0.527	0.564	0.602	15.9	48	2	2	4	4	0.477	0.374	0.374	47.7
4	3	2	3	2	0.595	0.685	0.737	7.2	49	3	2	4	4	0.512	0.451	0.460	23.7
5	4	2	3	2	0.681	0.826	0.885	3.5	50	4	2	4	4	0.551	0.540	0.556	12.9
6	2	2	4	2	0.506	0.469	0.525	24.4	51	5	3	4	4	0.590	0.649	0.656	7.4
7	3	2	4	2	0.556	0.574	0.646	11.4	52	4	4	4	4	0.546	0.558	0.558	12.6
8	4	2	4	2	0.618	0.695	0.776	5.8	53	5	4	4	4	0.589	0.656	0.657	7.3
9	2	2	5	2	0.493	0.401	0.467	34.5	54	6	4	4	4	0.639	0.759	0.761	4.4
10	3	2	5	2	0.532	0.494	0.576	16.5	55	2	2	5	4	0.469	0.328	0.341	62.0
11	4	2	5	2	0.580	0.599	0.693	8.7	56	3	2	5	4	0.498	0.398	0.420	31.1
12	2	2	2	3	0.527	0.602	0.564	15.9	57	4	2	5	4	0.531	0.478	0.507	17.2
13	3	2	2	3	0.598	0.713	0.690	7.3	58	3	3	5	4	0.497	0.406	0.422	30.7
14	4	2	2	3	0.685	0.847	0.837	3.6	59	4	3	5	4	0.528	0.487	0.507	17.0
15	2	2	3	3	0.503	0.491	0.491	24.7	60	5	3	5	4	0.563	0.574	0.598	10.0
16	3	2	3	3	0.555	0.591	0.602	11.7	61	5	4	5	4	0.562	0.578	0.599	10.0
17	4	2	3	3	0.616	0.708	0.727	6.0	62	6	4	5	4	0.603	0.670	0.693	6.2
18	3	3	3	3	0.553	0.607	0.607	11.5	63	2	2	6	4	0.463	0.292	0.315	78.1
19	4	3	3	3	0.612	0.724	0.728	5.9	64	3	2	6	4	0.488	0.356	0.388	39.5
20	5	3	3	3	0.687	0.850	0.858	3.2	65	4	2	6	4	0.516	0.429	0.468	22.0
21	2	2	4	3	0.488	0.415	0.437	35.2	66	3	3	6	4	0.487	0.362	0.389	39.1
22	3	2	4	3	0.529	0.504	0.536	17.0	67	4	3	6	4	0.514	0.435	0.468	21.9
23	4	2	4	3	0.576	0.607	0.647	9.0	68	5	3	6	4	0.544	0.514	0.551	13.1
24	3	3	4	3	0.528	0.514	0.540	16.8	69	5	4	6	4	0.543	0.517	0.552	13.1
25	4	3	4	3	0.573	0.617	0.647	8.9	70	6	4	6	4	0.577	0.600	0.638	8.2
26	5	3	4	3	0.628	0.728	0.762	5.0	71	2	2	7	4	0.459	0.263	0.292	95.8
27	2	2	5	3	0.478	0.360	0.394	47.4	72	3	2	7	4	0.481	0.323	0.360	48.8
28	3	2	5	3	0.512	0.440	0.485	23.3	73	4	2	7	4	0.505	0.390	0.435	27.5
29	4	2	5	3	0.550	0.531	0.585	12.6	74	3	3	7	4	0.480	0.327	0.362	48.4
30	3	3	5	3	0.511	0.446	0.488	23.1	75	4	3	7	4	0.503	0.394	0.435	27.3
31	4	3	5	3	0.548	0.538	0.585	12.5	76	5	3	7	4	0.529	0.466	0.512	16.6
32	5	3	5	3	0.591	0.636	0.688	7.2	77	4	4	7	4	0.503	0.395	0.436	27.2
33	2	2	6	3	0.471	0.318	0.360	61.3	78	5	4	7	4	0.529	0.468	0.513	16.5
34	3	2	6	3	0.499	0.390	0.444	30.5	79	6	4	7	4	0.558	0.543	0.592	10.6
35	4	2	6	3	0.531	0.472	0.535	16.7	80	2	2	2	5	0.493	0.467	0.401	34.5
36	3	3	6	3	0.499	0.394	0.446	30.3	81	3	2	2	5	0.540	0.544	0.493	17.1
37	4	3	6	3	0.530	0.477	0.535	16.6	82	4	2	2	5	0.593	0.641	0.601	9.1
38	5	3	6	3	0.565	0.565	0.629	9.8	83	2	2	3	5	0.478	0.394	0.360	47.4
39	2	2	2	4	0.506	0.525	0.469	24.4	84	3	2	3	5	0.516	0.466	0.442	23.8
40	3	2	2	4	0.562	0.616	0.575	11.7	85	4	2	3	5	0.558	0.554	0.537	12.9
41	4	2	2	4	0.627	0.730	0.700	6.0	86	3	3	3	5	0.511	0.488	0.446	23.1
42	2	2	3	4	0.488	0.437	0.415	35.2	87	4	3	3	5	0.550	0.574	0.536	12.7
43	3	2	3	4	0.531	0.521	0.509	17.2	88	5	3	3	5	0.596	0.668	0.636	7.4
44	4	2	3	4	0.581	0.621	0.618	9.2	89	2	2	4	5	0.469	0.341	0.328	62.0
45	3	3	3	4	0.528	0.540	0.514	16.8	90	3	2	4	5	0.500	0.409	0.403	31.4

TABLE 7.16 (*continued*)

No.	a_1	a_2	b	c	π_{11}	ξ_{11}	ξ_{21}	V	No.	a_1	a_2	b	c	π_{11}	ξ_{11}	ξ_{21}	V
91	4	2	4	5	0.535	0.488	0.488	17.4	136	6	5	8	5	0.528	0.463	0.499	16.3
92	3	3	4	5	0.497	0.422	0.406	30.7	137	2	2	3	6	0.471	0.360	0.318	61.3
93	4	3	4	5	0.529	0.502	0.487	17.1	138	3	2	3	6	0.506	0.423	0.391	31.3
94	5	3	4	5	0.565	0.587	0.577	10.1	139	4	2	3	6	0.543	0.501	0.476	17.4
95	4	4	4	5	0.528	0.509	0.490	16.9	140	3	3	3	6	0.499	0.446	0.394	30.3
96	5	4	4	5	0.563	0.597	0.577	10.0	141	4	3	3	6	0.533	0.522	0.474	17.0
97	6	4	4	5	0.603	0.688	0.670	6.2	142	5	3	3	6	0.573	0.604	0.563	10.1
98	2	2	5	5	0.462	0.302	0.302	78.5	143	2	2	4	6	0.463	0.315	0.292	78.1
99	3	2	5	5	0.489	0.364	0.371	40.0	144	3	2	4	6	0.492	0.374	0.359	40.1
100	4	2	5	5	0.518	0.436	0.449	22.4	145	4	2	4	6	0.524	0.445	0.436	22.5
101	3	3	5	5	0.487	0.373	0.373	39.3	146	4	3	4	6	0.516	0.461	0.435	22.1
102	4	3	5	5	0.514	0.446	0.449	22.1	147	5	3	4	6	0.548	0.537	0.515	13.3
103	5	3	5	5	0.544	0.524	0.530	13.3	148	4	4	4	6	0.514	0.469	0.437	21.8
104	5	4	5	5	0.543	0.530	0.530	13.2	149	5	4	4	6	0.544	0.548	0.516	13.1
105	6	4	5	5	0.576	0.613	0.614	8.3	150	6	4	4	6	0.579	0.630	0.598	8.3
106	5	5	5	5	0.542	0.531	0.531	13.2	151	2	2	5	6	0.457	0.280	0.271	96.6
107	6	5	5	5	0.576	0.615	0.615	8.3	152	3	2	5	6	0.482	0.336	0.333	49.9
108	2	2	6	5	0.457	0.271	0.280	96.6	153	4	2	5	6	0.509	0.401	0.403	28.2
109	3	2	6	5	0.481	0.328	0.345	49.5	154	3	3	5	6	0.479	0.347	0.335	48.8
110	4	2	6	5	0.506	0.395	0.417	28.0	155	4	3	5	6	0.503	0.412	0.403	27.8
111	3	3	6	5	0.479	0.335	0.347	48.8	156	5	3	5	6	0.531	0.483	0.476	17.0
112	4	3	6	5	0.502	0.402	0.416	27.7	157	4	4	5	6	0.502	0.418	0.404	27.5
113	5	3	6	5	0.529	0.473	0.491	16.8	158	5	4	5	6	0.528	0.490	0.477	16.8
114	5	4	6	5	0.528	0.477	0.492	16.7	159	6	4	5	6	0.557	0.565	0.553	10.8
115	6	4	6	5	0.556	0.552	0.569	10.7	160	5	5	5	6	0.527	0.492	0.478	16.7
116	5	5	6	5	0.527	0.478	0.492	16.7	161	6	5	5	6	0.556	0.569	0.554	10.7
117	6	5	6	5	0.556	0.554	0.570	10.7	162	3	3	6	6	0.001	0.591	0.591	100.3
118	2	2	7	5	0.453	0.246	0.262	116.5	163	4	3	6	6	0.001	0.721	0.591	79.5
119	3	2	7	5	0.474	0.299	0.322	60.0	164	5	3	6	6	0.001	0.856	0.592	68.5
120	4	2	7	5	0.496	0.360	0.389	34.1	165	4	4	6	6	0.001	0.722	0.722	59.2
121	3	3	7	5	0.473	0.304	0.324	59.3	166	5	4	6	6	0.001	0.859	0.723	48.6
122	4	3	7	5	0.494	0.366	0.389	33.9	167	6	4	6	6	0.001	0.999	0.723	42.4
123	5	3	7	5	0.517	0.431	0.459	20.8	168	5	5	6	6	0.001	0.861	0.861	38.1
124	5	4	7	5	0.516	0.434	0.459	20.7	169	6	5	6	6	0.001	0.999	0.862	32.1
125	6	4	7	5	0.541	0.503	0.531	13.4	170	6	6	6	6	0.001	0.999	0.999	26.2
126	5	5	7	5	0.516	0.435	0.460	20.7	171	3	3	7	6	0.001	0.507	0.591	118.9
127	6	5	7	5	0.540	0.504	0.532	13.4	172	4	3	7	6	0.001	0.618	0.592	90.8
128	3	2	8	5	0.469	0.275	0.303	71.4	173	5	3	7	6	0.001	0.735	0.592	76.0
129	4	2	8	5	0.489	0.332	0.366	40.9	174	4	4	7	6	0.001	0.620	0.723	70.3
130	3	3	8	5	0.468	0.279	0.305	70.7	175	5	4	7	6	0.001	0.738	0.724	55.9
131	4	3	8	5	0.487	0.336	0.366	40.6	176	6	4	7	6	0.001	0.859	0.724	47.6
132	4	4	8	5	0.487	0.337	0.367	40.5	177	5	5	7	6	0.001	0.739	0.861	45.3
133	5	4	8	5	0.507	0.399	0.432	25.0	178	6	5	7	6	0.001	0.861	0.862	37.2
134	6	4	8	5	0.528	0.463	0.499	16.3	179	6	6	7	6	0.001	0.862	0.999	31.2
135	5	5	8	5	0.507	0.399	0.432	25.0	180	3	3	8	6	0.001	0.444	0.591	140.2

TABLE 7.16 (*continued*)

No.	a_1	a_2	b	c	π_{11}	ξ_{11}	ξ_{21}	V	No.	a_1	a_2	b	c	π_{11}	ξ_{11}	ξ_{21}	V
181	4	3	8	6	0.001	0.542	0.592	103.8	220	5	5	7	7	0.001	0.739	0.739	52.6
182	5	3	8	6	0.001	0.644	0.593	84.7	221	6	5	7	7	0.001	0.861	0.740	44.3
183	4	4	8	6	0.001	0.543	0.723	82.9	222	6	6	7	7	0.001	0.862	0.862	36.2
184	5	4	8	6	0.001	0.646	0.724	64.3	223	4	4	8	7	0.001	0.543	0.620	94.3
185	6	4	8	6	0.001	0.753	0.724	53.5	224	5	4	8	7	0.001	0.646	0.621	75.4
186	5	5	8	6	0.001	0.647	0.862	53.6	225	6	4	8	7	0.001	0.753	0.621	64.4
187	6	5	8	6	0.001	0.754	0.863	43.0	226	5	5	8	7	0.001	0.647	0.739	61.0
188	6	6	8	6	0.001	0.755	0.999	36.9	227	6	5	8	7	0.001	0.754	0.740	50.2
189	2	2	4	7	0.001	0.703	0.401	171.3	228	6	6	8	7	0.001	0.755	0.862	42.0
190	3	2	4	7	0.486	0.346	0.324	49.8	229	2	2	5	8	0.001	0.563	0.351	236.7
191	3	3	4	7	0.480	0.362	0.327	48.4	230	3	2	5	8	0.474	0.292	0.276	72.6
192	4	3	4	7	0.506	0.426	0.393	27.7	231	3	3	5	8	0.468	0.305	0.279	70.7
193	5	3	4	7	0.535	0.495	0.466	16.9	232	4	3	5	8	0.489	0.360	0.335	41.0
194	4	4	4	7	0.503	0.436	0.395	27.2	233	5	3	5	8	0.512	0.419	0.397	25.5
195	5	4	4	7	0.530	0.508	0.466	16.6	234	4	4	5	8	0.487	0.367	0.337	40.5
196	6	4	4	7	0.560	0.581	0.541	10.7	235	5	4	5	8	0.507	0.428	0.397	25.1
197	2	2	5	7	0.453	0.262	0.246	116.5	236	6	4	5	8	0.531	0.492	0.461	16.5
198	3	2	5	7	0.477	0.312	0.302	60.7	237	5	5	5	8	0.507	0.432	0.399	25.0
199	4	2	5	7	0.503	0.372	0.366	34.7	238	6	5	5	8	0.529	0.498	0.462	16.3
200	3	3	5	7	0.473	0.324	0.304	59.3	239	3	2	7	8	0.462	0.247	0.247	99.6
201	4	3	5	7	0.495	0.384	0.366	34.1	240	3	3	7	8	0.459	0.255	0.248	97.8
202	5	3	5	7	0.520	0.448	0.433	21.0	241	4	3	7	8	0.476	0.304	0.298	57.1
203	4	4	5	7	0.493	0.390	0.368	33.7	242	5	3	7	8	0.494	0.356	0.353	35.9
204	5	4	5	7	0.516	0.457	0.433	20.8	243	4	4	7	8	0.474	0.307	0.300	56.7
205	6	4	5	7	0.542	0.526	0.503	13.5	244	5	4	7	8	0.491	0.361	0.353	35.6
206	5	5	5	7	0.516	0.460	0.435	20.7	245	6	4	7	8	0.509	0.416	0.409	23.6
207	6	5	5	7	0.541	0.531	0.504	13.4	246	5	5	7	8	0.491	0.362	0.354	35.5
208	3	3	6	7	0.001	0.591	0.507	118.9	247	6	5	7	8	0.508	0.419	0.410	23.5
209	4	3	6	7	0.001	0.721	0.507	97.7	248	6	6	7	8	0.508	0.419	0.410	23.4
210	5	3	6	7	0.001	0.856	0.507	86.5	249	3	3	8	8	0.001	0.444	0.444	181.1
211	4	4	6	7	0.001	0.723	0.620	70.3	250	4	3	8	8	0.001	0.542	0.445	143.8
212	5	4	6	7	0.001	0.860	0.620	59.4	251	5	3	8	8	0.001	0.644	0.445	124.0
213	6	4	6	7	0.001	0.999	0.621	53.1	252	4	4	8	8	0.001	0.543	0.543	107.3
214	5	5	6	7	0.001	0.861	0.739	45.3	253	5	4	8	8	0.001	0.646	0.544	88.2
215	6	5	6	7	0.001	0.999	0.739	39.2	254	6	4	8	8	0.001	0.753	0.544	77.0
216	6	6	6	7	0.001	0.999	0.862	31.2	255	5	5	8	8	0.001	0.647	0.647	69.4
217	4	4	7	7	0.001	0.620	0.620	81.5	256	6	5	8	8	0.001	0.754	0.648	58.6
218	5	4	7	7	0.001	0.738	0.621	66.9	257	6	6	8	8	0.001	0.755	0.755	47.9
219	6	4	7	7	0.001	0.859	0.621	58.4									

the usage rate in original units. Then the two-variable test plans for the Weibull and lognormal distributions are immediately applicable if we specify the censoring usages—not the censoring times. In many applications, censoring times are predetermined for convenient management of test resources. Then the censoring usages depend on the respective usage rates, which are to be optimized. This results in a small change in the optimization models (G. Yang, 2005). But the test plans given in Tables 7.14 and 7.16 are still applicable by using a_{ij}, b, and c calculated from

$$a_{ij} = \frac{1}{\sigma}[\ln(\eta_{ij}\, f_2) - \mu_{22}] = \frac{1}{\sigma}\left[\ln(\eta_{ij}\, f_2) - \mu_{20} - B\ln\left(\frac{f_2}{f_0}\right)\right],$$

$$b = \frac{1}{\sigma}(\mu_{00} - \mu_{20}), \tag{7.109}$$

$$c = \frac{1}{\sigma}\left[\mu_{00} - \mu_{02} + \ln\left(\frac{f_2}{f_0}\right)\right] = \frac{1}{\sigma}(1 - B)\ln\left(\frac{f_2}{f_0}\right),$$

where η_{ij} is the censoring time, B the usage rate effect parameter in (7.21), and f_0 and f_2 the usual and maximum allowable usage rates, respectively. Note that the units of usage rate should be in accordance with those of the censoring time. For example, if the usage rate is in cycles per hour, the censoring time should be in hours.

Example 7.20 In Example 7.15 we presented the actual Yang compromise plan for testing the compact electromagnetic relays at higher temperatures and switching rates. Develop the test plan for which the necessary data were given in Example 7.15; that is, the use temperature is 30°C, the maximum allowable temperature is 125°C, the usual switching rate is 5 cycles per minute, the maximum allowable switching rate is 30 cycles per minute, the sample size is 120, the censoring time at 125°C is 96 hours, and the censoring time at the low temperature (to be optimized) is 480 hours.

SOLUTION The test of similar relays at 125°C and 5 cycles per minute showed that the cycles to failure can be modeled with the Weibull distribution with shape parameter 1.2 and characteristic life 56,954 cycles. These estimates approximate the shape parameter and characteristic life of the compact relays under study. Thus we have $\sigma = 1/1.2 = 0.83$ and $\mu_{20} = \ln(56, 954) = 10.95$. Using the reliability prediction handbook MIL-HDBK-217F (U.S. DoD, 1995), we preestimate the failure rates to be 1.39×10^{-4} failures per hour or 0.46×10^{-6} failure per cycle at a switching rate of 5 cycles per minute, and 14.77×10^{-4} failures per hour or 0.82×10^{-6} failures per cycle at a switching rate of 30 cycles per minute. The preestimates of the location parameters of the log life are obtained from (7.103) as $\mu_{00} = 14.66$ and $\mu_{02} = 14.08$. From (7.21), the preestimate of B is

$$B = \frac{\mu_{00} - \mu_{02}}{\ln(f_0) - \ln(f_2)} = \frac{14.66 - 14.08}{\ln(5) - \ln(30)} = -0.324.$$

Since $B = -0.324 < 1$, increasing switching rate shortens the test length.

Using the preestimates and the censoring times given, we have

$$a_1 = a_{11} = a_{12} = \frac{1}{0.83}[\ln(480 \times 30 \times 60) - 10.95 + 0.324\ln(30/5)] = 3.98,$$

$$a_2 = a_{21} = a_{22} = 2.04,$$

$$b = \frac{1}{0.83}(14.66 - 10.95) = 4.47, \qquad c = \frac{1}{0.83}(1 + 0.324)\ln(30/5) = 2.86.$$

As in Example 7.18, Table 7.14 and repeated linear interpolations can yield the test plan for (4, 2, 4.47, 2.86). Here we calculate the optimization model (7.107) directly and obtain $\pi_{11} = 0.604$, $\xi_{11} = 0.57$, and $\xi_{21} = 0.625$. The standardized test plan is then transformed to the actual test plan using (7.108). The actual plan is shown in Table 7.9.

7.9 HIGHLY ACCELERATED LIFE TESTS

In contrast to the quantitative ALTs presented in previous sections, the highly accelerated life test (HALT) is qualitative. It is not intended for estimating product reliability. Rather, HALT is used in the early design and development phase to reveal the potential failure modes that would probably occur in field operation. Any failures observed in testing are treated seriously. Failure analysis is performed to determine the causes. Then appropriate corrective actions are developed and implemented, followed by confirmation of the effectiveness of the remedy. Once a failure mode is eliminated, the reliability of the product is increased to a higher level. Clearly, the primary purpose of HALT is not to measure, but to improve reliability.

HALT is aimed at stimulating failures effectively and efficiently. To accomplish this, one often uses the most effective test stresses, which may or may not be seen in service. In other words, it is not necessary to duplicate the field operating stresses. For instance, a pacemaker can be subjected to a wide range of thermal cycles in testing to produce a disconnection of solder joints, although it never encounters such stressing in the human body. Stresses frequently used include temperature, thermal cycling, humidity, vibration, voltage, and any other stimulus highly capable of generating failures in a short time. The stresses are applied in step-up fashion; that is, the severity of the stresses is increased progressively until a failure occurs. Unlike the quantitative ALT, HALT allows a stress to exceed the operational limit and reach the destruction limit, as the purpose is to discover potential failure modes. Once a failure is produced, a failure analysis is performed, followed by the development and implementation of the corrective action. The improved product is then subjected to the next HALT for confirming the effectiveness of the fix and identifying new failure modes. Apparently, a complete HALT process is a test–fix–test process, but it is considerably more powerful than the traditional test–fix–test cycle in terms of efficiency. For detailed description of HALT, see Hobbs (2000) and O'Connor (2001). Examples of HALT application are given in Becker and Ruth (1998), Silverman (1998), and Misra and Vyas (2003).

Due to the lack of direct correlation between the test stresses and the actual operational stresses, there is no guarantee that HALT-produced failure modes would occur in the field. Although eradication of all observed failure modes certainly improves reliability, it can lead to overdesign, which leads to unnecessary design expenditure and time. Therefore, it is critical to review every failure induced by HALT, and to identify and fix those that will affect customers. The work is difficult, but essential. Some design reliability techniques such as FMEA are helpful for this purpose.

TABLE 7.17 Differences Between HALT and Quantitative ALT

Aspect	HALT	Quantitative ALT
Purpose	Improve reliability	Measure reliability
When to conduct	Design and development phase	Design and development phase and after
Type of stress	Any, as long as effective	Field stress
Severity of stress	Up to destruction limit	Below operational limit
Stress loading pattern	Step stress	Constant, step, progressive, cyclic, or random stress
Test time	Short	Long
Censoring	No	Possible
Sample size	Small	Larger
Acceleration model	Not useful	Required
Life data analysis	No	Yes
Failure analysis	Yes	Optional
Failure modes to occur in field	Uncertain	Yes
Failure due mainly to:	Design mistake	Lack of robustness and design mistake
Confirmation test	Required	No

We have seen above that the stress levels used in a HALT can exceed the operational limit, and the resulting failure modes may be different from those in the field. Such a test may violate the fundamental assumptions for quantitative ALT. Thus, the lifetimes from a HALT cannot be extrapolated to design stresses for reliability estimation. Conversely, the quantitative test plans described earlier are inept for HALT development or debugging. The differences between the two approaches are summarized in Table 7.17.

PROBLEMS

7.1 Develop an accelerated life test plan for estimating the reliability of a product of your choice (e.g., paper clip, light bulb, hair drier). The plan should include the acceleration method, stress type (constant temperature, thermal cycling, voltage, etc.), stress levels, sample size, censoring times, failure definition, data collection method, acceleration model, and data analysis method, which should be determined before data are collected. Justify the test plan. Write this plan as a detailed proposal to your management.

7.2 Explain the importance of possessing understanding of the effects of a stress to be applied when one is planning an accelerated test. What types of stresses may be suitable for accelerating metal corrosion? What stresses accelerate conductor electromigration?

7.3 List the stresses that can accelerate fatigue failure, and explain the fatigue mechanism under each stress.

TABLE 7.18 Life Data of the Electromechanical Assembly

Group	Mechanical Load (kgf)	Operating Rate (cycles/min)	Number of Test Units	Life (10^3 cycles)
1	8	10	16	20.9, 23.8, 26.8, 39.1, 41.5, 51.5, 55.9, 69.4, 70, 70.5, 79.6, 93.6, 131.9, 154, 155.9, 229.9
2	16	10	10	4.8, 6.8, 8.7, 14.8, 15.7, 16.6, 18.9, 21.3, 27.6, 58.6
3	8	20	10	10.5, 12.7, 18, 22.4, 26.2, 28, 29.3, 34.9, 61.5, 82.4
4	16	20	10	0.9, 1.8, 3.1, 5, 6.3, 6.6, 7.4, 8.7, 10.3, 12.1

7.4 An electromechanical assembly underwent testing at elevated mechanical load and higher operating rates. The test conditions and fatigue life data are given in Table 7.18.

   (a) Plot the interaction graph for the mechanical load and operating rate where the response is the log median life. Comment on the interaction.

   (b) Plot the main effects for the mechanical load and operating rate where the response is the log median life. Do the two stresses have significant effects on the life?

   (c) Suppose that both the life–mechanical load and life–operating rate relations can be modeled adequately with the inverse power relationship. Write down the acceleration model: the relationship between the median life and the mechanical load and operating rate.

   (d) Estimate the relationship parameters, and comment on the adequacy of the relationship.

   (e) Calculate the acceleration factor between the median life at the test condition (16 kgf and 20 cycles per minute) and the median life at the use condition (2.5 kgf and 5 cycles per minute).

   (f) What is the estimate of median life at the use condition?

7.5 Explain the differences between the failure mechanisms described by the Coffin–Manson relationship and the Norris–Landzberg relationship.

7.6 Explain why increasing a usage rate does not necessarily reduce the test time. What are the possible consequences of ignoring the usage rate effects?

7.7 Propose methods for estimating the parameters of the Eyring relationship (7.8) and the generalized Eyring relationship (7.27) through the linear regression analysis.

7.8 Refer to Problem 7.4.

   (a) Plot the life data of each test group on lognormal probability paper.

   (b) Plot the life data of each test group on Weibull probability paper.

(c) Decide which life distribution fits better. Explain your choice.

(d) Estimate the parameters of the selected distribution for each test group.

(e) Suppose that the test units of group 1 were censored at 1×10^5 cycles. Plot the life data of this group on the probability paper selected in part (c). Does this distribution still look adequate? Estimate the distribution parameters and compare them with those obtained in part (d).

7.9 A preliminary test on a valve was conducted to obtain a preestimate of the life distribution, which would be used for the subsequent optimal design of accelerated life tests. In the test, 10 units were baked at the maximum allowable temperature and yielded the following life data (10^3 cycles): 67.4, 73.6*, 105.6, 115.3, 119.3, 127.5, 170.8, 176.2, 200.0*, 200.0*, where an asterisk implies censored. Historical data suggest that the valve life is adequately described by the Weibull distribution.

(a) Plot the life data on Weibull probability paper.

(b) Comment on the adequacy of the Weibull distribution.

(c) Estimate the Weibull parameters.

(d) Calculate the B_{10} life.

(e) Estimate the probability of failure at 40,000 cycles.

(f) Use an acceleration factor of 35.8 between the test temperature and the use temperature, and estimate the characteristic life at the use temperature.

7.10 Refer to Problem 7.9.

(a) Do Problem 7.9 (c)–(f) using the maximum likelihood method.

(b) Comment on the differences between the results from part (a) and those from Problem 7.9.

(c) Calculate the two-sided 90% confidence intervals for the Weibull parameters, B_{10} life, and the probability of failure at 40,000 cycles at the test temperature.

7.11 Refer to Example 7.3.

(a) Plot the life data of the three groups on the same Weibull probability paper.

(b) Plot the life data of the three groups on the same lognormal probability paper.

(c) Does the Weibull or lognormal distribution fit better? Select the better one to model the life.

(d) Comment on the parallelism of the three lines on the probability paper of the distribution selected.

(e) For each test voltage, estimate the (transformed) location and scale parameters.

(f) Estimate the common (transformed) scale parameter.

TABLE 7.19   Test Conditions and Life Data for the GaAs pHEMT Switches

Temperature (°C)	Relative Humidity (%)	Number of Devices Censored	Censoring Time (h)	Failure Times (h)
115	85	1	340	74, 200, 200, 290, 305
130	85	3	277	105, 160, 245
130	95	2	120	58, 65, 90, 105
145	85	0	181	8.5, 5 × [20, 70], 105, 110, 110, 140, 181

    (g) Estimate the distribution parameters at the rated voltage of 50 V.

    (h) Estimate the MTTF at 50 V, and compare the result with the mean life estimate obtained in Example 7.3.

7.12 Repeat Problem 7.11 (e)–(h) using the maximum likelihood method.

7.13 Ersland et al. (2004) report the results of dc biased life tests performed on GaAs pHEMT switches under elevated temperature and humidity. The approximate life data are shown in Table 7.19, where 5 × [20, 70] means that 5 failures occurred between 20 and 70 hours.

    (a) Plot the life data of the four groups on the same lognormal probability paper. Comment on the adequacy of the lognormal distribution.

    (b) Is it evident that the lognormal shape parameter is dependent on the test condition? What factor may contribute to the peculiarity?

    (c) Write down an acceleration relationship for the test.

    (d) Using a constant shape parameter, work out the total sample log likelihood function.

    (e) Estimate the model parameters.

    (f) Calculate the two-sided 90% confidence intervals for the activation energy and the relative humidity exponent.

    (g) Are the activation energy and relative humidity exponent statistically different from the empirical ones, which Hallberg and Peck (1991) report: 0.9 eV and −3?

    (h) Estimate the 10th percentile at the use condition 30°C and 45% relative humidity.

    (i) Calculate the two-sided 90% confidence interval for the 10th percentile at the use condition.

7.14 Accelerated life tests are to be conducted to evaluate the reliability of a type of laser diode at a use current of 30 mA. A sample of 85 units is available for testing at three elevated levels of current. The high level is the maximum allowable current of 220 mA. The censoring times are 500 hours at the high level, 880 hours at the middle level, and 1050 hours at the low level of current. The life of the diodes is modeled with the lognormal distribution with shape parameter 1.1. The scale parameters at

30 and 220 mA are preestimated to be 6.2 and 11.7, respectively. Design the Yang compromise test plan.

7.15 Refer to Example 7.3. Suppose that the test data have to be analyzed at 2000 hours and that the maximum allowable voltage is 120 V.

(a) Develop the Yang compromise test plan. Use the data given in the example to calculate the unknowns for the test plan.

(b) Compute the standardized variance for the test plan used in the example.

(c) What is the variance increase ratio?

(d) If a test unit were lost at the low level during testing, what would be the variance increase ratios for the two test plans? Which test plan is more sensitive to the loss of a test unit?

7.16 A sample of 65 hydraulic components is to be tested at elevated temperature and pressure in order to estimate the reliability at a use condition. The use temperature is $50°C$ and the normal pressure is 9.18×10^6 Pa, while the maximum allowable ones are $130°C$ and 32.3×10^6 Pa. Data analysis for the prior-generation product indicates that the life can be modeled with a Weibull distribution with shape parameter 2.3. The characteristic lives are 12,537, 1085, and 3261 hours at the use condition, $130°C$ and 9.18×10^6 Pa, and $50°C$ and 32.3×10^6 Pa, respectively.

(a) Write down the acceleration relationship, assuming no interaction between the temperature and pressure.

(b) Preestimate the characteristic life at $130°C$ and 32.3×10^6 Pa.

(c) Design the Yang compromise test plan, given that the test at low temperature is censored at 960 hours and the test at high temperature is censored at 670 hours.

7.17 Describe the HALT process, and explain the role of a HALT in planning a quantitative ALT.

8

DEGRADATION TESTING AND ANALYSIS

8.1 INTRODUCTION

As explained in Chapter 7, the accelerated life test is an important task in nearly all effective reliability programs. In today's competitive business environment, the time allowed for testing is continuously reduced. On the other hand, products become more reliable thanks to advancements in technology and manufacturing capability. So it is not uncommon that accelerated life tests yield no or few failures at low stress levels. In these situations it is difficult or impossible to analyze the life data and make meaningful inferences about product reliability. For some products whose performance characteristics degrade over time, a failure is said to have occurred when a performance characteristic crosses a specified threshold. Such a failure, called a *soft failure*, is discussed in Chapters 2 and 5. Degradation of the characteristic indicates a deterioration in reliability. The measurements of the characteristic contain much useful and credible information about product reliability. Therefore, it is possible to infer reliability by analyzing the degradation data.

A degradation test has several advantages over a life test. Reliability analysis using degradation data directly relates reliability to physical characteristics. This distinction enables reliability to be estimated even before a test unit fails and thus greatly shortens the test time. Degradation analysis often yields more accurate estimates than those from life data analysis, especially when a test is highly censored. As shown in Chapter 7, an unfailed unit affects an estimate only through its censoring time. The time from censoring to failure, or the remaining

Life Cycle Reliability Engineering, by Guangbin Yang
Copyright © 2007 John Wiley & Sons, Inc.

life, is unknown and not taken into account in life data analysis. In contrast, a degradation test measures the performance characteristics of an unfailed unit at different times, including the censoring time. The degradation process and the distance between the last measurement and a specified threshold are known. In degradation analysis, such information is also utilized to estimate reliability. Certainly, degradation analysis has drawbacks and limitations. For example, it usually requires intensive computations.

In this chapter we describe different techniques for reliability estimation from degradation data, which may be generated from nondestructive or destructive inspections. The principle and method for accelerated degradation test with tightened thresholds are also presented. We also give a brief survey of the optimal design of degradation test plans.

8.2 DETERMINATION OF THE CRITICAL PERFORMANCE CHARACTERISTIC

The performance of a product is usually measured by multiple characteristics. Typically, a small component such as a capacitor or resistor has three or more characteristics, and a complicated system such as the automobile can have dozens, if not hundreds. Each characteristic reflects to some degree the level of product reliability. In degradation analysis, quantitatively relating reliability to all characteristics is difficult, if not impossible. In fact, most characteristics are neither independent of each other nor equally important. In many applications, there is one critical characteristic, which describes the dominant degradation process. This one can be used to characterize product reliability. The robust reliability design discussed in Chapter 5 is based on such a characteristic.

In many applications, the critical characteristic is fairly obvious and can be identified using physical knowledge, customer requirement, or experience. For a component to be installed in a system, the critical characteristic of the component is often the one that has the greatest impact on system performance. The drift and variation in this characteristic cause the system performance to deteriorate remarkably. The relationship between component characteristics and system performance may be explored using a design of experiment. For a commercial product, the critical characteristic is the one that most concerns customers. The degradation in this characteristic closely reflects customer dissatisfaction on the product. Determination of the critical characteristic may be aided by using the quality deployment function, which translates customer expectations to product performance characteristics. This technique was described in detail in Chapter 3.

The critical characteristic chosen to characterize product reliability must be increasing or decreasing monotonically over time. That is, the degradation in the characteristic is irreversible. This requirement is satisfied in most applications. For example, attrition of automobile tires always increases with mileage, and the bond strength of solder joints always decreases with age. For some electronic products, however, the characteristic may not be monotone in early life, due to *burn-in effects*. The characteristic fluctuates in a short period of time and then becomes monotone afterward. An example is light-emitting diodes, the luminous power of which may increase in the first few hundred hours of operation and then decreases over time. For these products it is important that the degradation analysis excludes observations collected in the burn-in period.

8.3 RELIABILITY ESTIMATION FROM PSEUDOLIFE

Suppose that a degradation test uses a sample of n units. During testing, each unit is inspected periodically to measure the critical performance characteristic y. The inspection is nondestructive, meaning that the unit is not damaged by the inspection and resumes its function after inspection. Let y_{ij} denote the measurement of y on unit i at time t_{ij}, where $i = 1, 2, \ldots, n$, $j = 1, 2, \ldots, m_i$, and m_i is the number of measurements on unit i. The degradation path can be modeled by

$$y_{ij} = g(t_{ij}; \beta_{1i}, \beta_{2i}, \ldots, \beta_{pi}) + e_{ij}, \tag{8.1}$$

where $g(t_{ij}; \beta_{1i}, \beta_{2i}, \ldots, \beta_{pi})$ is the true degradation of y of unit i at time t_{ij} and e_{ij} is the error term. Often, the error term is independent over i and j and is modeled with the normal distribution with mean zero and standard deviation σ_e, where σ_e is constant. Although measurements are taken on the same unit, the potential autocorrelation among e_{ij} may be ignored if the readings are widely spaced. In (8.1), $\beta_{1i}, \beta_{2i}, \ldots, \beta_{pi}$ are unknown degradation model parameters for unit i and should be estimated from test data, and p is the number of such parameters.

During testing, the inspections on unit i yield the data points (t_{i1}, y_{i1}), (t_{i2}, y_{i2}), $\ldots, (t_{im_i}, y_{im_i})$. Since $e_{ij} \sim N(0, \sigma_e^2)$, the log likelihood L_i for the measurement data of unit i is

$$L_i = -\frac{m_i}{2} \ln(2\pi) - m_i \ln(\sigma_e) - \frac{1}{2\sigma_e^2} \sum_{j=1}^{m_i} [y_{ij} - g(t_{ij}; \beta_{1i}, \beta_{2i}, \ldots, \beta_{pi})]^2. \tag{8.2}$$

The estimates $\hat{\beta}_{1i}, \hat{\beta}_{2i}, \ldots, \hat{\beta}_{pi}$ and $\hat{\sigma}_e$ are obtained by maximizing L_i directly.

The parameters may also be estimated by the least squares method. This is done by minimizing the sum of squares of the deviations of the measurements from the true degradation path, which is given by

$$\text{SSD}_i = \sum_{j=1}^{m_i} e_{ij}^2 = \sum_{j=1}^{m_i} [y_{ij} - g(t_{ij}; \beta_{1i}, \beta_{2i}, \ldots, \beta_{pi})]^2, \tag{8.3}$$

where SSD_i is the sum of squares of deviations for unit i. Note that the maximum likelihood estimates are the same as the least squares estimates.

Once the estimates $\hat{\beta}_{1i}, \hat{\beta}_{2i}, \ldots, \hat{\beta}_{pi}$ are obtained, we can calculate the pseudolife. If a failure occurs when y crosses a specified threshold, denoted G, the life of unit i is given by

$$\hat{t}_i = g^{-1}(G; \hat{\beta}_{1i}, \hat{\beta}_{2i}, \ldots, \hat{\beta}_{pi}), \tag{8.4}$$

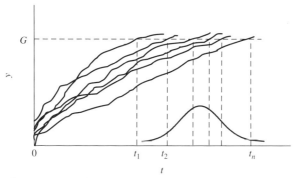

FIGURE 8.1 Relation of degradation path, pseudolife, and life distribution

where g^{-1} is the inverse of g. Applying (8.4) to each test unit yields the lifetime estimates $\hat{t}_1, \hat{t}_2, \ldots, \hat{t}_n$. Apparently, pseudolifetimes are complete exact data. In Chapter 7 we described probability plotting for this type of data. By using the graphical method, we can determine a life distribution that fits these life data adequately and estimate the distribution parameters. As explained in Chapter 7, the maximum likelihood method should be used for estimation of distribution parameters and other quantities of interest when commercial software is available. Figure 8.1 depicts the relation of degradation path, pseudolife, and life distribution.

For some products, the true degradation path is simple and can be written in a linear form:

$$g(t) = \beta_{1i} + \beta_{2i}t, \tag{8.5}$$

where $g(t), t$, or both may represent a log transformation. Some examples follow. The wear of an automobile tire is directly proportional to mileage and $\beta_{1i} = 0$. Tseng et al. (1995) model the log luminous flux of the fluorescent lamp as a linear function of time. K. Yang and Yang (1998) establish a log-log linear relationship between the variation ratio of luminous power and the aging time for a type of infrared light-emitting diodes. The MOS field-effect transistors have a linear relationship between the log current and log time, according to J. Lu et al. (1997).

For (8.5), the least squares estimates of β_{1i} and β_{2i} are

$$\hat{\beta}_{1i} = \bar{y}_i - \hat{\beta}_{2i}\bar{t}_i,$$

$$\hat{\beta}_{2i} = \frac{m_i \sum_{j=1}^{m_i} y_{ij}t_{ij} - \sum_{j=1}^{m_i} y_{ij} \sum_{j=1}^{m_i} t_{ij}}{m_i \sum_{j=1}^{m_i} t_{ij}^2 - \left(\sum_{j=1}^{m_i} t_{ij}\right)^2},$$

$$\bar{y}_i = \frac{1}{m_i} \sum_{j=1}^{m_i} y_{ij}, \qquad \bar{t}_i = \frac{1}{m_i} \sum_{j=1}^{m_i} t_{ij}.$$

Then the pseudolife of unit i is

$$\hat{t}_i = \frac{G - \hat{\beta}_{1i}}{\hat{\beta}_{2i}}.$$

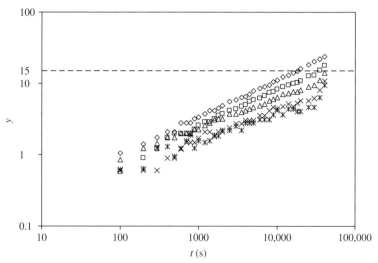

FIGURE 8.2   Percent transconductance degradation over time

Example 8.1   J. Lu et al. (1997) give the percent transconductance degradation data taken at different times for five units of an MOS field-effect transistor. The testing was censored at 40,000 seconds. The failure criterion is the percent transconductance degradation greater than or equal to 15%. The data are shown in Table 8.11 of Problem 8.10 and plotted on the log-log scale in Figure 8.2, where the vertical axis is the percent transconductance degradation and the horizontal axis is the time in seconds. The plot suggests a log-log linear degradation model:

$$\ln(y_{ij}) = \beta_{1i} + \beta_{2i}\ln(t) + e_{ij}.$$

The degradation model above is fitted to each degradation path. Simple linear regression analysis suggests that the degradation model is adequate. The least squares estimates for the five paths are shown in Table 8.1. After obtaining the estimates, we calculate the pseudolifetimes. For example, for unit 5 we have

$$\ln(\hat{t}_5) = \frac{\ln(15) + 2.217}{0.383} = 12.859 \quad \text{or} \quad \hat{t}_5 = 384,285 \text{ seconds.}$$

TABLE 8.1   Least Squares Estimates of Model Parameters

Estimate	Unit				
	1	2	3	4	5
$\hat{\beta}_{1i}$	−2.413	−2.735	−2.056	−2.796	−2.217
$\hat{\beta}_{2i}$	0.524	0.525	0.424	0.465	0.383

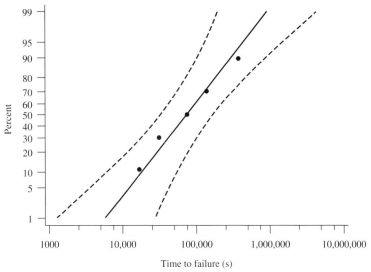

FIGURE 8.3   Lognormal plot, ML fits, and percentile confidence intervals for the pseudolife data of Example 8.1

Similarly, the pseudolifetimes for the other four units are $\hat{t}_1 = 17{,}553$, $\hat{t}_2 = 31{,}816$, $\hat{t}_3 = 75{,}809$, and $\hat{t}_4 = 138{,}229$ seconds. Among the commonly used life distributions, the lognormal provides the best fit to these data. Figure 8.3 shows the lognormal plot, ML fit, and two-sided 90% confidence interval for percentiles. The ML estimates of the scale and shape parameters are $\hat{\mu} = 11.214$ and $\hat{\sigma} = 1.085$.

8.4   DEGRADATION ANALYSIS WITH RANDOM-EFFECT MODELS

8.4.1   Random-Effect Models

From Example 8.1 it can be seen that the values of the degradation model parameters β_1 and β_2 are different for each unit. In general, the values of some or all of
the model parameters $\beta_1, \beta_2, \ldots, \beta_p$ in (8.1) vary from unit to unit. They can be
considered to be drawn randomly from the respective populations. The random
effects exist because of material property change, manufacturing process variation, stress variation, and many other uncontrollable factors. The parameters are
not independent; they usually form a joint distribution. In practice, a multivariate
normal distribution with mean vector $\boldsymbol{\mu_\beta}$ and variance–covariance matrix $\boldsymbol{\Sigma_\beta}$
is often used for simplicity. For example, Ahmad and Sheikh (1984) employ a
bivariate normal model to study the tool-wear problem. Using the multivariate
normal distribution and (8.1), the likelihood function l for all measurement data
of the n units can be written as

$$l = \prod_{i=1}^{n} \int_{-\infty}^{\infty} \cdots \int_{-\infty}^{\infty} \frac{1}{(2\pi)^{(p+m_i)/2} \sigma_e^{m_i} |\boldsymbol{\Sigma_\beta}|^{1/2}}$$

$$\cdot \exp \left\{ -\frac{1}{2} \left[\sum_{j=1}^{m_i} z_{ij}^2 + (\boldsymbol{\beta}_i - \boldsymbol{\mu}_{\boldsymbol{\beta}})' \boldsymbol{\Sigma}_{\boldsymbol{\beta}}^{-1} (\boldsymbol{\beta}_i - \boldsymbol{\mu}_{\boldsymbol{\beta}}) \right] \right\} \, d\beta_{1i} \cdots d\beta_{pi}, \quad (8.6)$$

where $z_{ij} = [y_{ij} - g(t_{ij}; \beta_{1i}, \beta_{2i}, \dots, \beta_{pi})]/\sigma_e$, $\boldsymbol{\beta}_i = [\beta_{1i}, \beta_{2i}, \dots, \beta_{pi}]$, and $|\boldsymbol{\Sigma}_{\boldsymbol{\beta}}|$ is the determinant of $\boldsymbol{\Sigma}_{\boldsymbol{\beta}}$. Conceptually, the model parameters, including the mean vector $\boldsymbol{\mu}_{\boldsymbol{\beta}}$, the variance–covariance matrix $\boldsymbol{\Sigma}_{\boldsymbol{\beta}}$, and the standard deviation of error σ_e, can be estimated by directly maximizing the likelihood. In practice, however, the calculation is extremely difficult unless the true degradation path takes a simple linear form such as in (8.5).

Here we provide a multivariate approach to estimating the model parameters $\boldsymbol{\mu}_{\boldsymbol{\beta}}$ and $\boldsymbol{\Sigma}_{\boldsymbol{\beta}}$. The approach is approximately accurate, yet very simple. First, we fit the degradation model (8.1) to each individual degradation path and calculate the parameter estimates $\hat{\beta}_{1i}, \hat{\beta}_{2i}, \dots, \hat{\beta}_{pi}$ ($i = 1, 2, \dots, n$) by maximizing the log likelihood (8.2) or by minimizing the sum of squares of the deviations (8.3). The estimates of each parameter are considered as a sample of n observations. The sample mean vector is

$$\overline{\boldsymbol{\beta}} = [\overline{\beta}_1, \overline{\beta}_2, \dots, \overline{\beta}_p], \quad (8.7)$$

where

$$\overline{\beta}_j = \frac{1}{n} \sum_{i=1}^{n} \hat{\beta}_{ji}, \qquad j = 1, 2, \dots, p. \quad (8.8)$$

The sample variance–covariance matrix is given by

$$\mathbf{S} = \begin{bmatrix} s_{11} & s_{12} & \cdots & s_{1p} \\ s_{21} & s_{22} & \cdots & s_{2p} \\ \cdots & \cdots & \cdots & \cdots \\ s_{p1} & s_{p2} & \cdots & s_{pp} \end{bmatrix}, \quad (8.9)$$

where

$$s_{kj} = \frac{1}{n} \sum_{i=1}^{n} (\hat{\beta}_{ki} - \overline{\beta}_k)(\hat{\beta}_{ji} - \overline{\beta}_j), \quad (8.10)$$

for $k = 1, 2, \dots, p$ and $j = 1, 2, \dots, p$. $s_{kj} (k \neq j)$ is the covariance of β_k and β_j. When $k = j$, s_{kk} is the sample variance of β_k. When the sample size is small, say $n \leq 15$, the variance–covariance component is corrected by replacing n with $n - 1$ to obtain the unbiased estimate. The computations can be done easily using a statistical software package such as Minitab.

The association between β_k and β_j can be measured by the correlation coefficient ρ_{kj}, which is defined as

$$\rho_{kj} = \frac{s_{kj}}{\sqrt{s_{kk}} \sqrt{s_{jj}}}. \quad (8.11)$$

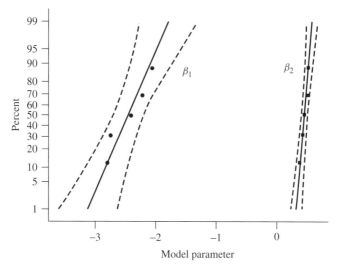

FIGURE 8.4 Normal fits to estimates of β_1 and β_2

$\overline{\beta}$ and **S** are estimates of μ_β and Σ_β, respectively. The estimates are useful in calculating degradation reliability, as we will see in Section 8.4.2.

Example 8.2 In Example 8.1 we have fitted the log-log linear model to each degradation path and obtained the estimates $\hat{\beta}_1$ and $\hat{\beta}_2$ for each path. Now β_1 and β_2 are considered as random variables, and we want to estimate μ_β and Σ_β by using the multivariate approach.

SOLUTION   First we plot the five estimates of β_1 and β_2 on normal probability paper, as shown in Figure 8.4. Loosely, the plot indicates that β_1 and β_2 have joint normal distribution. The mean of β_1 is $\overline{\beta}_1 = (-2.413 - 2.735 - 2.056 - 2.796 - 2.217)/5 = -2.443$. Similarly, $\overline{\beta}_2 = 0.464$. Then the estimate of the mean vector is $\overline{\beta} = [\overline{\beta}_1, \overline{\beta}_2] = [-2.443, 0.464]$. The variance estimate of β_1 is

$$s_{11} = \frac{(-2.413 + 2.443)^2 + \cdots + (-2.217 + 2.443)^2}{5 - 1} = 0.1029.$$

Similarly, $s_{22} = 0.00387$. The estimate of the covariance of β_1 and β_2 is

$$s_{12} = s_{21} = \frac{1}{5 - 1}[(-2.413 + 2.443)(0.524 - 0.464) + \cdots$$
$$+ (-2.217 + 2.443)(0.383 - 0.464)] = -0.01254.$$

The estimate of the variance–covariance matrix is

$$\mathbf{S} = \begin{bmatrix} s_{11} & s_{12} \\ s_{21} & s_{22} \end{bmatrix} = \begin{bmatrix} 0.1029 & -0.01254 \\ -0.01254 & 0.00387 \end{bmatrix}.$$

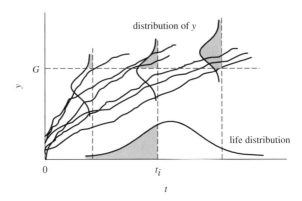

FIGURE 8.5   Relation of degradation to failure

The correlation coefficient between β_1 and β_2 is

$$\rho_{12} = \frac{s_{12}}{\sqrt{s_{11}}\sqrt{s_{22}}} = \frac{-0.01254}{\sqrt{0.1029}\sqrt{0.00387}} = -0.628.$$

The large absolute value suggests that the correlation between the two parameters cannot be ignored, while the negative sign indicates that β_1 increases as β_2 decreases, and vice versa. In other words, in this particular case, a unit with a smaller degradation percentage early in the test time ($t = 1$ second) will have a greater degradation rate.

8.4.2   Relation of Degradation to Failure

As we saw earlier, some or all of the parameters in the true degradation path $g(t; \beta_1, \beta_2, \ldots, \beta_p)$ are random variables. As such, the degradation amount at a given time varies from unit to unit, and some units may have crossed a specified threshold by this time. The probability of a critical characteristic crossing a threshold at a given time equals the probability of failure at that time. As the time proceeds, these probabilities increase. Figure 8.5 depicts the relation of the distribution of characteristic y to the life distribution of the product for the case where failure is defined in terms of $y \geq G$, where G is a specified threshold. In Figure 8.5, the shaded fraction of y distribution at time t_i is equal to the shaded fraction of the life distribution at that time, representing that the probability of $y(t_i) \geq G$ equals the probability of $T \leq t_i$, where T denotes the time to failure. In general, the probability of failure at time t can be expressed as

$$F(t) = \Pr(T \leq t) = \Pr[y(t) \geq G] = \Pr[g(t; \beta_1, \beta_2, \ldots, \beta_p) \geq G]. \quad (8.12)$$

For a monotonically decreasing performance characteristic, the probability of failure is calculated by replacing $y \geq G$ in (8.12) with $y \leq G$. In some simple cases that will be presented later, (8.12) may lead to a closed form for $F(t)$. In most applications, however, (8.12) has to be evaluated numerically for the given distributions of the model parameters.

As (8.12) indicates, the probability of failure depends on distribution of the model parameters, which, in turn, is a function of stress level. As such, an accelerated test is often conducted at an elevated stress level to generate more failures or a larger amount of degradation before the test is censored. Sufficient degradation reduces the statistical uncertainty of the estimate of the probability of failure. On the other hand, (8.12) also indicates that the probability of failure is influenced by the threshold. Essentially, a threshold is subjective and may be changed in specific applications. For a monotonically increasing performance characteristic, the smaller the threshold, the shorter the life and the larger the probability of failure. In this sense, a threshold can be considered as a stress; tightening a threshold accelerates the test. This acceleration method was mentioned in Chapter 7 and is discussed in detail in this chapter.

8.4.3   Reliability Estimation by Monte Carlo Simulation

Once the estimates of μ_β and Σ_β are obtained for $\beta_1, \beta_2, \ldots, \beta_p$, we can use a Monte Carlo simulation to generate a large number of degradation paths. The probability of failure $F(t)$ is approximated by the percentage of simulated degradation paths crossing a specified threshold at the time of interest. The steps for evaluating $F(t)$ are as follows:

1. Generate n (a large number, say 100,000) sets of realizations of $\beta_1, \beta_2, \ldots, \beta_p$ from a multivariate normal distribution with mean vector $\overline{\beta}$ and variance–covariance matrix $\mathbf{S}$. The n sets of realizations are denoted by $\beta_{1i}', \beta_{2i}', \ldots, \beta_{pi}'$, where $i = 1, 2, \ldots, n$.
2. For each i, calculate the true degradation y_i at the given time t by substituting $\beta_{1i}', \beta_{2i}', \ldots, \beta_{pi}'$ into $g(t; \beta_1, \beta_2, \ldots, \beta_p)$.
3. Count the number of y_i $(i = 1, 2, \ldots, n)$ crossing the threshold. Let r denote this number.
4. The probability of failure at time t is approximated by $F(t) \approx r/n$. Then the reliability is $1 - F(t)$.

Example 8.3   In Example 8.2 we computed the mean vector $\overline{\beta}$ and the variance–covariance matrix $\mathbf{S}$ for an MOS field-effect transistor. Now we want to evaluate the probability of failure at 1000, 2000, 3000, ..., 900,000 seconds by Monte Carlo simulation.

SOLUTION   Using Minitab we generated 65,000 sets of β_1' and β_2' from the bivariate normal distribution with mean vector $\overline{\beta}$ and the variance–covariance matrix $\mathbf{S}$ calculated in Example 8.2. At a given time (e.g., $t = 40,000$), we computed the percent transconductance degradation for each set (β_1', β_2'). Then count the number of degradation percentages greater than 15%. For $t = 40,000$, the number is $r = 21,418$. The probability of failure is $F(40,000) \approx$

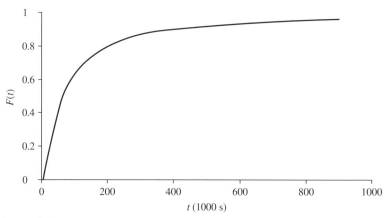

FIGURE 8.6 Failure probabilities computed from the Monte Carlo simulation data

$21,418/65,000 = 0.3295$. Repeat the calculation for other times, and plot the probabilities of failure (Figure 8.6). In the next subsection this plot is compared with others obtained using different approaches.

8.4.4 Reliability Estimation with a Bivariate Normal Distribution

We use a bivariate normal distribution to model the joint distribution of β_1 and β_2, where the means are μ_{β_1} and μ_{β_2}, the variances are $\sigma_{\beta_1}^2$ and $\sigma_{\beta_2}^2$, and the correlation coefficient is ρ_{12}. For the linear model (8.5), if a failure is said to have occurred when $y \geq G$, the probability of failure can be written as

$$F(t) = \Pr[g(t) \geq G] = \Pr(\beta_1 + \beta_2 t \geq G)$$

$$= \Phi \left[\frac{\mu_{\beta_1} + t\mu_{\beta_2} - G}{(\sigma_{\beta_1}^2 + t^2\sigma_{\beta_2}^2 + 2t\rho_{12}\sigma_{\beta_1}\sigma_{\beta_2})^{1/2}} \right], \tag{8.13}$$

where $\Phi(\cdot)$ is the cumulative distribution function (cdf) of the standard normal distribution. To evaluate $F(t)$, the means, variances, and correlation coefficient in (8.13) are substituted with their estimates.

Example 8.4 In Example 8.2 we computed the mean vector $\overline{\beta}$ and the variance–covariance matrix $\mathbf{S}$ for the MOS field-effect transistor. Now we use (8.13) to evaluate the probabilities of failure at 1000, 2000, 3000, ..., 900,000 seconds.

SOLUTION Because the degradation path is linear on the log-log scale as shown in Example 8.1, both G and t in (8.13) must take the log transformation. For $t = 40,000$, for example, the probability of failure is

$$F(40,000) =$$

$$\Phi \left[\frac{-2.443 + 10.5966 \times 0.464 - 2.7081}{(0.1029 + 10.5966^2 \times 0.00387 - 2 \times 10.5966 \times 0.6284 \times \sqrt{0.1029}\sqrt{0.00387})^{1/2}} \right]$$

$$= \Phi(-0.4493) = 0.3266.$$

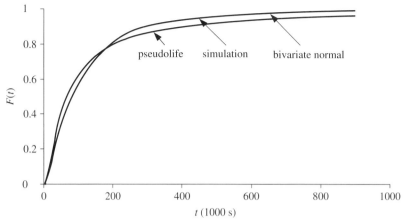

FIGURE 8.7 Probabilities of failure calculated using different methods

The probabilities of failure at other times are computed similarly. Figure 8.7 shows probabilities at various times calculated from (8.13). For comparison, probabilities using Monte Carlo simulation and pseudolife calculation are also shown in Figure 8.7. The probability plots generated from (8.13) and the Monte Carlo simulation cannot be differentiated visually, indicating that estimates from the two approaches are considerably close. In contrast, the pseudolife calculation gives significantly different results, especially when the time is greater than 150,000 seconds. Let's look at the numerical differences at the censoring time $t = 40,000$ seconds. Using (8.13), the probability of failure is $F(40,000) = 0.3266$. The Monte Carlo simulation gave the probability as $F(40,000) = 0.3295$, as shown in Example 8.3. The percentage difference is only 0.9%. Using the pseudolife approach, the probability is

$$F(40,000) = \Phi \left[\frac{\ln(40,000) - 11.214}{1.085} \right] = 0.2847.$$

It deviates from $F(40,000) = 0.3295$ (the Monte Carlo simulation result) by 13.6%. In general, compared with the other two methods, the pseudolife method provides less accurate results. But its simplicity is an obvious appeal.

8.4.5 Reliability Estimation with a Univariate Normal Distribution

As in (8.5), let's consider the simple linear model $g(t) = \beta_1 + \beta_2 t$, where both $g(t)$ and t are on the original scale (no log transformation). Suppose that the parameter β_1 is fixed and β_2 varies from unit to unit. That is, β_1 representing the initial degradation amount at time zero is common to all units, whereas β_2 representing the degradation rate is a random variable. An important special case is β_1 equal to zero. For example, automobile tires do not wear before use. Suppose that β_2 has a normal distribution with mean μ_{β_2} and standard deviation σ_{β_2}. For a monotonically increasing characteristic, the probability of failure can be written as

$$F(t) = \Pr[g(t) \geq G] = \Pr(\beta_1 + \beta_2 t \geq G) = \Pr\left(\beta_2 \geq \frac{G - \beta_1}{t}\right)$$

$$= \Phi\left[\frac{\mu_{\beta_2}/(G - \beta_1) - 1/t}{\sigma_{\beta_2}/(G - \beta_1)}\right]. \tag{8.14}$$

Now let's consider the case where $\ln(\beta_2)$ can be modeled with a normal distribution with mean μ_{β_2} and standard deviation σ_{β_2}; that is, β_2 has a lognormal distribution with scale parameter μ_{β_2} and shape parameter σ_{β_2}. For a monotonically increasing characteristic, the probability of failure can be expressed as

$$F(t) = \Pr[g(t) \geq G] = \Phi\left\{\frac{\ln(t) - [\ln(G - \beta_1) - \mu_{\beta_2}]}{\sigma_{\beta_2}}\right\}. \tag{8.15}$$

This indicates that the time to failure also has a lognormal distribution; the scale parameter is $\ln(G - \beta_1) - \mu_{\beta_2}$ and the shape parameter is σ_{β_2}. Substituting the estimates of the degradation model parameters and G into (8.15) gives an estimate of the probability of failure.

Example 8.5 A solenoid valve is used to control the airflow at a desired rate. As the valve ages, the actual airflow rate deviates from the rate desired. The deviation represents the performance degradation of the valve. A sample of 11 valves was tested, and the percent deviation of the airflow rate from that desired was measured at different numbers of cycles. Figure 8.8 plots the degradation paths of the 11 units. Assuming that the valve fails when the percent deviation is greater than or equal to 20%, estimate the reliability of the valve at 50,000 cycles.

SOLUTION   As shown in Figure 8.8, the degradation data suggest a linear relationship between the percent deviation and the number of cycles. Since measurements of percent deviation at the first inspection time (1055 cycles) are very

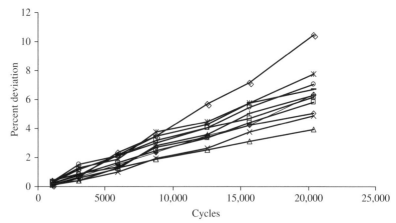

FIGURE 8.8   Degradation paths of the solenoid valves

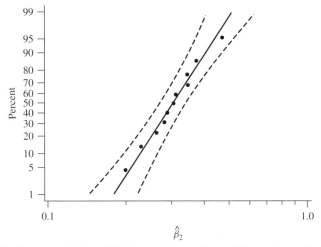

FIGURE 8.9 Lognormal plot, ML fits, and percentile confidence intervals for the estimates of β_2

small, deviations at time zero are negligible. Thus, the true degradation path can be modeled by $g(t) = \beta_2 t$, where t is in thousands of cycles. The simple linear regression analyses yielded estimates of β_2 for the 11 test units. The estimates are 0.2892, 0.2809, 0.1994, 0.2303, 0.3755, 0.3441, 0.3043, 0.4726, 0.3467, 0.2624, and 0.3134. Figure 8.9 shows the lognormal plot of these estimates with ML fit and two-sided 90% confidence intervals for percentiles. It is seen that β_2 can be approximated adequately by a lognormal distribution with $\hat{\mu}_{\beta_2} = -1.19406$ and $\hat{\sigma}_{\beta_2} = 0.22526$.

From (8.15), the cycles to failure of the valve also have a lognormal distribution. The probability of failure at 50,000 cycles is

$$F(50) = \Phi\left\{\frac{\ln(50) - [\ln(20) + 1.19406]}{0.22526}\right\} = 0.1088.$$

The reliability at this time is 0.8912, indicating that about 89% of the valves will survive 50,000 cycles of operation.

8.5 DEGRADATION ANALYSIS FOR DESTRUCTIVE INSPECTIONS

For some products, inspection for measurement of performance characteristics (often, monotonically decreasing strengths) must damage the function of the units either completely or partially. Such units cannot restart the same function as before the inspection and are discontinued from testing. Thus, each unit of the products can be inspected only once during testing and generates one measurement. For example, a solder joint has to be sheared or pulled off to get its joint strength, and measuring the dielectric strength of an insulator must break down the insulator. For such products, the degradation analysis methods described earlier in the chapter are not applicable. In this section we present the random-process method for degradation testing and data analysis. It is worth noting that this method is equally applicable to nondestructive products and is especially suitable for cases when degradation models are complicated. Examples of such application include K. Yang and Xue (1996), K. Yang and Yang (1998), and W. Wang and Dragomir-Daescu (2002).

8.5.1 Test Methods

The test uses n samples and m destructive inspections. The inspection times are $t_1, t_2, \ldots, t_m$. At t_1 (which may be time zero), n_1 units are destructed and yield the measurements y_{i1} ($i = 1, 2, \ldots, n_1$). Then the degradation test continues (starts) on the remaining $n - n_1$ units until t_2, at which point n_2 units are inspected. The inspection gives the measurements $y_{i2}(i = 1, 2, \ldots, n_2)$. The process is repeated until $(n - n_1 - n_2 - \cdots - n_{m-1})$ units are measured at t_m. The last inspection produces the measurements $y_{im}(i = 1, 2, \ldots, n - n_1 - n_2 - \cdots - n_{m-1})$. Obviously, $\sum_{j=1}^{m} n_j = n$. The n test units may be allocated equally to each inspection time; however, this allocation may not be statistically efficient. A better allocation should consider the shape of the degradation path as well as the change in performance dispersion over time. A degradation path may be linear, convex, or concave. On the other hand, the measurement dispersion may be constant or depend on time. Figure 8.10 shows examples of the three shapes of the degradation paths and the decreasing performance dispersion for the linear path. Decreasing dispersion for other shapes is possible but is not shown in Figure 8.10. Principles for better sample allocation are given below. Optimal allocation as a part of test planning deserves further study.

- If a degradation path is linear and performance dispersion is constant over time, the n test units should be apportioned equally to each inspection.
- If a degradation path is linear and performance dispersion decreases with time, more units should be allocated to low time inspections.

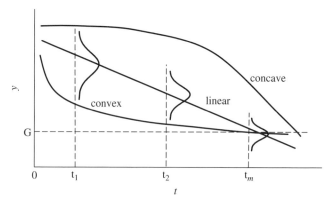

FIGURE 8.10   Three shapes of degradation paths and decreasing performance dispersion

- For the convex decreasing degradation path in Figure 8.10, the degradation rate becomes smaller as time increases. For the degradation amount between two consecutive inspection times to be noticeable, more units should be allocated to high time inspections regardless of the performance dispersion. The effect of unit-to-unit variability is usually less important than the aging effect.

- For the concave decreasing degradation path in Figure 8.10, the degradation rate is flat at a low time. More units should be assigned to low time inspections. This principle applies to both constant and decreasing performance dispersion.

8.5.2 Data Analysis

The following assumptions are needed for analyzing destructive degradation data:

- The (log) performance characteristic y at any given time has a (transformed) location-scale distribution (Weibull, normal, or lognormal). Other distributions are less common, but may be used.

- The location parameter μ_y is a function of time t and is denoted by $\mu_y(t; \boldsymbol{\beta})$, where $\boldsymbol{\beta} = [\beta_1, \beta_2, \ldots, \beta_p]$ is the vector of p unknown parameters. $\mu_y(t; \boldsymbol{\beta})$ represents a (transformed) "typical" degradation path. Its specific form is known from experience or test data. The simplest case is μ_y as a linear function of (log) time t:

$$\mu_y(t; \beta_1, \beta_2) = \beta_1 + \beta_2 t. \tag{8.16}$$

- If the scale parameter σ_y is a function of time t, the specific form of the function $\sigma_y(t; \boldsymbol{\theta})$ is known from experience or test data, where $\boldsymbol{\theta} = [\theta_1, \theta_2, \ldots, \theta_k]$ is the vector of k unknown parameters. Often, σ_y is assumed to be independent of time. For example, Nelson (1990, 2004) uses a constant σ_y for insulation breakdown data, and W. Wang and Dragomir-Daescu (2002) model degradation of the relative orbit area of an induction motor with an invariable σ_y.

As described in the test method, the inspection at time t_j yields n_j measurements y_{ij}, where $j = 1, 2, \ldots, m$ and $i = 1, 2, \ldots, n_j$. Now we want to use the measurements y_{ij} to estimate the reliability at a given time. As described earlier, the performance characteristic at any time has a (transformed) location-scale distribution. Let $f_y(y; t)$ denote the conditional probability density function (pdf) of the distribution at time t. $f_y(y; t)$ contains the unknown parameters $\mu_y(t; \boldsymbol{\beta})$ and $\sigma_y(t; \boldsymbol{\theta})$, which, in turn, depend on the unknown parameters β and θ. These parameters can be estimated using the maximum likelihood method. Because the observations y_{ij} come from different test units, they are independent of each other. The total sample log likelihood can be written as

$$L(\boldsymbol{\beta}, \boldsymbol{\theta}) = \sum_{j=1}^{m} \sum_{i=1}^{n_j} \ln[f_y(y_{ij}; t_j)]. \tag{8.17}$$

Maximizing (8.17) directly gives the estimates of the model parameters $\boldsymbol{\beta}$ and $\boldsymbol{\theta}$. If σ_y is constant and $\mu_y(t; \boldsymbol{\beta})$ is a linear function of (log) time as given in (8.16), then (8.17) will be greatly simplified. In this special case, commercial software packages such as Minitab and Reliasoft ALTA for accelerated life test analysis can apply to estimate the model parameters β_1, β_2, and σ_y. This is done by treating (8.16) as an acceleration relationship, where t_j is considered a stress level. Once the estimates are computed, the conditional cdf $F_y(y; t)$ is readily available. Now let's consider the following cases.

Case 1: Weibull Performance   Suppose that the performance characteristic y has a Weibull distribution with shape parameter β_y and characteristic life α_y, where β_y is constant and $\ln(\alpha_y) = \beta_1 + \beta_2 t$. Since the measurements are complete exact data, from (7.59) and (8.17) the total sample log likelihood is

$$L(\beta_1, \beta_2, \beta_y) = \sum_{j=1}^{m} \sum_{i=1}^{n_j} \left[\ln(\beta_y) - \beta_y(\beta_1 + \beta_2 t_j) \right.$$
$$\left. + (\beta_y - 1) \ln(y_{ij}) - \left(\frac{y_{ij}}{e^{\beta_1 + \beta_2 t_j}} \right)^{\beta_y} \right]. \qquad (8.18)$$

The estimates $\hat{\beta}_1$, $\hat{\beta}_2$, and $\hat{\beta}_y$ can be calculated by maximizing (8.18) directly. As in accelerated life test data analysis, commercial software can be employed to obtain these estimates. In computation, we treat the performance characteristic y as life, the linear relationship $\ln(\alpha_y) = \beta_1 + \beta_2 t$ as an acceleration model, the inspection time t as a stress, m as the number of stress levels, and n_j as the number of units tested at "stress level" t_j. If y is a monotonically decreasing characteristic such as strength, the effect of time on y is analogous to that of stress on life. If y is a monotonically increasing characteristic (mostly in non-destructive cases), the effects are exactly opposite. Such a difference does not impair the applicability of the software to this type of characteristic. In this case, the parameter β_2 is positive.

The conditional cdf for y can be written as

$$F_y(y; t) = 1 - \exp\left[-\left(\frac{y}{e^{\hat{\beta}_1 + \hat{\beta}_2 t}} \right)^{\hat{\beta}_y} \right]. \qquad (8.19)$$

Case 2: Lognormal Performance   If the performance characteristic y has a lognormal distribution with scale parameter μ_y and shape parameter σ_y, $\ln(y)$ has the normal distribution with mean μ_y and standard deviation σ_y. If σ_y is constant and $\mu_y = \beta_1 + \beta_2 t$ is used, the total sample log likelihood can be obtained easily from (7.73) and (8.17). As in the Weibull case, the model parameters may be calculated by maximizing the likelihood directly or by using the existing commercial software.

The conditional cdf for y can be expressed as

$$F_y(y; t) = \Phi\left[\frac{\ln(y) - \hat{\beta}_1 - \hat{\beta}_2 t}{\hat{\sigma}_y} \right]. \qquad (8.20)$$

Case 3: Normal Performance with Nonconstant σ_y Sometimes the performance characteristic y can be modeled with a normal distribution with mean μ_y and standard deviation σ_y, where

$$\ln(\mu_y) = \beta_1 + \beta_2 t \quad \text{and} \quad \ln(\sigma_y) = \theta_1 + \theta_2 t. \tag{8.21}$$

From (7.73), (8.17), and (8.21), the total sample log likelihood is

$$L(\beta_1, \beta_2, \theta_1, \theta_2) = \sum_{j=1}^{m} \sum_{i=1}^{n_j} \left\{ -\frac{1}{2} \ln(2\pi) - \theta_1 - \theta_2 t_j \right.$$
$$\left. -\frac{1}{2} \left[\frac{y_{ij} - \exp(\beta_1 + \beta_2 t_j)}{\exp(\theta_1 + \theta_2 t_j)} \right]^2 \right\}. \tag{8.22}$$

The existing commercial software do not handle nonconstant σ_y cases; the estimates $\hat{\beta}_1$, $\hat{\beta}_2$, $\hat{\theta}_1$, and $\hat{\theta}_2$ are calculated by maximizing (8.22) directly. This will be illustrated in Example 8.6.

The conditional cdf for y is

$$F_y(y; t) = \Phi \left[\frac{y - \exp(\hat{\beta}_1 + \hat{\beta}_2 t)}{\exp(\hat{\theta}_1 + \hat{\theta}_2 t)} \right]. \tag{8.23}$$

The three cases above illustrate how to determine the conditional cdf for the performance characteristic. Now we want to relate the performance distribution to a life distribution. Similar to the nondestructive inspection case described in Section 8.4.2, the probability of failure at a given time is equal to the probability of the performance characteristic crossing a threshold at that time. In particular, if a failure is defined in terms of $y \leq G$, the probability of failure at time t equals the probability of $y(t) \leq G$: namely,

$$F(t) = \Pr(T \leq t) = \Pr[y(t) \leq G] = F_y(G; t). \tag{8.24}$$

In some simple cases, it is possible to express $F(t)$ in a closed form. For example, for case 2, the probability of failure is given by

$$F(t) = \Phi \left[\frac{\ln(G) - \hat{\beta}_1 - \hat{\beta}_2 t}{\hat{\sigma}_y} \right] = \Phi \left\{ \frac{t - [\ln(G) - \hat{\beta}_1]/\hat{\beta}_2}{-\hat{\sigma}_y/\hat{\beta}_2} \right\}. \tag{8.25}$$

This indicates that the time to failure has a normal distribution with mean $[\ln(G) - \hat{\beta}_1]/\hat{\beta}_2$ and standard deviation $-\hat{\sigma}_y/\hat{\beta}_2$. Note that β_2 is negative for a monotonically decreasing characteristic, and thus $-\hat{\sigma}_y/\hat{\beta}_2$ is positive.

Example 8.6 In Section 5.13 we presented a case study on the robust reliability design of IC wire bonds. The purpose of the study was to select a setting of bonding parameters that maximizes robustness and reliability. In the experiment, wire bonds were generated with different settings of bonding parameters according to the experimental design. The bonds generated with the same setting were divided into two groups each with 140 bonds. One group underwent level 1 thermal cycling and the other group was subjected to level 2. For each group a sample of 20 bonds were sheared at 0, 50, 100, 200, 300, 500, and 800 cycles, respectively, for the measurement of bonding strength. In this example, we want to estimate the reliability of the wire bonds after 1000 cycles of level 2 thermal cycling, where the bonds were generated at the optimal setting of the bonding parameters. The optimal setting is a stage temperature of 150°C, ultrasonic power of 7 units, bonding force of 60 gf, and bonding time of 40 ms. As described in Section 5.13, the minimum acceptable bonding strength is 18 grams.

SOLUTION   The strength measurements [in grams (g)] at each inspection time can be modeled with a normal distribution. Figure 8.11 shows normal fits to the strength data at, for example, 0, 300, and 800 cycles. In Section 5.13 we show that the normal mean and standard deviation decrease with the number of thermal cycles. Their relationships can be modeled using (8.21). Figure 8.12 plots the relationships for the wire bonds that underwent level 2 thermal cycling. Simple linear regression analysis gives $\hat{\beta}_1 = 4.3743$, $\hat{\beta}_2 = -0.000716$, $\hat{\theta}_1 = 2.7638$, and $\hat{\theta}_2 = -0.000501$. Substituting these estimates into (8.23) and (8.24) can yield an estimate of the probability of failure at a given time. To improve the accuracy of the estimate, we use the maximum likelihood method. The parameters are estimated by maximizing (8.22), where y_{ij} are the strength measurements, $m = 7$, and $n_j = 20$ for all j. The estimates obtained from linear regression analysis serve as the initial values. The maximum likelihood estimates are $\hat{\beta}_1 = 4.3744$, $\hat{\beta}_2 = -0.000712$, $\hat{\theta}_1 = 2.7623$, and $\hat{\theta}_2 = -0.000495$. From (8.23) and (8.24), the estimate of reliability at 1000 cycles is

$$R(1000) = 1 - \Phi \left[\frac{18 - \exp(4.3744 - 0.000712 \times 1000)}{\exp(2.7623 - 0.000495 \times 1000)} \right] = 0.985.$$

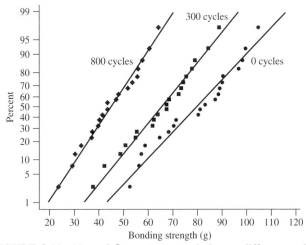

FIGURE 8.11   Normal fits to the strength data at different cycles

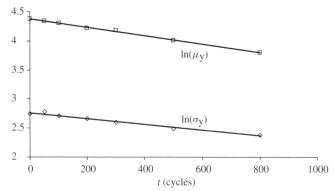

FIGURE 8.12 $\ln(\mu_y)$ and $\ln(\sigma_y)$ versus the number of cycles

8.6   STRESS-ACCELERATED DEGRADATION TESTS

In the preceding sections we described techniques for degradation analysis at a single stress level, which may be a use stress level or an elevated level. In many situations, testing a sample at a use stress level yields a small amount of degradation in a reasonable length of time. Insufficient degradation inevitably provides biased reliability estimates. To overcome this problem, as in accelerated life testing, two or more groups of test units are subjected to higher-than-use stress levels. The degradation data so obtained are extrapolated to estimate the reliability at a use stress level. In this section we present methods for accelerated degradation analysis.

8.6.1   Pseudo Accelerated Life Test Method

Suppose that n_k units are tested at stress level S_k, where $k = 1, 2, \ldots, q$, and q is the number of stress levels. Tested at S_k, unit i ($i = 1, 2, \ldots, n_k$) is inspected at time t_{ijk} and produces the measurement y_{ijk}, where $j = 1, 2, \ldots, m_{ik}$ and m_{ik} is the number of measurements on unit i tested at S_k. Having obtained the measurements, we calculate the approximate life $\hat{t}_{ik}$ of unit i tested at S_k using the pseudolife method described in Section 8.3. Treat the life data $\hat{t}_{1k}, \hat{t}_{2k}, \ldots, \hat{t}_{n_k k}$ as if they came from an accelerated life test at S_k. Then the life distribution at the use stress level can be estimated from q sets of such life data by using the graphical or maximum likelihood method presented in Section 7.7. The calculation is illustrated in Example 8.7.

Example 8.7   Stress relaxation is the loss of stress in a component subjected to a constant strain over time. The contacts of electrical connectors often fail due to excessive stress relaxation. The relationship between stress relaxation and aging time at temperature T can be expressed as

$$\frac{\Delta s}{s_0} = At^B \exp\left(-\frac{E_a}{kT}\right),$$
(8.26)

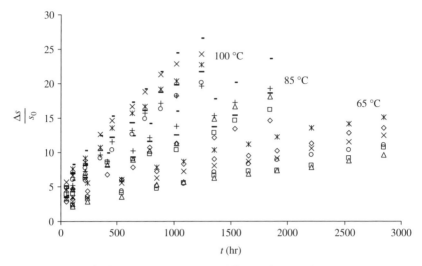

FIGURE 8.13   Stress relaxation at different times and temperatures

where Δs is the stress loss by time t, s_0 the initial stress, the ratio $\Delta s/s_0$ the stress relaxation (in percent), A and B are unknowns, and other notation is as in (7.4). Here A usually varies from unit to unit, and B is a fixed-effect parameter. At a given temperature, (8.26) can be written as

$$\ln\left(\frac{\Delta s}{\Delta s_0}\right) = \beta_1 + \beta_2 \ln(t),$$ (8.27)

where $\beta_1 = \ln(A) - E_a/kT$ and $\beta_2 = B$. To estimate the reliability of a type of connectors, a sample of 18 units was randomly selected and divided into three equal groups, which were tested at 65, 85, and 100°C. The stress relaxation data measured during testing are plotted in Figure 8.13. The electrical connector is said to have failed if the stress relaxation exceeds 30%. Estimate the probability of failure of the connectors operating at 40°C (use temperature) for 15 years (design life).

SOLUTION   First we fit (8.27) to each degradation path, and estimate the parameters β_1 and β_2 for each unit using the least squares method. Figure 8.14 plots the fits of the degradation model to the data and indicates that the model is adequate. Then we calculate the approximate lifetime of each test unit using $\hat{t} = \exp\{[\ln(30) - \hat{\beta}_1]/\hat{\beta}_2\}$. The resulting lifetimes are 15,710, 20,247, 21,416, 29,690, 41,167, and 42,666 hours at 65°C; 3676, 5524, 7077, 7142, 10,846, and 10,871 hours at 85°C; and 1702, 1985, 2434, 2893, 3343, and 3800 hours at 100°C. The life data are plotted on the lognormal probability paper, as shown in Figure 8.15. It is seen that the lifetimes at the three temperatures are reasonably lognormal with a common shape parameter σ.

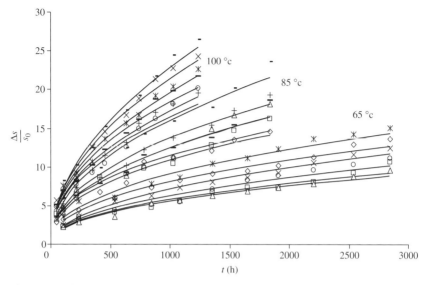

FIGURE 8.14 Degradation model fitted to the measurement data of stress relaxation

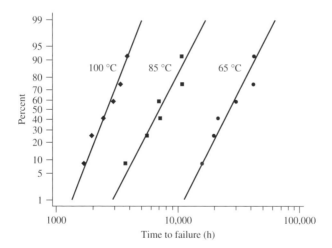

FIGURE 8.15 Lognormal fits to the pseudolifetimes at different temperatures

Since a failure occurs when $\Delta s/s_0 \geq 30$, from (8.26) the nominal life can be written as

$$t = \left(\frac{30}{A}\right)^{1/B} \exp\left(\frac{E_a}{kBT}\right).$$

Thus, it is legitimate to assume that the lognormal scale parameter μ is a linear function of $1/T$: namely, $\mu = \gamma_0 + \gamma_1/T$, where γ_0 and γ_1 are constants.

By using the maximum likelihood method for accelerated life data analysis as described in Section 7.7, we obtain the ML estimates as $\hat{\gamma}_0 = -14.56$, $\hat{\gamma}_1 = 8373.35$, and $\hat{\sigma} = 0.347$. The estimate of the scale parameter at 40°C is $\hat{\mu} = -14.56 + 8373.35/313.15 = 12.179$. Then the probability of failure of the connectors operating at 40°C for 15 years (131,400 hours) is

$$F(131,400) = \Phi \left[\frac{\ln(131,400) - 12.179}{0.347} \right] = 0.129 \quad \text{or} \quad 12.9\%.$$

That is, an estimated 12.9% of the connectors will fail by 15 years when used at 40°C.

8.6.2 Random-Effect Method

In Section 8.4 we delineated use of the random-effect model to analyze degradation data obtained at a single stress level. Now we extend the analysis to multiple stress levels. The analysis still consists of two steps: estimating degradation model parameters and evaluating the probability of failure. As in Section 8.4, the model parameters can be estimated using the maximum likelihood method, but the calculation requires much effort. For computational simplicity, readers may use the data analysis method for destructive inspection described in Section 8.6.3 or the pseudo accelerated life test method presented in Section 8.6.1.

Estimating the Model Parameters   The degradation model $g(t; \beta_1, \beta_2, \ldots, \beta_p)$ describes the true degradation path at a given stress level. The stress effect is implicitly embedded into one or more parameters among $\beta_1, \beta_2, \ldots, \beta_p$. In other words, at least one of these parameters depends on the stress level. For example, β_1 in Example 8.7 depends on temperature, whereas β_2 does not. The temperature effect on stress relaxation can be described by $\ln(\Delta s/s_0) = \ln(A) - E_a/kT + \beta_2 \ln(t)$, where the parameters A and β_2 vary from unit to unit, and the activation energy E_a is usually of fixed-effect type. In general, because of the additional stress term, the total number of model parameters increases, but is still denoted by p here for notional convenience. Thus, the true degradation path can be written as $g(t, S; \beta_1, \beta_2, \ldots, \beta_p)$, where S denotes the (transformed) stress, and the model parameters are free of stress effects. The mean vector $\boldsymbol{\mu}_\beta$ and the variance–covariance matrix $\boldsymbol{\Sigma}_\beta$ may be estimated directly by maximizing the likelihood (8.6), where the true degradation path is replaced with $g(t_{ij}, S_i; \beta_{1i}, \beta_{2i}, \ldots, \beta_{pi})$ and S_i is the stress level of unit i. Note that the stress parameters are constant and the corresponding variances and covariance elements are zero. As pointed out earlier, estimating $\boldsymbol{\mu}_\beta$ and $\boldsymbol{\Sigma}_\beta$ from (8.6) is computationally intensive.

Evaluating the Probability of Failure   Similar to (8.12), for a monotonically increasing characteristic, the probability of failure at the use stress level S_0 can be expressed as

$$F(t) = \Pr[g(t, S_0; \beta_1, \beta_2, \ldots, \beta_p) \geq G]. \qquad (8.28)$$

In some simple cases, $F(t)$ can be expressed in closed form. Let's consider the true degradation path

$$g(t, S) = \beta_1 + \gamma S + \beta_2 t,$$

where β_1 and β_2 have bivariate normal distribution and γ is a constant. The probability of failure at the use stress level S_0 is

$$F(t) = \Pr[g(t, S_0) > G] = \Pr(\beta_1 + \gamma S_0 + \beta_2 t \geq G)$$

$$= \Phi \left[\frac{\mu_{\beta_1} + \gamma S_0 + t\mu_{\beta_2} - G}{(\sigma_{\beta_1}^2 + t^2 \sigma_{\beta_2}^2 + 2t\rho_{12}\sigma_{\beta_1}\sigma_{\beta_2})^{1/2}} \right], \tag{8.29}$$

which is a closed-form expression for $F(t)$. In general, if $F(t)$ can be expressed in closed form, the evaluation is done simply by substituting estimates of μ_β and Σ_β into the $F(t)$ expression. Otherwise, we use the Monte Carlo simulation method described in Section 8.4.

8.6.3 Random-Process Method

In Section 8.5 we described the random-process method for degradation analysis of destructive and nondestructive measurements from a single stress level. As shown below, the method can be extended to multiple stress levels. The extension uses the following assumptions:

- The (log) performance characteristic y at a given time t and stress S has a (transformed) location-scale distribution (Weibull, normal, or lognormal). Other distributions are less common, but can be used.
- The location parameter μ_y is a monotone function of t and S. The function is denoted by $\mu_y(t, S; \beta)$, where $\beta = [\beta_1, \beta_2, \ldots, \beta_p]$. The specific form of $\mu_y(t, S; \beta)$ is known from experience or test data. The simplest case is μ_y as a linear function of (log) time and (transformed) stress:

$$\mu_y(t, S; \beta) = \beta_1 + \beta_2 t + \beta_3 S. \tag{8.30}$$

- If the scale parameter σ_y is a function of t and S, the specific form of the function $\sigma_y(t, S; \theta)$ is known from experience or test data, where $\theta = [\theta_1, \theta_2, \ldots, \theta_k]$. Often, σ_y is assumed to be independent of t and S.

Estimating the Model Parameters For an accelerated degradation test with destructive inspections, the test method described in Section 8.5 applies to each of k stress levels. Let y_{ijk} denote measurement at time t_{jk} on unit i of stress level S_k, where $i = 1, 2, \ldots, n_{jk}, j = 1, 2, \ldots, m_k$, and $k = 1, 2, \ldots, q; n_{jk}$ is the number of units inspected at t_{jk}; m_k is the number of inspections at S_k; and q is the number of stress levels. Obviously, $\sum_{k=1}^{q} \sum_{j=1}^{m_k} n_{jk} = n$, where n is the total sample size. Let $f_y(y; t, S)$ denote the pdf of y distribution conditional on t and S. Similar to (8.17), the total sample log likelihood can be expressed as

$$L(\boldsymbol{\beta}, \boldsymbol{\theta}) = \sum_{k=1}^{q} \sum_{j=1}^{m_k} \sum_{i=1}^{n_{jk}} \ln[f_y(y_{ijk}; t_{jk}, S_k)]. \tag{8.31}$$

For nondestructive inspections, the notations above are slightly different. y_{ijk} denotes the measurement at time t_{ijk} on unit i of stress level S_k, where $i = 1, 2, \ldots, n_k$, $j = 1, 2, \ldots, m_{ik}$, and $k = 1, 2, \ldots, q$; n_k is the number of units at S_k; and m_{ik} is the number of inspections on unit i of stress level S_k. Clearly, $\sum_{k=1}^{q} n_k = n$. The total sample log likelihood is similar to (8.31), but the notation changes accordingly. Note that the likelihood may be approximately correct for nondestructive inspections because of potential autocorrelation among the measurements. To reduce the autocorrelation, inspections should be widely spaced.

Maximizing the log likelihood directly yields estimates of the model parameters $\boldsymbol{\beta}$ and $\boldsymbol{\theta}$. If σ_y is constant and $\mu_y(t, S; \boldsymbol{\beta})$ is a linear function of (log) time and (transformed) stress given by (8.30), then (8.31) will be greatly simplified. In this special case, commercial software packages for accelerated life test analysis can be used to estimate the model parameters $\beta_1, \beta_2, \beta_3$, and σ_y. In calculation, we treat (8.30) as a two-variable acceleration relationship, where the time t is considered as a stress.

Evaluating the Probability of Failure After obtaining the estimates $\hat{\boldsymbol{\beta}}$ and $\hat{\boldsymbol{\theta}}$, we can calculate the conditional cdf for y, denoted by $F_y(y; t, S)$. If a failure is defined in terms of $y \leq G$, the probability of failure at time t and use stress level S_0 is given by

$$F(t, S_0) = F_y(G; t, S_0). \tag{8.32}$$

For example, if y has the lognormal distribution with μ_y modeled by (8.30) and constant σ_y, the estimate of the probability of failure at t and S_0 is

$$F(t, S_0) = \Phi\left[\frac{\ln(G) - \hat{\beta}_1 - \hat{\beta}_2 t - \hat{\beta}_3 S_0}{\hat{\sigma}_y} \right]$$

$$= \Phi\left\{ \frac{t - [\ln(G) - \hat{\beta}_1 - \hat{\beta}_3 S_0]/\hat{\beta}_2}{-\hat{\sigma}_y/\hat{\beta}_2} \right\}. \tag{8.33}$$

Note that (8.33) is similar to (8.25).

The following example illustrates application of the random-process method to nondestructive inspections.

Example 8.8 Refer to Example 8.7. Using the random-process method, estimate the probability of failure of connectors operating at $40°C$ (use temperature) for 15 years (design life).

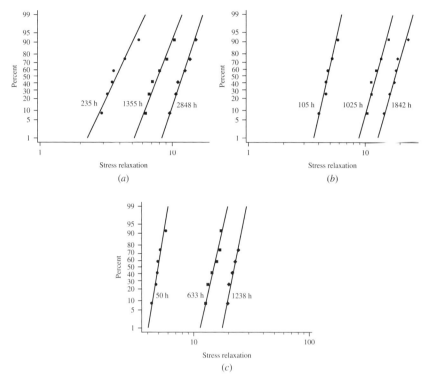

FIGURE 8.16 Lognormal fits to stress relaxation measurements at (*a*) 65°C, (*b*) 85°C, and (*c*) 100°C

SOLUTION   Figures 8.16 shows examples of lognormal fits to measurements of stress relaxation at different times and temperatures. It is seen that the stress relaxation can be approximated adequately using lognormal distribution and the shape parameter is reasonably independent of time and temperature. From (8.26), the scale parameter can be written as (8.30), where t is the log time and $S = 1/T$. In this example we have $q = 3$, $n_1 = n_2 = n_3 = 6$, and $m_{i1} = 11$, $m_{i2} = 10$, $m_{i3} = 10$ for $i = 1, 2, \ldots, 6$. The total sample log likelihood is

$$L(\boldsymbol{\beta}, \sigma_y) = \sum_{k=1}^{3} \sum_{i=1}^{6} \sum_{j=1}^{m_{ik}} \left\{ -\frac{1}{2} \ln(2\pi) - \ln(\sigma_y) \right.$$

$$\left. - \frac{1}{2\sigma_y^2} \left[\ln(y_{ijk}) - \beta_1 - \beta_2 \ln(t_{ijk}) - \frac{\beta_3}{T_k} \right]^2 \right\}.$$

Directly maximizing the likelihood yields the estimates $\hat{\beta}_1 = 9.5744$, $\hat{\beta}_2 = 0.4519$, $\hat{\beta}_3 = -3637.75$, and $\hat{\sigma}_y = 0.1532$. The calculation was performed using the Solver feature of Microsoft Excel. Alternatively, as described earlier, we may treat the measurement data as if they came from an accelerated life test. The pseudotest involves two stresses (temperature and time), and the acceleration model

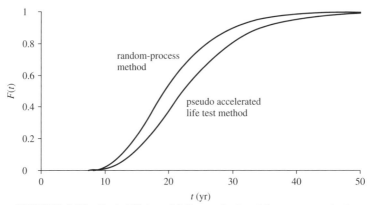

FIGURE 8.17   Probabilities of failure calculated from two methods

combines the Arrhenius relationship and the inverse power relationship. Minitab
gave the estimates $\hat{\beta}_1 = 9.5810$, $\hat{\beta}_2 = 0.4520$, $\hat{\beta}_3 = -3640.39$, and $\hat{\sigma}_y = 0.1532$,
which are close to these from the Excel calculation and are used for subsequent
analysis.

Because stress relaxation is a monotonically increasing characteristic, the probability of failure at t and S_0 is the complement of the probability given in (8.33)
and can be written as

$$F(t) = \Phi\left\{\frac{\ln(t) - [\ln(G) - \hat{\beta}_1 - \hat{\beta}_3 S_0]/\hat{\beta}_2}{\hat{\sigma}_y/\hat{\beta}_2}\right\} = \Phi\left[\frac{\ln(t) - 12.047}{0.3389}\right]. \quad (8.34)$$

This shows that the time to failure has a lognormal distribution with scale parameter 12.047 and shape parameter 0.3389. Note that, in Example 8.7, the pseudo
accelerated life test method resulted in a lognormal distribution with scale and
shape parameters equal to 12.179 and 0.347. At the design life of 15 years
(131,400 hours), from (8.34) the probability of failure is 0.2208. This estimate
should be more accurate than that in Example 8.7. For comparison, the probabilities at different times calculated from the two methods are plotted in Figure 8.17.
It is seen that the random-process method always gives a higher probability of
failure than the other method in this case. In general, the random-process method
results in more accurate estimates.

8.7   ACCELERATED DEGRADATION TESTS WITH TIGHTENED THRESHOLDS

As shown earlier, degradation analysis often involves complicated modeling and
intensive computation. In contrast, as presented in Chapter 7, methods for the
analysis of accelerated life test data are fairly easy to implement. The methods,
however, become inefficient when there are no or few failures at low stress
levels. Such difficulties may be overcome by measuring performance degradation
in accelerated life testing and tightening a failure threshold. Because life is
a function of threshold, tightening a threshold produces more failures. Here,

tightening a threshold refers to reducing the threshold for a monotonically increasing characteristic and increasing the threshold for a monotonically decreasing one. In this section we discuss the relationship between threshold and life and describe a test method and life data analysis.

8.7.1 Relationship Between Life and Threshold

A degrading product is said to have failed if a performance characteristic crosses its failure threshold. The life is the time at which the characteristic reaches the threshold. Apparently, the life of a unit depends on the value of the threshold. For a monotonically increasing characteristic, the smaller the value, the shorter the life, and vice versa. A unit that does not fail at a usual threshold may have failed at a tightened one. Thus, a tightened threshold yields more failures in a censored test. In this sense, life can be accelerated by tightening a threshold. Such an acceleration method does not induce failure modes different from those at the usual threshold because failures at different values of a threshold are caused by the same degradation mechanisms. The relationship between life and threshold for a monotonically increasing characteristic is depicted in Figure 8.18, where G_0 is the usual threshold, G_1 and G_2 are reduced thresholds, and $f(t|G_i)$ is the pdf of life conditional on G_i.

A quantitative relationship between life and threshold is useful and important. The threshold for a component is usually specified by the manufacturer of the component to meet functional requirements in most applications. The specification is somewhat subjective and arbitrary. Certainly, it can be changed in specific system designs. In other words, the value of a threshold for a component depends on the system specifications. For example, a component installed in a critical military system is often required to have a tighter threshold than when installed in a commercial product. Provided with the relationship, system designers can estimate the reliability of the component at a desirable value of the threshold without repeating reliability testing. The life–threshold relationship finds applications in the design of automobile catalytic converters, which convert the engine exhausts (hydrocarbon and NO_x, a group of highly reactive gases containing nitrogen and oxygen in varying amounts) to less harmful gases.

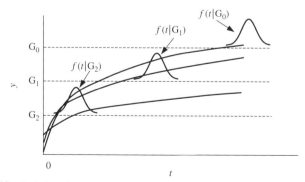

FIGURE 8.18 Relationship between life and threshold for an increasing characteristic

The performance of the catalytic converter is characterized by the index ratio, which increases with mileage. A smaller threshold for the index ratio is needed to reduce the actual emission level, but that increases warranty repairs. In the early design phase, design engineers often rely on the life–threshold relationship to evaluate the impact on warranty costs of lowering the index ratio threshold.

Example 8.9 In Example 8.7 the usual threshold for stress relaxation is 30%. Reducing the threshold shortens the time to failure. Determine the relationship between the time to failure and the threshold, and evaluate the effect of threshold on the probability of failure.

SOLUTION   From (8.34), the estimate of the mean log life (location parameter μ_t) is

$$\hat{\mu}_t = \frac{\ln(G) - \hat{\beta}_1 - \hat{\beta}_3 S_0}{\hat{\beta}_2} = 4.5223 + 2.2124 \ln(G),$$

where $\hat{\beta}_1$, $\hat{\beta}_2$, and $\hat{\beta}_3$ are as obtained in Example 8.8. It is seen that the mean log life is a linear function of the log threshold. The influence of the threshold on life is significant because of the large slope. Figure 8.19 plots the probabilities of failure with thresholds 30%, 25%, and 20%. It is seen that reducing the threshold greatly increases the probability of failure. For instance, at a design life of 15 years, the probabilities of failure at the three thresholds are 0.2208, 0.6627, and 0.9697, respectively.

8.7.2   Test Method

The test uses a sample of size n, q stress levels and m tightened thresholds. A group of n_k units is tested at stress level S_k ($k = 1, 2, \ldots, q$) until time η_k. Here $\sum_{k=1}^{q} n_k = n$. The times are recorded when the performance measurement of a unit reaches the tightened thresholds $G_1, G_2, \ldots, G_m$, where G_1 is closest to and G_m is farthest from G_0 (the usual threshold). As a result, a test unit

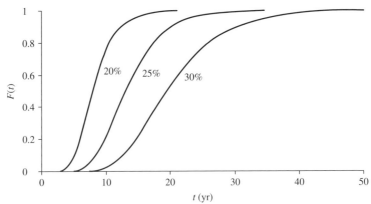

FIGURE 8.19   Probabilities of failure with different thresholds

can have m "life" observations. Let t_{ijk} denote the "failure" time of unit i at S_k and at G_j, where $i = 1, 2, \ldots, n_k$, $j = 1, 2, \ldots, m$, and $k = 1, 2, \ldots, q$. The life distribution at the use stress level S_0 and the usual threshold G_0 is estimated by utilizing these "life" data.

The most severe threshold G_m should be as tight as possible to maximize the threshold range and the number of failures at G_m, but it must not fall in the fluctuating degradation stage caused by the burn-in effects at the beginning of a test. The space between two thresholds should be as wide as possible to reduce the potential autocorrelation among the failure times. For this purpose, we usually use $m \leq 4$. In Section 8.8 we describe optimal design of the test plans.

8.7.3 Life Data Analysis

The life data analysis uses the following assumptions:

- The (log) time to failure at a given threshold has a (transformed) location-scale distribution (Weibull, normal, or lognormal). Other distributions are less common, but may be used.
- The scale parameter σ does not depend on stress and threshold.
- The location parameter μ is a linear function of (transformed) stress and threshold: namely,

$$\mu(S, G) = \beta_1 + \beta_2 S + \beta_3 G. \tag{8.35}$$

where β_1, β_2, and β_3 are parameters to be estimated from test data. Nonlinear relationships may be used, but the analysis will be greatly complicated.

The life data analysis is to estimate the life distribution at the use stress level and the usual threshold. This can be done by using the graphical or maximum likelihood method described in Section 7.7. The analysis is illustrated in Example 8.10.

Example 8.10 Infrared light-emitting diodes (IRLEDs) are high-reliability opto-electronic devices used widely in communication systems. The devices studied here are GaAs/GaAs IRLEDs, the wavelength of which is 880 nm and the design operating current, 50 mA. The performance of the devices is measured mainly by the variation ratio of luminous power. A failure is said to have occurred if the ratio is greater than 30%. To estimate the reliability at the operating current of 50 mA, 40 units were sampled and divided into two groups. A group of 25 units was tested at 170 mA; the group of 15 units was tested at 320 mA. Two reduced thresholds (10% and 19%) were used in testing. Table 8.2 summarizes the test plan.

The test units were inspected for luminous power before and during testing. The inspections were scheduled at 0, 24, 48, 96, 155, 368, 768, 1130, 1536, 1905, 2263, and 2550 hours for the 170 mA group, and at 0, 6, 12, 24, 48, 96, 156, 230, 324, 479, and 635 hours for the 320 mA group. Figure 8.20 shows

TABLE 8.2 Accelerated Degradation Test Plan for the IRLEDs

Group	Sample Size	Current (mA)	Threshold (%)	Censoring Time (h)
1	15	320	19	635
2	15	320	10	635
3	25	170	19	2550
4	25	170	10	2550

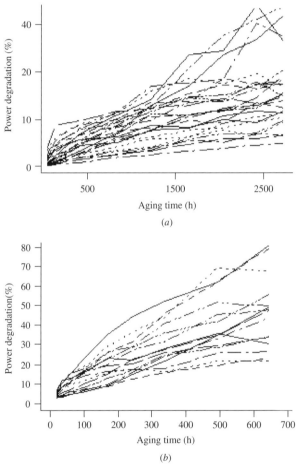

FIGURE 8.20 Degradation of luminous power at (*a*) 170 mA and (*b*) 320 mA

the values of the variation ratio at each inspection time. The data are shown in Tables 8.9 and 8.10 of Problem 8.8.

The variation ratio of luminous power is a function of time and current, and can be written as

$$y = \frac{A}{I^B} t^C, \tag{8.36}$$

TABLE 8.3   Failure Times (Hours) at Different Levels of Current and Threshold

Test	320 mA		170 mA	
Unit	19%	10%	19%	10%
1	176	39	1751	1004
2	544	243	2550^a	735
3	489	117	2550^a	2550^a
4	320	183	2550^a	2550^a
5	108	35	1545	867
6	136	47	2550^a	1460
7	241	113	2550^a	1565
8	73	27	1480	710
9	126	59	2550^a	1389
10	92	35	2550^a	2550^a
11	366	74	2550^a	1680
12	158	58	2550^a	2550^a
13	322	159	2550^a	873
14	220	64	2550^a	2550^a
15	325	195	2550^a	2138
16			2550^a	840
17			2550^a	2550^a
18			2550^a	2550^a
19			2550^a	1185
20			2550^a	898
21			2550^a	636
22			2550^a	2329
23			2550^a	1965
24			2550^a	2550^a
25			1611	618

^{a}Censored.

where y is the variation ratio, I the current, A and C random-effect parameters, and B a fixed-effect constant.

To calculate the approximate failure times at a threshold of 10%, we fitted the log transformed (8.36) to each of the degradation paths whose y values are greater than or equal to 10% at their censoring times. Interpolating y to 10% gives the failure times at this threshold. The units with y value less than 10% are considered censored. To calculate the failure times at the threshold of 19%, we interpolated y to 19% for the units with y values greater than or equal to 19% at their censoring times. The remaining units are considered censored. The resulting exact failure times are shown in Table 8.3. Alternatively, the failure times can be expressed as interval data by using the measurements and corresponding inspection times. This approach does not require fitting degradation model and interpolation. For example, the measurement on unit 1 of 170 mA was 16% at the inspection time of 1536 hours, and 22.5% at 1905 hours. If the reduced threshold is 19%, the

unit failed between the two consecutive inspections. Therefore, the failure time
is within the interval [1536,1905]. In contrast, the exact time from interpolation
is 1751.

Figure 8.21 plots the lognormal fits to the failure times listed in Table 8.3. It
is shown that the shape parameter σ is approximately independent of current and
threshold. In addition, from (8.36), the scale parameter μ is a linear function of
the log current and the log threshold: namely,

$$\mu(I, G) = \beta_1 + \beta_2 \ln(I) + \beta_3 \ln(G). \qquad (8.37)$$

The estimates of the model parameters were computed using Minitab as $\hat{\beta}_1 =$
28.913, $\hat{\beta}_2 = -4.902$, $\hat{\beta}_3 = 1.601$, and $\hat{\sigma} = 0.668$. The mean log life at the
operating current of 50 mA and the usual threshold of 30% is $\hat{\mu} = 15.182$. The
probability of failure in 10 years (87,600 hours) is negligible, so the device has
ultrahigh reliability.

8.8   ACCELERATED DEGRADATION TEST PLANNING

Like accelerated life tests, degradation tests merit careful planning. In this section
we present compromise test plans for accelerated degradation tests with tightened
thresholds and survey briefly other degradation test plans.

8.8.1   Compromise Plans for Tightened Threshold Tests

The test plans here use the 2^2 full factorial design in Figure 7.21, where S_1
is a two-level stress and S_2 is a two-level threshold. The sample is divided
into two groups. One group is tested at the low level of S_1, and the other is
subjected to the high level. The two groups employ two levels of threshold. Since

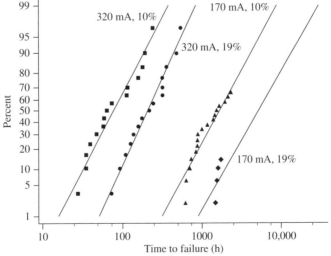

FIGURE 8.21   Lognormal fits to the failure times at different levels of current and
threshold

the threshold is thought of as a stress, the accelerated degradation tests can be perceived of as accelerated life tests with two accelerating variables. As such, with the assumptions stated in Section 8.7.3, the compromise test plans for accelerated degradation tests can be developed using the methods for accelerated life tests described in Sections 7.8.4 and 7.8.5. We use the notation in Sections 7.8.4, and choose ξ_{11} (low stress level), ξ_{21} (low threshold level), and π_{11} (proportion of sample size allocated to ξ_{11}), to minimize the asymptotic variance of the MLE of the mean log life at the use stress and the usual threshold. Note the uniqueness of the test plans here: $\pi_{11} = \pi_{12}$, $\pi_{21} = \pi_{22}$, and $\pi_{11} + \pi_{22} = 1$. G. Yang and Yang (2002) describe in detail development of the test plans.

Tables 8.4 and 8.5 list the test plans for Weibull and lognormal distributions, respectively. In the tables, a_1 is the standardized censoring time at S_1, and a_2 at S_2. It is seen that a few plans have ξ_{11}, ξ_{21}, or both at the use (usual) level. In these cases, the tests become partially accelerated degradation tests. To find a plan from the tables, one looks up the value of c first, then b, a_2, and a_1 in order. Linear interpolation may be needed for a combination (a_1, a_2, b, c) not given in the tables. Extrapolation outside the tables is not valid. Instead, use a numerical algorithm for solution. After obtaining the standardized values, we convert them to the actual values using the formulas given in Section 7.8.4.

The test plans depend on the values of μ_{00}, μ_{22}, μ_{02}, μ_{20}, and σ. They are unknown at the test planning stage. Generally, μ_{00} can be approximated by using experience, similar data, or a reliability prediction handbook such as MIL-HDBK-217F (U.S. DoD, 1995), as described in Chapter 7. μ_{22}, μ_{20}, and σ can be preestimated using a preliminary degradation test conducted at a high level of stress. Since G_2 is the tightest threshold, most test units will fail at this threshold when the test is terminated. The failure times are used to preestimate μ_{22} and σ. On the other hand, the preestimate of μ_{20} can be calculated by employing the approximate pseudolife method described earlier in the chapter. The unknown μ_{02} is estimated by $\mu_{00} + \mu_{22} - \mu_{20}$, provided that (8.35) holds.

Example 8.11 A type of new tire has a nominal tread depth of $\frac{10}{32}$ inch and must be retreaded when the tread is worn to $\frac{2}{32}$ inch. To evaluate the wear life of the tires under a load of 1350 pounds, a sample of 35 units is to be placed in test. The sample is divided into two groups, one loaded with the maximum allowable load of 2835 pounds, and the other with a lighter load. Each group will be run for 13,000 miles and subject to two thresholds, with one being $\frac{8}{32}$ inch (the tightest threshold). The wear life (in miles) can be modeled with the Weibull distribution with $\mu(P, G) = \beta_1 + \beta_2 \ln(P) + \beta_3 \ln(G)$, where μ is the log Weibull characteristic life, P the load, and G the threshold. A preliminary test was conducted on six tires loaded with 2835 pounds of pressure and run for 6500 miles. The tread loss was measured every 500 miles until test termination. The degradation data yielded the preestimates $\mu_{22} = 8.56$, $\mu_{20} = 9.86$, and $\sigma = 0.45$. The fleet test data of the prior-generation tires resulted in the preestimate $\mu_{00} = 11.32$. Develop a compromise test plan for the experiment.

TABLE 8.4 Compromise Test Plans for a Weibull Distribution

No.	a_1	a_2	b	c	π_{11}	ξ_{11}	ξ_{21}	V	No.	a_1	a_2	b	c	π_{11}	ξ_{11}	ξ_{21}	V
1	2	2	2	2	0.542	1.000	1.000	5.8	46	4	3	3	4	0.399	1.000	0.775	7.8
2	3	2	2	2	0.842	0.893	1.000	3.1	47	5	3	3	4	0.792	0.756	0.746	4.5
3	2	2	3	2	0.559	0.807	1.000	10.1	48	2	2	4	4	0.429	0.617	0.587	31.1
4	3	2	3	2	0.669	0.775	1.000	5.4	49	3	2	4	4	0.456	0.599	0.604	19.6
5	4	2	3	2	0.850	0.843	1.000	2.9	50	4	2	4	4	0.675	0.541	0.599	11.5
6	2	2	4	2	0.575	0.625	0.930	16.3	51	5	3	4	4	0.714	0.645	0.687	6.6
7	3	2	4	2	0.622	0.664	1.000	8.5	52	4	4	4	4	0.450	0.992	0.995	7.9
8	4	2	4	2	0.745	0.710	1.000	4.6	53	5	4	4	4	0.432	1.000	0.987	6.1
9	2	2	5	2	0.577	0.506	0.812	24.2	54	6	4	4	4	0.782	0.749	0.790	4.0
10	3	2	5	2	0.612	0.555	0.918	12.4	55	2	2	5	4	0.458	0.502	0.557	41.7
11	4	2	5	2	0.689	0.614	0.960	6.8	56	3	2	5	4	0.468	0.516	0.588	25.2
12	2	2	2	3	0.416	1.000	0.821	10.0	57	4	2	5	4	0.633	0.479	0.568	15.2
13	3	2	2	3	0.726	0.770	0.787	6.2	58	3	3	5	4	0.468	0.636	0.757	20.1
14	4	2	2	3	0.908	0.891	0.894	3.1	59	4	3	5	4	0.438	0.732	0.779	13.5
15	2	2	3	3	0.456	0.806	0.781	15.6	60	5	3	5	4	0.666	0.566	0.644	8.9
16	3	2	3	3	0.556	0.714	0.777	9.6	61	5	4	5	4	0.635	0.589	0.665	8.8
17	4	2	3	3	0.771	0.716	0.808	5.2	62	6	4	5	4	0.723	0.652	0.732	5.5
18	3	3	3	3	0.473	1.000	1.000	7.2	63	2	2	6	4	0.478	0.423	0.524	54.0
19	4	3	3	3	0.501	1.000	0.984	5.0	64	3	2	6	4	0.482	0.447	0.566	31.7
20	5	3	3	3	0.858	0.852	0.932	2.8	65	4	2	6	4	0.603	0.432	0.545	19.3
21	2	2	4	3	0.492	0.621	0.731	23.1	66	3	3	6	4	0.488	0.534	0.735	26.4
22	3	2	4	3	0.522	0.632	0.765	13.5	67	4	3	6	4	0.463	0.617	0.765	17.2
23	4	2	4	3	0.695	0.611	0.752	7.8	68	5	3	6	4	0.634	0.507	0.610	11.6
24	3	3	4	3	0.509	0.791	1.000	10.9	69	5	4	6	4	0.602	0.533	0.639	11.5
25	4	3	4	3	0.487	0.901	1.000	7.1	70	6	4	6	4	0.686	0.578	0.686	7.3
26	5	3	4	3	0.757	0.721	0.852	4.4	71	2	2	7	4	0.492	0.365	0.490	68.0
27	2	2	5	3	0.512	0.505	0.672	32.4	72	3	2	7	4	0.495	0.392	0.540	39.1
28	3	2	5	3	0.529	0.535	0.728	18.3	73	4	2	7	4	0.582	0.394	0.525	23.8
29	4	2	5	3	0.648	0.538	0.712	10.7	74	3	3	7	4	0.502	0.460	0.707	33.8
30	3	3	5	3	0.530	0.640	0.960	15.5	75	4	3	7	4	0.480	0.534	0.746	21.5
31	4	3	5	3	0.513	0.738	0.996	9.8	76	5	3	7	4	0.612	0.460	0.582	14.5
32	5	3	5	3	0.698	0.628	0.794	6.2	77	4	4	7	4	0.514	0.575	0.977	18.5
33	2	2	6	3	0.524	0.425	0.614	43.4	78	5	4	7	4	0.489	0.676	0.998	12.7
34	3	2	6	3	0.538	0.459	0.682	23.9	79	6	4	7	4	0.467	0.783	1.000	9.4
35	4	2	6	3	0.618	0.480	0.680	14.0	80	2	2	2	5	0.280	1.000	0.516	22.1
36	3	3	6	3	0.541	0.536	0.905	21.2	81	3	2	2	5	0.672	0.600	0.501	15.3
37	4	3	6	3	0.528	0.621	0.956	13.0	82	4	2	2	5	0.811	0.698	0.581	8.1
38	5	3	6	3	0.661	0.558	0.749	8.3	83	2	2	3	5	0.331	0.800	0.502	30.2
39	2	2	2	4	0.335	1.000	0.635	15.5	84	3	2	3	5	0.546	0.538	0.483	21.0
40	3	2	2	4	0.388	0.875	0.622	10.9	85	4	2	3	5	0.721	0.568	0.534	11.8
41	4	2	2	4	0.844	0.778	0.703	5.3	86	3	3	3	5	0.337	1.000	0.638	14.3
42	2	2	3	4	0.384	0.803	0.613	22.3	87	4	3	3	5	0.332	1.000	0.633	11.2
43	3	2	3	4	0.535	0.620	0.597	14.8	88	5	3	3	5	0.757	0.685	0.625	6.6
44	4	2	3	4	0.737	0.630	0.641	8.2	89	2	2	4	5	0.379	0.613	0.487	40.2
45	3	3	3	4	0.394	1.000	0.790	10.5	90	3	2	4	5	0.417	0.563	0.494	26.9

TABLE 8.4 (*continued*)

No.	a_1	a_2	b	c	π_{11}	ξ_{11}	ξ_{21}	V	No.	a_1	a_2	b	c	π_{11}	ξ_{11}	ξ_{21}	V
91	4	2	4	5	0.665	0.489	0.501	15.9	136	6	5	8	5	0.470	0.710	0.992	12.1
92	3	3	4	5	0.384	0.782	0.629	19.3	137	2	2	3	6	0.291	0.798	0.425	39.2
93	4	3	4	5	0.346	0.894	0.640	14.1	138	3	2	3	6	0.558	0.480	0.408	28.2
94	5	3	4	5	0.690	0.587	0.580	9.2	139	4	2	3	6	0.712	0.520	0.460	15.9
95	4	4	4	5	0.392	0.988	0.799	10.4	140	3	3	3	6	0.294	1.000	0.535	18.8
96	5	4	4	5	0.370	1.000	0.800	8.3	141	4	3	3	6	0.285	1.000	0.534	15.1
97	6	4	4	5	0.740	0.686	0.670	5.6	142	5	3	3	6	0.737	0.629	0.539	9.1
98	2	2	5	5	0.413	0.498	0.470	52.0	143	2	2	4	6	0.339	0.610	0.415	50.5
99	3	2	5	5	0.421	0.498	0.488	33.4	144	3	2	4	6	0.416	0.504	0.413	35.3
100	4	2	5	5	0.627	0.434	0.476	20.4	145	4	2	4	6	0.661	0.449	0.432	20.8
101	3	3	5	5	0.419	0.632	0.620	25.3	146	4	3	4	6	0.303	0.890	0.537	18.5
102	4	3	5	5	0.383	0.726	0.635	17.8	147	5	3	4	6	0.675	0.542	0.503	12.1
103	5	3	5	5	0.647	0.519	0.545	12.0	148	4	4	4	6	0.347	0.985	0.667	13.2
104	5	4	5	5	0.598	0.551	0.567	11.8	149	5	4	4	6	0.323	1.000	0.670	10.9
105	6	4	5	5	0.691	0.601	0.625	7.5	150	6	4	4	6	0.712	0.637	0.583	7.5
106	5	5	5	5	0.439	0.971	0.980	8.1	151	2	2	5	6	0.376	0.495	0.404	63.5
107	6	5	5	5	0.408	1.000	0.991	6.6	152	3	2	5	6	0.387	0.480	0.415	42.6
108	2	2	6	5	0.438	0.420	0.450	65.6	153	4	2	5	6	0.625	0.399	0.411	26.1
109	3	2	6	5	0.438	0.435	0.477	40.7	154	3	3	5	6	0.378	0.629	0.523	31.0
110	4	2	6	5	0.599	0.393	0.456	25.3	155	4	3	5	6	0.340	0.722	0.534	22.7
111	3	3	6	5	0.444	0.531	0.608	32.2	156	5	3	5	6	0.636	0.481	0.474	15.4
112	4	3	6	5	0.411	0.612	0.628	22.0	157	4	4	5	6	0.384	0.792	0.666	16.9
113	5	3	6	5	0.618	0.466	0.518	15.1	158	5	4	5	6	0.345	0.929	0.672	13.1
114	5	4	6	5	0.418	0.781	0.805	12.8	159	6	4	5	6	0.323	1.000	0.670	10.9
115	6	4	6	5	0.388	0.904	0.807	10.1	160	5	5	5	6	0.393	0.967	0.816	10.1
116	5	5	6	5	0.463	0.810	0.982	10.5	161	6	5	5	6	0.362	1.000	0.825	8.4
117	6	5	6	5	0.433	0.944	0.987	8.0	162	3	3	6	6	0.406	0.528	0.516	38.6
118	2	2	7	5	0.456	0.364	0.429	81.0	163	4	3	6	6	0.370	0.608	0.530	27.3
119	3	2	7	5	0.454	0.384	0.462	48.9	164	5	3	6	6	0.609	0.434	0.452	19.1
120	4	2	7	5	0.577	0.361	0.441	30.5	165	4	4	6	6	0.411	0.662	0.664	21.2
121	3	3	7	5	0.462	0.458	0.594	40.2	166	5	4	6	6	0.375	0.778	0.672	16.0
122	4	3	7	5	0.433	0.530	0.619	26.8	167	6	4	6	6	0.344	0.900	0.674	12.8
123	5	3	7	5	0.598	0.425	0.495	18.6	168	5	5	6	6	0.420	0.807	0.817	12.7
124	5	4	7	5	0.439	0.672	0.803	15.8	169	6	5	6	6	0.388	0.940	0.821	10.0
125	6	4	7	5	0.412	0.778	0.806	12.1	170	6	6	6	6	0.432	0.958	0.972	8.1
126	5	5	7	5	0.481	0.695	0.983	13.2	171	3	3	7	6	0.428	0.455	0.508	47.3
127	6	5	7	5	0.454	0.810	0.990	9.9	172	4	3	7	6	0.393	0.526	0.525	32.6
128	3	2	8	5	0.467	0.342	0.445	58.0	173	5	3	7	6	0.589	0.396	0.433	23.0
129	4	2	8	5	0.561	0.335	0.428	36.1	174	4	4	7	6	0.433	0.569	0.662	26.2
130	3	3	8	5	0.476	0.402	0.578	49.3	175	5	4	7	6	0.398	0.669	0.671	19.2
131	4	3	8	5	0.449	0.467	0.607	32.1	176	6	4	7	6	0.369	0.775	0.674	15.2
132	4	4	8	5	0.484	0.502	0.784	27.4	177	5	5	7	6	0.441	0.693	0.818	15.7
133	5	4	8	5	0.456	0.590	0.801	19.2	178	6	5	7	6	0.411	0.807	0.823	12.1
134	6	4	8	5	0.431	0.684	0.806	14.5	179	6	6	7	6	0.452	0.822	0.974	10.1
135	5	5	8	5	0.494	0.609	0.984	16.4	180	3	3	8	6	0.444	0.400	0.498	56.9

TABLE 8.4 (*continued*)

No.	a_1	a_2	b	c	π_{11}	ξ_{11}	ξ_{21}	V	No.	a_1	a_2	b	c	π_{11}	ξ_{11}	ξ_{21}	V
181	4	3	8	6	0.413	0.463	0.519	38.4	220	5	5	7	7	0.407	0.691	0.700	18.4
182	5	3	8	6	0.574	0.366	0.417	27.2	221	6	5	7	7	0.375	0.805	0.704	14.5
183	4	4	8	6	0.449	0.499	0.659	31.7	222	6	6	7	7	0.416	0.819	0.833	11.9
184	5	4	8	6	0.417	0.587	0.670	22.9	223	4	4	8	7	0.419	0.498	0.567	36.4
185	6	4	8	6	0.389	0.680	0.673	17.7	224	5	4	8	7	0.384	0.585	0.575	27.0
186	5	5	8	6	0.457	0.607	0.819	19.1	225	6	4	8	7	0.355	0.678	0.578	21.4
187	6	5	8	6	0.429	0.707	0.824	14.5	226	5	5	8	7	0.425	0.605	0.701	22.1
188	6	6	8	6	0.468	0.719	0.977	12.3	227	6	5	8	7	0.395	0.705	0.705	17.1
189	2	2	4	7	0.307	0.607	0.361	62.0	228	6	6	8	7	0.434	0.717	0.834	14.3
190	3	2	4	7	0.466	0.423	0.350	44.7	229	2	2	5	8	0.317	0.491	0.313	90.1
191	3	3	4	7	0.308	0.777	0.456	30.1	230	3	2	5	8	0.352	0.434	0.317	64.6
192	4	3	4	7	0.269	0.887	0.463	23.5	231	3	3	5	8	0.317	0.625	0.398	44.4
193	5	3	4	7	0.667	0.505	0.445	15.4	232	4	3	5	8	0.278	0.715	0.404	34.3
194	4	4	4	7	0.312	0.983	0.573	16.4	233	5	3	5	8	0.626	0.423	0.379	23.3
195	5	4	4	7	0.287	1.000	0.576	13.7	234	4	4	5	8	0.319	0.787	0.501	24.4
196	6	4	4	7	0.693	0.596	0.517	9.6	235	5	4	5	8	0.281	0.923	0.504	19.7
197	2	2	5	7	0.344	0.493	0.353	76.2	236	6	4	5	8	0.258	1.000	0.504	17.0
198	3	2	5	7	0.363	0.460	0.360	53.0	237	5	5	5	8	0.325	0.962	0.610	14.7
199	4	2	5	7	0.626	0.371	0.363	32.4	238	6	5	5	8	0.295	1.000	0.616	12.6
200	3	3	5	7	0.345	0.627	0.452	37.4	239	3	2	7	8	0.368	0.358	0.312	85.0
201	4	3	5	7	0.306	0.718	0.460	28.2	240	3	3	7	8	0.371	0.452	0.391	63.2
202	5	3	5	7	0.629	0.449	0.421	19.2	241	4	3	7	8	0.333	0.520	0.400	46.1
203	4	4	5	7	0.349	0.789	0.572	20.5	242	5	3	7	8	0.581	0.352	0.348	32.9
204	5	4	5	7	0.310	0.926	0.576	16.2	243	4	4	7	8	0.373	0.566	0.499	35.2
205	6	4	5	7	0.287	1.000	0.576	13.7	244	5	4	7	8	0.335	0.664	0.504	27.1
206	5	5	5	7	0.356	0.964	0.699	12.3	245	6	4	7	8	0.305	0.769	0.506	22.3
207	6	5	5	7	0.325	1.000	0.706	10.4	246	5	5	7	8	0.378	0.689	0.612	21.3
208	3	3	6	7	0.374	0.526	0.448	45.7	247	6	5	7	8	0.345	0.802	0.615	17.1
209	4	3	6	7	0.336	0.604	0.458	33.3	248	6	6	7	8	0.385	0.818	0.727	13.8
210	5	3	6	7	0.603	0.407	0.402	23.4	249	3	3	8	8	0.391	0.397	0.387	74.1
211	4	4	6	7	0.378	0.660	0.570	25.2	250	4	3	8	8	0.354	0.458	0.398	52.9
212	5	4	6	7	0.340	0.775	0.576	19.4	251	5	3	8	8	0.566	0.327	0.336	38.1
213	6	4	6	7	0.308	0.897	0.578	16.0	252	4	4	8	8	0.393	0.496	0.497	41.5
214	5	5	6	7	0.384	0.805	0.699	15.2	253	5	4	8	8	0.356	0.583	0.504	31.4
215	6	5	6	7	0.351	0.937	0.702	12.1	254	6	4	8	8	0.326	0.675	0.506	25.4
216	6	6	6	7	0.394	0.956	0.831	9.7	255	5	5	8	8	0.397	0.604	0.612	25.3
217	4	4	7	7	0.401	0.567	0.569	30.5	256	6	5	8	8	0.366	0.703	0.616	19.9
218	5	4	7	7	0.364	0.666	0.576	23.0	257	6	6	8	8	0.405	0.716	0.728	16.4
219	6	4	7	7	0.334	0.772	0.578	18.5									

SOLUTION   From the preestimates we obtain $\mu_{02} = 11.32 + 8.56 - 9.86 = 10.02$, $a_1 = a_2 = 2.03$, $b = 3.24$, and $c = 2.89$. Make the approximations $a_1 = a_2 \approx 2$ and $c \approx 3$. Since the desired value of b is not included in Table 8.4, linear interpolation is needed. First find the plans for $(a_1, a_2, b, c) = (2, 2, 3, 3)$ and $(2, 2, 4, 3)$. Then interpolation to $(2, 2, 3.24, 3)$ gives $\pi_{11} = 0.465$, $\xi_{11} = 0.762$, $\xi_{21} - 0.769$, and $V = 17.4$. For comparison, we calculated the optimization model

TABLE 8.5 Compromise Test Plans for a Lognormal Distribution

No.	a_1	a_2	b	c	π_{11}	ξ_{11}	ξ_{21}	V	No.	a_1	a_2	b	c	π_{11}	ξ_{11}	ξ_{21}	V
1	2	2	2	2	0.478	1.000	1.000	5.6	46	4	3	3	4	0.677	0.685	0.659	6.5
2	3	2	2	2	0.846	0.879	1.000	2.6	47	5	3	3	4	0.805	0.781	0.765	3.7
3	2	2	3	2	0.472	0.837	1.000	9.8	48	2	2	4	4	0.379	0.696	0.678	26.7
4	3	2	3	2	0.704	0.711	0.897	4.8	49	3	2	4	4	0.572	0.489	0.514	16.7
5	4	2	3	2	0.850	0.844	1.000	2.4	50	4	2	4	4	0.681	0.559	0.603	9.4
6	2	2	4	2	0.495	0.654	1.000	15.4	51	5	3	4	4	0.722	0.667	0.703	5.4
7	3	2	4	2	0.646	0.593	0.814	7.5	52	4	4	4	4	0.417	1.000	1.000	7.2
8	4	2	4	2	0.747	0.699	0.952	3.9	53	5	4	4	4	0.697	0.678	0.709	5.4
9	2	2	5	2	0.537	0.442	0.645	21.5	54	6	4	4	4	0.786	0.771	0.810	3.3
10	3	2	5	2	0.616	0.507	0.747	10.8	55	2	2	5	4	0.409	0.561	0.676	35.9
11	4	2	5	2	0.691	0.600	0.874	5.8	56	3	2	5	4	0.552	0.427	0.482	21.8
12	2	2	2	3	0.374	1.000	0.823	9.4	57	4	2	5	4	0.641	0.489	0.563	12.5
13	3	2	2	3	0.751	0.767	0.753	5.1	58	3	3	5	4	0.501	0.467	0.510	20.9
14	4	2	2	3	0.923	0.910	0.896	2.5	59	4	3	5	4	0.600	0.510	0.570	12.2
15	2	2	3	3	0.395	0.912	0.883	14.0	60	5	3	5	4	0.674	0.584	0.655	7.4
16	3	2	3	3	0.646	0.630	0.681	8.1	61	5	4	5	4	0.657	0.592	0.661	7.3
17	4	2	3	3	0.779	0.732	0.802	4.3	62	6	4	5	4	0.725	0.675	0.752	4.6
18	3	3	3	3	0.423	1.000	1.000	7.0	63	2	2	6	4	0.431	0.470	0.671	46.6
19	4	3	3	3	0.731	0.753	0.809	4.2	64	3	2	6	4	0.398	0.563	0.675	29.7
20	5	3	3	3	0.867	0.868	0.936	2.3	65	4	2	6	4	0.615	0.436	0.531	15.9
21	2	2	4	3	0.432	0.693	0.878	20.7	66	3	3	6	4	0.507	0.404	0.476	26.6
22	3	2	4	3	0.600	0.534	0.627	11.7	67	4	3	6	4	0.584	0.452	0.537	15.6
23	4	2	4	3	0.703	0.618	0.734	6.4	68	5	3	6	4	0.644	0.520	0.616	9.6
24	3	3	4	3	0.554	0.573	0.656	11.3	69	5	4	6	4	0.633	0.525	0.621	9.5
25	4	3	4	3	0.667	0.638	0.741	6.3	70	6	4	6	4	0.687	0.600	0.706	6.1
26	5	3	4	3	0.763	0.735	0.853	3.6	71	2	2	7	4	0.447	0.405	0.665	58.8
27	2	2	5	3	0.457	0.560	0.868	28.9	72	3	2	7	4	0.417	0.487	0.671	36.6
28	3	2	5	3	0.578	0.462	0.583	15.9	73	4	2	7	4	0.597	0.394	0.503	19.7
29	4	2	5	3	0.659	0.537	0.681	8.9	74	3	3	7	4	0.511	0.358	0.448	32.8
30	3	3	5	3	0.549	0.487	0.605	15.5	75	4	3	7	4	0.573	0.406	0.509	19.3
31	4	3	5	3	0.632	0.553	0.688	8.7	76	5	3	7	4	0.623	0.468	0.584	12.0
32	5	3	5	3	0.705	0.638	0.790	5.2	77	4	4	7	4	0.567	0.410	0.514	19.2
33	2	2	6	3	0.473	0.470	0.852	38.6	78	5	4	7	4	0.615	0.473	0.587	11.9
34	3	2	6	3	0.565	0.407	0.547	20.7	79	6	4	7	4	0.660	0.541	0.667	7.7
35	4	2	6	3	0.631	0.475	0.638	11.7	80	2	2	2	5	0.263	1.000	0.541	19.0
36	3	3	6	3	0.546	0.423	0.564	20.2	81	3	2	2	5	0.684	0.622	0.505	12.0
37	4	3	6	3	0.611	0.487	0.644	11.5	82	4	2	2	5	0.826	0.732	0.605	6.5
38	5	3	6	3	0.669	0.564	0.739	6.9	83	2	2	3	5	0.293	0.925	0.551	25.1
39	2	2	2	4	0.308	1.000	0.658	13.8	84	3	2	3	5	0.596	0.522	0.466	16.9
40	3	2	2	4	0.706	0.685	0.604	8.2	85	4	2	3	5	0.728	0.597	0.554	9.5
41	4	2	2	4	0.859	0.808	0.721	4.3	86	3	3	3	5	0.319	1.000	0.710	12.7
42	2	2	3	4	0.337	0.920	0.679	19.1	87	4	3	3	5	0.642	0.632	0.558	9.2
43	3	2	3	4	0.614	0.570	0.552	12.2	88	5	3	3	5	0.769	0.714	0.649	5.5
44	4	2	3	4	0.746	0.655	0.653	6.7	89	2	2	4	5	0.338	0.699	0.551	33.7
45	3	3	3	4	0.368	1.000	0.882	9.5	90	3	2	4	5	0.295	0.828	0.551	24.2

TABLE 8.5 (*continued*)

No.	a_1	a_2	b	c	π_{11}	ξ_{11}	ξ_{21}	V	No.	a_1	a_2	b	c	π_{11}	ξ_{11}	ξ_{21}	V
91	4	2	4	5	0.669	0.512	0.514	12.8	136	6	5	8	5	0.621	0.464	0.554	12.1
92	3	3	4	5	0.351	0.883	0.698	16.5	137	2	2	3	6	0.260	0.929	0.463	31.9
93	4	3	4	5	0.319	1.000	0.710	12.7	138	3	2	3	6	0.229	1.000	0.457	25.1
94	5	3	4	5	0.697	0.614	0.601	7.6	139	4	2	3	6	0.719	0.551	0.482	12.7
95	4	4	4	5	0.563	0.564	0.533	12.1	140	3	3	3	6	0.281	1.000	0.594	16.3
96	5	4	4	5	0.659	0.630	0.607	7.4	141	4	3	3	6	0.285	1.000	0.546	13.9
97	6	4	4	5	0.746	0.712	0.693	4.7	142	5	3	3	6	0.748	0.661	0.565	7.5
98	2	2	5	5	0.370	0.562	0.550	43.8	143	2	2	4	6	0.304	0.701	0.463	41.5
99	3	2	5	5	0.329	0.671	0.552	30.2	144	3	2	4	6	0.261	0.833	0.464	30.9
100	4	2	5	5	0.631	0.452	0.483	16.5	145	4	2	4	6	0.663	0.475	0.450	16.7
101	3	3	5	5	0.384	0.710	0.700	21.6	146	4	3	4	6	0.577	0.509	0.454	16.1
102	4	3	5	5	0.349	0.854	0.708	15.9	147	5	3	4	6	0.682	0.571	0.526	10.0
103	5	3	5	5	0.319	1.000	0.710	12.7	148	4	4	4	6	0.528	0.537	0.469	15.6
104	5	4	5	5	0.372	1.000	0.870	9.2	149	5	4	4	6	0.631	0.591	0.532	9.7
105	6	4	5	5	0.364	1.000	0.884	8.1	150	6	4	4	6	0.718	0.665	0.608	6.3
106	5	5	5	5	0.417	1.000	1.000	7.2	151	2	2	5	6	0.338	0.563	0.463	52.5
107	6	5	5	5	0.404	1.000	1.000	6.4	152	3	2	5	6	0.295	0.673	0.465	37.6
108	2	2	6	5	0.395	0.471	0.549	55.4	153	4	2	5	6	0.626	0.421	0.424	21.0
109	3	2	6	5	0.356	0.565	0.552	37.0	154	3	3	5	6	0.349	0.712	0.587	26.2
110	4	2	6	5	0.606	0.405	0.457	20.6	155	4	3	5	6	0.311	0.856	0.592	19.9
111	3	3	6	5	0.467	0.397	0.420	33.4	156	5	3	5	6	0.281	1.000	0.594	16.2
112	4	3	6	5	0.563	0.425	0.462	20.0	157	4	4	5	6	0.361	0.869	0.718	14.6
113	5	3	6	5	0.627	0.484	0.532	12.5	158	5	4	5	6	0.331	1.000	0.727	11.6
114	5	4	6	5	0.609	0.492	0.537	12.4	159	6	4	5	6	0.323	1.000	0.738	10.4
115	6	4	6	5	0.662	0.560	0.611	8.0	160	5	5	5	6	0.377	1.000	0.864	8.9
116	5	5	6	5	0.605	0.494	0.539	12.3	161	6	5	5	6	0.367	1.000	0.888	8.0
117	6	5	6	5	0.656	0.562	0.614	8.0	162	3	3	6	6	0.375	0.595	0.587	32.6
118	2	2	7	5	0.415	0.405	0.546	68.5	163	4	3	6	6	0.339	0.716	0.595	24.1
119	3	2	7	5	0.378	0.488	0.552	44.6	164	5	3	6	6	0.309	0.843	0.597	19.3
120	4	2	7	5	0.588	0.368	0.435	25.0	165	4	4	6	6	0.388	0.727	0.720	18.3
121	3	3	7	5	0.480	0.347	0.394	40.6	166	5	4	6	6	0.358	0.859	0.726	14.1
122	4	3	7	5	0.554	0.383	0.440	24.4	167	6	4	6	6	0.332	0.996	0.728	11.6
123	5	3	7	5	0.608	0.438	0.505	15.4	168	5	5	6	6	0.402	0.865	0.859	11.2
124	5	4	7	5	0.595	0.444	0.509	15.2	169	6	5	6	6	0.377	1.000	0.864	8.9
125	6	4	7	5	0.640	0.506	0.580	10.0	170	6	6	6	6	0.417	1.000	1.000	7.2
126	5	5	7	5	0.593	0.445	0.511	15.2	171	3	3	7	6	0.396	0.512	0.588	39.9
127	6	5	7	5	0.636	0.508	0.582	10.0	172	4	3	7	6	0.362	0.616	0.597	28.8
128	3	2	8	5	0.395	0.430	0.551	53.0	173	5	3	7	6	0.332	0.725	0.599	22.6
129	4	2	8	5	0.576	0.337	0.416	29.7	174	4	4	7	6	0.408	0.625	0.721	22.5
130	3	3	8	5	0.487	0.311	0.373	48.4	175	5	4	7	6	0.380	0.738	0.728	17.0
131	4	3	8	5	0.548	0.349	0.421	29.1	176	6	4	7	6	0.355	0.856	0.731	13.7
132	4	4	8	5	0.541	0.353	0.425	28.9	177	5	5	7	6	0.422	0.744	0.861	13.8
133	5	4	8	5	0.585	0.405	0.486	18.3	178	6	5	7	6	0.399	0.862	0.866	10.8
134	6	4	8	5	0.624	0.463	0.553	12.1	179	6	6	7	6	0.437	0.868	1.000	9.0
135	5	5	8	5	0.584	0.406	0.488	18.3	180	3	3	8	6	0.412	0.449	0.588	48.0

TABLE 8.5 (*continued*)

No.	a_1	a_2	b	c	π_{11}	ξ_{11}	ξ_{21}	V	No.	a_1	a_2	b	c	π_{11}	ξ_{11}	ξ_{21}	V
181	4	3	8	6	0.381	0.540	0.598	34.1	220	5	5	7	7	0.390	0.745	0.740	16.1
182	5	3	8	6	0.352	0.637	0.602	26.2	221	6	5	7	7	0.365	0.863	0.744	12.9
183	4	4	8	6	0.424	0.548	0.722	27.2	222	6	6	7	7	0.403	0.868	0.864	10.5
184	5	4	8	6	0.398	0.647	0.730	20.3	223	4	4	8	7	0.396	0.549	0.621	31.1
185	6	4	8	6	0.375	0.751	0.734	16.1	224	5	4	8	7	0.368	0.648	0.627	23.7
186	5	5	8	6	0.437	0.652	0.863	16.8	225	6	4	8	7	0.343	0.751	0.630	19.2
187	6	5	8	6	0.416	0.757	0.869	13.0	226	5	5	8	7	0.408	0.653	0.741	19.3
188	6	6	8	6	0.452	0.762	1.000	10.9	227	6	5	8	7	0.384	0.757	0.746	15.2
189	2	2	4	7	0.277	0.702	0.400	50.1	228	6	6	8	7	0.420	0.761	0.866	12.7
190	3	2	4	7	0.233	0.836	0.401	38.5	229	2	2	5	8	0.285	0.552	0.356	72.5
191	3	3	4	7	0.285	0.888	0.504	25.1	230	3	2	5	8	0.243	0.676	0.353	55.0
192	4	3	4	7	0.251	1.000	0.511	20.3	231	3	3	5	8	0.294	0.714	0.443	36.7
193	5	3	4	7	0.673	0.535	0.469	12.7	232	4	3	5	8	0.256	0.859	0.447	29.2
194	4	4	4	7	0.498	0.516	0.420	19.4	233	5	3	5	8	0.227	1.000	0.448	24.8
195	5	4	4	7	0.609	0.559	0.475	12.3	234	4	4	5	8	0.303	0.873	0.542	20.8
196	6	4	4	7	0.699	0.625	0.542	8.1	235	5	4	5	8	0.272	1.000	0.548	17.1
197	2	2	5	7	0.311	0.564	0.400	62.1	236	6	4	5	8	0.263	1.000	0.554	15.7
198	3	2	5	7	0.266	0.675	0.401	45.9	237	5	5	5	8	0.313	1.000	0.651	12.8
199	4	2	5	7	0.625	0.395	0.379	25.9	238	6	5	5	8	0.302	1.000	0.665	11.7
200	3	3	5	7	0.319	0.713	0.505	31.2	239	3	2	7	8	0.294	0.489	0.353	73.5
201	4	3	5	7	0.281	0.857	0.509	24.3	240	3	3	7	8	0.345	0.513	0.444	52.4
202	5	3	5	7	0.251	1.000	0.511	20.3	241	4	3	7	8	0.308	0.617	0.449	39.8
203	4	4	5	7	0.329	0.871	0.618	17.6	242	5	3	7	8	0.277	0.726	0.451	32.4
204	5	4	5	7	0.299	1.000	0.625	14.2	243	4	4	7	8	0.353	0.626	0.544	29.9
205	6	4	5	7	0.290	1.000	0.633	12.9	244	5	4	7	8	0.323	0.739	0.548	23.6
206	5	5	5	7	0.342	1.000	0.743	10.8	245	6	4	7	8	0.296	0.857	0.550	19.7
207	6	5	5	7	0.331	1.000	0.760	9.7	246	5	5	7	8	0.363	0.745	0.649	18.6
208	3	3	6	7	0.346	0.596	0.506	38.2	247	6	5	7	8	0.337	0.864	0.652	15.1
209	4	3	6	7	0.309	0.717	0.511	28.9	248	6	6	7	8	0.374	0.869	0.758	12.2
210	5	3	6	7	0.279	0.844	0.512	23.6	249	3	3	8	8	0.364	0.449	0.445	61.5
211	4	4	6	7	0.357	0.728	0.619	21.6	250	4	3	8	8	0.328	0.541	0.450	45.8
212	5	4	6	7	0.326	0.860	0.623	17.0	251	5	3	8	8	0.297	0.637	0.452	36.7
213	6	4	6	7	0.299	0.997	0.625	14.2	252	4	4	8	8	0.372	0.549	0.545	35.3
214	5	5	6	7	0.369	0.866	0.739	13.2	253	5	4	8	8	0.342	0.648	0.549	27.4
215	6	5	6	7	0.342	1.000	0.743	10.8	254	6	4	8	8	0.316	0.752	0.551	22.5
216	6	6	6	7	0.381	1.000	0.864	8.6	255	5	5	8	8	0.382	0.653	0.650	21.9
217	4	4	7	7	0.379	0.626	0.620	26.1	256	6	5	8	8	0.357	0.758	0.653	17.6
218	5	4	7	7	0.349	0.739	0.625	20.2	257	6	6	8	8	0.392	0.761	0.759	14.5
219	6	4	7	7	0.323	0.857	0.627	16.6									

for directly (2.03, 2.03, 3.24, 2.89) and got $\pi_{11} = 0.477$, $\xi_{11} = 0.758$, $\xi_{21} = 0.800$, and $V = 16.1$. In this case, the approximation and interpolation yield fairly accurate results. The standardized plan is then converted to the actual plan, as shown in Table 8.6. In implementing this plan, the six tires used for the preliminary test should continue being tested until 13,000 miles as part of the group at 2,835 pounds. Thus, this group requires only an additional 12 units.

TABLE 8.6   Actual Test Plan for the Tires

Group	Sample Size	Load (pounds)	Threshold (inches)	Censoring Time (miles)
1	17	1611	3/32	13,000
2	17	1611	8/32	13,000
3	18	2835	3/32	13,000
4	18	2835	8/32	13,000

8.8.2   Survey of Degradation Test Plans

As stated earlier, a degradation test may be conducted at a use stress level or higher than a use stress level. For a test at a use stress level, we need to choose the sample size, inspection times (inspection frequency), and test termination time. In addition to these variables, we need to determine the stress levels and sample size allocated to each stress level for a constant-stress accelerated degradation test. The common design of the test plans selects the optimum values of these variables to minimize the statistical error, total test cost, or both. In particular, the test plans can be formulated as one of the following optimization problems:

- Minimize the asymptotic variance (or mean square error) of the estimate of a life percentile or other quantity at a use stress level, subject to a prespecified cost budget.
- Minimize the total test cost, subject to the allowable statistical error.
- Minimize the asymptotic variance (or mean square error) and the total test cost simultaneously.

Nelson (2005) provides a nearly complete bibliography of test plans. Most plans are formulated as a first optimization problem. Yu (2003) designs optimum accelerated degradation test plans using the nonlinear integer programming technique. The plans are designed to choose the optimal combination of sample size, inspection frequency, and test termination time at each stress level by minimizing the mean square error of the estimated $100p$th percentile of the product life distribution at a use stress level. The optimization is subject to the constraint that the total test cost must not exceed a given budget. In an earlier work, Yu (1999) presents similar test plans but minimizes the variance of the estimate. Q. Li and Kececioglu (2003) present a four-step approach to the calculation of sample size, test stress levels, respective sample allocation assigned to each stress level, measurement times, and test termination time. The decision variables are optimized to minimize the mean square error of the estimate of the mean life subject to a cost budget. Using a simple constant rate relationship between the applied stress and product performance, Park and Yum (1997) devise optimum test plans for destructive inspections. The plans also determine the stress levels, proportion of test units allocated to each stress level, and measurement times, which minimize the asymptotic variance of the MLE of the mean life at a use stress level. Park and Yum (1999) compare numerically plans they developed for accelerated life and degradation tests. Unsurprisingly, they conclude that accelerated degradation test plans provide more accurate estimates of life percentiles, especially when the probabilities of failure are small. Boulanger and Escobar (1994) design optimum accelerated degradation test plans for a particular degradation model that may be suitable to describe a degradation process in which the amount of degradation over time levels off toward a stress-dependent plateau (maximum degradation). The design consists of three steps, the first of which is to determine the stress levels and corresponding proportions of test units by minimizing the variance of the weighted least squares estimate of the mean of the log plateaus at the use condition. The second step is to optimize the times at which to measure the units at a selected stress level, then the results of the two steps are combined to determine the total number of test units.

For degradation tests at a single constant-stress level, S. Wu and Chang (2002) propose an approach to the determination of sample size, inspection frequency, and test termination time. The optimization criterion is to minimize the variance of a life percentile estimate subject to total test cost. Marseguerra et al. (2003) develop test plans similar to those of S. Wu and Chang (2002), and consider additionally simultaneous minimization of the variance and total test cost. The latter part deals with the third optimization problem described above.

Few publications deal with the second optimization problem. Yu and Chiao (2002) design optimal plans for fractional factorial degradation experiments with the aim of improving product reliability. The plans select the inspection frequency, sample size, and test termination time at each run by minimizing the total test cost subject to a prespecified correct decision probability. Tang et al. (2004) conduct the optimal design of step-stress accelerated degradation tests. The objective of the design is to minimize the total test cost subject to a variance constraint. The minimization yields the optimal sample size, number of inspections at each intermediate stress level, and number of total inspections.

Generally speaking, the optimal design of accelerated degradation tests is considerably more difficult than that of accelerated life tests, mainly because the former involves complicated degradation models and more decision variables. It is not surprising that there is scant literature on this subject. As we may have observed, degradation testing and analysis is a promising and rewarding technique. Increasingly wide applications of this technique require more practically useful test plans.

PROBLEMS

8.1 A product usually has more than one performance characteristic. Describe the general approaches to determination of the critical characteristic. Explain why the critical characteristic selected must be monotone.

8.2 Discuss the advantages and disadvantages of pseudolife analysis. Compared with the traditional life data analysis, do you expect a pseudolife analysis to give a more accurate estimate? Why?

TABLE 8.7   Valve Recession Data (Inches)

	Valve						
Time (h)	1	2	3	4	5	6	7
0	0	0	0	0	0	0	0
15	0.001472	0.001839	0.001472	0.001839	0.001839	0.001839	0.002575
45	0.002943	0.004047	0.003311	0.002943	0.003311	0.002943	0.003679
120	0.005886	0.00699	0.005886	0.004415	0.005518	0.005886	0.005886
150	0.006254	0.008093	0.006622	0.005150	0.006254	0.006990	0.007726
180	0.008461	0.009933	0.008093	0.006622	0.007726	0.008461	0.010301

8.3 K.Yang and Xue (1996) performed degradation analysis of the exhaust valves installed in a certain internal combustion engine. The degradation is the valve recession, representing the amount of wear in a valve over time; the valve recession data at different inspection times are shown in Table 8.7. Develop a model to describe the degradation paths. If the valve is said to have failed when the recession reaches 0.025 inch, estimate the probability of failure at 500 hours through pseudolife analysis.

8.4 Refer to Problem 8.3. Suppose that the degradation model parameters for the valve recession have random effects and have a bivariate normal distribution. Calculate the mean vector and variance–covariance matrix using the multivariate approach. Estimate the probability of failure of the valve at 500 hours through Monte Carlo simulation.

8.5 Refer to Problem 8.4. Calculate the probability of failure of the valve at 500 hours by using (8.13). Compare the result with those in Problems 8.3 and 8.4.

8.6 In Section 8.5.1 we describe methods for sample allocation to destructive inspections. Explain how the methods improve the statistical accuracy of a reliability estimate.

8.7 A type of new polymer was exposed to the alkaline environment at elevated temperatures to evaluate the long-term reliability of the material. The experiment tested standard bars of the material at 50, 65, and 80°C, each with 25 units. Five units were inspected destructively for tensile strength at each inspection time during testing. The degradation performance is the ratio of the tensile strength to the original standard strength. The material is said to have failed when the ratio is less than 60%. Table 8.8 shows the values of the ratio at different inspection times (in days) and temperatures.

(a) For each combination of temperatures and inspection times, plot the data and ML fits on lognormal paper. Comment on the adequacy of the lognormal distribution.

(b) Does the shape parameter change with time?

TABLE 8.8 Polymer Degradation Data

	Temperature (°C)		
Days	50	65	80
8	98.3 94.2 96.5 98.1 96.0	87.5 85.2 93.3 90.0 88.4	80.8 82.3 83.7 86.6 81.1
25	92.4 88.1 90.5 93.4 90.2	83.2 80.5 85.7 86.3 84.2	73.3 72.3 71.9 74.5 76.8
75	86.2 82.7 84.2 86.1 85.5	77.0 73.2 79.8 75.4 76.2	67.4 65.4 64.3 65.3 64.5
130	82.3 78.5 79.4 81.8 82.3	73.9 70.1 75.8 72.3 71.7	64.3 60.4 58.6 58.9 59.7
180	77.7 74.6 76.1 77.9 79.2	68.7 65.3 69.8 67.4 66.6	60.4 55.3 56.7 57.3 55.7

TABLE 8.9 Degradation Data for the IRLEDs at 170 mA

	Time (h)										
Unit	24	48	96	155	368	768	1130	1536	1905	2263	2550
1	0.1	0.3	0.7	1.2	3.0	6.6	12.1	16.0	22.5	25.3	30.0
2	2.0	2.3	4.7	5.9	8.2	9.3	12.6	12.9	17.5	16.4	16.3
3	0.3	0.5	0.9	1.3	2.2	3.8	5.5	5.7	8.5	9.8	10.7
4	0.3	0.5	0.8	1.1	1.5	2.4	3.2	5.1	4.7	6.5	6.0
5	0.2	0.4	0.9	1.6	3.9	8.2	11.8	19.5	26.1	29.5	32.0
6	0.6	1.0	1.6	2.2	4.6	6.2	10.5	10.2	11.2	11.6	14.6
7	0.2	0.4	0.7	1.1	2.4	4.9	7.1	10.4	10.8	13.7	18.0
8	0.5	0.9	1.8	2.7	6.5	10.2	13.4	22.4	23.0	32.2	25.0
9	1.4	1.9	2.6	3.4	6.1	7.9	9.9	10.2	11.1	12.2	13.1
10	0.7	0.8	1.4	1.8	2.6	5.2	5.7	7.1	7.6	9.0	9.6
11	0.2	0.5	0.8	1.1	2.5	5.6	7.0	9.8	11.5	12.2	14.2
12	0.2	0.3	0.6	0.9	1.6	2.9	3.5	5.3	6.4	6.6	9.2
13	2.1	3.4	4.1	4.9	7.2	8.6	10.8	13.7	13.2	17.0	13.9
14	0.1	0.2	0.5	0.7	1.2	2.3	3.0	4.3	5.4	5.5	6.1
15	0.7	0.9	1.5	1.9	4.0	4.7	7.1	7.4	10.1	11.0	10.5
16	1.8	2.3	3.7	4.7	6.1	9.4	11.4	14.4	16.2	15.6	16.6
17	0.1	0.2	0.5	0.8	1.6	3.2	3.7	5.9	7.2	6.1	8.8
18	0.1	0.1	0.2	0.3	0.7	1.7	2.2	3.0	3.5	4.2	4.6
19	0.5	0.7	1.3	1.9	4.8	7.7	9.1	12.8	12.9	15.5	19.3
20	1.9	2.3	3.3	4.1	5.2	8.9	11.8	13.8	14.1	16.2	17.1
21	3.7	4.8	7.3	8.3	9.0	10.9	11.5	12.2	13.5	12.4	13.8
22	1.5	2.2	3.0	3.7	5.1	5.9	8.1	7.8	9.2	8.8	11.1
23	1.2	1.7	2.0	2.5	4.5	6.9	7.5	9.2	8.5	12.7	11.6
24	3.2	4.2	5.1	6.2	8.3	10.6	14.9	17.5	16.6	18.4	15.8
25	1.0	1.6	3.4	4.7	7.4	10.7	15.9	16.7	17.4	28.7	25.9

(c) Develop a degradation model for a given temperature.

(d) For each temperature, estimate the degradation model parameters using the least squares method. Which model parameter(s) depend on temperature?

(e) For each temperature, estimate the probability of failure at the test termination time.

(f) Repeat part (e) for a design life of 10 years.

8.8 Example 8.10 describes the degradation analysis for the IRLEDs. The degradation data of the device at different measurement times (in hours) and current levels (in mA) are shown in Tables 8.9 and 8.10.

(a) Fit the degradation model (8.36) to each degradation path, and estimate the model parameters using the least squares method.

(b) Compute the pseudolife of each unit.

(c) For each current, plot the pseudolife data and the ML fits on lognormal probability paper. Comment on the adequacy of the lognormal distribution.

(d) Is it evident that the shape parameter depends on current?

(e) Use the inverse power relationship for current, and estimate the lognormal scale parameter and the probability of failure at a use current of 50 mA. Comment on the difference in results from those in Example 8.10.

8.9 Refer to Problem 8.7.

(a) Is it evident that the shape parameter depends on temperature?

(b) Develop a degradation model to describe the relationship between the mean log ratio and time and temperature.

TABLE 8.10   Degradation Data for the IRLEDs at 320 mA

Unit	\multicolumn{10}{c}{Time (h)}
	6	12	24	48	96	156	230	324	479	635
1	4.3	5.8	9.5	10.2	13.8	20.6	19.7	25.3	33.4	27.9
2	0.5	0.9	1.4	3.3	5.0	6.1	9.9	13.2	17.0	20.7
3	2.6	3.6	4.6	6.9	9.5	13.0	15.3	13.5	19.0	19.5
4	0.2	0.4	0.9	2.4	4.5	7.1	13.4	21.2	30.7	41.7
5	3.7	5.6	8.0	12.8	16.0	23.7	26.7	38.4	49.2	47.2
6	3.2	4.3	5.8	9.9	15.2	20.3	26.2	33.6	39.5	53.2
7	0.8	1.7	2.8	4.6	7.9	12.4	20.2	24.8	32.5	45.4
8	4.3	6.5	7.8	13.0	21.7	33.0	42.1	49.9	59.9	78.6
9	1.4	2.7	5.0	7.8	14.5	23.3	29.0	43.3	59.8	77.4
10	3.4	4.6	7.8	13.0	16.8	26.8	34.1	41.5	67.0	65.5
11	3.6	4.7	6.2	9.1	11.7	13.8	14.5	15.5	23.1	24.0
12	2.3	3.7	5.6	8.8	13.7	17.2	24.8	29.1	42.9	45.3
13	0.5	0.9	1.9	3.5	5.9	10.0	14.4	22.0	26.0	31.8
14	2.6	4.4	6.0	8.7	14.6	16.8	17.9	23.2	27.0	31.3
15	0.1	0.4	0.7	2.0	3.5	6.6	12.2	18.8	32.3	47.0

TABLE 8.11 Degradation Data for an MOS Field-Effect Transistor

Time (s)	Unit				
	1	2	3	4	5
100	1.05	0.58	0.86	0.6	0.62
200	1.4	0.9	1.25	0.6	0.64
300	1.75	1.2	1.45	0.6	1.25
400	2.1	1.75	1.75	0.9	1.3
500	2.1	2.01	1.75	0.9	0.95
600	2.8	2	2	1.2	1.25
700	2.8	2	2	1.5	1.55
800	2.8	2	2	1.5	1.9
900	3.2	2.3	2.3	1.5	1.25
1,000	3.4	2.6	2.3	1.7	1.55
1,200	3.8	2.9	2.6	2.1	1.5
1,400	4.2	2.9	2.8	2.1	1.55
1,600	4.2	3.2	3.15	1.8	1.9
1,800	4.5	3.6	3.2	2.1	1.85
2,000	4.9	3.8	3.2	2.1	2.2
2,500	5.6	4.2	3.8	2.4	2.2
3,000	5.9	4.4	3.8	2.7	2.5
3,500	6.3	4.8	4	2.7	2.2
4,000	6.6	5	4.2	3	2.8
4,500	7	5.6	4.4	3	2.8
5,000	7.8	5.9	4.6	3	2.8
6,000	8.6	6.2	4.9	3.6	3.1
7,000	9.1	6.8	5.2	3.6	3.1
8,000	9.5	7.4	5.8	4.2	3.1
9,000	10.5	7.7	6.1	4.6	3.7
10,000	11.1	8.4	6.3	4.2	4.4
12,000	12.2	8.9	7	4.8	3.7
14,000	13	9.5	7.2	5.1	4.4
16,000	14	10	7.6	4.8	4.4
18,000	15	10.4	7.7	5.3	4.1
20,000	16	10.9	8.1	5.8	4.1
25,000	18.5	12.6	8.9	5.7	4.7
30,000	20.3	13.2	9.5	6.2	4.7
35,000	22.1	15.4	11.2	8	6.4
40,000	24.2	18.1	14	10.9	9.4

Source: J. Lu et al. (1997).

(c) Write down the total sample log likelihood.

(d) Estimate the model parameters.

(e) Calculate the probability of failure of the polymer after 10 years of continuous use at 40°C.

8.10 Refer to Example 8.1. The percent transconductance degradation data taken at different times (in seconds) for five units of an MOS field-effect transistor are shown in Table 8.11.

    (a) Without fitting a degradation model, determine the failure time intervals for each unit at thresholds 3%, 8%, and 13%, respectively.

    (b) Estimate the life distributions at each threshold.

    (c) Does the scale parameter depend on threshold?

    (d) Develop a relationship between the distribution location parameter and the threshold.

    (e) Estimate the distribution location parameter at the usual threshold of 15%. Compare the result with that in Example 8.1.

8.11 To estimate the reliability of a resistor at a use temperature of 50°C, the manufacturer plans to sample 45 units and divide them into two groups, each tested at an elevated temperature. The failure of the resistor is defined in terms of a resistance drift greater than 500 ppm (parts per million). The tightest failure criterion is 100 ppm, and the maximum allowable temperature is 175°C. The time to failure of the resistor has a Weibull distribution with shape parameter 1.63. Preestimates of the log characteristic life are $\mu_{00} = 12.1$, $\mu_{20} = 8.3$, and $\mu_{22} = 5.9$. Each group is tested for 2350 hours or until all units fail, whichever is sooner. Develop a compromise test plan for the experiment.

9

RELIABILITY VERIFICATION TESTING

9.1 INTRODUCTION

In the design and development phase of a product life cycle, reliability can be designed into products proactively using the techniques presented in earlier chapters. The next task is to verify that the design meets the functional, environmental, reliability, and legal requirements specified in the product planning phase. This task is often referred to in industry as *design verification* (DV). Reliability verification testing is an integral part of DV testing and is aimed particularly at verifying design reliability. If during testing a design fails to demonstrate the reliability required, it must be revised following rigorous failure analysis. Then the redesigned product is resubjected to verification testing. The process of test–fix–test is continued until the reliability required is achieved. The repetitive process jeopardizes the competitiveness of the product in the marketplace, due to the increased cost and time to market. Nowadays, most products are designed with the aim of passing the first DV testing. Therefore, it is vital to design-in reliability and to eliminate potential failure modes even before prototypes are built.

A design is released to production if it passes DV testing successfully. The production process is then set up to manufacture products that meet all requirements with minimum variation. As we know, the designed-in or inherent reliability level is always degraded by process variation. This is also true for product functionality and other performances. Therefore, prior to full production, the process must pass a qualification test, usually called *process validation* (PV) in industry. Its purpose

Life Cycle Reliability Engineering, by Guangbin Yang
Copyright © 2007 John Wiley & Sons, Inc.

is to validate that the established process is capable of manufacturing products that meet the functional, environmental, reliability, and legal requirements specified in the product planning phase. Like DV testing, PV testing must include a reliability verification test to demonstrate that the final product achieves the required reliability level. If the process is validated, full production may begin. Otherwise, process changes must be implemented following root cause analysis. Failure to pass PV testing results in expensive test–fix–test repetition. Therefore, it is critical to perform a careful process planning by using proactive techniques such as process FMEA, process capability study, and statistical process control charts.

In this chapter we present techniques for reliability verification testing performed in the DV and PV stages. The techniques include:

- *Bogey testing*: widely implemented in industry
- *Sequential life testing*: used widely for military and governmental products
- *Degradation testing*: aimed at verifying the reliability of highly reliable products

9.2 PLANNING RELIABILITY VERIFICATION TESTS

9.2.1 Types of Tests

There are four types of test for reliability verification: bogey (here *bogey* means requirement) testing, sequential life testing, test-to-failure testing, and degradation testing. In test planning we need to determine the type of test appropriate for the product and test purpose.

Bogey testing is used to test a sample of predetermined size for a certain period of time. The reliability required is verified if no failures occur in the testing. The sample size and test time are determined by the methods described in Sections 9.3 and 9.4. This type of test is easy to implement; it does not require failure monitoring and performance measurement during testing. Thus, it is widely favored by industry. For example, automobile manufacturers and their suppliers are loyal practitioners of this test method. However, the pros and cons of this test method are exactly the same. Failure or degradation observations during testing are neither necessary nor useful in arriving at a conclusion about the reliability. Consequently, the test method requires a large sample size and/or extensive test length, as we show in Section 9.3.

Sequential life testing tests samples one unit at a time until failure or until a prespecified period of time has elapsed. The accumulated test results are compared with the predetermined decision rules to conclude whether (1) the reliability required is achieved, (2) the reliability required is not achieved, or (3) the test is to be continued. Because of this dynamic nature, the sample size is not fixed. This test method is appealing in that it needs a smaller sample size than that required in a bogey test. In contrast to a bogey test, which only protects consumer risk, a sequential test considers both consumer and producer risks. It is often applied to military and governmental products.

Test-to-failure testing, testing samples until they fail, often takes longer; however, it requires fewer samples and generates considerably more information. The actual reliability level can also be estimated from the test. The test is usually conducted under accelerating conditions, and thus needs an appropriate acceleration relationship. For some products, failure is defined in terms of a performance characteristic crossing a threshold. As described in Chapter 8, degradation measurements of products can be used to estimate reliability. Thus, it is not necessary to test such products until failure. This advantage may make a *degradation test* suitable for highly reliable products.

9.2.2 Test Samples

In the DV stage, a reliability verification test is conducted to demonstrate the reliability of design. The prototypes for testing should be representative of the final product in every detail of the design, including, for example, structure, function, connection, materials, components, and housing. The use of surrogate parts must be avoided under all circumstances. On the other hand, the tooling and assembly process of prototypes must be as close as possible to the actual production process. Although the production process is not fully set up in the DV stage, process planning usually makes appreciable progress in the determination of process steps, tooling machines, assembly process, and others, which form the base of the prototype building process. Let's consider an automobile body control module as an example. The DV prototypes should be built to represent the final modules to be installed in automobiles. This is accomplished by using the same circuit design, the same materials for the printed circuit board and module package, and the same electronic components and solders from the manufacturers selected. The tooling steps and process parameters, such as the components populating sequence and wave-soldering temperature and time, are determined from the process planning outputs.

Reliability verification testing in the PV stage is to validate that the production process is capable of manufacturing products that achieve the reliability level required. By this step, the process has been set up and is intended for production at full capacity. Thus, the test samples are the products that customers will see in the market. In other words, the samples and the final products are not differentiable because both use the same materials, components, production processes, and process monitoring and measuring techniques. For the automobile body control module discussed above, the DV and PV samples are basically the same except that the PV test units are built using the actual production process, which is slightly different from that for DV prototyping in some process parameters. Strictly speaking, PV samples cannot fully characterize a full-production population, because the test units are built in a short period of time and thus do not contain much lot-to-lot variability.

9.2.3 Test Stresses

Reliability verification testing in both the DV and PV stages should use the same test stress types and levels. Stress types may include temperature, mechanical

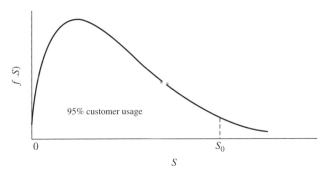

FIGURE 9.1   Real-world usage profile

vibration, thermal cycling, humidity, mechanical force, electrical load, radiation, high altitude, salt spray, and many others. The type and magnitude of the stresses to be applied in testing should be determined from the real-world usage profile, which defines the customer operational frequency, load, and environment. The profile, also known as *stress distribution*, is shown in Figure 9.1, where S is the stress and $f(S)$ is the probability density function (pdf) of S. The test stress level is usually chosen to be a high percentile to represent a large percentage of customer usage. For example, S_0 in Figure 9.1 is the 95th percentile of the profile. The profile essentially describes the distribution of an external noise factor. If a reliability verification test is conducted on a subsystem or component of the product, the profile should be translated to the load applied to the subsystem or component under study. Then the load must be superimposed on the internal noise factors. The total stress determines the test stress level. External and internal noise factors were discussed in Chapter 5.

When customer real-world usage profiles are unattainable, test stresses are selected from appropriate engineering standards. For example, MIL-HDBK-781 (U.S. DoD, 1996) provides information on how to assess test environments and to design tests for military equipment. In private industries, most sectors have test standards; large-scale manufacturers also publish test guidelines suitable for their own products. Suppliers of components or subsystems are often required to adopt the guidelines or other engineering standards endorsed by contractors. These documents specify test stress profiles as well as test durations and sample sizes. Alternation of any items specified must be justified carefully and ratified by contractors.

As shown in subsequent sections, verifying high reliability at a high confidence level requires a large sample size or long test time, which is unaffordable in most applications. Accelerated testing is a natural choice. However, care must be exercised to ensure that elevated stress levels do not produce failure modes different from those in the real world.

9.2.4   Test Time

Basically, the test time is dictated by the time associated with the reliability required. The time may be the warranty period, design life, or others, depending on the reliability specifications. The test time is, however, usually too long to be affordable. For example, testing an automotive component to a design life of 100,000 miles is economically prohibitive and time impermissible. As shown later, the test duration is further prolonged if a reduction in sample size is essential. Because the test time has a direct influence on the total cost and time to market, it is a major concern in planning a reliability verification test. Test planners often seek ways to shorten the test time, and consider accelerated testing a natural choice. As stated earlier, elevated stress levels must not cause failure modes that differ from those in the filed.

If an increase in sample size is acceptable, the test time may be compressed by testing a larger sample. The reduction in test time, however, has implications that test planners must consider. If the failure modes have an increasing hazard rate, failures are caused by wear-out mechanisms, which progress over time. Thus, sufficient test duration is required to induce a significant amount of wear out. Most mechanical and some electronic components belong to this category. If the failure modes display a nonincreasing (constant or decreasing) hazard rate, testing more samples for a shorter time is effective in precipitating failures. Most electronic components and systems have such a characteristic; their test time can safely be shortened by increasing the sample size.

9.3 BOGEY TESTING

A bogey test is a one-shot test in which a fixed number of samples are run simultaneously for a predetermined length of time under specified test environments. If no failures occur, we conclude that the required reliability is achieved at the given confidence level. A bogey test is characterized simply by the sample size, test time, and test stresses. Since the test time and stresses are often prespecified as described in Section 9.2, in this section we present methods for calculating the sample size. The methods deal with the binomial and Weibull distributions.

9.3.1 Binomial Bogey Testing

Suppose that we want to demonstrate the reliability R_L at a $100C\%$ confidence level. The task is equivalent to testing the hypotheses

$$H_0: R(t_L) \geq R_L, \qquad H_1: R(t_L) < R_L,$$

where $R(t_L)$ is the true reliability of the population at time t_L.

A random sample of size n is drawn from a large population. Each unit is tested until the specified time t_L, unless it fails sooner. H_0 is rejected if $r > c$, where r is the number of failures in test and c is the critical value. Because each test unit has a binary outcome (i.e., either success or failure), r has a binomial distribution given by

$$p(r) = C_n^r p^r (1 - p)^{n-r}, \qquad r = 0, 1, \ldots, n,$$

where p is the probability of failure.

The probability that the number of failures r is less than or equal to the critical value c is

$$\Pr(r \leq c) = \sum_{i=0}^{c} C_n^i p^i (1 - p)^{n-i}, \tag{9.1}$$

It is desirable to have a type II error (consumer's risk) of less than or equal to $1 - C$ when $p = 1 - R_L$. Hence, we have

$$\Pr(r \leq c | p = 1 - R_L) \leq 1 - C. \tag{9.2}$$

Combining (9.1) and (9.2) gives

$$\sum_{i=0}^{c} C_n^i (1 - R_L)^i R_L^{n-i} \leq 1 - C. \tag{9.3}$$

If c, R_L, and C are given, (9.3) can be solved for the minimum sample size. When $c = 0$, which is the case in bogey testing, (9.3) reduces to

$$R_L^n \leq 1 - C. \tag{9.4}$$

From (9.4), the minimum sample size is

$$n = \frac{\ln(1 - C)}{\ln(R_L)}. \tag{9.5}$$

If a sample of size n (the minimum sample size) produces zero failures in testing, we conclude that the product achieves the required reliability R_L at a $100C\%$ confidence level. Figure 9.2 plots the minimum sample sizes for various values of C and R_L. It is shown that the sample size increases with the reliability required given a confidence level, or with the confidence level given a required reliability. It rises sharply when the required reliability approaches 1.

Example 9.1 Determine the minimum sample size to demonstrate R90/C90, which in industry commonly denotes 90% reliability at a 90% confidence level. What is the minimum sample size for verifying R99/C90?

SOLUTION   The minimum sample size for verifying R90/C90 is

$$n = \frac{\ln(1 - 0.9)}{\ln(0.9)} = 22.$$

If $R = 99\%$ and $C = 90\%$, the minimum sample size is $n = 230$.

In some applications, we may be interested in the lower reliability bound when testing a sample of size n. If no failures occur in t_L, the lower-bound reliability

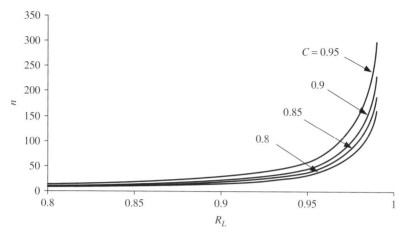

FIGURE 9.2   Sample sizes at different values of C and R_L

at a $100C\%$ confidence level can be calculated from (9.5) as

$$R_L = (1 - C)^{1/n}. \tag{9.6}$$

Example 9.2   A random sample of 30 units was tested for 15,000 cycles and produced no failures. Calculate the lower 90% confidence bound on reliability.

SOLUTION   From (9.6), the lower 90% confidence bound on reliability is

$$R_L = (1 - 0.9)^{1/30} = 0.926.$$

Note that this reliability is at 15,000 cycles under the test conditions.

9.3.2   Weibull Bogey Testing

We have seen in Section 9.3.1 that the minimum sample size becomes too large to be affordable when a high reliability is to be verified. As Example 9.1 shows, 230 units are needed to demonstrate 99% reliability at a 90% confidence level. The sample size may be reduced if we have some information about the product life from historical data. Suppose that the product life has a Weibull distribution with scale parameter α and shape parameter β, and β is known. The task is still to demonstrate the lower-bound reliability R_L at a $100C\%$ confidence level. To perform a hypothesis test, a sample of size n_0 is drawn at random and undergoes bogey testing for a specified period of time t_0. The reliability at t_0 is

$$R(t_0) = \exp\left[-\left(\frac{t_0}{\alpha}\right)^{\beta}\right]. \tag{9.7}$$

The probability of the sample of size n_0 yielding zero failures is obtained from (9.1) as

$$\Pr(r = 0) = \exp\left[-n_0 \left(\frac{t_0}{\alpha}\right)^\beta\right]. \tag{9.8}$$

Similarly, a sample of size n tested for t_L without failures has

$$\Pr(r = 0) = \exp\left[-n \left(\frac{t_L}{\alpha}\right)^\beta\right]. \tag{9.9}$$

Equating (9.8) to (9.9) gives

$$n_0 = n\pi^{-\beta}, \tag{9.10}$$

where $\pi = t_0/t_L$, and is called the *bogey ratio*.
Substituting (9.5) into (9.10) yields

$$n_0 = \frac{\ln(1 - C)}{\ln(R_L)\pi^\beta}. \tag{9.11}$$

Equation (9.11) collapses to (9.5) when the bogey ratio equals 1 and indicates that the sample size can be reduced by increasing the bogey ratio (i.e., extending the test time). The magnitude of reduction depends on the value of β. The larger the value, the greater the reduction. Table 9.1 shows the sample sizes for different values of R_L, C, π, and β.

Equation (9.11) can be derived through another approach due partially to C. Wang (1991). Suppose that a sample of size n_0 is tested for time t_0 without failures. According to Nelson (1985), the lower $100C\%$ confidence bound on the Weibull scale parameter α is

$$\alpha_L = \left(\frac{2n_0 t_0^\beta}{\chi_{C,2}^2}\right)^{1/\beta}, \tag{9.12}$$

where $\chi_{C,2}^2$ is the $100C$th percentile of the χ^2 distribution with 2 degrees of freedom. The lower bound on reliability at t_L is

$$R_L = \exp\left[-\left(\frac{t_L}{\alpha_L}\right)^\beta\right] = \exp\left(-\frac{t_L^\beta \chi_{C,2}^2}{2t_0^\beta n_0}\right). \tag{9.13}$$

Let $\pi = t_0/t_L$. Then the minimum sample size can be written as

$$n_0 = -\frac{\chi_{C,2}^2}{2\pi^\beta \ln(R_L)}. \tag{9.14}$$

Considering that $\chi_{C,2}^2 = -2\ln(1 - C)$, (9.14) reduces to (9.11).

TABLE 9.1 Sample Size for Bogey Test of a Weibull Distribution

β	π	100C	80					90					95				
		$100R_L$	90	92.5	95	97.5	99	90	92.5	95	97.5	99	90	92.5	95	97.5	99
1.25	1		16	21	32	64	161	22	30	45	91	230	29	39	59	119	299
	1.5		10	13	19	39	97	14	18	28	55	139	18	24	36	72	180
	2		7	9	14	27	68	10	13	19	39	97	12	17	25	50	126
	2.5		5	7	10	21	51	7	10	15	29	73	10	13	19	38	95
	3		4	6	8	17	41	6	8	12	24	59	8	10	15	30	76
	3.5		4	5	7	14	34	5	7	10	19	48	6	9	13	25	63
	4		3	4	6	12	29	4	6	8	17	41	6	7	11	21	53
1.5	1		16	21	32	64	161	22	30	45	91	230	29	39	59	119	299
	1.5		9	12	18	35	88	12	17	25	50	125	16	21	32	65	163
	2		6	8	12	23	57	8	11	16	33	82	11	14	21	42	106
	2.5		4	6	8	17	41	6	8	12	24	58	8	10	15	30	76
	3		3	4	7	13	31	5	6	9	18	45	6	8	12	23	58
	3.5		3	4	5	10	25	4	5	7	14	35	5	6	9	19	46
	4		2	3	4	8	21	3	4	6	12	29	4	5	8	15	38
1.75	1		16	21	32	64	161	22	30	45	91	230	29	39	59	119	299
	1.5		8	11	16	32	79	11	15	23	45	113	14	19	29	59	147
	2		5	7	10	19	48	7	9	14	28	69	9	12	18	36	89
	2.5		4	5	7	13	33	5	6	10	19	47	6	8	12	24	60
	3		3	4	5	10	24	4	5	7	14	34	5	6	9	18	44
	3.5		2	3	4	8	18	3	4	6	11	26	4	5	7	14	34
	4		2	2	3	6	15	2	3	4	9	21	3	4	6	11	27
2	1		16	21	32	64	161	22	30	45	91	230	29	39	59	119	299
	1.5		7	10	14	29	72	10	14	20	41	102	13	18	26	53	133
	2		4	6	8	16	41	6	8	12	23	58	8	10	15	30	75
	2.5		3	4	6	11	26	4	5	8	15	37	5	7	10	19	48
	3		2	3	4	8	18	3	4	5	11	26	4	5	7	14	34
	3.5		2	2	3	6	14	2	3	4	8	19	3	4	5	10	25
	4		1	2	2	4	11	2	2	3	6	15	2	3	4	8	19
2.25	1		16	21	32	64	161	22	30	45	91	230	29	39	59	119	299
	1.5		7	9	13	26	65	9	12	19	37	93	12	16	24	48	120
	2		4	5	7	14	34	5	7	10	20	49	6	9	13	25	63
	2.5		2	3	4	9	21	3	4	6	12	30	4	5	8	16	38
	3		2	2	3	6	14	2	3	4	8	20	3	4	5	10	26
	3.5		1	2	2	4	10	2	2	3	6	14	2	3	4	8	18
	4		1	1	2	3	8	1	2	2	5	11	2	2	3	6	14
2.5	1		16	21	32	64	161	22	30	45	91	230	29	39	59	119	299
	1.5		6	8	12	24	59	8	11	17	34	84	11	14	22	43	109
	2		3	4	6	12	29	4	6	8	17	41	6	7	11	21	53
	2.5		2	3	4	7	17	3	3	5	10	24	3	4	6	12	31
	3		1	2	3	5	11	2	2	3	6	15	2	3	4	8	20
	3.5		1	1	2	3	7	1	2	2	4	10	2	2	3	6	14
	4		1	1	1	2	6	1	1	2	3	8	1	2	2	4	10
2.75	1		16	21	32	64	161	22	30	45	91	230	29	39	59	119	299
	1.5		6	7	11	21	53	8	10	15	30	76	10	13	20	39	98
	2		3	4	5	10	24	4	5	7	14	35	5	6	9	18	45
	2.5		2	2	3	6	13	2	3	4	8	19	3	4	5	10	24
	3		1	2	2	4	8	2	2	3	5	12	2	2	3	6	15

TABLE 9.1 (*continued*)

	100C	80					90					95				
β	π	100R_L	90	92.5	95	97.5	99	90	92.5	95	97.5	99	90	92.5	95	97.5	99
	3.5		1	1	2	3	6	1	1	2	3	8	1	2	2	4	10
	4		1	1	1	2	4	1	1	1	3	6	1	1	2	3	7
3	1		16	21	32	64	161	22	30	45	91	230	29	39	59	119	299
	1.5		5	7	10	19	48	7	9	14	27	68	9	12	18	36	89
	2		2	3	4	8	21	3	4	6	12	29	4	5	8	15	38
	2.5		1	2	3	5	11	2	2	3	6	15	2	3	4	8	20
	3		1	1	2	3	6	1	2	2	4	9	2	2	3	5	12
	3.5		1	1	1	2	4	1	1	2	3	6	1	1	2	3	7
	4		1	1	1	1	3	1	1	1	2	4	1	1	1	2	5

Example 9.3 An engineer plans to demonstrate that an electronic sensor achieves a lower 90% confidence bound reliability of 95% at 15,000 cycles. Historical data analysis indicated that the life distribution is approximately Weibull with a shape parameter between 1.5 and 2. The engineer wants to reduce the sample size by testing the sensors for 33,000 cycles. Determine the minimum sample size.

SOLUTION   The bogey ratio is $\pi = 33{,}000/15{,}000 = 2.2$. To be conservative, the value of the shape parameter is chosen to be 1.5. When $R_L = 0.95$, $C = 0.9$, and $\beta = 1.5$, the sample size is 16 for $\pi = 2$ and 12 for $\pi = 2.5$ from Table 9.1. Linear interpolation gives the required sample size of 14. Direct calculation from (9.11) also yields $n_0 = 14$. Now the bogey testing is to test 14 units of the sensor for 33,000 cycles. If no failures appear, the reliability of 95% at 15,000 cycles is demonstrated at a 90% confidence level.

Example 9.4 Refer to Example 9.3. Suppose that the maximum allowable sample size is 10. Calculate the test time required.

SOLUTION   From (9.11) the bogey ratio is

$$\pi = \left[\frac{\ln(1 - C)}{n_0 \ln(R_L)}\right]^{1/\beta} = \left[\frac{\ln(1 - 0.9)}{10 \ln(0.95)}\right]^{1/1.5} = 2.72.$$

The test time required is $t_0 = \pi t_L = 2.72 \times 15{,}000 = 40{,}800$ cycles.

As shown in Examples 9.3 and 9.4, reduction in sample size is at the expense of increased test time. In many situations it is impossible to prolong a test. Instead, elevation of test stress levels is feasible and practical. If the acceleration factor A_f is known between the elevated and use stress levels, the actual test time t_a is

$$t_a = \frac{t_0}{A_f}. \tag{9.15}$$

In Chapter 7 we described methods for calculating A_f.

9.4 SAMPLE SIZE REDUCTION BY TAIL TESTING

9.4.1 Test Method

The inherent product life is determined by the physical characteristics of the product, such as the dimensions and material properties; there exist relationships between the life and the characteristics. The relationships are also called *transfer functions*, meaning that the life is transferred from the characteristics. Nelson (1990, 2004) provides examples in conductor and dielectrics whose life is highly correlated to the size of the parts. The examples include:

- The life of a capacitor dielectric is predominated by the area of the dielectric.
- The life of cable insulation is determined largely by the cable length.
- The life of conductor in microelectronics is proportional to its length.
- The life of electrical insulation is dominated by its thickness.

Allmen and Lu (1994) also observe that:

- The fatigue life of the connecting rod of an automotive engine is proportional to the hardness of material.
- The life of a vehicle suspension arm is dominated by the stock thickness.

When a physical characteristic is transferred to life, variation in the physical characteristic may be depressed, amplified, or unchanged, depending on their transfer function. Figures 9.3 illustrates the transfer functions for these three cases, where t is the life and y is a larger-the-better physical characteristic. It can be seen that units at the weakest extreme of the characteristic have the shortest lives. If the most vulnerable units can pass a bogey test, it is safe to conclude that the remainder would have survived the test. Indeed, there is no need to test stronger units. Then the reliability verification becomes the task of selecting and testing the samples from the weak tail of the characteristic distribution. Specifically, the task consists of the following steps:

1. Determine the critical physical characteristic y.
2. Establish a relationship between life and y (i.e., the transfer function) through the analysis of historical data and/or analytical studies.
3. Draw a large sample for measurements of y.
4. Estimate the y distribution.
5. Compute the life of each unit measured using the transfer function.
6. Estimate the life distribution.
7. Choose a fraction q of the life distribution, and calculate the $100q$th percentile t_q. We must have $q > 1 - R_L$ and $t_q > t_L$, where t_L is the time at which R_L is required.
8. Convert t_q to the $(100q')$th percentile $y_{q'}$ of the y distribution using the transfer function, where q' is a fraction of the y distribution.

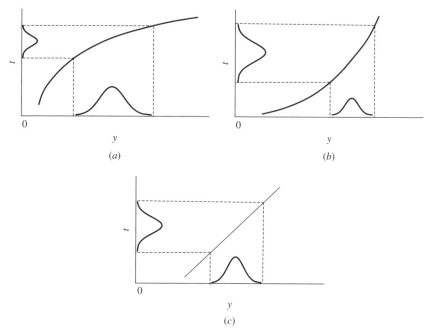

FIGURE 9.3 Transfer function yielding (*a*) a dampened life distribution, (*b*) an amplified life distribution, and (*c*) an unchanged life distribution

9. Draw a sample of size n_q from the population; each unit must fall in the tail area defined by $y_{q'}$.

10. Test the n_q units until t_L. If no failures occur, R_L is demonstrated at a $100C\%$ confidence level.

In this approach, the sample size n_q and tail fraction q are two important quantities, which are discussed in the following subsections.

9.4.2 Determination of the Sample Size

A sample of size n_q is drawn from the characteristic distribution's weak tail defined by $y_{q'}$. Each of the n_q units is tested until t_L unless it fails sooner. If the $100q$th percentile t_q of the life distribution is greater than the test time t_L, the reliability of the units at t_L can be written as

$$R(t_L|t_q) = \Pr(t \geq t_L|t \leq t_q) = \frac{\Pr(t_L \leq t \leq t_q)}{\Pr(t \leq t_q)} = 1 - \frac{1 - R(t_L)}{q}, \quad (9.16)$$

where $R(t_L)$ is the reliability of a unit randomly selected from the entire population. Equation (9.16) shows that $R(t_L|t_q)$ decreases with the value of q, and $R(t_L|t_q) < R(t_L)$ when $q < 1$. When $q = 1$, that is, the test units are randomly drawn from the entire population, (9.16) is reduced to $R(t_L|t_q) = R(t_L)$.

The original task of reliability verification is to demonstrate that the entire population achieves R_L at a $100C\%$ confidence level. When the samples are drawn from the lower tail of the population, the task is equivalent to testing the hypotheses

$$H_0:\ R(t_L|t_q) \geq 1 - \frac{1 - R_L}{q}, \qquad H_1:\ R(t_L|t_q) < 1 - \frac{1 - R_L}{q}.$$

Similar to (9.4), if no failures are allowed in the bogey test, the type II error can be obtained from the binomial distribution: namely,

$$\left(1 - \frac{1 - R_L}{q}\right)^{n_q} \leq 1 - C. \tag{9.17}$$

Then the minimum sample size is

$$n_q = \frac{\ln(1 - C)}{\ln(R_L + q - 1) - \ln(q)}, \tag{9.18}$$

where $q > 1 - R_L$. When $q = 1$, (9.18) collapses to (9.5). When $1 - R_L < q < 1$, the required sample size is smaller than n of (9.5). Figure 9.4 plots the ratios of n_q to n for different values of R_L and q. It can be seen that for a given value of R_L, the smaller the q value, the greater the reduction in sample size. However, the value of q should not be very small, considering the variability of the transfer function. Generally, we select $q \geq 0.3$.

9.4.3 Risks of Tail Testing

The risks associated with tail testing come from the variability in transfer function. In practice, a transfer function is derived from historical data and thus contains statistical uncertainty and/or model error. As a result, the transfer function may yield an underestimated or overestimated life. Figure 9.5 shows three

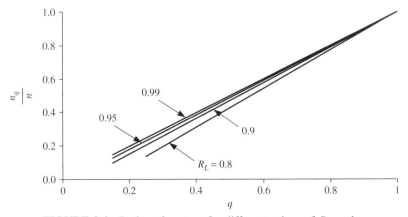

FIGURE 9.4 Ratios of n_q to n for different values of R_L and q

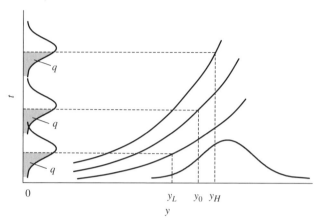

FIGURE 9.5 Life variation due to variability of transfer function

nonlinear transfer functions and the resulting life distributions. In this figure, the middle transfer function is assumed to be the correct one, and the other two contain deviations. The correct one results in middle life distribution on the t-axis. The upper transfer function produces longer lives, whereas the lower transfer function yields shorter failure times. In the case of overestimation, the $100q$th percentile of the erroneous life distribution would require sampling below y_H, which is larger than the correct y_0, as shown in Figure 9.5. Obviously, the test result is optimistic. Better said, the population reliability may not achieve the required retiability at the confidence level specified even if no failures occur in the test. In the case of underestimation, the $100q$th percentile of the incorrect life distribution is transferred to y_L, which is lower than the correct y_0. Consequently, the test result is pessimistic. Indeed, there is a possibility that the population reliability meets the reliability requirement even if the sample fails to pass the test.

The stated risks vanish when the characteristic has a normal distribution and the transfer function is linear. Suppose that y has a normal distribution $N(\mu_y, \sigma_y^2)$, and the transfer function is

$$t = ay + b, \tag{9.19}$$

where a and b are constants. Then the life is also normal with mean $\mu_t = a\mu_y + b$ and standard deviation $\sigma_t = a\sigma_y$. The $100q$th percentile of the life distribution is

$$t_q = \mu_t + z_q\sigma_t = a\mu_y + az_q\sigma_y + b, \tag{9.20}$$

where z_q is the $100q$th percentile of the standard normal distribution.

The $(100q')$th percentile of y distribution corresponding to t_q is obtained from (9.19) and (9.20) as

$$y_{q'} = \frac{t_q - b}{a} = \frac{a\mu_y + az_q\sigma_y + b - b}{a} - \mu_y + z_q\sigma_y, \tag{9.21}$$

which indicates that the $(100q')$th percentile of y distribution does not depend on the values of a and b. Therefore, variation in the estimates of a and b does not impose risks to tail testing. Furthermore, (9.21) also shows that $q' = q$.

9.4.4   Application Example

Example 9.5   A bogey test is designed to verify that a shaft meets a lower 90% confidence bound reliability of 95% at 5×10^5 cycles under the cyclic loading specified. From (9.5) the test requires 45 samples, which is too large in this case, due to the cost and test time restrictions. So the tail-testing technique is considered here. Determine the sample size for the test.

SOLUTION   Calculation of the tail-testing sample size follows the steps described in Section 9.4.1.

1. Choose the shaft characteristic to describe the fatigue life. It is known that the fatigue life is influenced dramatically by material properties, surface finish, and diameter. Since the variability in the first two factors is well under control, diameter is the predominant characteristic and thus is selected to characterize the fatigue life.

2. Develop the transfer function that relates the fatigue life to shaft diameter. From the theory of material strength and the $S-N$ curve (a curve plotting the relationship between mechanical stress and the number of cycles to failure), we derived the transfer function as

$$L = ay^b, \tag{9.22}$$

   where L is the fatigue life, y is the diameter (in millimeters), and a and b are constants depending on the material properties. To estimate a and b, the historical data of a similar part made of the same material were analyzed. Figure 9.6 shows the fatigue life of the part at various values of diameter, and the fit of (9.22) to the data. Simple linear regression analysis gives the estimates $\hat{a} = 3 \times 10^{-27}$ and $\hat{b} = 24.764$.

3. Draw 45 samples randomly from the entire population and take the measurements of the diameter. Probability plot indicates that the diameter has the normal distribution $N(21.21, 0.412^2)$.

4. Calculate the lives of the 45 units by using (9.22) with the estimates $\hat{a}$ and $\hat{b}$. The life can be modeled adequately with the lognormal distribution $LN(14.56, 0.479^2)$.

5. Choose $q = 0.3$. The 30th percentile is $t_{0.3} = 1.64 \times 10^6$ cycles, which is obtained from the lognormal distribution.

6. Convert $t_{0.3}$ to the $(100q')$th percentile of the diameter distribution and get $y_{q'} = 20.989$ mm. From the normal distribution, we obtain the lower tail fraction of the diameter as $q' = 0.296$, which is close to $q = 0.3$.

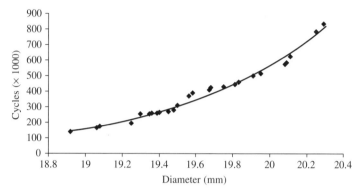

FIGURE 9.6 Fatigue life of a similar part at various values of diameter

7. The sample size is calculated from (9.18) as

$$n_q = \frac{\ln(1 - 0.9)}{\ln(0.95 + 0.3 - 1) - \ln(0.3)} = 13.$$

This ample size is considerably smaller than the 45 derived from (9.5).

Now the bogey test plan is to draw 13 units from the lower tail of the diameter distribution such that the measurements of the diameter are less than $y_{q'} = 20.989$ mm, and test the 13 units until 5×10^5 cycles under the specified cyclic loading condition. If no failures occur, we conclude that at a 90% confidence level, the shaft population achieves 95% reliability at 5×10^5 cycles.

9.5 SEQUENTIAL LIFE TESTING

Sequential life testing is to test one unit at a time until it fails or until a predetermined period of time has elapsed. As soon as a new observation becomes available, an evaluation is made to determine if (1) the required reliability is demonstrated, (2) the required reliability is not demonstrated, or (3) the test should be continued. Statistically speaking, sequential life testing is a hypothesis-testing situation in which the test statistic is reevaluated as a new observation is available and then compared against the decision rules. When rejection or acceptance rules are satisfied, the test is discontinued and the conclusion is arrived at. Otherwise, the test should continue. It can be seen that the sample size required to reach a conclusion is a random number and cannot be predetermined. Because of the sequential nature, the test method needs fewer samples than a bogey test.

9.5.1 Theoretical Basics

Consider the hypotheses

$$H_0: \theta = \theta_0, \qquad H_1: \theta = \theta_1,$$

where θ is a parameter of the life distribution (e.g., an exponential MTTF or Weibull scale parameter) and θ_0 and θ_1 are the values specified for θ. Loosely, θ_0 represents the upper limit of reliability requirement above which the lot of the product should be accepted; θ_1 is the lower limit of reliability requirement below which the lot of product should be rejected. The ratio

$$d = \frac{\theta_0}{\theta_1}, \qquad (9.23)$$

is called the *discrimination ratio*.

Let X be the random variable with the pdf given by $f(x;\theta)$. Suppose that a sequential life testing generates $x_1, x_2, \ldots, x_n$, which are n independent observations of X. As presented in Chapter 7, the likelihood of the n observations is

$$L(x_1, x_2, \ldots, x_n; \theta) = \prod_{i=1}^{n} f(x_i; \theta). \qquad (9.24)$$

We define the ratio of the likelihood at θ_1 to that at θ_0 as

$$LR_n = \frac{L(x_1, x_2, \ldots, x_n; \theta_1)}{L(x_1, x_2, \ldots, x_n; \theta_0)}. \qquad (9.25)$$

LR_n is also called the *probability ratio* because the sample likelihood is the joint pdf for the sample as shown in (9.24). Given a data set $x_1, x_2, \ldots, x_n$, the likelihood depends only on the value of θ. The maximum likelihood principle indicates that the likelihood is maximized when the value of θ takes the true value. We can reason that a value of θ closer to the true one would result in a larger value of the likelihood. Following the same reasoning, if θ_0 is closer to the true value of θ than θ_1, $L(x_1, x_2, \ldots, x_n; \theta_0)$ is greater than $L(x_1, x_2, \ldots, x_n; \theta_1)$, and LR_n is less than 1. LR_n would become smaller when θ_0 approaches, and θ_1 leaves, the true value. It is reasonable to find a bound, say A, such that if $LR_n \leq A$, we would accept H_0. Similarly, we may also determine a bound, say B, such that if $LR_n \geq B$, we would reject H_0. If LR_n is between the bounds, we would fail to accept or reject H_0; thus, the test should be continued to generate more observations. The decision rules are as follows:

- Accept H_0 if $LR_n \leq A$.
- Reject H_0 if $LR_n \geq B$.
- Draw one more unit and continue the test if $A < LR_n < B$.

By following the decision rules above and the definitions of type I and type II errors, we can determine the bounds as

$$A = \frac{\beta}{1 - \alpha}, \qquad (9.26)$$

$$B = \frac{1 - \beta}{\alpha}, \qquad (9.27)$$

where α is the type I error (producer's risk) and β is the type II error (consumer's risk).

In many applications it is computationally more convenient to use the log likelihood ratio: namely,

$$\ln(LR_n) = \sum_{i=1}^{n} \ln\left[\frac{f(x_i;\theta_1)}{f(x_i;\theta_0)}\right]. \tag{9.28}$$

Then the continue test region becomes

$$\ln\left(\frac{\beta}{1-\alpha}\right) < \ln(LR_n) < \ln\left(\frac{1-\beta}{\alpha}\right). \tag{9.29}$$

It is worth noting that the true values of the two types of errors are not exactly equal to the specified values of α and β. It is difficult to calculate the true errors, but they are bounded by

$$\alpha' \le 1/B \quad \text{and} \quad \beta' \le A,$$

where α' and β' denote the true values of α and β, respectively. For example, if a test specifies $\alpha = 0.1$ and $\beta = 0.05$, the true errors are bounded by $\alpha' \le 0.105$ and $\beta' \le 0.056$. It can be seen that the upper bounds are slightly higher than the specified values. Generally, the maximum relative error of α' to α is

$$\frac{\alpha' - \alpha}{\alpha} = \frac{1/B - \alpha}{\alpha} = \frac{\beta}{1-\beta}.$$

The maximum relative error of β' to β is

$$\frac{\beta' - \beta}{\beta} = \frac{A - \beta}{\beta} = \frac{\alpha}{1-\alpha}.$$

The *operating characteristic* (O.C.) *curve* is useful in hypothesis testing. It plots the probability of accepting H_0 when H_0 is true for different true values of θ. The probability, denoted by $Pa(\theta)$, can be written as

$$Pa(\theta) = \frac{B^h - 1}{B^h - A^h}, \quad h \ne 0, \tag{9.30}$$

where h is a constant related to the value of θ. The relationship between h and θ is defined by

$$\int_{-\infty}^{\infty} \left[\frac{f(x;\theta_1)}{f(x;\theta_0)}\right]^h f(x;\theta)dx = 1. \tag{9.31}$$

Solving (9.31) gives $\theta(h)$. Then we can use the following steps to generate the O.C. curve:

1. Set a series of arbitrary numbers for h which may be between, for example, -3 and 3.

2. Calculate $\theta(h)$ at the values of h specified.
3. Calculate $Pa(\theta)$ at the values of h using (9.30).
4. Generate the O.C. curve by plotting $Pa(\theta)$ versus $\theta(h)$.

Let's consider two special cases of (9.30). When $h = 1$, (9.30) becomes

$$Pa(\theta) = \frac{B-1}{B-A} = 1 - \alpha. \qquad (9.32)$$

When $h = -1$, (9.30) is reduced to

$$Pa(\theta) = \frac{B^{-1} - 1}{B^{-1} - A^{-1}} = \beta. \qquad (9.33)$$

Example 9.6 Consider a sequential life test for the exponential distribution. Suppose that $\theta_0 = 2000$, $\theta_1 = 1000$, $\alpha = 0.1$, and $\beta = 0.1$. Develop the decision bounds and O.C. curve for the test.

SOLUTION The decision bounds are

$$A = \frac{\beta}{1 - \alpha} = \frac{0.1}{1 - 0.1} = 0.111,$$

$$B = \frac{1 - \beta}{\alpha} = \frac{1 - 0.1}{0.1} = 9.$$

Thus, if a sequential test of n units results in $LR_n \le 0.111$, the null hypothesis $\theta_0 = 2000$ is accepted. If $LR_n \ge 9$, the null hypothesis is rejected. If $0.111 < LR_n < 9$, take one more unit and continue the test. To construct the O.C. curve for the test, we first solve (9.31) for the exponential distribution, where

$$f(x;\theta) = \frac{1}{\theta} \exp\left(-\frac{x}{\theta}\right), \qquad x \ge 0.$$

From (9.31) we have

$$\int_0^\infty \left[\frac{\theta_0 \exp(-x/\theta_1)}{\theta_1 \exp(-x/\theta_0)} \right]^h \frac{1}{\theta} \exp\left(-\frac{x}{\theta}\right) dx = 1.$$

Solving the equation gives

$$\theta = \frac{(\theta_0/\theta_1)^h - 1}{h(1/\theta_1 - 1/\theta_0)}. \qquad (9.34)$$

From (9.34), if $\theta = \theta_0$, then $h = 1$. From (9.32) we have $Pa(\theta_0) = 1 - \alpha$. That is, if θ_0 is the true MTTF, the probability of accepting the lot equals $1 - \alpha$.

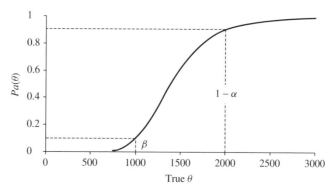

FIGURE 9.7   O.C. curve for the sequential test plan of Example 9.6

Similarly, if $\theta = \theta_1$, then $h = -1$. From (9.33) we obtain $Pa(\theta_1) = \beta$; that is, if θ_1 is the true MTTF, the probability of accepting the lot is equal to β.

To construct the O.C. curve, set $h = -2, -1.9, -1.8, \ldots, 1.8, 1.9, 2$, and calculate the corresponding values of θ from (9.34), and of $Pa(\theta)$ from (9.30). Then the O.C. curve is the plot of the sets of $Pa(\theta)$ and θ values, as shown in Figure 9.7. From the curve we see that if the true value $\theta = 2000$, the probability of accepting the lot is 0.9, which equals $1 - \alpha$, and if $\theta = 1000$, the probability is 0.1, which equals β.

9.5.2   Binomial Sequential Life Testing

As in bogey testing, we are sometimes interested in whether a test unit fails in a fixed period of time in sequential testing. The outcome of the test is either failure or success. Thus, the probability of an occurrence is described by a binomial distribution: namely,

$$p(x) = p^x (1 - p)^{1-x}, \qquad x = 0, 1, \tag{9.35}$$

where p is the probability of failure, $x = 0$ if no failure occurs, and $x = 1$ if a failure occurs.

Suppose that p_0 is the lower limit of failure probability below which the lot of product should be accepted and p_1 is the upper limit of failure probability above which the lot should be rejected. Clearly, $p_0 < p_1$. Then the sequential testing is equivalent to testing the hypotheses

$$H_0: p = p_0, \qquad H_1: p = p_1.$$

For n observations, the log likelihood ratio given by (9.28) can be written as

$$\ln(LR_n) = r \ln \left[\frac{p_1(1 - p_0)}{p_0(1 - p_1)} \right] - n \ln \left(\frac{1 - p_0}{1 - p_1} \right), \tag{9.36}$$

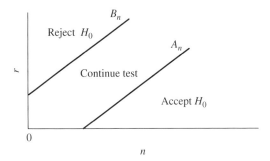

FIGURE 9.8   Graphical binomial sequential test plan

where r is the total number of failures in n trials and $r = \sum_{i=1}^{n} x_i$.

The continue-test region can be obtained by substituting (9.36) into (9.29). Further simplification gives

$$A_n < r < B_n, \tag{9.37}$$

where

$$A_n = C \ln\left(\frac{\beta}{1-\alpha}\right) + nC \ln\left(\frac{1-p_0}{1-p_1}\right), \qquad B_n = C \ln\left(\frac{1-\beta}{\alpha}\right)$$
$$+ nC \ln\left(\frac{1-p_0}{1-p_1}\right), \qquad C = \ln^{-1}\left[\frac{p_1(1-p_0)}{p_0(1-p_1)}\right].$$

A_n and B_n are the bounds of the test. According to the decision rules, we accept H_0 if $r \leq A_n$, reject H_0 if $r \geq B_n$, and draw one more unit and continue the test if $A_n < r < B_n$. A_n and B_n are two parallel straight lines, as shown in Figure 9.8. The cumulative number of failures can be plotted on the graph to show the current decision and track the test progress.

To construct the O.C. curve for this test, we first solve (9.31) for the binomial distribution defined by (9.35) and obtain

$$p = \frac{1 - \left(\dfrac{1-p_1}{1-p_0}\right)^h}{\left(\dfrac{p_1}{p_0}\right)^h - \left(\dfrac{1-p_1}{1-p_0}\right)^h}. \tag{9.38}$$

The probability of accepting H_0 when p is the true probability of failure is obtained from (9.30) by setting $\theta = p$: namely,

$$Pa(p) = \frac{B^h - 1}{B^h - A^h}, \qquad h \neq 0. \tag{9.39}$$

Then the O.C. curve can be generated by following the steps described earlier.

In test planning we may be interested in the minimum number of trials leading to acceptance of H_0. The fastest path to the decision takes place when no failures occur in the trails. The minimum number n_a is given by

$$A_n = C \ln \left(\frac{\beta}{1-\alpha} \right) + n_a C \ln \left(\frac{1-p_0}{1-p_1} \right) = 0$$

or

$$n_a = \frac{\ln \left(\dfrac{1-\alpha}{\beta} \right)}{\ln \left(\dfrac{1-p_0}{1-p_1} \right)}. \tag{9.40}$$

Similarly, the minimum number of trials leading to rejection of H_0 occurs when all trails fail. The minimum number n_r is given by

$$B_n = C \ln \left(\frac{1-\beta}{\alpha} \right) + n_r C \ln \left(\frac{1-p_0}{1-p_1} \right) = n_r$$

or

$$n_r = \frac{C \ln \left(\dfrac{1-\beta}{\alpha} \right)}{1 - C \ln \left(\dfrac{1-p_0}{1-p_1} \right)}. \tag{9.41}$$

The expected number of trials $E(n|p)$ to reach an accept or reject decision is given by

$$E(n|p) = \frac{Pa(p) \ln \left(\dfrac{1}{A} \right) + [1 - Pa(p)] \ln \left(\dfrac{1}{B} \right)}{p \ln \left(\dfrac{p_0}{p_1} \right) + (1-p) \ln \left(\dfrac{1-p_0}{1-p_1} \right)}, \tag{9.42}$$

which indicates that $E(n|p)$ is a function of the true p, which is unknown. In calculation it can be replaced with an estimate.

Example 9.7   An automotive supplier wants to demonstrate the reliability of a one-shot airbag at a specified time and test condition. Suppose that the contract for the airbag specifies $p_0 = 0.001$, $p_1 = 0.01$, $\alpha = 0.05$, and $\beta = 0.1$. Develop a sequential test plan.

SOLUTION   Substituting the given data into (9.37), we obtain the continue-test region $0.0039n - 0.9739 < r < 0.0039n + 1.2504$. Following our decision rules, we accept H_0 (the probability of failure is less than or equal to 0.001 at the specified time and test condition) if $r \le 0.0039n - 0.9739$, reject H_0 if $r \ge 0.0039n + 1.2504$, and take an additional unit for test if $0.0039n - 0.9739 < r < 0.0039n + 1.2504$.

The minimum number of trials that lead to acceptance of H_0 is determined from (9.40) as $n_a = 249$. The minimum number of trials resulting in rejection of H_0 is calculated from (9.41) as $n_r = 2$.

Now we compute the expected number of trials for the test. The supplier was confident that the airbag achieves the required reliability based on the accelerated test data of a similar product and has $p = 0.0008$. Substituting the value of p into (9.38) gives $h = 1.141$. With the given α and β values, we obtain $A = 0.1053$ and $B = 18$. From (9.39), $Pa(p) = 0.9658$. Then the expected number of trials is calculated from (9.42) as $E(n|p) = 289$. The test plan is plotted in Figure 9.9. The minimum numbers can also be read from the graph.

To construct an O.C. curve for the test, set h to various numbers between -3 and 3. Then calculate the corresponding values of p from (9.38) and of $Pa(p)$ from (9.39). The plot of $Pa(p)$ versus p is the O.C. curve, shown in Figure 9.10. It is seen that the probability of accepting H_0 decreases sharply as the true p increases when it is less than 0.005. That is, the test plan is sensitive to the change in p in the region.

To compare the sequential life test with the bogey test, we determine the sample size for the bogey test that demonstrates 99.9% reliability at a 90% confidence level, which is equivalent to $p_0 = 0.001$ and $\beta = 0.1$ in this example. From (9.5) we obtain $n = 2302$. The sample size is substantially larger than 289 (the expected number of trials in the sequential life test).

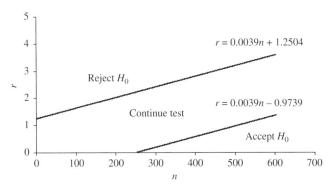

FIGURE 9.9 Sequential life test plan for Example 9.7

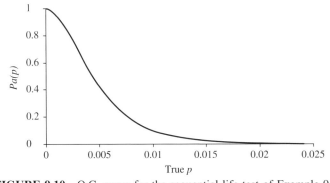

FIGURE 9.10 O.C. curve for the sequential life test of Example 9.7

9.5.3 Exponential Sequential Life Testing

Exponential distribution can approximate the life distribution of some products such as the flash random access memory. Because of its simplicity, the distribution is widely used and perhaps misused. In this subsection we present sequential life test for this distribution. The exponential pdf is

$$f(t) = \frac{1}{\theta} \exp\left(-\frac{t}{\theta}\right),$$

where t is the lifetime and θ is the mean time to failure. The sequential life testing is to test the hypotheses

$$H_0: \theta = \theta_0, \qquad H_1: \theta = \theta_1,$$

where $\theta_0 > \theta_1$. In addition to θ_0 and θ_1, the test also specifies α and β.

To perform the hypothesis test, we construct the log likelihood ratio using (9.28) and obtain

$$\ln(LR_n) = \sum_{i=1}^{n} \ln\left[\frac{(1/\theta_1)\exp(-t_i/\theta_1)}{(1/\theta_0)\exp(-t_i/\theta_0)}\right] = n \ln\left(\frac{\theta_0}{\theta_1}\right) - T\left(\frac{1}{\theta_1} - \frac{1}{\theta_0}\right), \quad (9.43)$$

where n is the total number of trials and T is the total time to failure of the n units ($T = \sum_{i=1}^{n} t_i$).

From (9.29) and (9.43), the continue-test region is

$$A_n < T < B_n, \qquad (9.44)$$

where

$$A_n = C \ln\left(\frac{\alpha}{1-\beta}\right) + nC \ln\left(\frac{\theta_0}{\theta_1}\right), \qquad B_n = C \ln\left(\frac{1-\alpha}{\beta}\right)$$

$$+ nC \ln\left(\frac{\theta_0}{\theta_1}\right), \qquad C = \frac{\theta_0 \theta_1}{\theta_0 - \theta_1}.$$

Note that the observation in the test is the time to failure. The decision variable is the total time to failure, not the total number of failures. Thus, the decision rules are that we accept H_0 if $T \geq B_n$, reject H_0 if $T \leq A_n$, and take an additional unit and continue the test if $A_n < T < B_n$. The shortest route to the reject decision is testing

$$n = \frac{\ln[(1-\beta)/\alpha]}{\ln(\theta_0/\theta_1)}$$

units which fail at time zero. The shortest route to the accept decision is testing one unit that survives at least the time given by

$$B_1 = C \ln \left(\frac{1 - \alpha}{\beta} \right) + C \ln \left(\frac{\theta_0}{\theta_1} \right).$$

The O.C. curve for the test plan can be developed using (9.30) and (9.34). The procedure was described in Example 9.6.

The use of (9.44) requires testing units individually to failure. Compared with a truncation test, the test method reduces sample size and increases test time. This is recommended when accelerated testing is applicable. Sometimes we may be interested in simultaneous testing of a sample of sufficient size. The decision rules and test plans are described in, for example, Kececioglu (1994) and MIL-HDBK-781 (U.S. DoD, 1996).

Example 9.8   A manufacturer was obligated to demonstrate the MTTF of a new electronic product not less than 5000 hours. Suppose that the unacceptable MTTF lower limit is 3000 hours, $\alpha = 0.05$, and $\beta = 0.1$. Determine a sequential test plan. An accelerated life test of 5 units yielded the failure times: 196.9, 15.3, 94.2, 262.6, and 111.6 hours. Suppose that the acceleration factor is 55. Make a decision as to whether to continue the test from the test data.

SOLUTION   The continue-test region is calculated from (9.44) as $-21677.8 + 3831.2n < T < 16884.7 + 3831.2n$. According to the decision rules, we would conclude that the MTTF of the product meets the requirement of 5000 hours if $T \geq 16884.7 + 3831.2n$ but does not meet the requirement if $T \leq -21677.8 + 3831.2n$. Otherwise, take one more unit and continue the test.

To make a decision on whether to continue the test, we convert the failure times to those at the use stress level by multiplying the acceleration factor. The equivalent total failure time is

$$T = 55 \times (196.9 + 15.3 + 94.2 + 262.6 + 111.6) = 37,433.$$

The decision bounds are

$$A_5 = -21677.8 + 3831.2 \times 5 = -2521.8 \quad \text{and} \quad B_5 = 16884.7$$
$$+ 3831.2 \times 5 = 36040.7.$$

Since $T > B_5$, we conclude that the product achieves the MTTF of 5000 hours. The sequential test results and decision process are plotted in Figure 9.11. It is seen that the accumulated test time crosses the B_n bound after a test of 5 units.

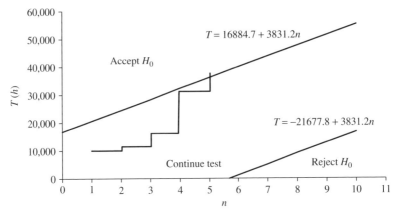

FIGURE 9.11   Sequential life test plan and results of Example 9.8

9.5.4   Weibull Sequential Life Testing

A Weibull distribution is used most commonly because of its flexibility in shape, as delineated in Chapter 2. In this subsection we present a sequential life test for this distribution. The Weibull pdf is

$$f(t) = \frac{m}{\eta} \left(\frac{t}{\eta} \right)^{m-1} \exp\left[-\left(\frac{t}{\eta} \right)^m \right], \qquad t \geq 0,$$

where m is the shape parameter and η is the scale parameter.

If we define $y = t^m$ where m is assumed known, y has an exponential distribution with scale parameter (mean) $\theta = \eta^m$. Then the sequential test plan for a Weibull distribution can be obtained by modifying the plan for the exponential distribution, which was described earlier. Suppose that we want to demonstrate the scale parameter of a Weibull distribution such that if $\eta = \eta_0$ the probability of accepting the lot is $1 - \alpha$, and if $\eta = \eta_1$ where $\eta_1 < \eta_0$, the probability of acceptance is β. This is equivalent to testing the exponential hypotheses

$$H_0\!: \theta = \theta_0, \qquad H_1\!: \theta = \theta_1,$$

where $\theta_0 = \eta_0^m$ and $\theta_1 = \eta_1^m$.

From (9.44), the continue-test region is defined by

$$A_n < T < B_n, \tag{9.45}$$

where

$$T = \sum_{i=1}^{n} t_i^m, \qquad A_n = C \ln\left(\frac{\alpha}{1-\beta} \right) + nmC \ln\left(\frac{\eta_0}{\eta_1} \right),$$

$$B_n = C \ln\left(\frac{1-\alpha}{\beta} \right) + nmC \ln\left(\frac{\eta_0}{\eta_1} \right), \qquad C = \frac{(\eta_0 \eta_1)^m}{\eta_0^m - \eta_1^m}.$$

We accept H_0 if $T \geq B_n$, reject H_0 if $T \leq A_n$, and continue the test otherwise. The O.C. curve can be constructed by using the formulas and procedures for the exponential distribution with the transformations $\theta_0 = \eta_0^m$ and $\theta_1 = \eta_1^m$.

The test method uses a known shape parameter of the Weibull distribution. In practice, it may be estimated from the accelerated test data obtained in an earlier design and development stage or from historical data on a similar product. If such data are not available, the shape parameter can be estimated from the sequential test itself. But the test plan needs to be modified accordingly as the updated estimate becomes available. The procedures are similar to those of Harter and Moore (1976) and are described as follows:

1. Test at least three units, one at a time, until all have failed.
2. Estimate the shape and scale parameters from the test data.
3. Calculate A_n and B_n using the estimate of the shape parameter.
4. Apply the decision rules to the failure times in the order in which they were observed. If a reject or accept decision is made, stop the test. Otherwise, go to step 5.
5. Take one more unit and continue the test until it fails or until a decision to accept is reached. If it fails, go to step 2.

Although the test data provide a better estimate of the shape parameter, the estimate may still have a large deviation from the true value. The deviation, of course, affects actual type I and type II errors. Therefore, it is recommended that the sensitivity of the test plan be assessed to the uncertainty of the estimate. Sharma and Rana (1993) and Hauck and Keats (1997) present formulas for examining the response of $Pa(\eta)$ to misspecification of the shape parameter and conclude that the test plan is not robust against a change in the shape parameter.

Example 9.9 The life of a mechanical component can be modeled with a Weibull distribution with $m = 1.5$. The manufacturer is required to demonstrate that the scale parameter meets the standard of 55,000 cycles. Given $\eta_1 = 45,000$ cycles, $\alpha = 0.05$, and $\beta = 0.1$, develop a sequential test plan.

SOLUTION   Substituting the given data into (9.45), we obtain the continue-test region as

$$-106 \times 10^6 + 11.1 \times 10^6 n < \sum_{i=1}^{n} t_i^{1.5} < 82.7 \times 10^6 + 11.1 \times 10^6 n.$$

The test plan is plotted in Figure 9.12. Note that the vertical axis T is the total transformed failure time. The O.C. curve is constructed by using (9.30) and (9.34) and the transformation $\theta_i = \eta_i^m$ ($i = 0, 1$). Figure 9.13 shows an O.C. curve that plots the probability of acceptance at different true values of the Weibull scale parameter, where $\eta = \theta^{1/m}$.

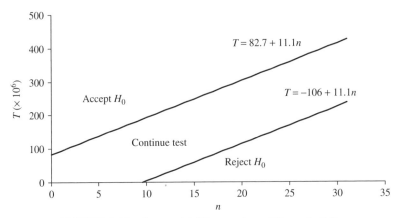

FIGURE 9.12 Sequential life test plan of Example 9.9

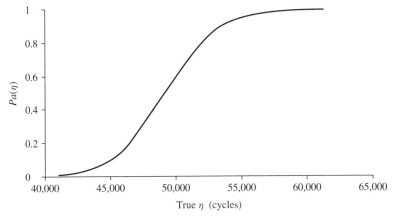

FIGURE 9.13 O.C. curve for the sequential life test of Example 9.9

9.6 RELIABILITY VERIFICATION USING PRIOR INFORMATION

The bogey tests or sequential life tests described in preceding sections may require a large sample size when high reliability is to be demonstrated at a high confidence level. The sample size can be reduced by using the Bayesian method if there exists known prior information about the life parameter to be verified. Fortunately, such information is sometimes available from the accelerated tests conducted earlier in the design and development phase, and/or from the failure data of prior-generation products. Incorporation of prior information into the development of test plans can be accomplished using the Bayesian method, which involves intensive statistical computations. The Bayesian procedures are as follows:

1. Determine the prior pdf $\rho(\theta)$ of the life parameter θ to be demonstrated. For example, θ is the reliability in binomial bogey testing and the MTTF

in exponential sequential life testing. This step involves the collection of prior failure data, selection of the prior distribution, and estimation of the distribution parameters.

2. Choose a probabilistic distribution to model the distribution of the test outcomes, say x, given parameter θ. The distribution is conditional on θ and is denoted here as $h(x/\theta)$. For example, if the test outcomes are either success or failure, the distribution is binomial.

3. Calculate the conditional joint pdf of the n independent observations from the test given parameter θ. This is the likelihood in Section 9.5.1 and can be written as

$$L(x_1, x_2, \ldots, x_n|\theta) = \prod_{i=1}^{n} h(x_i|\theta).$$

4. Calculate the joint pdf of the n independent observations from the test and of parameter θ. This is done by multiplying the conditional joint pdf and the prior pdf: namely,

$$f(x_1, x_2, \ldots, x_n; \theta) = L(x_1, x_2, \ldots, x_n|\theta)\rho(\theta).$$

5. Determine the marginal pdf of the n observations $k(x_1, x_2, \ldots, x_n)$ by integrating the joint pdf with respect to parameter θ over its entire range. That is,

$$k(x_1, x_2, \ldots, x_n) = \int f(x_1, x_2, \ldots, x_n; \theta)\, d\theta.$$

6. Using Bayes' rule,

$$g(\theta|x_1, x_2, \ldots, x_n) = \frac{L(x_1, x_2, \ldots, x_n|\theta)\rho(\theta)}{\int L(x_1, x_2, \ldots, x_n|\theta)\rho(\theta)\, d\theta},$$

find the posterior pdf of parameter θ. It is computed by dividing the joint pdf by the marginal pdf of the n observations:

$$g(\theta|x_1, x_2, \ldots, x_n) = \frac{f(x_1, x_2, \ldots, x_n; \theta)}{k(x_1, x_2, \ldots, x_n)}.$$

7. Devise a test plan by using the posterior pdf of parameter θ and the type I and type II errors specified.

The procedures above are illustrated below through application to the development of a bogey test. Although complicated mathematically, sequential life tests using the Bayesian method have been reported in the literature. Interested readers may consult Sharma and Rana (1993), Deely and Keats (1994), B. Lee (2004), and F. Wang and Keats (2004).

As described in Section 9.3.1, a binomial bogey testing is to demonstrate R_L at a $100C\%$ confidence level. The sampling reliability, say R, is a random variable, and its prior distribution is assumed known. It is well accepted by, for example, Kececioglu (1994), Kleyner et al. (1997), and Guida and Pulcini (2002), that the prior information on R can be modeled with a beta distribution given by

$$\rho(R) = \frac{R^{a-1}(1-R)^{b-1}}{\beta(a, b)}, \qquad 0 \le R \le 1,$$

where $\beta(a, b) = \Gamma(a)\Gamma(b)/\Gamma(a + b)$, $\Gamma(\cdot)$ is the gamma function, and a and b are unknown parameters to be estimated from past data. Martz and Waller (1982) provide methods for estimating a and b.

Because a bogey test generates a binary result (either success or failure), the test outcome is described by the binomial distribution with a given R. The pdf is

$$h(x|R) = (1 - R)^x R^{1-x},$$

where $x = 0$ if a success occurs and $x = 1$ if a failure occurs.

By following the Bayesian procedures, we obtain the posterior pdf of R for n success outcomes (i.e., no failures are allowed in testing):

$$g(R|x_1 = 0, x_2 = 0, \ldots, x_n = 0) = \frac{R^{a+n-1}(1-R)^{b-1}}{\beta(a + n, b)}. \qquad (9.46)$$

Note that the posterior distribution is also the beta distribution, but the parameters are $a + n$ and b.

The bogey test plan with no failures allowed is to determine the sample size required to demonstrate R_L at a $100C\%$ confidence level. This is equivalent to selecting n such that the probability of R not less than R_L is equal to C. Then we have

$$\int_{R_L}^{1} g(R|x_1 = 0, x_2 = 0, \ldots, x_n = 0) \, dR = C$$

or

$$\int_{R_L}^{1} \frac{R^{a+n-1}(1-R)^{b-1}}{\beta(a + n, b)} dR = C. \qquad (9.47)$$

Equation (9.47) is solved numerically for n. The sample size is smaller than that given by (9.5).

9.7 RELIABILITY VERIFICATION THROUGH DEGRADATION TESTING

So far in this chapter we have discussed reliability verification through bogey testing and sequential life testing. These test methods may require extensive test time to reach a conclusion, especially when high reliability is to be proved. As described in Chapter 8, the failure of some products is defined in terms of a performance characteristic exceeding a specified threshold. Degradation of the performance characteristic is highly correlated to reliability. Therefore, it is possible to verify reliability by analyzing performance measurement data. For example, Sohn and Jang (2001) present an acceptance sampling plan for computer keyboards based on degradation data. Although reliability verification through degradation testing has numerous benefits, we find few application examples in industry mainly because there are no systematic approaches that are ready for implementation. Research on this topic is also scarce in the literature. Here we describe two approaches.

The first approach is to verify the one-sided lower-bound reliability at a given confidence level. Suppose that as in bogeying testing, we want to demonstrate a required reliability R_L at time t_L at a $100C\%$ confidence level. The procedures for calculating the lower-bound reliability presented below are based on the destructive degradation analysis, which is equally applicable to nondestructive data, as pointed out in Chapter 8. The assumptions and notation described in Section 8.5 are used here for data analysis. Other methods, such as pseudolife analysis and random-effect analysis, may be employed, but the resulting accuracy and amount of computational work should be considered. The procedures are as follows:

1. Calculate the MLE of β and θ by maximizing the sample log likelihood given by (8.17).
2. Compute the estimate of reliability at t_L, denoted $\hat{R}(t_L)$, by employing (8.24) for $y \leq G$ and the like for $y \geq G$.
3. Calculate the variance–covariance matrix for β and θ by evaluating the inverse of the local Fisher information matrix given in Section 7.6.2.
4. Calculate the variance of $\hat{R}(t_L)$, denoted $\hat{V}\text{ar}[\hat{R}(t_L)]$, using (7.48).
5. The one-sided lower $100C\%$ confidence bound reliability based on a normal approximation is

$$\hat{R}(t_L) - z_C\sqrt{\hat{V}\text{ar}[\hat{R}(t_L)]}.$$

If the lower bound is greater than R_L, we conclude that the product meets the reliability requirement at a $100C\%$ confidence level.

The calculation above uses the known specific forms of $\mu_y(t; \beta)$ and $\sigma_y(t; \theta)$. In practice, they are often unknown but can be determined from test data. First we estimate the location and scale parameters at each inspection time. Then the linear or nonlinear regression analysis of the estimates derives the specific functions. Nonlinear regression analysis is described in, for example, Seber and Wild (2003). In many applications, the scale parameter is constant. This greatly simplifies subsequent analysis.

The approach described above is computationally intensive. Here we present an approximate yet simple method. Suppose that a sample of size n is tested until t_0, where $t_0 < t_L$. If the test were terminated at t_L, the sample would yield r failures. Then $\hat{p} = r/n$ estimates the probability of failure p at t_L. The number of failures r is unknown; it may be calculated from the pseudolife method described in Chapter 8. In particular, a degradation model is fitted to each degradation path; then the degradation characteristic at t_L is estimated from the model. If the resulting characteristic reaches the threshold, the unit is said to have failed. Loosely, r is binomial.

The reliability verification is equivalent to testing the hypotheses

$$H_0: p \leq 1 - R_L, \qquad H_1: p > 1 - R_L.$$

When the sample size is relatively large and p is not extremely close to zero or one, the test statistic

$$Z_0 = \frac{r - n(1 - R_L)}{\sqrt{n R_L (1 - R_L)}} \tag{9.48}$$

can be approximated with the standard normal distribution. The decision rule is that we accept H_0 at a $100C\%$ confidence level if $Z_0 \leq z_C$, where z_C is the $100C$th percentile of the standard normal distribution.

Example 9.10 In an electrical welding process, the failure of an electrode is said to have occurred when the diameter of a weld spot is less than 4 mm. The diameter decreases with the number of spots welded by the electrode. To demonstrate that a newly designed electrode meets the lower 95% confidence bound reliability of 92.5% at 50,000 spots, 75 electrodes were sampled, and each was tested until 35,000 spots were welded. Degradation analysis showed that five units would fail if the test were continued until 50,000 spots. Determine if the electrode meets the R92.5/C95 requirement.

SOLUTION   From (9.48) we have

$$Z_0 = \frac{5 - 75 \times (1 - 0.925)}{\sqrt{75 \times 0.925 \times (1 - 0.925)}} = -0.274.$$

Since $Z_0 < z_{0.95} = 1.645$, we conclude that electrode meets the specified reliability requirement.

PROBLEMS

9.1 Describes the pros and cons of bogey test, sequential life test, and degradation test for reliability verification.

9.2 Find the minimum sample size to demonstrate R95/C95 by bogey testing. If the sample size is reduced to 20, what is the confidence level? If a test uses a sample of 25 units and generates no failures, what is the lower-bound reliability demonstrated at a 90% confidence level?

9.3 A manufacturer wants to demonstrate that a new micro relay achieves a lower 90% confidence bound reliability of 93.5% at 25,000 cycles. The relay has a Weibull distribution with shape parameter 1.8. How many units shall be tested for 25,000 cycles? If the test schedule can accommodate 35,000 cycles, what is the resulting sample size? If only 12 units are available for the test, how many cycles should be run?

9.4 Redo Problem 9.3 for cases in which the shape parameter has a $\pm 20\%$ deviation from 1.8. Compare the results with those of Problem 9.3.

9.5 Explain the rationale of tail testing. Discuss the benefits, risks, and limitations of the test method.

9.6 A manufacturer wants to demonstrate the reliability of a new product by sequential life testing. The required reliability is 0.98, and the minimum acceptable reliability is 0.95. Develop a binomial sequential test plan to verify the reliability at $\alpha = 0.05$ and $\beta = 0.1$. How many units on average does the test need to reach a reject or accept decision? What is the probability of accepting the product when the true reliability is 0.97?

9.7 The life of an electronic system has the exponential distribution. The system is designed to have an MTBF of 2500 hours with a minimum acceptable MTBF of 1500 hours. The agreed-upon producer and consumer risks are 10%. Develop and plot the sequential life test plan and the O.C. curve. Suppose that the test has yielded two failures in 600 and 2300 hours. What is the decision at this point?

9.8 A mechanical part has a Weibull distribution with shape parameter 2.2. The manufacturer is required to demonstrate the characteristic life of 5200 hours with a minimum acceptable limit of 3800 hours. The probability is 0.95 of accepting the part that achieves the specified characteristic life, while the probability is 0.9 of rejecting the part that has 3800 hours. What are the decision rules for the test? Develop an O.C. curve for the test plan. What are the probabilities of accepting the part when the true values of the characteristic life are 5500 and 3500 hours?

9.9 Redo Problem 9.8 for cases in which the shape parameter has a $\pm 20\%$ deviation from 2.2. Comment on the differences due to the changes in shape parameter.

9.10 Derive the formulas for evaluating the sensitivity of Pa to misspecification of the shape parameter of the Weibull distribution.

9.11 To demonstrate that a product achieved the lower 90% confidence bound reliability of 90% at 15,000 hours, a sample of 55 units was subjected to degradation testing. The test lasted 2500 hours and yielded no failures. Degradation analysis gave the reliability estimate as

$$R(t) = 1 - \Phi \left[\frac{\ln(t) - 10.9}{1.05} \right].$$

Does the product meet the reliability requirement specified?

10

STRESS SCREENING

10.1 INTRODUCTION

A production process must be validated successfully before full production can begin. As described in Chapter 9, process validation testing includes reliability verification testing to demonstrate that the final products achieve the reliability target required. Nevertheless, we should not expect that each batch from the production line will meet the reliability requirement. Indeed, due to process variation, material flaws, and inadequate design, some products may bear latent defects which cannot be detected by functional tests. If shipped to customers, the defective products will manifest themselves to failures in an unexpectedly short time in the field. Such failures, known as *infant mortality*, are critical concerns of many electronic products. Stress screening is to reduce, if not eliminate, defects by subjecting all products manufactured to an elevated stress level for a certain period of time. The stressing process causes the latent defects to be detectable, thus preventing defective products from being delivered to customers. Although reactive (in contrast to proactive reliability improvement in the design and development phase), stress screening is an effective means of increasing the field reliability. In fact, screening is the last measure that a manufacture can take to improve the field reliability before products enter the market.

In this chapter we describe the concept of different screening techniques and the design of screening plans. Then the principle of degradation screening is discussed and applied to part-level screening. This is followed by discussions on life-based module-level stress screening. Combining part- and module-level

Life Cycle Reliability Engineering, by Guangbin Yang
Copyright © 2007 John Wiley & Sons, Inc.

screening, we also present the development of two-level optimal screen plans that minimize an important segment of the life cycle cost and meet the field reliability requirement.

10.2 SCREENING TECHNIQUES

The screening techniques currently being implemented in industry may be classified into five categories: (1) burn-in, (2) environmental stress screening (ESS), (3) highly accelerated stress screening (HASS), (4) discriminator screening, and (5) degradation screening. Although these techniques differ in many aspects, the objective remains the same, that is, to weed out the defective products before shipment to customers. Each of the screening techniques is described briefly below.

The burn-in technique, which originated in the defense industry, was the earliest screening approach used to precipitate defective electronic components and is still implemented widely in industry. In a burn-in, products are powered up and subjected to a constant stress level within the specification limits for a certain length of time. For example, the burn-in strategies for the microcircuits specified in MIL-STD-883F (U.S. DoD, 2004) require electrically biasing the devices at a minimum of 125°C for 168 hours. The burn-in strategies are effective in weeding out surface and metallization defects and weak bonds. In general, a burn-in condition is lenient and is capable of precipitating significant latent defects. Products with subtle defects will certainly escape a burn-in process. These products will be delivered to customers and fail prematurely. Jensen and Peterson (1982), and Kuo et al. (1998) describe burn-in techniques.

Similar to burn-in, ESS is also a screening method that subjects all products to an elevated stress level for a predetermined duration. ESS differs from burn-in in that it exposes products to environmental stresses outside the specification limits. The most commonly used stresses are thermal cycling, random vibration, power cycling, temperature, and humidity. In applications, the combination of two or more stresses is often used to enhance screening effectiveness. MIL-HDBK-344A (U.S. DoD, 1993) well documents techniques for planning and evaluating ESS programs for military electronic products. In many occasions, ESS is mistakenly interexchanged with burn-in. Although both have a great similarity, the distinctions are obvious. First, an ESS applies environmental stresses, which are often time dependent, while a burn-in frequently uses a constant temperature. Environmental stimuli are more powerful in precipitating latent defects to patent failures. For example, thermal cycling is more stressful than a constant temperature in removing the weak solder joints of electronic components. Second, an ESS stresses products outside the specification limits. A high stress level greatly accelerates the aging process of defects. Third, ESS and burn-in often find different flaws, mainly because they use different stresses. For the distinctions above, it is widely recognized that an ESS is more effective than a burn-in.

A HASS is a more stressful ESS. In a HASS, the applied stresses may not necessarily be the ones that would be experienced in the field. That is, any stress may be employed as long as it is effective in precipitating relevant defects that would occur in normal use. The stress levels should be as high as possible to turn latent defects into patent failures swiftly in order to compress the screen duration. As we know, while applied stresses accelerate the failure of defective products, they also cause damage to defective-free products. Here *damage* refers to a reduction in life time. For example, a high screen stress level degrades the performance of a good product and reduces the remaining life. Therefore, the screen stress levels must be optimized such that defective-free products have sufficient remaining life. The initial values of the stress levels may be ascertained by analyzing the HALT data; the optimal values are determined in subsequent HASS implementation. Hobbs (2000) describes the theory and application of HASS in a great detail. There are many practical examples in the literature; some are given in Silverman (1998), Rahe (2000), and Misra and Vyas (2003).

Discriminator screening utilizes a discriminator, which is comprised of a parameter or a weighted combination of several parameters of the product under screening, to identify products with intrinsic defects. If the value of the discriminator for a product crosses a specified threshold, a defect is detected and the product should be weeded out; otherwise, the product is defect-free. It is worth noting that the parameters are not limited to the product performance characteristics and can be any properties that allow one to discriminate defective products from good ones. For example, low-frequency $1/f$ noise can be used as a discriminator to signify the oxide-trap charge-related defects for discrete MOS devices and small-scale circuits, according to Fleetwood et al. (1994). When a discriminator consists of only one parameter, a normal value of the parameter indicates flawlessness directly, and vice versa. That is, a product is said to be defect-free if the value of the parameter is within a prespecified threshold. Otherwise, the product is considered defective and subject to removal. Such a simple relationship does not exist if several parameters are used to form a discriminator. Specifically, all individual parameters being within their thresholds may lead to a discriminator being outside the normal limit. In application, the parameters selected must be readily measurable and sensitive to the magnitude of a defect. Since the parameters are measured at use conditions, the screening is nondestructive on both flaw and good products. Because of this advantage, discriminator screening requires short screen time and does not damage or degrade good products. In practice, however, a sensitive and reliable discriminator is not easy to find, and misdiscrimination often occurs. In other words, a good product may be classified as a bad one and removed from the population; this is type I error. On the other hand, a defective product may escape the screening and be shipped to customers; this is the type II error. A good discriminator should be one that achieves minimum errors of the two types. Jensen (1995) describes in detail the theory and application examples of this screening technique.

Like ESS, degradation screening subjects all products to an elevated stress level for a prespecified length of time. During screening, the performance characteristic of defective products degrades rapidly over time while the good ones deteriorate gradually. The substantial difference between degradation rates allows

TABLE 10.1   Differences and Similarities Between Screening Methods

Aspect	Burn-in	ESS	HASS	Discriminator Screening	Degradation Screening
Stress type	Mostly temperature	Thermal cycling, random vibration, power cycling, temperature, humidity, etc.	Thermal cycling, random vibration, power cycling, temperature, humidity, etc.	No stresses applied	Thermal cycling, random vibration, power cycling, temperature, humidity, etc.
Stress level	Low and constant	High and variable	Very high and variable	Not applicable	High and variable
Screen duration	Long	Short	Very short	Very short	Very short
Cost	High	Low	Low	Low	Low
Damage to good products	Little	Large	Very large	Almost no	Little
Type I error	Very low	Low	Very high	High	Low
Type II error	High	Low	Very low	High	Low
Defect criterion	Functional demise	Functional demise	Functional demise	Discriminator crossing a threshold	Characteristic crossing a threshold

defective products to be identified by measuring the performance characteristic. A product is said to be defective if its measurement crosses a specified threshold at the end of screening; otherwise, it is a good one and survives the screening. In application, the threshold can be tighter than the usual one that defines a failure in the field. From the perspective of defect identification, degradation screening is similar to discriminator screening, but it is more effective in detecting bad products. On the other hand, this screening method causes less damage to good products than does the ESS. In subsequent sections we discuss this method in detail.

As we may have noted, the five screening techniques described above have great differences and similarities, which are summarized in Table 10.1.

10.3   DESIGN OF SCREEN PLANS

10.3.1   Characterization of Screen Plans

A screen plan is characterized by the screening technique, stress type, stress levels, screen duration, and defect criteria. They must be determined prior to screening. Selection of a screening method is most critical and has a fundamental impact on the screening effectiveness, time, cost, field reliability, and other factors. The selection can be aided by referring to Table 10.1. If a reliable and sensitive discriminator can be formed and the measurement of the parameters does not require expensive equipment, the discriminator screening method should be the best choice. The next-most-preferable technique is degradation screening if a performance characteristic is highly indicative of the latent defect of concern. Nowadays, HASS is widely implemented to compress the time to market; however, we should not overlook the fact that the screen stress may cause damage to good products. Using a HASS, some defect-free products at the low tail of the life distribution may be destroyed and thus removed. The surviving products may accumulate a large amount of degradation during screening, which greatly shortens the remaining life. Therefore, choosing between HASS and ESS should be well justified based on various criteria, including time, cost, field reliability, and other factors. In many situations, burn-in is not effective for components; instead, large-scale systems favor this method.

Once the screening technique is determined, we should choose the stress type. In general, the stress type selected should effectively stimulate and accelerate the failure mechanisms governing the early failures that would be observed in the field. Often, preliminary tests may be conducted to discover the failure modes caused by latent defects and to examine the effectiveness of the stresses selected in precipitating the defects. ESS or degradation screening usually employs stresses that a product will encounter in normal operation. In contrast, one may use any stress type in a HASS as long as it is effective.

For a burn-in, the stress levels should be within the specification limits. The burn-in conditions for some types of electronic products can be found in relevant engineering standards. For example, MIL-STD-883F (U.S. DoD, 2004) gives different stress levels and durations for burning-in microcircuits. In contrast to a burn-in, an ESS or a degradation screening exposes products to a stress level outside the specification limits and inside the operational limits at which the products can perform a full function. Thus, continuously monitoring the functionality is possible during screening. In a HASS, products are stressed outside the operational limits and thus fail to perform a full function during screening. The stress level must be lowered below operational limits to detect defects. In general, a screening should use a high stress level to reduce the screen duration. But the stress level should not induce failure modes that differ from those of early failures in the field.

The defect criteria for a screen define what constitutes a defect. For a defect leading to a hard failure, complete cession of product function indicates a defect. If a soft failure is concerned, a product is defective if its degradation reaches a threshold at the end of screening. As we will see later, the threshold for degradation screening may be tighter than that used in the field to define a failure. For a discriminator screening, a defect is said to have been detected if the value of the discriminator crosses a critical value, which is often chosen to minimize type I and type II errors.

10.3.2 Optimization of Screen Plans

Once the screening technique, stress types, stress levels, and defect criteria are determined, the screen duration should be chosen. The duration has conflicting impacts on the cost, reliability, and requirement for screening equipment capacity. Insufficient screening saves screening cost and equipment capacity but reducese field reliability and increases field repair cost. On the other hand, excessive screening eliminates nearly all infantile failures and decreases the field repair cost but incurs high screening expense and requires ample screening equipment capacity. Apparently, an optimal duration should be chosen to make the best trade-off among these conflicting factors.

To obtain the optimal duration, an optimization model is needed. In the literature, much of the effort toward establishing the optimal duration derives from minimizing the total cost. Some examples are given in Reddy and Dietrich (1994), Pohl and Dietrich (1995a, b), Kar and Nachlas (1997), Yan and English (1997), C. L. Wu and Su (2002), and Sheu and Chien (2004, 2005). Here the total cost is formulated as a segment of the life cycle cost and may consist of the following cost components:

- Cost of screen setup, which is fixed
- Cost of screening for a specified duration which is variable and depends on the screen duration
- Cost of good products being weeded out
- Cost of repair in subsequent higher-level (e.g., board-level) screening and in the field
- Cost of reputation loss due to infantile failures

To develop more realistic screening strategies, minimization of the total cost is often subject to constraints on the field reliability and screening equipment capacity, which are formulated as a function of the screen duration. The constrained optimization problems are discussed in, for example, Chi and Kuo (1989), Mok and Xie (1996), T. Kim and Kuo (1998), and G. Yang (2002).

In situations where the field reliability is critical, the screen duration may be optimized to maximize the reliability while the constraints on total cost and other factors, if any, are satisfied. K. Kim and Kuo (2005) present a study of this type which determines the optimal burn-in period by maximizing system reliability.

10.4 PRINCIPLE OF DEGRADATION SCREENING

For a product whose performance degrades over time, a failure is said to have occurred if a performance characteristic (say, y) crosses a specified threshold. The faster the degradation, the shorter the life. Thus, the life is determined by the degradation rate of y. A population of the products usually contains a fraction of both good and substandard units. In practice, good products overwhelmingly outnumber substandard ones. Stressed at an elevated level during screening, the substandard products degrade rapidly, resulting in early failures, while the good ones degrade gradually, causing random or wear-out failures. For example, Henderson and Tutt (1997) report that under biased electrical current and thermal stress, the bad units of the GaAs-based heterojunction bipolar transistors have a considerably larger collector current than the good units. Croes et al. (1998) also present experimental results showing that substandard units of metal film resistors suffer greater resistance drift than the good units do when subjected to a high temperature. The important difference between degradation rates yields a bimodal distribution of y, where the main and weak distributions are dominated by good and substandard products, respectively. Figure 10.1 illustrates the degradation paths of a sample containing good and substandard units, and the bimodal distribution of y resulting from the difference between degradation rates.

The degradation rates of good and bad products depend on the screen stress level. The higher the stress level, the greater the degradation rates. To shorten the screen duration, a high stress level is often applied. However, the stress level should not induce degradation modes that differ from those of early failures in the field. If y is a monotonically increasing characteristic, the bimodal cumulative distribution function (cdf) of y at a given stress level can be written as

$$\Pr(y \geq G_0) = \alpha_1 \Pr[y_1(t) \geq G_0] + \alpha_2 \Pr[y_2(t) \geq G_0] = \alpha_1 F_1(t) + \alpha_2 F_2(t),$$
(10.1)

where the subscripts 1 and 2 denote substandard and good subpopulations, α the fraction of a subpopulation, F the cdf of the life distribution of a subpopulation, and G_0 the usual threshold. Equation (10.1) indicates that the bimodal life distribution, which is hypothesized in Jensen and Petersen (1982) and has been used extensively, is the result of the bimodal distribution of y. The relationship between the bimodal distributions of y and life is depicted in Figure 10.2, where pdf(t) is the probability density function (pdf) of life, pdf[$y(t)$] the pdf of y at time t, and G^* a threshold smaller than G_0. In Figure 10.2 the shaded area of pdf(t) represents the probability of failure by time t_2, which equals $\Pr[y(t_2) \geq G_0]$ represented by the shaded area of pdf[$y(t_2)$].

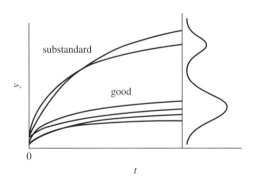

FIGURE 10.1   Difference between degradation rates causing a bimodal distribution

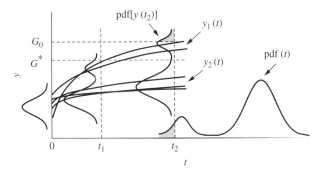

FIGURE 10.2 Relationship between the bimodal distributions of y and life

Before screening $(t = 0)$, y of a population does not display an apparent bimodal distribution, as illustrated in Figure 10.2. The reason is that the products with patent defects have been removed in either the manufacturing process or the product final test; latent defects do not appreciably degrade the value of y. During screening, y_1 (i.e., y of a substandard product) degrades rapidly due to the growth of latent defects, whereas y_2 (i.e., y of a good product) degrades gradually only because of the aging effect. The difference yields a bimodal distribution of y. The parameters of the distribution are determined by the measurements of y_1 and y_2, which in turn depend on the screen stress and duration. At a given stress level, y_1 and y_2 degrade at different rates with screen time. Thus, the location parameters of y_1 and y_2 distributions degrade, and their difference is widened. Therefore, the two modes of the y distribution shift apart as the screen time increases, as shown in Figure 10.2. When the two modes are apart far enough, it is possible to find an appropriate G^*, which is tighter than G_0, to discriminate between the substandard and good products. A tighter threshold G^* refers to a threshold that is smaller than G_0 for a monotonically increasing y and larger than G_0 for a monotonically decreasing y. If the value of y of a product crosses G^* at the end of screening, the product is said to be *substandard* and is removed from the population; otherwise, the product is defect-free and survives the screening.

Tightening a threshold from G_0 to G^* shortens the screen duration and thus alleviates the aging effects of screen stress on the good products. The design of a screen plan includes the selection of G^*. The selection is constrained by, for example, the screen stress level, screen duration, and field repair cost. A higher field repair cost imposes a tighter threshold to weed out more defects. A screen at a lower stress level for a shorter time also requires a tighter threshold to increase the screen effectiveness.

10.5 PART-LEVEL DEGRADATION SCREENING

In practice, one often conducts two-level screening; that is, part- and module-level screening, where a module may be a board, subsystem, or system. The purpose of part-level screening is to weed out the substandard parts by subjecting the part population to an elevated stress level. The screened parts are then assembled into a module. Because the assembly process may introduce defects, the module is then screened for a specified duration. In this section we focus on part-level degradation screening.

Part-level degradation screening stresses products at an elevated level. Let t_p' denote the screen duration at this stress level. Then the equivalent time t_p at the use stress level is

$$t_p = A_p t_p', \tag{10.2}$$

where A_p is the acceleration factor and can be estimated using the theory of accelerated testing discussed in Chapter 7. For example, if temperature is the screening stress, the Arrhenius relationship may be applied to determine the value of A_p.

A part, good or bad, subjected to degradation screening may fail to pass or survive screening. For a part having a monotonically increasing performance characteristic, the probability p_0 of the part passing the screen can be written as

$$p_0 = \alpha_1 \Pr[y_1(t_p) \le G^*] + \alpha_2 \Pr[y_2(t_p) \le G^*]. \tag{10.3}$$

The probability p_1 that a part passing the screen is from the substandard subpopulation is

$$p_1 = \frac{\alpha_1}{p_0} \Pr[y_1(t_p) \le G^*]. \tag{10.4}$$

The probability p_2 that a part passing the screen is from the good subpopulation is

$$p_2 = \frac{\alpha_2}{p_0} \Pr[y_2(t_p) \le G^*]. \tag{10.5}$$

From (10.3) through (10.5), we have $p_1 + p_2 = 1$.

All parts from the screened population have survived the screening process. However, due to type I and type II errors, the screening process might not be perfect. Thus, the screened population might still contain some substandard parts. A part selected at random from the screened population is either good or substandard. The field reliability $R_p(t)$ at time t of a part from the screened population can be written as

$$R_p(t) = p_1 R_{p1}(t|t_p) + p_2 R_{p2}(t|t_p), \tag{10.6}$$

where $R_{p1}(t|t_p)$ is the field reliability at t of a substandard part from the screened population and $R_{p2}(t|t_p)$ is the field reliability at t of a good part from the screened population. Because

$$R_{pi}(t|t_p) = \frac{R_{pi}(t+t_p)}{R_{pi}(t_p)} = \frac{\Pr[y_i(t+t_p) \le G_0]}{\Pr[y_i(t_p) \le G_0]}, \qquad i = 1, 2,$$

(10.6) can be rewritten as

$$R_p(t) = p_1 \frac{\Pr[y_1(t + t_p) \leq G_0]}{\Pr[y_1(t_p) \leq G_0]} + p_2 \frac{\Pr[y_2(t + t_p) \leq G_0]}{\Pr[y_2(t_p) \leq G_0]}. \tag{10.7}$$

Substituting (10.4) and (10.5) into (10.7) gives

$$R_p(t) = \theta_1 \Pr[y_1(t + t_p) \leq G_0] + \theta_2 \Pr[y_2(t + t_p) \leq G_0], \tag{10.8}$$

where

$$\theta_i = \frac{\alpha_i \Pr[y_i(t_p) \leq G^*]}{p_0 \Pr[y_i(t_p) \leq G_0]}, \qquad i = 1, 2. \tag{10.9}$$

The cdf of a part from a screened population is

$$F_p(t) = 1 - \theta_1 \Pr[y_1(t + t_p) \leq G_0] - \theta_2 \Pr[y_2(t + t_p) \leq G_0]. \tag{10.10}$$

The associated pdf is given by

$$f_p(t) = \theta_1 f_{p1}(t + t_p) + \theta_2 f_{p2}(t + t_p), \tag{10.11}$$

where

$$f_{pi}(t + t_p) = -\frac{d\Pr[y_i(t + t_p) \leq G_0]}{dt}, \qquad i = 1, 2.$$

Now let's consider a special case where the performance characteristic of a part has a normal or lognormal distribution. As explained in Chapter 8, this case is frequently encountered in practice. Because lognormal data can be transformed into normal data, only the normal distribution is discussed here. The following assumptions are made:

- The performance characteristic y is monotonically increasing and has a bimodal normal distribution; $y_1(y_2)$ has a location parameter $\mu_{y_1}(\mu_{y_2})$ and scale parameter $\sigma_{y_1}(\sigma_{y_2})$.
- Neither σ_{y_1} nor σ_{y_2} depend on screen stress and time.
- The location parameter is a (transformed) linear function of screen time for both substandard and good parts: namely,

$$\mu_{y_i}(t) = \beta_{1i} + \beta_{2i}t, \qquad i = 1, 2, \tag{10.12}$$

where β_{1i} is the mean of the initial values of y_i before screening and β_{2i} is the degradation rate of y_i. These parameters can be estimated from preliminary test data.

From (10.11) and (10.12), the pdf of a part from the screened population is

$$f_p(t) = \frac{\theta_1}{\sigma_{t_1}} \phi \left(\frac{t - \mu_{t_1}}{\sigma_{t_1}} \right) + \frac{\theta_2}{\sigma_{t_2}} \phi \left(\frac{t - \mu_{t_2}}{\sigma_{t_2}} \right), \tag{10.13}$$

where $\phi(\cdot)$ is the pdf of the standard normal distribution and

$$\mu_{t_i} = \frac{G_0 - \beta_{1i} - \beta_{2i} t_p}{\beta_{2i}}, \qquad \sigma_{t_i} = \frac{\sigma_{y_i}}{\beta_{2i}}, \qquad i = 1, 2. \tag{10.14}$$

The cdf of a part from the screened population is

$$F_p(t) = \theta_1 \Phi \left(\frac{t - \mu_{t_1}}{\sigma_{t_1}} \right) + \theta_2 \Phi \left(\frac{t - \mu_{t_2}}{\sigma_{t_2}} \right), \tag{10.15}$$

where $\Phi(\cdot)$ is the cdf of the standard normal distribution. Equation (10.15) indicates that the life distribution of a substandard or good part from the screened population has a normal distribution with mean μ_{t_i} and standard deviation σ_{t_i}.

Example 10.1 A part supplier screens a type of electronic component it produces at 175°C for 120 hours. A component is defective and removed from the population if the performance characteristic drifts more than 12% by the end of screening. The survival components are shipped to system manufacturers and will operate at 35°C. The system specification requires the component degradation to be less than 25%. The acceleration factor between the screen and the use temperatures is 22. A preliminary test of 180 components at the screen temperature identified 12 defective units. Degradation analysis of the test data shows that the degradation percentage has a normal distribution with $\mu_{y_1} = 0.23t'$, $\sigma_{y_1} = 4.5$, $\mu_{y_2} = 0.018t'$, and $\sigma_{y_2} = 3.2$, where the prime denotes the screen condition. Calculate the probability of failure at a design life of 20,000 hours for the screened components at 35°C. How much reduction in probability of failure at the design life does the screen achieve?

SOLUTION Considering the temperature acceleration effect, the mean estimates of the degradation percentage at 175°C are converted to those at 35°C: namely,

$$\mu_{y_1} = \frac{0.23}{22} t = 0.0105t, \qquad \mu_{y_2} = \frac{0.018}{22} t = 0.8182 \times 10^{-3} t.$$

The equivalent screen time at 35°C is $t_p = 22 \times 120 = 2640$. From the data given, the fractions of the substandard and good subpopulations are $\hat{\alpha}_1 = 12/180 = 0.0667$ and $\hat{\alpha}_2 = 1 - 0.0667 = 0.9333$.

The estimate of the probability of a defective component escaping the screen is

$$\Pr[y_1(t_p) \le G^*] = \Phi \left(\frac{G^* - \mu_{y_1}}{\sigma_{y_1}} \right) = \Phi \left(\frac{12 - 0.0105 \times 2640}{4.5} \right) = 0.0002.$$

The estimate of the probability of a defect-free component surviving the screen is

$$\Pr[y_2(t_p) \leq G^*] = \Phi\left(\frac{12 - 0.8182 \times 10^{-3} \times 2640}{3.2}\right) = 0.9989.$$

The estimate of the reliability of a defective component at the end of screening is

$$\Pr[y_1(t_p) \leq G_0] = \Phi\left(\frac{G_0 - \mu_{y_1}}{\sigma_{y_1}}\right) = \Phi\left(\frac{25 - 0.0105 \times 2640}{4.5}\right) = 0.2728.$$

The estimate of the reliability of a defect-free component at the end of screening is

$$\Pr[y_2(t_p) \leq G_0] = \Phi\left(\frac{25 - 0.8182 \times 10^{-3} \times 2640}{3.2}\right) \approx 1.$$

From (10.3), the estimate of the probability of a component passing the screen is

$$p_0 = 0.0667 \times 0.0002 + 0.9333 \times 0.9989 = 0.9323.$$

Substituting into (10.9) the necessary data obtained above gives

$$\theta_1 = \frac{0.0667 \times 0.0002}{0.9323 \times 0.2728} = 0.5245 \times 10^{-4},$$

$$\theta_2 = \frac{0.9333 \times 0.9989}{0.9323 \times 1} = 0.99997.$$

From (10.14), after screening, the estimates of the mean and standard deviation of the life distribution of the defective components are

$$\mu_{t_1} = \frac{25 - 0.0105 \times 2640}{0.0105} = -259 \quad \text{and} \quad \sigma_{t_1} = \frac{4.5}{0.0105} = 429.$$

In Problem 10.7 we ask for an explanation of the negative mean life.

Similarly, after screening, the estimates of the mean and standard deviation of the life distribution of the good components are

$$\mu_{t_2} = \frac{25 - 0.8182 \times 10^{-3} \times 2640}{0.8182 \times 10^{-3}} = 27,915,$$

$$\sigma_{t_2} = \frac{3.2}{0.8182 \times 10^{-3}} = 3911.$$

After operating 20,000 hours at 35°C, the screened components are estimated using (10.15) to have a probability of failure

$$\hat{F}_p(20,000) = 0.5245 \times 10^{-4} \times \Phi\left(\frac{20,000 + 259}{429}\right)$$

$$+ 0.99997 \times \Phi\left(\frac{20,000 - 27,915}{3911}\right) = 0.0215.$$

From (10.1), if the component population were not screened, the probability of failure at 20,000 hours would be

$$\Pr(y \geq 25) = 0.0667 \times \left[1 - \Phi\left(\frac{25 - 0.0105 \times 20,000}{4.5}\right)\right]$$

$$+ 0.9333 \times \left[1 - \Phi\left(\frac{25 - 0.8182 \times 10^{-3} \times 20,000}{3.2}\right)\right]$$

$$= 0.07.$$

Therefore, the screen reduces the probability of failure at the time of 20,000 hours by $0.07 - 0.0215 = 0.0485$. Figure 10.3 plots the probabilities of failure at different times for both the screened and unscreened populations. It can be seen that the improvement retains until the time reaches 22,500 hours, after which the probability of failure is exacerbated by the screening. This is understandable. Nearly all defective components would fail before 22,500 hours. After this time, the failure is dominated by the good components. Because of the screen stress effects, a screened good component has a greater degradation percentage than an unscreened good component, causing the higher probability of failure.

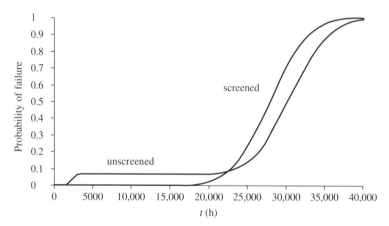

FIGURE 10.3   Probabilities of failure for screened and unscreened components

10.6   MODULE-LEVEL SCREENING

The screened parts are assembled into a module according to the design configuration, where a module may refer to a board, subsystem, or system, as described earlier. The assembly process usually consists of multiple steps, each of which may introduce different defects, including, for example, weak solder joints, loose connections, and crack wire bonds. Most of the defects are latent and cannot be detected in final production tests. When stressed in the field, they will fail in early life. Therefore, it is often desirable to precipitate and remove such defects before customer delivery. This can be accomplished by performing module-level screening. During the screening, defective connections dominate the failure. Meanwhile, the already-screened parts may also fail. So in this section we model failure of both parts and connections.

10.6.1   Part Failure Modeling

Screened parts may fail during module-level screening and field operation. Upon failure, the failed parts are replaced by new parts from the screened population. The replacement process is a typical renewal process. According to renewal theory (Cox, 1962; Tijms, 1994), the renewal density function $h_p(t)$ can be written as

$$h_p(t) = f_p(t) + \int_0^t h_p(t-s) f_p(s)\, ds. \tag{10.16}$$

Then the expected number of renewals $N_p(t)$ within time interval $[0, t]$ is

$$N_p(t) = \int_0^t h_p(t)\, dt. \tag{10.17}$$

To calculate $N_p(t)$, we first need to determine the Laplace transform of $f_p(t)$, denoted $f_p^*(s)$. Then the Laplace transform of the renewal density function is given by

$$h_p(s) = \frac{f_p^*(s)}{1 - f_p^*(s)}. \tag{10.18}$$

The next step is to transform $h_p(s)$ inversely to the renewal density function in the time domain [i.e., $h_p(t)$]. Then $N_p(t)$ is calculated from (10.17). Unfortunately, the Laplace transform for most distributions (e.g., the Weibull) is intractable. In most situations, it is more convenient to use the following renewal equation to calculate $N_p(t)$. The renewal equation is

$$N_p(t) = F_p(t) + \int_0^t N_p(t-x) f_p(x)\, dx. \tag{10.19}$$

This renewal equation for $N_p(t)$ is a special case of a Volterra integral equation of the second kind, which lies in the field of numerical analysis. Many numerical methods have been proposed to solve the equation. However, these methods typically suffer from an accumulation of round-off error when t gets large. Using the basic concepts in the theory of Riemann–Stieltjes integration (Nielsen, 1997), Xie (1989) proposes a simple and direct solution method with good convergence properties. As introduced below, this method discretizes the time and computes recursively the renewal function on a grid of points.

For a fixed $t \geq 0$, let the time interval $[0, t]$ be partitioned to $0 = t_0 < t_1 < t_2 < \cdots < t_n = t$, where $t_i = id$ for a given grid size $d > 0$. For computational simplification set $N_i = N_p(id)$, $F_i = F_p[(i - 0.5)d]$, and $A_i = F_p(id)$, $1 \leq i \leq n$. The recursion scheme for computing N_i is

$$N_i = \frac{1}{1 - F_1} \left[A_i + \sum_{j=1}^{i-1}(N_j - N_{j-1})F_{i-j+1} - N_{i-1}F_1 \right], \qquad 1 \leq i \leq n,$$

(10.20)

starting with $N_0 = 0$. The recursion scheme is remarkable in resisting the accumulation of round-off error as t gets larger and gives surprisingly accurate results (Tijms, 1994). Implementation of the recursion algorithm needs a computer program, which is easy to code. In computation, the grid size d has a strong influence on the accuracy of the result. The selection depends on the accuracy required, the shape of $F_p(t)$, and the length of the time interval. A good way to determine whether the results are accurate enough is to do the computation for both grid sizes d and $d/2$. The accuracy is satisfactory if the difference between the two results is tolerable.

When t is remarkably longer than the mean of the life distribution, the expected number of renewals can be simply approximated by

$$N_p(t) \approx \frac{t}{\mu_t} + \frac{\sigma_t^2 - \mu_t^2}{2\mu_t^2},$$

(10.21)

where μ_t and σ_t are the mean and standard deviation of the life distribution $f_p(t)$. Note that (10.21) gives an exact result when μ_t and σ_t are equal. This is the case for the exponential distribution. In practice, the approximation has an adequate accuracy for a moderate value of t provided that $c_x^2 = \sigma_t^2/\mu_t^2$ is not too large or close to zero. Numerical investigations indicate that for practical purpose (10.21) can be used for $t \geq t_x$ (Tijms, 1994), where

$$t_x = \begin{cases} \dfrac{3}{2}c_x^2\mu_t, & c_x^2 > 1 \\ \mu_t, & 0.2 < c_x^2 \leq 1 \\ \dfrac{1}{2c_x^2}\mu_t, & 0 < c_x^2 \leq 0.2. \end{cases}$$

(10.22)

When $t \geq t_x$, the approximation by (10.21) usually results in a relative error of less than 5%.

The renewal equation (10.19) may also be solved by using an approximate approach. In the literature, much work has been done in developing approximations to the renewal functions for various life distributions, such as the Weibull, lognormal, and truncated normal. Examples of the work are given in, Baxter et al. (1982), Smeitink and Dekker (1990), Tijms (1994), Lomnicki (1996), and Garg and Kalagnanam (1998). To the best of the author's knowledge, the approximation for a mixed distribution such as $f_p(t)$ expressed in (10.11) has not yet been studied.

Example 10.2 Refer to Example 10.1. The electronic component is installed in a module and operates at 35°C. Once the component fails, it is replaced with a new one from the screened population. The replacement process is a renewal process. Calculate the expected number of renewals for the component in a design life of 20,000 hours. Redo the calculation for a service life of 50,000 hours.

SOLUTION Using the data in Example 10.1, we have $c_x^2 \approx 0.02$, and $t_x = 711{,}062$. Because $t = 20{,}000$ is considerably smaller than $t_x = 711{,}062$, (10.21) cannot approximate the expected number of renewals. In this case, (10.20) is used. The recursion scheme is coded in Visual Basic running on Excel. The source codes are given in Table 10.2 and can readily be modified for other distributions. The recursive calculation yields $N_p(20{,}000) = 0.0216$, which is nearly equal to the probability of failure at 20,000 hours obtained in Example 10.1. This is understandable. As shown in Figure 10.3, the component has an extremely low probability of failure within 20,000 hours, allowing (10.19) to be approximated by $N_p(t) \approx F_p(t)$. For a service life of 50,000 hours, the expected number of renewals is $N_p(50{,}000) = 1.146$, which is calculated by setting T0 = 50,000 in the computer program. In contrast, the probability of failure at this time is approximately 1. $N_p(t)$ is ploted in Figure 10.4 to illustrate how the expected number of renewals increases with time. It is seen that $N_p(t)$ becomes a plateau between 35,000 and 45,000 hours. In Problem 10.10 we ask for an explanation.

10.6.2 Connection Failure Modeling

Defective connections are the primary source of the early failure of modules. To precipitate and repair latent connection defects, modules are often screened at an elevated stress level. Upon failure, a connection is repaired. The repair is often a minimal repair, implying that a newly repaired connection has the same failure rate as immediately before failure. In Chapter 11 we discuss more about this repair strategy.

Suppose that a module consists of m types of connections, including, for example, surface-mounted technology (SMT) solder joints, plated-through-hole (PTH) solder joints, wire bonds, and die attachments. In the literature, much work reports that the life of a connection can be modeled with the Weibull distribution. Examples of the work include, Wen and Ross (1995), Yeo et al. (1996), Amagai (1999), and Strifas et al. (2002). Before a module-level screening, connections of each type contain both substandard and good connections, forming a mixed

TABLE 10.2   Visual Basic Codes for Example 10.2

```
Sub Np()
Dim N(5000), A(5000), F(5000)
TO = 20000
D = 10
M = TO/D
N(0) = 0
Mean1 = -259
Sigma1 = 429
Mean2 = 27915
Sigma2 = 3911
Theta1 = 0.00005245
Theta2 = 0.99997
For i = 0 To M
ZF1 = ((i - 0.5) * D - Mean1) / Sigma1
ZF2 = ((i - 0.5) * D - Mean2) / Sigma2
ZA1 = (i * D - Mean1) / Sigma1
ZA2 = (i * D - Mean2) / Sigma2
FP1 = Application.WorksheetFunction.NormSDist(ZF1)
FP2 = Application.WorksheetFunction.NormSDist(ZF2)
AP1 = Application.WorksheetFunction.NormSDist(ZA1)
AP2 = Application.WorksheetFunction.NormSDist(ZA2)
F(i) = Theta1 * FP1 + Theta2 * FP2
A(i) = Theta1 * AP1 + Theta2 * AP2
Next i
For i = 1 To M
Sum = 0
For j = 1 To i - 1
Sum = Sum + (N(j) - N(j - 1)) * F(i - j + 1)
Next j
N(i) = (A(i) + Sum - N(i - 1) * F(1)) / (1 - F(1))
Next i
Cells(1, 1) = N(M)
End Sub
```

population. Thus, the life of connections can be reasonably modeled using a mixed Weibull distribution (Chapter 2).

A module is usually subjected to an elevated stress level during module-level screening. Let t_c' denote the module-level screen duration at the elevated stress level. The equivalent screen time t_c for a connection at the use stress level is

$$t_c = A_c t_c',$$

where A_c is the acceleration factor between the module-level screen stress and the use stress for the connection. The value of A_c may vary with the type of connection, and can be estimated with the theory of accelerated testing described in Chapter 7.

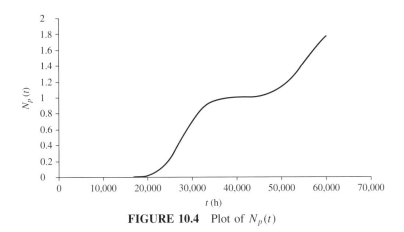

FIGURE 10.4 Plot of $N_p(t)$

When a module undergoes screening, all parts within the module are aged at the same time. The aging effects depend on the type of part. Some parts may be more sensitive than others to screening stress. Nevertheless, all parts suffer performance degradation during screening, causing permanent damage. For a particular part, the amount of degradation is determined by the screen stress level and duration. The equivalent aging time t_{cp} for a part at the use stress level is

$$t_{cp} = A_{cp} t_c',$$

where A_{cp} is the acceleration factor between the module-level screen stress and the use stress for the part.

Using the mixed Weibull distribution, the reliability of an unscreened connection of a certain type can be written as

$$R_c(t) = \rho_1 R_{c1}(t) + \rho_2 R_{c2}(t) = \rho_1 \exp\left[-\left(\frac{t}{\eta_1}\right)^{m_1}\right] + \rho_2 \exp\left[-\left(\frac{t}{\eta_2}\right)^{m_2}\right],$$
(10.23)

where

$R_c(t) =$ reliability of an unscreened connection,

$R_{c1}(t) =$ reliability of an unscreened substandard connection,

$R_{c2}(t) =$ reliability of an unscreened good connection,

$\rho_1 =$ fraction of the substandard connections,

$\rho_2 =$ fraction of the good connections,

$m_1 =$ Weibull shape parameter of the substandard connections,

$m_2 =$ Weibull shape parameter of the good connections,

$\eta_1 =$ Weibull characteristic life of the substandard connections,

$\eta_2 =$ Weibull characteristic life of the good connections.

The Weibull parameters m_i and η_i ($i = 1, 2$) can be estimated from accelerated life data using the graphical or maximum likelihood method described in Chapter 7.

Once a connection fails during module-level screening, it is repaired; then the screening continues until t'_c. Because a repaired connection has the same failure rate as right before failure, the expected number of repairs $N_c(t)$ to a connection in interval $[0, t]$ is

$$N_c(t) = \int_0^t \lambda_c(t)\, dt = -\ln[R_c(t)]$$

$$= -\ln\left\{\rho_1 \exp\left[-\left(\frac{t}{\eta_1}\right)^{m_1}\right] + \rho_2 \exp\left[-\left(\frac{t}{\eta_2}\right)^{m_2}\right]\right\}, \quad (10.24)$$

where $\lambda_c(t)$ is the failure rate of a connection at time t.

From (10.24), the expected number of repairs $N_f(\tau)$ to a screened connection by time τ (e.g., design life, warranty time) in the field is

$$N_f(\tau) = N_c(\tau + t_c) - N_c(t_c)$$

$$= -\ln\left\{\frac{\rho_1 \exp\left[-\left(\frac{\tau + t_c}{\eta_1}\right)^{m_1}\right] + \rho_2 \exp\left[-\left(\frac{\tau + t_c}{\eta_2}\right)^{m_2}\right]}{\rho_1 \exp\left[-\left(\frac{t_c}{\eta_1}\right)^{m_1}\right] + \rho_2 \exp\left[-\left(\frac{t_c}{\eta_2}\right)^{m_2}\right]}\right\}.$$

$$(10.25)$$

Example 10.3 A printed circuit board is populated with surface-mounted components; the connections are solder joints of the same type, which contain defects due to variation in the wave soldering process. The defective connections need to be precipitated and repaired before being delivered to customers. This is done by subjecting the boards to an accelerated thermal cycling profile for 12 cycles. The acceleration factor between the accelerating profile and the use profile is 31. A preliminary accelerated life test under the accelerating profile showed that the substandard and good solder joints have Weibull distributions and yielded the estimates $\hat{\rho}_1 = 0.04$, $\hat{\rho}_2 = 0.96$, $\hat{m}_1 = 0.63$, $\hat{m}_2 = 2.85$, $\hat{\eta}_1 = 238$ cycles, and $\hat{\eta}_2 = 12{,}537$ cycles. Calculate the expected number of repairs to a solder joint at the end of screening and in the warranty time of two years (equivalent to 1500 cycles). If the boards were not screened, what would be the reliability of the connection and the expected number of repairs to a solder joint at the end of the warranty period?

SOLUTION The equivalent number of cycles under the use profile is $t_c = 31 \times 12 = 372$. Substituting into (10.24) the value of t_c and the given data yields the expected number of repairs to a solder joint as

$$\hat{N}_c(372) = -\ln\left\{0.04 \times \exp\left[-\left(\frac{372}{238}\right)^{0.63}\right]\right.$$

$$\left. + 0.96 \times \exp\left[-\left(\frac{372}{12,537}\right)^{2.85}\right]\right\}$$

$$= 0.0299.$$

From (10.25), the expected number of repairs to a solder joint by the end of warranty time ($\tau = 1500$) is

$$\hat{N}_f(1500) = -\ln$$

$$\left\{\frac{0.04 \times \exp\left[-\left(\dfrac{1500 + 372}{238}\right)^{0.63}\right] + 0.96 \times \exp\left[-\left(\dfrac{1500 + 372}{12,537}\right)^{2.85}\right]}{0.04 \times \exp\left[-\left(\dfrac{372}{238}\right)^{0.63}\right] + 0.96 \times \exp\left[-\left(\dfrac{372}{12,537}\right)^{2.85}\right]}\right\}$$

$$= 0.0143.$$

For an unscreened solder joint, the reliability at the end of warranty time is calculated from (10.23) as

$$\hat{R}_c(1500) = 0.04 \times \exp\left[-\left(\frac{1500}{238}\right)^{0.63}\right] + 0.96 \times \exp\left[-\left(\frac{1500}{12,537}\right)^{2.85}\right]$$

$$= 0.9594.$$

From (10.24), if the board were not screened, the expected number of repairs to a solder joint by the end of warranty time would be

$$\hat{N}_c(1500) = -\ln[R_c(1500)] = -\ln(0.9594) = 0.0414.$$

The benefit from the module-level screening is obviously noted by comparing the values of $\hat{N}_f(1500)$ and $\hat{N}_c(1500)$.

10.7 MODULE RELIABILITY MODELING

It is usually required that the reliability of a screened module at time τ (e.g., design life or warranty time) is greater than a specified value R_0. The reliability of the module depends on the reliability of parts and connections of the module. From (10.8), after the two-level screening, the reliability of a part at time τ can be written as

$$R_p(\tau) = \theta_1 \Pr[y_1(\tau + t_{cp} + t_p) \le G_0] + \theta_2 \Pr[y_2(\tau + t_{cp} + t_p) \le G_0]. \quad (10.26)$$

Now we consider the reliability of a screened connection. From (10.23), the probability p_c of a connection passing the module-level screening is

$$p_c = \rho_1 R_{c1}(t_c) + \rho_2 R_{c2}(t_c) = \rho_1 \exp\left[-\left(\frac{t_c}{\eta_1}\right)^{m_1}\right] + \rho_2 \exp\left[-\left(\frac{t_c}{\eta_2}\right)^{m_2}\right].$$

The probability p_{c1} that a connection surviving the screening is substandard is given by

$$p_{c1} = \frac{\rho_1 R_{c1}(t_c)}{p_c} = \frac{\rho_1}{p_c} \exp\left[-\left(\frac{t_c}{\eta_1}\right)^{m_1}\right].$$

The probability p_{c2} that a connection surviving the screening is good is given by

$$p_{c2} = \frac{\rho_2 R_{c2}(t_c)}{p_c} = \frac{\rho_2}{p_c} \exp\left[-\left(\frac{t_c}{\eta_2}\right)^{m_2}\right].$$

Note that $p_{c1} + p_{c2} = 1$.

The reliability of a connection from the screened population at time τ is

$$R_c(\tau) = p_{c1} R_{c1}(\tau|t_c) + p_{c2} R_{c2}(\tau|t_c)$$

$$= p_{c1} \exp\left[-\left(\frac{\tau + t_c}{\eta_1}\right)^{m_1}\right] \Big/ \exp\left[-\left(\frac{t_c}{\eta_1}\right)^{m_1}\right]$$

$$+ p_{c2} \exp\left[-\left(\frac{\tau + t_c}{\eta_2}\right)^{m_2}\right] \Big/ \exp\left[-\left(\frac{t_c}{\eta_2}\right)^{m_2}\right]. \quad (10.27)$$

Suppose that a module ceases to function when any part or connection fails; that is, the parts and connections are in series. As presented in Chapter 4, if the parts and connections are independent of each other, the reliability of the module can be written as

$$R_m(\tau) = \prod_{i=1}^{n_P} [R_{pi}(\tau)]^{L_i} \prod_{j=1}^{n_C} [R_{cj}(\tau)]^{K_j}, \quad (10.28)$$

where

$$n_P = \text{number of types of parts,}$$

$$n_C = \text{number of types of connections,}$$

$$L_i = \text{number of parts of type } i,$$

$$K_j = \text{number of connections of type } j,$$

$$R_m(\tau) = \text{reliability of the module at time } \tau,$$

$$R_{pi}(\tau) = \text{reliability of a part of type } i \text{ at time } \tau,$$

$$R_{cj}(\tau) = \text{reliability of a connection of type } j \text{ at time } \tau.$$

TABLE 10.3 Component and Connection Information

Component or Connection	Type	Number	Reliability
Resistor			
10 kΩ	1	3	0.9995
390 Ω	2	1	0.9998
27 kΩ	3	2	0.9991
Capacitor	4	2	0.9986
LED	5	1	0.9995
Transistor	6	1	0.9961
SM connection	1	16	0.9999
PTH connection	2	7	0.9998

$R_{pi}(\tau)$ and $R_{cj}(\tau)$ are calculated from (10.26) and (10.27), respectively.

Example 10.4 An electronic module consists of six types of components and two types of connections. Table 10.3 summarizes information about components and connections of each type, where SM stands for surface-mounted, PTH for plated-through-hole, and LED for light-emitting diode. The reliabilities of the screened components and connections at 10,000 hours are shown in Table 10.3. Calculate the module reliability.

SOLUTION   From the data given, we have $n_P = 6, L_1 = 3, L_2 = 1, L_3 = 2,$ $L_4 = 2, L_5 = 1, L_6 = 1, n_C = 2, K_1 = 16,$ and $K_2 = 7.$ The module reliability at 10,000 hours is computed from (10.28) as

$$R_m(10,000) = \prod_{i=1}^{6} [R_{pi}(10,000)]^{L_i} \prod_{j=1}^{2} [R_{cj}(10,000)]^{K_j}$$

$$= 0.9995^3 \times 0.9998 \times 0.9991^2 \times 0.9986^2 \times 0.9995 \times 0.9961$$

$$\times 0.9999^{16} \times 0.9998^7$$

$$= 0.9864.$$

10.8   COST MODELING

As described earlier, parts are screened and assembled into modules. Then the modules are screened to precipitate connection defects. So the screening is a two-level screening. Screening at each level precipitates defects but incurs costs as well. A screen plan should be designed to minimize the relevant segment of the life cycle cost. This segment consists of the in-house screen cost and the field repair cost. In-house screen cost increases with screen duration, whereas field repair cost decreases with screen duration before it reaches a certain point in time. Since screen stresses degrade good parts, field repair cost turns to increase at a certain screen time. The critical time depends on the degradation rates of both defective and good parts, the fraction of substandard subpopulation, and the

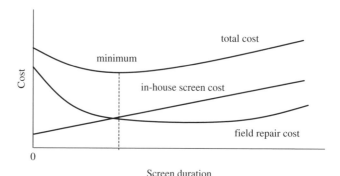

FIGURE 10.5   Costs as a function of screen duration

screen stress levels. Figure 10.5 shows in-house screen cost, field repair cost, and total cost as a function of screen duration. The in-house screen cost includes the part- and module-level screen costs, and the field repair cost involves the part replacement cost and connection repair cost.

The in-house screen and field repair costs incurred by parts consist of the following elements:

1. Cost of screen setup
2. Cost of screen for a specified duration
3. Cost of good parts being screened out
4. Cost of repair at the module-level screen and in the field

The part-cost model can be written as

$$\text{TP} = C_{sp} + \sum_{i=1}^{n_P} C_{pi} L_i t'_{pi} + \sum_{i=1}^{n_P} (C_{gpi} + C_{pi} t'_{pi}) L_i \alpha_2 \Pr[y_2(t_{pi}) \geq G_i^*]$$
$$+ \sum_{i=1}^{n_P} C_{phi} L_i N_{pi}(t_{cpi}) + \sum_{i=1}^{n_P} C_{pfi} L_i [N_{pi}(\tau + t_{cpi}) - N_{pi}(t_{cpi})],$$

$$(10.29)$$

where i denotes a type i part, and

$\quad$ TP $=$ total cost incurred by parts,

$\quad C_{sp}$ $=$ part-level screen setup cost,

$\quad C_{pi}$ $=$ screen cost per unit time (e.g., hour, cycle) for a part of type i,

$\quad C_{gpi}$ $=$ cost of a part of type i,

$\quad C_{phi}$ $=$ in-house repair cost due to the failure of a part of type i,

$\quad C_{pfi}$ $=$ field repair cost due to the failure of a part of type i.

Other notation is as given earlier.

Similarly, the in-house screen and field repair costs incurred by connections are comprised of cost elements 1, 2, and 4 given above. Then the connection-cost model can be written as

$$\text{TC} = C_{sc} + C_c t_c' + \sum_{j=1}^{n_C} C_{chj} K_j N_c(t_{cj}) + \sum_{j=1}^{n_C} C_{cfj} K_j [N_c(\tau + t_{cj}) - N_c(t_{cj})],$$
(10.30)

where j denotes a type j connection, and

$\quad$ TC $=$ total cost incurred by connections,

$\quad C_{sc} =$ module-level screen setup cost,

$\quad C_c =$ screen cost per unit time (e.g., hour, cycle) for a module,

$\quad C_{chj} =$ in-house repair cost due to the failure of a connection of type j,

$\quad C_{cfj} =$ field repair cost due to the failure of a connection of type j.

Other notation is as given earlier.

The total cost associated with the screen and repair is the sum of TP and TC:

$$\text{TM} = \text{TP} + \text{TC},$$
(10.31)

where TM is the total cost incurred by a module (parts and connection) due to the screen and repair. It represents an important segment of the life cycle cost of the module.

10.9 OPTIMAL SCREEN PLANS

As described earlier, screen plans are characterized by the screening technique, stress type, stress levels, screen duration, and defect criteria, among which the first three variables need to be prespecified. The remaining variables, including the screen durations for parts and module, as well as the thresholds of part performance characteristics, should be optimized. The optimization criterion is to minimize the total cost given in (10.31) subject to the constraint on module reliability. The optimization model can be formulated as

$$\text{Min(TM)},$$
(10.32a)

subject to

$$R_m(\tau) \geq R_0,$$
(10.32b)

$$G_i^* \leq G_{0i},$$
(10.32c)

$$G_i^* \geq y_{ai},$$
(10.32d)

$$t_{pi}', t_c', G_i^* \geq 0,$$
(10.32e)

where t'_c, t'_{pi}, and G^*_i ($i = 1, 2, \ldots, n_P$) are decision variables and y_{ai} is the minimum allowable threshold for a part of type i. Constraint (10.32c) is imposed to accelerate the screening process and to reduce the damage to good parts; (10.32d) is required for some parts whose degradation is not stable until y_{ai} is reached. The implications of other constraints are straightforward.

The actual number of decision variables in (10.32) depends on n_P, which may be large for a medium-scale module. In these situations, it is important to lump all similar parts and reduce the size of n_P. For example, the three types of resistors in Example 10.4 may be grouped into one type because their reliabilities and cost factors are close and the degradation thresholds (defined as the resistance drift percentage) are the same. Calculation of the optimization model can be accomplished using a nonlinear programming technique such as the Lagrangian approach and the penalization method. Bertsekas (1996), for example, provides a good description of the approaches.

Example 10.5 A printed circuit board (PCB), defined as a module here, is populated with eight components of the same type. The PCB has 16 solder joints for connections. It is required that the PCB has 90% reliability after five years (43,800 hours) of continuous use. The PCB fails if the performance characteristic y is greater than 100 for any component, or any solder joint disconnects. Suppose that y has a bimodal normal distribution, and

1. $\mu_{y_1}(t) = 10 + 0.01t$, $\mu_{y_2}(t) = 10 + 0.0012t$, $\sigma_{y_1} = \sigma_{y_2} = 6$, $y_a = 15$.
2. $m_1 = 0.5$, $\eta_1 = 5000$ hours; $m_2 = 2$, $\eta_2 = 10^7$ hours.
3. $C_{gp} = \$10$, $C_p = \$0.1$, $C_{sp} = \$100$, $C_{ph} = \$200$, $C_{pf} = \$2000$; $C_c = \$5$, $C_{sc} = \$200$, $C_{ch} = \$200$, $C_{cf} = \$2000$.
4. $\alpha_1 = 1\%$, $\alpha_2 = 99\%$; $\rho_1 = 1\%$, $\rho_2 = 99\%$.
5. $A_p = 50$, $A_c = 50$, $A_{cp} = 10$.

Choose optimal values of t'_p, t'_c, and G^* that minimize TM and meet the reliability requirement.

SOLUTION The optimization model for the problem is calculated using the penalization method and yields $t'_p = 51$ hours, $t'_c = 13.4$ hours, $G^* = 23.1$, and TM = \$631.68.

Now let's discuss the significance and implication of the optimal screen plan. First, reducing the threshold from the usual one ($G_0 = 100$) to the optimal value ($G^* = 23.1$) lowers the life cycle cost. To show this, Figure 10.6 plots TM for various values of G^*. TM is calculated by choosing optimal t'_p and t'_c at a given G^*. The minimum TM (\$631.68) is achieved at $G^* = 23.1$. If the usual threshold ($G_0 = 100$) were used in screening, the TM would be \$750. The saving due to use of the optimal G^* is $(750 - 631.68)/750 = 15.8\%$.

The optimal G^* also alleviates the aging effect of screen stress on good parts. Figure 10.7 shows the mean values μ_{y_2} of y_2 immediately after the module-level screening for various values of G^*. The μ_{y_2} decreases with G^*, indicating that

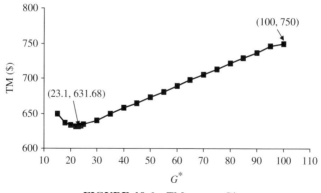

FIGURE 10.6 TM versus G^*

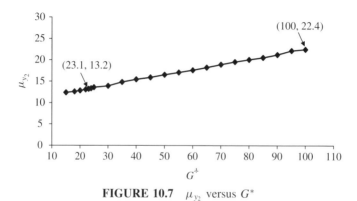

FIGURE 10.7 μ_{y_2} versus G^*

the degradation of a good part caused by the screen stress can be mitigated by use of a smaller G^*. If the usual threshold ($G_0 = 100$) were used in screening, μ_{y_2} would be 22.4. Use of the optimal tightened threshold ($G^* = 23.1$) reduces the degradation by $(22.4 - 13.2)/22.4 = 41.1\%$.

Now let's look at how the value of t'_p affects the cost elements. Figure 10.8 plots the following costs for various values of t'_p:

- *Cost 1*: cost of part repair at module-level screen and in the field
- *Cost 2*: cost of good parts being screened out
- *Cost 3*: part-level screen setup cost plus the cost of screen for t'_p
- TP

Cost 1 sharply decreases with the increase in t'_p before it reaches 51 hours. However, cost 1 increases with t'_p as it goes beyond 300 hours because excessive screen appreciably degrades good parts. This differs from the classical cost models, which ignore the aging effects of the screen stress on good parts. Cost 2

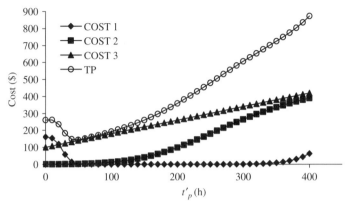

FIGURE 10.8 Costs versus t'_p

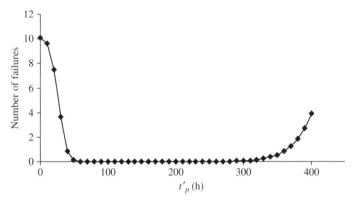

FIGURE 10.9 Number of field failures versus t'_p

increases with t'_p because the degradation of good parts increases the probability of screening out good parts. TP has an optimum value which achieves the best compromise among costs 1, 2, and 3.

The expected number of field failures per 1000 parts by the end of $\tau = 43,800$ hours is plotted in Figure 10.9 for various values of t'_p. As t'_p increases, the number of field failures decreases. When t'_p reaches 51 hours, the number of field failures begins to remain nearly constant. But as t'_p increases further beyond 300 hours, the number of field failures increases considerably, due to the degradation of good parts caused by the screen stress.

PROBLEMS

10.1 Describe the purposes of screening and the consequences of insufficient and excessive screening.

10.2 Explain the advantages and disadvantages of the commonly used screening techniques, including burn-in, ESS, HASS, discriminator screening, and degradation screening.

10.3 For a product that is said to have failed when its monotonically decreasing performance characteristic crosses a specified threshold, formulate and depict the relationship between the bimodal distributions of the life and characteristic.

10.4 A degradation screening requires products to be aged at an elevated stress level for a certain length of time. Explain why.

10.5. A type of part whose failure is defined in terms of $y \leq G_0$, is subjected to degradation screening at an elevated stress level for a length of time t_p. Develop formulas for calculating the following:

   (a) The probability of a part, substandard or good, passing the screen.
   (b) The probability that a part passing the screen is from the substandard subpopulation.
   (c) The probability that a part passing the screen is from the good subpopulation.
   (d) The reliability of a part from the screened population.
   (e) The pdf of a part from the screened population.

10.6 Refer to Problem 10.5. Suppose that the performance characteristic y can be modeled with the lognormal distribution. Calculate parts (a) through (e).

10.7 Revisit Example 10.1.

   (a) Explain why, after screening, the mean life of the defective components is negative.
   (b) Work out the pdf of the components before screening.
   (c) Calculate the pdf of the components after screening.
   (d) Plot on the same chart the pdfs of the components before and after screening. Comment on the shape of the pdf curves.

10.8 An electronic component is said to have failed if its performance characteristic exceeds 85. The component population contains 8% defective units and is subjected to degradation screening for 110 hours at an elevated stress level. The acceleration factor between the screen and use stress levels is 18. A unit is considered defective and is weeded out if the performance reaches 25 at the end of screening. Suppose that the performance is modeled using the normal distribution and that at the use stress level the degradation models are $\mu_{y_1} = 4.8 + 0.021t$, $\sigma_{y_1} = 5.5$, $\mu_{y_2} = 4.8 + 0.0018t$, and $\sigma_{y_2} = 3.7$.

   (a) Determine the equivalent screen time at the use stress level.
   (b) Calculate the probability of a defective component escaping the screen.
   (c) Compute the probability of a defect-free component surviving the screen.

(d) Estimate the reliability of a defective component at the end of screening.

(e) Calculate the probability of a component passing the screen.

(f) Compute the mean life and standard deviation of the defective components after screening.

(g) Calculate the mean life and standard deviation of the good components after screening.

(h) Work out the cdf of a component from the screened population.

(i) Work out the cdf of a component from an unscreened population.

(j) Plot the cdf's calculated in parts (h) and (i).

(k) Comment on the screen time.

10.9 Refer to Problem 10.8. The screened electronic components are installed in a system. Once the component fails, it is replaced with a new one from the screened population. Using a renewal process, calculate the expected number of renewals in 50,000 hours.

10.10 Refer to Example 10.2. The expected number of renewals for the electronic components is plotted in Figure 10.4, which shows a plateau between 35,000 and 45,000 hours. Explain why. Are plateaus expected to recur? If so, when is the next one?

10.11 Explain the concept of minimal repair and the reason why this repair strategy may be appropriate for connections.

10.12 A module contains both degradation parts and binary parts. Suppose that a two-level screening is conducted and that the defective and good binary parts form a mixed Weibull distribution.

(a) Develop the cdf and pdf of a binary part from the screened part population.

(b) Compute the expected number of renewals for a binary part.

(c) Calculate the reliability of the module.

(d) Work out the part-cost model for the degradation and binary parts.

(e) Write down the total cost model.

10.13 Refer to Problem 10.12. If the screen plan is to minimize the total cost and meet the reliability requirement simultaneously, write down the optimization model. What are the decision variables?

10.14 Refer to Problem 10.8. Six of the screened electronic components are assembled into a module. Containing 14 solder joints, the module is said to have failed if any solder joint disconnects or any component has a degradation exceeding 85. The module is subjected to module-level screening, and the reliability is required to be greater than 92% at 20,000 hours. Suppose that the life of the solder joints has a mixed Weibull distribution, and that $m_1 = 0.32$, $\eta_1 = 1080$ hours, $m_2 = 2.25$, $\eta_2 = 1.3 \times 10^7$ hours, $C_{gp} = \$12$, $C_p = \$0.15$, $C_{sp} = \$85$, $C_{ph} = \$185$, $C_{pf} = \$1230$; C_c

$= \$5.5$, $C_{sc} = \$175$, $C_{ch} = \$210$, $C_{cf} = \$2000$, $\rho_1 = 1\%$, $\rho_2 = 99\%$, A_c $= 35$, and $A_{cp} = 10$.

(a) Determine the optimal screen duration for module-level screening.

(b) Calculate the part-level screen cost.

(c) Compute the reliability of the module at 20,000 hours.

(d) Calculate the total cost.

(e) The part-level screen plan described in Problem 10.8 is not optimal and needs revision. Develop an optimal two-level screen plan, and redo parts (a) through (d).

(f) Compare the results from part (e) with those from parts (a) through (d). Comment on the differences.

10.15 Refer to Problem 10.14. Develop the optimal two-level screen plan by maximizing the module reliability, with the total cost not to exceed 1.8 times the minimum cost obtained in Problem 10.14(e).

11

WARRANTY ANALYSIS

11.1 INTRODUCTION

In the context of the product life cycle, warranty analysis is performed in the field deployment phase. In the earlier phases, including product planning, design and development, verification and validation, and production, a product team should have accomplished various well-orchestrated reliability tasks to achieve the reliability requirements in a cost-effective manner. However, it does not mean that the products would not fail in the field. In fact, some products would fail sooner than others for various reasons, such as improper operation, production process variation, and inadequate design. The failures not only incur costs to customers but often result in reputation and potential sales losses to manufacturers. Facing intense global competition, today most manufacturers offer warranty packages to customers to gain competitive advantage. The role of warranty in marketing a product is described in, for example, Murthy and Blischke (2005). In the marketplace, lengthy warranty coverage has become a bright sales point for many commercial products, especially for those that may incur high repair costs. Furthermore, it is often employed by manufacturers as a weapon to crack a new market. A recent example is that of South Korean automobiles, which entered North American markets with an unprecedented warranty plan covering the powertrain system for five years or 50,000 miles. This contrasts with the three-year, 36,000-mile plan offered by most domestic automakers. In addition to "voluntary" offers, government agencies may also mandate extended warranty coverage for certain products whose failures can result in severe consequences,

Life Cycle Reliability Engineering, by Guangbin Yang
Copyright © 2007 John Wiley & Sons, Inc.

such as permanent damage to the environment and loss of life. For instance, U.S. federal regulations require that automobile catalytic converters be warranted for eight years or 80,000 miles, since failure of the subsystem increases toxic emissions to the environment. In short, warranty offers have been popular in modern times, so warranty analysis has become increasingly important.

When products fail under warranty coverage, customers return their products for repair or replacement. The failure data, such as the failure time, failure mode, and use condition, are made known to manufacturers. Often, manufacturers maintain warranty databases to record and track these data. Such data contain precious and credible information about how well products perform in the field, and thus should be fully analyzed to serve different purposes. In general, warranty analyses are performed to:

- Determine monetary reserves for warranty.
- Estimate the number of warranty repairs or replacements.
- Estimate field reliability.
- Detect critical failure modes that would prompt product recalls.
- Identify unusual failure modes and failure probability to improve the product design and manufacturing process.
- Evaluate the effectiveness of fixes implemented in design, production, or warranty service.
- Determine whether it is necessary to buy back certain products that have generated numerous warranty claims.

In this chapter we present techniques for warranty analysis. In particular, we describe basic warranty policies, data mining strategies, reliability estimation from warranty data, warranty repair modeling and projection, field failure monitoring, and warranty cost reduction. For repairable systems, one may be interested in estimating mean cumulative function and repair rate from warranty data. Interested readers may consult Nelson (2003), and G. Yang et al. (2005), for example.

11.2 WARRANTY POLICIES

Webster's College Dictionary (Neufeldt and Guralnik, 1997) defines *warranty* as a seller's assurance to the purchaser that the goods or property is as represented and, if not, will be replaced or repaired. In the context of a product, a *seller* may be the manufacturer of the product or the dealer or retailer who sells the product. A *buyer* is a customer who pays for the product. The seller's assurance to the buyer can be considered to be a contractual agreement between the two parties and becomes effective upon the sale of the product. Indeed, in most situations, a warranty is a guaranteeing policy which a seller offers to a buyer at the time of sale, and is not subject to negotiation. Customers have no choice but to accept the policy if they decide to buy the product.

A warranty policy must be specific and definite. In particular, a warranty policy must specify (1) the warranty period (calendar time, usage, or other measure), (2) failure coverage, and (3) seller's and buyer's financial responsibility for a warranty repair or replacement. For a complicated product such as an automobile, the failure coverage may be too extensive to be exhausted. In these situations, the failures or parts not warranted are often stated instead. Three examples of warranty policies are given below.

1. *Automobile bumper-to-bumper warranty policy on Ford cars and light trucks.* Vehicle owners are not charged for covered warranty repairs during the bumper-to-bumper warranty period, which begins at the date of original purchase and lasts for 36 months or 36,000 miles, whichever comes first. Certified dealers will repair, replace, or adjust all parts on the vehicle that are defective in factory-supplied materials or workmanship. Exceptions include damage caused by accidents, collisions, fire, explosion, freezing, vandalism, abuse, neglect, improper maintenance, unapproved modifications, and others (truncated here for brevity). The warranty period covers both the time in service and mileage. Known as *two-dimensional coverage*, this is discussed later in detail.

2. *General Tire warranty policy on passenger tires.* This limited warranty coverage is for a maximum period of 72 months from the date of purchase, determined by the new vehicle registration date or new vehicle sales invoice showing the date of purchase. A covered unserviceable tire will be replaced with a comparable new tire according to the free replacement policy and the pro-rata replacement policy. The free replacement policy lasts 12 months or first $\frac{2}{32}$ of an inch (whichever comes first). After the free replacement policy expires, tire owners will receive a replacement pro-rata credit (excluding all applicable taxes) toward the purchase of a comparable new tire equal to the percentage of tread depth remaining down to the treadwear indicators ($\frac{2}{32}$ inch of tread remaining). The tire tread is worn out at this point and the pro-rata replacement policy ends regardless of the time period. The warranty policies do not cover failures caused by road hazard, improper operation or maintenance, intentional alternation, and others (truncated here for brevity).

3. *General Electric (GE) Company warranty policy on self-cleaning electric coil ranges.* For a period of one year from the date of the original purchase, GE will provide any part of the range that fails due to a defect in materials or workmanship. During this full one-year warranty, GE will also provide, free of charge, all labor and in-home service to replace the defective part. The warranty policy does not cover failures caused by improper installation, delivery, or maintenance, abuse, misuse, accident, fire, floods, acts of God, and others (truncated here for brevity).

As explained above, a warranty policy is comprised of three elements: warranty period, failure coverage, and seller's and buyer's financial responsibility for warranty service. Ideally, the warranty period should be expressed in terms of the time scale that describes the underlying failure process. For example, the warranty on the corrosion of automobile parts typically covers five years and unlimited mileage, because corrosion is closely related to calendar time, not to mileage. In general, the time scale may be the calendar time, usage (e.g., mileage or cycles), or others. For many products used intermittently, the failure process is often more closely related to usage than to calendar time. Thus, usage should serve as one of the scales for defining a warranty period. However, due to the difficulty in tracking usage for warranty purposes, calendar time is often employed instead. Washing machines, for instance, are warranted for a period of time and not for the cycles of use, although most failures result from use. When accumulated use is traceable, it is often used in conjunction with calendar time. A common example is the automobile bumper-to-bumper warranty policy described above, which specifies both the calendar time in service and mileage.

Among the three elements of a warranty policy, the warranty period is probably most influential on warranty costs. Lengthy warranty coverage erodes a large portion of revenues and deeply shrinks profit margins; however, it increases customer satisfaction and potential sales. Manufacturers often determine an optimal period by considering the effects of various factors, including, for example, product reliability, cost per repair, sales volume, unit price, legal requirements, and market competition. If the failure of a product can result in a substantial loss to society, the manufacturer would be impotent in making a warranty decision. Instead, governmental regulations usually mandate an extended warranty period for the product. Another important element of a warranty policy is failure coverage. Normally, a warranty covers all failures due to defective materials or workmanship. However, damage caused by conditions other than normal use, such as accident, abuse, or improper maintenance, are usually excluded. Failure coverage is often an industry standard; individual sellers would not like to override it. In contrast, sellers have greater room to manipulate the seller's and buyer's financial responsibility for warranty services. This results in different warranty policies. The most common ones are as follows:

1. *Free replacement policy.* When a product fails within the warranty period and failure coverage, it is repaired or replaced by the seller free of charge to the buyer. For a failure to be eligible for the policy, it must meet the warranty period and failure coverage requirements. Under this policy, the seller has to pay all costs incurred by the warranty service, including fees for materials, labor, tax, disposal, and others. Because of the substantial expenditure, sellers usually limit this policy to a short warranty length unless a longer period is stipulated by regulations.

2. *Pro-rata replacement policy.* When a product fails within the warranty period and failure coverage, it is repaired or replaced by the seller at a fraction of the repair or replacement cost to the buyer. The cost to the buyer is proportional to the age of the product at failure. The longer the product has been used, the more the buyer has to pay; this is reasonable. The cost is a function of the age in relative to the warranty length. Under this policy, customers may be responsible for tax and service charges. Let's consider the tire example given earlier. If a

tire is blown out in 15 months after the original purchase and the remaining tread depth is $\frac{6}{32}$ inch at failure, it is subject to the pro-rata replacement policy. Suppose that a comparable new tire has a tread depth of $\frac{11}{32}$ inch, and sells at $70. Then the cost to the customer is $(11/32 - 6/32)/(9/32) \times 70 = \$38.89$ plus applicable tax, where $\frac{9}{32}$ inch is the usable life.

3. *Combination free and pro-rata replacement policy.* This policy specifies two warranty periods, say t_1 and t_0 $(t_1 < t_0)$. If a product fails before t_1 expires and the failure is covered, it is repaired or replaced by the seller free of charge to the buyer. When a failure occurs in the interval between t_1 and t_0 and is under the failure coverage, the product is repaired or replaced by the seller at a fraction of the repair or replacement cost to the buyer.

The warranty policies described above are *nonrenewing*; that is, the repair or replacement of a failed product does not renew the warranty period. The repaired or replaced product assumes the remaining length of the original warranty period. The warranty policies for repairable products are often nonrenewing. In contrast, under *renewing policies*, the repaired or replaced products begin with a new warranty period. The policies cover mostly nonrepairable products.

A warranty period may be expressed in two dimensions. For most commercial products, the two dimensions represent calendar time and use. As soon as one of the two dimensions reaches its warranty limit, the warranty expires, regardless of the magnitude of the other dimension. If a product is operated heavily, the warranty will expire well before the warranty time limit is reached. On the other hand, if a product is subjected to light use, the warranty will expire well before the use reaches the warranty usage limit. The two-dimensional warranty policy greatly reduces the seller's warranty expenditure and conveys the costs to customers. This policy is depicted in Figure 11.1, where t and u are, respectively, the calendar time and usage, and the subscript 0 implies a warranty limit. Figure 11.1 shows that the failures occurring inside the window are covered and those outside are not. Let's revisit the automobile and tire warranty examples given earlier. The automobile bumper-to-bumper warranty is a two-dimensional (time in service and mileage) free replacement policy, where the warranty time and mileage limits are 36 months and 36,000 miles. The General Tire warranty on passenger tires is a two-dimensional combination free and pro-rata replacement policy, where the warranty periods t_1 and t_0 are two-dimensional vectors

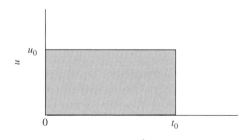

FIGURE 11.1 Two-dimensional warranty coverage

with t_1 equal to 12 months and first $\frac{2}{32}$ inch, and t_0 equal to 72 months and $\frac{2}{32}$ inch of tread remaining.

11.3 WARRANTY DATA MINING

11.3.1 Warranty Data

Products failed under warranty coverage are repaired or replaced by the manufacturer or its authorized service providers. When failures are claimed, information about the failed products is disclosed to the manufacturer. Such information is precious and credible and should be analyzed thoroughly to support business and engineering decision making. As such, most manufacturers maintain warranty databases to store the warranty data. In general, warranty data contain three pieces of information, which are as described here.

Product Data   The data often include product serial number, production date, plant identification, sales date, sales region, price, accumulated use, warranty repair history, and others. Some of these data may be read directly from the failed products, whereas others need to be extracted from serial numbers. The data may be analyzed for different purposes. For example, the data are useful in identification of unusual failure patterns in certain production lots, evaluation of relationship between field reliability and sales region (use environment), study of customer use, and determination of the time from production to sales. Manufacturers often utilize product data, along with failure data and repair data (discussed below), to perform buyback analysis, which supports a decision as to whether it is profitable for manufacturers to buy back from customers certain products that have generated numerous warranty claims.

Failure Data   When a failure is claimed, the repair service provider should record the data associated with the failure, such as the customer complaint symptoms, use conditions at failure, and accumulated use. After the failure is fixed, the diagnosis findings, failure modes, failed part numbers, causes, and postfix test results must be documented. It is worth noting that the failure modes observed by repair technicians usually are not the same as the customer complaint symptoms, since customers often lack product knowledge and express what they observed in nontechnical terms. However, the symptom description is helpful in diagnosing and isolating a failure correctly and efficiently.

Repair Data   Such data should contain the labor time and cost, part numbers serviced, costs of parts replaced, technician work identification and affiliation, date of repair, and others. In warranty repair practice, an unfailed part close to a failure state may be adjusted, repaired, or replaced. Thus, it is possible that the parts serviced may outnumber the parts that failed. The repair data should be analyzed on a regular basis to track warranty spending, identify the opportunity for improving warranty repair procedures, and increase customer satisfaction. In addition, manufacturers often utilize the data to estimate cost per repair, warranty cost per unit, and total warranty cost.

11.3.2 Warranty Data Mining Strategy

Data mining, sometimes called *knowledge discovery*, is a computer-assisted process of searching and analyzing enormous amount of data and then extracting the meaning of the data. Data mining uses a variety of tools, including statistical analysis, decision tree, neural net, principal component and factor analysis, and many other techniques. Readers interested in detailed description of data mining definitions, methodologies, tools, and applications may consult Han and Kamber (2000), and Ye (2003). In warranty data mining, we confine ourselves to a discussion on retrieving warranty data from databases, analyzing the data with statistical tools, and making business and engineering decisions based on the analysis results. In this subsection we focus on strategies for searching the warranty data necessary for subsequent analyses; in the sections to follow we deal with the other two tasks.

Manufacturers usually maintain warranty databases to store product data, failure data, and repair data of all warranty claims made against their products. The claim data need to be kept for a long period of time, ranging from years to decades, for financial, engineering, legal, or other purposes. Time is not the only factor that complicates and swells the database. A manufacturer often makes various products, each comprising numerous parts, and each part may fail in different modes and mechanisms. Failed products may be claimed at a variety of locations and result in a wide range of repair costs. A database capable of accommodating these variables and sorting data by one or more of these variables is complicated in nature. Searching such a database for desired data requires an effective and efficient approach as well as careful implementation. Although the approach depends on a specific product and database, the following generic strategy is useful. The strategy consists of four steps.

1. *Define the objective of the warranty data analysis.* The objective includes, but is not limited to, determination of monetary reserves for warranty, projection of warranty repairs or replacements to the end of warranty period, estimation of field reliability, identification of critical failure modes, manufacturing process improvement, and evaluation of fix effectiveness. This step is critical, because the type of data to be retrieved vary with the objective. For example, the estimation of field reliability uses first failure data, whereas the warranty repair projection includes repeat repairs.

2. *Determine the data scope.* A warranty database usually contains three categories of data, including the product data, failure data, and repair data, as described earlier. In this step, one should clearly define what specific warranty data in each category are needed to achieve the objective. For example, if the objective is to evaluate the effectiveness of a part design change, the products must be grouped into two subpopulations, one before and one after the time when the design change is implemented in production, for the purpose of comparison. The grouping may be done by specifying the production dates of the two subpopulations.

3. *Create data search filters and launch the search.* In this step, one has to interrogate the warranty database by creating search filters. In the context of warranty database, a *filter* is the characteristic of a product, failure, or repair. The filters are established such that only the data defined in the data scope are extracted from the database. Upon establishment of the filters, a data search may be initiated. The time a search takes can vary considerably, depending on the size of the database, the complexity of the filters, and the speed of the computers.

4. *Format the data representation.* When data search is completed, one may download the data sets and orchestrate them in a format with which subsequent data analyses are efficient. Some comprehensive databases are equipped with basic statistical tools to generate graphical charts, descriptive statistics, probability plots, and others. A preliminary analysis using such tools is good preparation for a more advanced study.

Example 11.1   Automobiles are installed with an on-board diagnostic (OBD) system to monitor the failure of the emission-related exhaust gas recirculation (EGR) system. When a failure occurs, the OBD system should detect the failure, illuminate a light (e.g., "Service Engine Soon") on the instrument panel cluster to alert the driver to the need for repair, and store the diagnostic trouble codes corresponding to the failure on the powertrain control module. When no failure occurs, the OBD system should not perform these functions. However, due to noise disturbance, the OBD system sometimes malfunctions and commits α and β error. The α error, a false alarm, is measured by the probability that the diagnostic system detects a failure given that no failure occurred. The β error is measured by the probability that the diagnostic system fails to detect a failure given that a failure has occurred. For a detailed description of the system, see Section 5.12. According to G. Yang and Zaghati (2004), the reliability $R(t)$ of an OBD system can be written as

$$R(t) = 1 - F_\alpha(t) - F_\beta(t), \tag{11.1}$$

where $F_\alpha(t)$ is the probability that no failure occurs and the OBD system detects a failure, and $F_\beta(t)$ is the probability that a failure occurs and the OBD system does not detect the failure. The objective of warranty data analysis is to estimate the reliability of the OBD system installed in Ford Motor Company's vehicle A of model year B. Determine the data mining strategy.

SOLUTION   To calculate $R(t)$ using (11.1), we first need to work out $F_\alpha(t)$ and $F_\beta(t)$ from the warranty data. The probabilities are estimated from the times to first failure of the vehicles that generate the α and β errors. The life data can be obtained by searching Ford's warranty database, called the Analytic Warranty System. The data search strategy is as follows.

1. By definition, the warranty claims for α error are those that show a trouble light and result in no part repair or replacement. Hence, the filters for retrieving such claims specify: Part Quantity = 0, Material Cost = 0,

Vehicle Line = Vehicle A, Model Year = B, Customer Concern Code = Light On, and OBD System = EGR.

2. The warranty claims for β error are those that result in EGR part repair or replacement but lack a trouble light. Thus, the filters are: Part Number = all EGR parts, Vehicle Line = Vehicle A, Model Year = B, Customer Concern Code $\neq$ Light On.

3. In regard to an automobile warranty, life is often measured by month to failure. The life data extracted from the database are grouped by month to failure and month in service of vehicles (see Table 11.5, for example). Such a data arrangement is convenient for the calculation of $F_\alpha(t)$ and $F_\beta(t)$. It is worth noting that the life data are right-censored interval data (Chapter 7).

11.3.3 Limitations of Warranty Data

Warranty data contain credible information about how well products function in the real world. Warranty data are more realistic than laboratory data and advantageous in various aspects. Nevertheless, warranty data are sometimes criticized as being "dirty" because of the following deficiencies:

1. A population of products may work in a wide variety of use conditions. For example, automobiles may be operated in hot and dry, hot and humid, or cold and dry environments. Driving habits vary from vehicle to vehicle and affect vehicle reliability. The warranty data from different use conditions often are not differentiable. The intermixture results in biased estimates of reliability, warranty cost, and other quantities of interest.

2. Product reliability of different lots may vary considerably due to production process variation, especially when the process occasionally gets out of control. It is not unusual to see that the newly launched products have higher failure rates than those made later. In many situations, a gradual or abrupt change in the production process is unknown to warranty analysts, and thus the failure data of different lots are mixed and treated from a homogeneous population for subsequent analyses.

3. When a soft failure (due to excessive degradation) occurs, a customer may not bring in the product immediately for warranty repair unless the warranty is about to expire. Frequently, the customer makes a warranty claim at the earliest convenience or when the failure can no longer be tolerated. A delay prolongs the time to failure artificially and biases warranty analysis. Rai and Singh (2004) consider this type of delay. Another type of delay often happens between the time a claim is made and the time it is entered into the database being used for analysis. Such delays result from the process of warranty claim reporting and ratification. This case is especially true when the manufacturer is not the warranty service provider. In this case, the count of recent claims underestimates the number of claims actually made. Kalbfleisch et al. (1991) discuss this type of delay.

4. The warranty claiming process is strongly influenced by customers' subjectivity. During the warranty period customers often tend to be picky about a

product and do not tolerate much performance degradation. Warranty claims are frequently made against products that have degraded significantly but have not failed technically. In today's tough business climate, many such products will be repaired or replaced to increase customer satisfaction. Even if there is neither repair nor replacement, premature claims still incur diagnostic costs. Hence, such claims result in pessimistic estimates of reliability, warranty cost, and other quantities of interest.

5. The population of products in service decreases with time, and the number of reductions is often unknown to manufacturers. In warranty analysis, the sales volume is assumed to be the working population and clearly overestimates the number of units actually in service. For example, automakers do not have accurate knowledge of the number of vehicles that are under warranty and have been salvaged due to devastating accidents. Such vehicles are still counted in many warranty analyses.

11.4 RELIABILITY ESTIMATION FROM WARRANTY CLAIM TIMES

Although laboratory tests including design verification and process validation tests have confirmed that products meet specified reliability requirements, manufacturers are often interested in reliability performance in the real world. Reliability estimation from warranty data provides realistic answers. However, due to the limitations of warranty data, the estimation must be based on some assumptions. As we understand, warranty data usually contain repeat repairs; that is, a product can be repaired more than once for the same problem during the warranty period. In reliability analysis, we consider only first failures. Subsequent failures, if any, have to be ignored. In addition, the times to first failure are assumed to be identically and independently distributed. Because of the warranty data limitations described earlier, this assumption may be partially violated. If it is strongly believed that this is the case, it is vital to take actions to mitigate the effects of limitations. For example, if a population is heterogeneous, we should break it into multiple homogeneous subpopulations and analyze each subpopulation separately.

11.4.1 Warranty Data Structure

In practice, products are sold continually, and their times in service are unequal. Manufacturers often track the quantity of units that are sold in any given period and the quantity of units from that period that are returned for warranty repair in subsequent time periods. The period may be a week, a month, a quarter, or other lengths. A monthly period is probably the most common measure; for example, automobile manufacturers use the month in service. For simplicity, products sold in the same time period are considered to have the same time in service, and products that fail in the same period are said to have the same lifetime. Apparently, the failure time of a product cannot exceed its time in service. Let n_i be the number of products sold in time period i and r_{ij} be the number of failures occurring in time period j to the units sold in time period i, where $i = 1, 2, \ldots, k$, $j = 1, 2, \ldots, k$, and k is the maximum time in service. In the automotive industry, k is often referred to as the *maturity*. Note that products sold in the first time period have the maximum time in service. In general, the data can be tabulated as in Table 11.1, where TTF stands for time to failure and TIS for time in service, and $r_{i.} = \sum_{j=1}^{i} r_{ij}$, $r_{.j} = \sum_{i=j}^{k} r_{ij}$, and $r_{..} = \sum_{i=1}^{k} r_{i.} = \sum_{j=1}^{k} r_{.j}$. Here $r_{i.}$ is the total number of failures among n_i units, $r_{.j}$ the total number of failures in j periods, and $r_{..}$ the total number of failures among all products sold. Since the failure time of a product is less than or equal to its time in service, the failure data are populated diagonally in Table 11.1.

11.4.2 Reliability Estimation

The data in Table 11.1 are multiply right-censored data (Chapter 7). In particular, $n_1 - r_{1.}$ units are censored at the end of period 1, $n_2 - r_{2.}$ units are censored at the end of period 2, and so on. For the sake of data analysis, the life data are rearranged and shown in Table 11.2.

Commercial software packages, such as Minitab and Weibull++ of Reliasoft, can perform life data analysis. The analysis usually starts with probability plotting to select an appropriate life distribution. Then the maximum likelihood method

TABLE 11.1 Warranty Data Structure

		TTF					Sales
TIS	1	2	3	...	k	Total	Volume
1	r_{11}					$r_{1.}$	n_1
2	r_{21}	r_{22}				$r_{2.}$	n_2
3	r_{31}	r_{32}	r_{33}			$r_{3.}$	n_3
⋮	⋮	⋮	⋮			⋮	⋮
k	r_{k1}	r_{k2}	r_{k3}	...	r_{kk}	$r_{k.}$	n_k
Total	$r_{.1}$	$r_{.2}$	$r_{.3}$	...	$r_{.k}$	$r_{..}$	

TABLE 11.2 Multiply Right-Censored Data

TTF	Number of Failures	Number of Censored Units
1	$r_{.1}$	$n_1 - r_{1.}$
2	$r_{.2}$	$n_2 - r_{2.}$
3	$r_{.3}$	$n_3 - r_{3.}$
⋮	⋮	⋮
k	$r_{.k}$	$n_k - r_{k.}$

is applied to estimate distribution parameters and other quantities of interest. The analysis was described in detail in Chapter 7 and is illustrated in the following example.

Example 11.2 A manufacturer sold 22,167 units of a washing machine during the past 13 consecutive months. The product is warranted for 12 months in service under a free replacement policy. The first failure data and sales volumes are shown in Table 11.3, where TTF and TIS are in months. Note that any failures that occur after the warranty expires arc unknown to the manufacturer and are not included in Table 11.3. Estimate the probability of failure at the end of the warranty time and the number of upcoming warranty claims.

SOLUTION   Because the products are warranted for 12 months, the failure data for the 765 units in service for 13 months are available only up to 12 months. These units are treated as having 12 months in service and are combined with the 1358 units having 12 months in service to calculate the number of survivals. The life data are shown in Table 11.4.

The life data in Table 11.4 were analyzed using Minitab. Graphical analysis indicates that the lognormal distribution fits the data adequately. Figure 11.2 shows the lognormal probability plot, least squares fits, and the two-sided 90% percentile confidence intervals. Further analysis using the maximum likelihood method yields estimates of the scale and shape parameters as $\hat{\mu} = 6.03$ and $\hat{\sigma} = 1.63$. The estimate of the probability of failure at the end of warranty time is

$$\hat{F}(12) = \Phi\left[\frac{\ln(12) - 6.03}{1.63}\right] = 0.0148.$$

TABLE 11.3   Warranty Data of Washing Machines

| TIS | \multicolumn{12}{c}{TTF} | Total | Sales Volume |
	1	2	3	4	5	6	7	8	9	10	11	12		
1	0												0	568
2	0	1											1	638
3	0	1	1										2	823
4	0	0	1	1									2	1,231
5	0	0	1	1	0								2	1,863
6	1	0	1	0	1	2							5	2,037
7	1	1	3	2	1	4	8						20	2,788
8	2	3	2	6	2	1	6	4					26	2,953
9	1	2	0	3	4	2	2	3	6				23	3,052
10	1	3	2	2	3	4	3	5	4	8			35	2,238
11	0	1	2	0	4	2	1	2	0	4	3		19	1,853
12	1	0	3	1	3	2	1	3	3	2	1	2	22	1,358
13	2	0	0	2	1	0	0	2	1	0	2	1	11	765
Total	9	12	16	18	19	17	21	19	14	14	6	3	168	22,167

TABLE 11.4 Life Data of Washing Machines

TTF (months)	Number of Failures	Number of Censored Units
1	9	568
2	12	637
3	16	821
4	18	1229
5	19	1861
6	17	2032
7	21	2768
8	19	2927
9	14	3029
10	14	2203
11	6	1834
12	3	2090

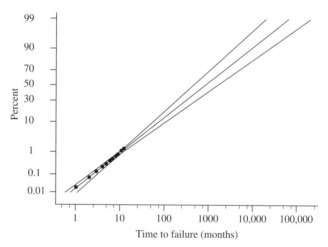

FIGURE 11.2 Lognormal plot, least squares fits, and 90% confidence intervals for the washing machine data

The number of units that would fail by the end of the warranty period is $0.0148 \times 22,167 = 328$. Since 168 units have failed up to the current month, there would be an additional $328 - 168 = 160$ warranty claims.

11.5 TWO-DIMENSIONAL RELIABILITY ESTIMATION

As explained in Section 11.2, some products are subject to two-dimensional warranty policies, which cover a specified period of time and amount of use (whichever comes first). The common example is automobiles, which are typically warranted for 36 months in service and 36,000 miles (whichever occurs sooner) in the United States. The failure of such products is time and usage dependent; in other words, the reliability of products is a function of time and use. Modeling two-dimensional reliability provides more realistic estimates. Such models are needed by manufacturers to evaluate reliability, to predict warranty claims and costs, and to assess customer satisfaction.

In this section we describe a practical approach to modeling and estimating two-dimensional reliability from warranty data. More statistical methods for this topic are given in, for example, Blischke and Murthy (1994, 1996), Lawless et al. (1995), Eliashberg et al. (1997), S. Yang et al. (2000), H. Kim and Rao (2000), G. Yang and Zaghati (2002), and Jung and Bai (2006).

11.5.1 Two-Dimensional Probabilities of Failure

Products accumulate use at different rates in the field and may fail at any combination of usage and time. Figure 11.3 shows the usage accumulation processes of four failed products on a time–usage plane. For usage m and time t (e.g., the warranty limits), m and t partition the plane into four regions I, II, III, and IV, as shown in Figure 11.3. Region I embraces failures of usage less than m and time less than t. If a product at failure has a usage greater than m and a time less than t, the failure occurs in region II. Region III contains failures at usage of less than m and time greater than t. If a product survives both m and t, the failure occurs in region IV.

The probability of failure at usage m and time t is given by

$$F_{M,T}(m, t) = \Pr(M \le m, T \le t) = \int_0^t \int_0^m f_{M,T}(m, t)\, dm\, dt, \qquad (11.2)$$

where M denotes the usage to failure, T the time to failure, and $f_{M,T}(m, t)$ the joint probability density function (pdf) of M and T.

The reliability is the probability that a product survives both usage m and time t, and can be written as

$$R(m, t) = \Pr(M \ge m, T \ge t) = \int_t^\infty \int_m^\infty f_{M,T}(m, t)\, dm\, dt. \qquad (11.3)$$

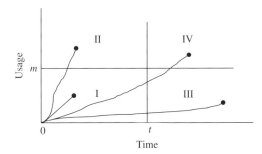

FIGURE 11.3 Time–usage plane partitioned into four regions

Note that this two-dimensional reliability is not the complement of the probability of failure. This is because failures may occur in regions II and III. The probability of failure in region II is

$$\Pr(M \geq m, T \leq t) = \int_0^t \int_m^\infty f_{M,T}(m, t) \, dm \, dt. \tag{11.4}$$

The probability of failure in region III is

$$\Pr(M \leq m, T \geq t) = \int_t^\infty \int_0^m f_{M,T}(m, t) \, dm \, dt. \tag{11.5}$$

Apparently, the probabilities of failure in the four regions add to 1. Then the two-dimensional reliability is

$$R(m, t) = 1 - \Pr(M \leq m, T \leq t) - \Pr(M \geq m, T \leq t) - \Pr(M \leq m, T \geq t), \tag{11.6}$$

which can be written as

$$R(m, t) = 1 - F_T(t) - F_M(m) + F_{M,T}(m, t), \tag{11.7}$$

where $F_T(t)$ and $F_M(m)$ are, respectively, the marginal probabilities of failure of T and M, and

$$F_T(t) = \Pr(M \leq m, T \leq t) + \Pr(M > m, T \leq t), \tag{11.8}$$

$$F_M(m) = \Pr(M \leq m, T \leq t) + \Pr(M \leq m, T > t). \tag{11.9}$$

The calculation of probabilities in the four regions requires estimation of the joint pdf. It can be written as

$$f_{M,T}(m, t) = f_{M|T}(m) f_T(t), \tag{11.10}$$

where $f_{M|T}(m)$ is the conditional pdf of M at a given time T and $f_T(t)$ is the marginal pdf of T. In the following subsections we present methods for estimating the two pdf's from warranty data.

11.5.2 Usage Accumulation Modeling

As shown in Figure 11.1, a two-dimensional warranty policy covers a time period t_0 and usage u_0, whichever comes first. This policy implies that a product is not warranted when the time in service is greater than t_0 although the usage does not exceed u_0. Conversely, the warranty expires when the usage surpasses u_0 even though the product age is less than t_0. Since products accumulate usage at different rates, the population usages are usually distributed widely. The fast accumulators are out of warranty coverage due to the usage exceeding u_0. The number of such accumulators increases with time. Figure 11.4 shows the usage

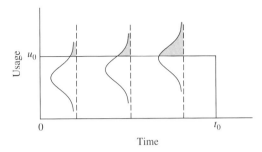

FIGURE 11.4   Usage distributions at various times

distributions at different times in service, where the shaded areas represent the fractions of the population falling outside the warranty usage limit. In the context of life testing, the two-dimensional warranty policy is equivalent to a dual censoring. Such censoring biases the estimation of $f_T(t)$ and $f_{M|T}(m)$ because failures occurring at a usage greater than u_0 are unknown to manufacturers. The bias is exacerbated as T increases toward t_0 and more products exceed the warranty usage limit. In reliability analysis, it is important to correct the bias. This can be accomplished by using a usage accumulation model, which describes the relationship between the usage and time. Here we present two approaches to modeling usage accumulation: the linear accumulation method and the sequential regression analysis.

Linear Accumulation Method   If a product accumulates usage linearly over time, the usage accumulation rate ρ can be written as

$$\rho = \frac{u}{t},\tag{11.11}$$

where u is the usage accumulated up to time t. Often, ρ is constant for the same product and varies from product to product. It may be modeled with the lognormal distribution with scale parameter μ_ρ and shape parameter σ_ρ. When automobiles are concerned, ρ is the mileage accumulation rate. As indicated in Lawless et al. (1995), M. Lu (1998), and Krivtsov and Frankstein (2004), the mileage accumulation rate is nearly constant for the same vehicle and is described adequately by the lognormal distribution.

Equation (11.11) can be written as

$$\ln(u) = \ln(\rho) + \ln(t).$$

Because ρ is modeled using lognormal distribution, $\ln(\rho)$ has a normal distribution with mean μ_ρ and standard deviation σ_ρ. Then u has the lognormal distribution with scale parameter $\mu_\rho + \ln(t)$ and shape parameter σ_ρ. It is clear that the usage distribution at a given time depends on the values of μ_ρ and σ_ρ. Ideally, μ_ρ and σ_ρ should be estimated from carefully planned survey data, because such data are not subject to warranty censoring. However, survey data are expensive to obtain and not available in most situations. As an alternative, we may use recall data to estimate these parameters. In general, recalls usually cover products that are no longer under warranty coverage. So recall data are also free of warranty censoring. If recall size is large (say, 60% or more of the population), the estimates are competent in terms of statistical accuracy. When both survey data and recall data are unavailable, we shall use sequential regression analysis (discussed next) to model the usage accumulation.

Once μ_ρ and σ_ρ are estimated, we can calculate the fraction of the population exceeding the warranty usage limit at a given time. The fraction is given by

$$\Pr(U \geq u_0|t) = 1 - \Phi\left[\frac{\ln(u_0) - \ln(t) - \mu_\rho}{\sigma_\rho}\right], \tag{11.12}$$

where U denotes the usage.

Example 11.3   A sport utility vehicle (SUV) is subject to a bumper-to-bumper warranty covering 36 months or 36,000 miles, whichever comes first. A safety concern about the vehicle prompted a large-scale recall and resulted in the estimates $\hat{\mu}_\rho = 6.85$ and $\hat{\sigma}_\rho = 0.72$. Estimate the fraction of the vehicles that exceed 36,000 miles at 36 months in service.

SOLUTION   Substituting the data into (11.12) yields

$$\Pr(U \geq 36,000|t = 36) = 1 - \Phi\left[\frac{\ln(36,000) - \ln(36) - 6.85}{0.72}\right] = 0.468.$$

This indicates that 46.8% of the vehicles will exceed the warranty mileage limit although the vehicle age is still within the warranty time limit; these vehicles are no longer warranted. To show how the warranty dropout increases over time, Figure 11.5 plots the fractions at various times up to 36 months. It is seen that the fraction may be negligible before 12 months in service, after which the fraction increases rapidly.

Sequential Regression Analysis   The linear accumulation model described above works well when there is a large amount of survey data or recall data. When such data are not obtainable, we may perform sequential regression analysis of the warranty repair data to establish a usage accumulation model.

Manufacturers have failure numbers and sales volume (Table 11.1), the usage to failure of each claimed product and the associated failure modes. The products usually fail in a large variety of failure modes. Each failure mode has a usage to failure distribution. This distribution may be, and often is, different from the usage distribution of a population, because some failure modes tend to occur on high or low usage accumulators. Mixing the usage data of all failure modes greatly mitigates or eliminates the effects of failure occurrence patterns. Therefore, the distribution of usage to failure of all failure modes at a given time approximately estimates the distribution of usage at that time.

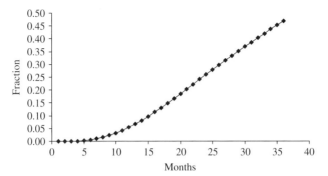

FIGURE 11.5 Fractions of vehicles exceeding warranty mileage limit at various months in service

We use the following assumptions:

1. Product infant mortality ceases by time period t_1. This assumption is necessary for products that suffer significant early failures. It ensures that the usage distribution is not dominated and distorted by the failure modes of early failures.

2. The warranty dropout fraction is negligible up to time period t_2, where $t_2 > t_1$. For example, we may choose $t_2 = 12$ months for the SUV in Example 11.3.

The usage data at time period t ($t = t_1, t_1 + 1, \ldots, t_2$) can be considered free of censoring based on assumption 2. An appropriate distribution is fitted to the usage data at time periods $t_1, t_1 + 1, \ldots, t_2$, respectively. The distribution may be a location-scale distribution (e.g., lognormal, normal, or Weibull). Note that the distribution is conditional on time period t. Let $f_{U|T}(u)$ denote the conditional pdf and $\mu_u(t)$ and $\sigma_u(t)$ denote the location and scale parameters of $f_{U|T}(u)$. Graphical analysis or maximum likelihood estimation of the usage data at each time period yields the estimates $\hat{\mu}_u(t)$ and $\hat{\sigma}_u(t)$. Regression analysis on data sets $[t_1, \hat{\mu}_u(t_1)], [t_1 + 1, \hat{\mu}_u(t_1 + 1)], \ldots, [t_2, \hat{\mu}_u(t_2)]$ establishes the dependence of location parameter on time, which can be written as

$$\mu_u(t) = g_1(\hat{\boldsymbol{\theta}}_1, t), \tag{11.13}$$

where $\boldsymbol{\theta}_1$ is the model parameter vector, $^\wedge$ denotes an estimate, and g_1 represents a function. Similarly, the scale parameter can be expressed as

$$\sigma_u(t) = g_2(\hat{\boldsymbol{\theta}}_2, t), \tag{11.14}$$

where the notation is similar to that in (11.13).

The next step is to use (11.13) and (11.14) to project the location and scale parameters at time period $(t_2 + 1)$, denoted $\mu'_u(t_2 + 1)$ and $\sigma'_u(t_2 + 1)$, respectively. At this time period, some products start to be dropped out due to usage exceeding u_0. The number of products that are dropped out and failed, denoted r'_{t_2+1}, is estimated by

$$r'_{t_2+1} = r_{\cdot(t_2+1)} \frac{1 - p_0}{p_0}, \tag{11.15}$$

where $r_{\cdot(t_2+1)}$ is the number of warranted products that fail at time period $t_2 + 1$, and

$$p_0 = \Pr[U \leq u_0 | \mu'_u(t_2 + 1), \sigma'_u(t_2 + 1)].$$

Then the usage data at time period $t_2 + 1$ can be viewed as the censoring data. The usage data of $r_{\cdot(t_2+1)}$ failed products are known, and the unrecorded r'_{t_2+1} failed units are considered to be right censored at u_0. The same type of distribution selected earlier is fitted to the censoring data, and we obtain the new estimates $\hat{\mu}_u(t_2 + 1)$ and $\hat{\sigma}_u(t_2 + 1)$. The distribution with these new estimates has a better fit than the projected distribution, and thus $\hat{\mu}_u(t_2 + 1)$ and $\hat{\sigma}_u(t_2 + 1)$ are added to the existing data series $\hat{\mu}_u(t)$ and $\hat{\sigma}_u(t)$ ($t = t_1, t_1 + 1, \ldots, t_2$) to update estimates of the regression model parameters θ_1 and θ_2. This projection and regression process repeats until the maximum time period k is reached. Equations (11.13) and (11.14) of the last update constitute the final usage accumulation model and are used to calculate $\Pr(U \geq u_0 | t)$. The projection and regression process is explained pictorially in Figure 11.6.

11.5.3 Estimation of Marginal Life Distribution

The marginal life(time) distribution $f_T(t)$ is estimated using *the hazard plotting method*. It is an alternative to, and sometimes more convenient than, the probability plotting method. This method is based on the one-to-one relationship between the cumulative distribution function (cdf) $F(t)$ and the cumulative hazard function $H(t)$. The relationship can be written as

$$F(t) = 1 - \exp[-H(t)]. \tag{11.16}$$

The cdf of the Weibull distribution with shape parameter β and scale parameter α is

$$F(t) = 1 - \exp\left[-\left(\frac{t}{\alpha}\right)^\beta\right], \qquad t > 0. \tag{11.17}$$

Combining (11.16) and (11.17) gives

$$H(t) = \left(\frac{t}{\alpha}\right)^\beta. \tag{11.18}$$

This can be linearized as

$$\ln[H(t)] = -\beta \ln(\alpha) + \beta \ln(t), \tag{11.19}$$

which indicates that the Weibull cumulative hazard rate is a linear function of time t on a log-log scale. If a data set plotted on this log-log scale is close to a straight line, the life can be modeled with a Weibull distribution. The Weibull

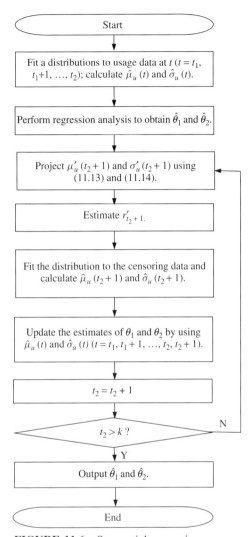

FIGURE 11.6   Sequential regression process

shape parameter equals the slope of the straight line; the scale parameter is calculated from the slope and intercept.

Similarly, the exponential distribution with hazard rate λ has

$$H(t) = \lambda t. \tag{11.20}$$

For a normal distribution with mean μ and standard deviation σ, we have

$$\Phi^{-1}[1 - e^{-H(t)}] = -\frac{\mu}{\sigma} + \frac{1}{\sigma}t. \tag{11.21}$$

A lognormal distribution with scale parameter μ and shape parameter σ has

$$\Phi^{-1}[1 - e^{-H(t)}] = -\frac{\mu}{\sigma} + \frac{1}{\sigma}\ln(t). \tag{11.22}$$

Nelson (1972, 1982) describes hazard plotting in detail.

To utilize the transformed linear relationships (11.19) through (11.22) to estimate the life distribution, one first must calculate the hazard rate. For a continuous nonnegative random variable T representing time to failure, the hazard function $h(t)$ is defined by

$$h(t) = \lim_{\Delta t \to 0} \frac{\Pr(t < T \le t + \Delta t | T > t)}{\Delta t},$$

where Δt is a very small time interval. This equation can be written as

$$h(t) = \lim_{\Delta t \to 0} \frac{N(t) - N(t + \Delta t)}{N(t)\Delta t}, \tag{11.23}$$

where $N(t)$ is the number of surviving products at time t. In warranty analysis, Δt is often considered to be one time period, such as one month, and thus (11.23) can be rewritten as

$$h(t) = \frac{\text{number of failures during time } t \text{ and } t + 1}{\text{number of survivals at time } t}. \tag{11.24}$$

Since the products are subject to two-dimensional censoring, the failures occurring beyond u_0 are not recorded in warranty databases. Thus, the numerator in (11.24) is unknown. For (11.24) to be applicable, we substitute "number of failures during time t and $t + 1$" by "number of first warranty repairs during time t and $t + 1$" and adjust the denominator accordingly, to the risk set, which contains only the surviving products whose usage is less than or equal to u_0 at time t. Then (11.24) can be rewritten as

$$h(t) = \frac{\Delta r(t)}{\Pr(U \le u_0 | t) N(t)}, \tag{11.25}$$

where $\Delta r(t)$ is the number of first warranty repairs during time t and $t + 1$. Applying (11.25) to the warranty data in Table 11.1, we obtain the estimate of the hazard rate in time period j as

$$\hat{h}_j = \frac{r_{\cdot j}}{\Pr(U \le u_0 | j) N_j}, \tag{11.26}$$

where

$$N_j = \sum_{i=j}^{k} \left(n_i - \sum_{l=1}^{j-1} r_{il} \right). \tag{11.27}$$

The cumulative hazard rate up to time period j is estimated by

$$\hat{H}_j = \sum_{i=0}^{j} \hat{h}_i, \tag{11.28}$$

where $\hat{h}_0 = 0$. After calculating $\hat{H}_j$ $(j = 1, 2, \ldots, k)$, we can fit the transformed linear relationships (11.19) through (11.22) to data sets $(j, \hat{H}_j)$. Like probability plotting, the relationship that gives the straightest plot is probably the best distribution. For example, if (11.19) provides the best fit, the Weibull distribution is chosen. More important, the selection should be justified by the physics of failure. The following example illustrates the use of hazard plotting method to estimate the marginal time distribution of an automobile component. Krivtsov and Frankstein (2004) discuss the estimation of marginal mileage distribution of automobile components.

Example 11.4 A mechanical assembly is installed in the sport utility vehicles dealt with in Example 11.3. The vehicles had a maturity of 11 months at the time of data analysis, meaning that the vehicles sold in the first month had 11 months in service, or $k = 11$. The sales volumes and first failure counts of the assembly are summarized in Table 11.5. Estimate the marginal life distribution of the assembly and calculate the reliability at the end of the warranty period (36 months).

SOLUTION The total number of first failures in each month is given in Table 11.5 and repeated in Table 11.6 for the sake of calculation. The number of surviving vehicles at month j is calculated from (11.27). For example, at month 3, the number is

$$N_3 = \sum_{i=3}^{11} \left(n_i - \sum_{l=1}^{2} r_{il} \right) = [20{,}806 - (1+3)] + [18{,}165 - (3+2)] + \cdots$$
$$+ [2868 - (1+0)] = 120{,}986.$$

The numbers of surviving vehicles for $j = 1, 2, \ldots, 11$ are calculated and summarized in Table 11.6.

As shown in Example 11.3, the mileage distribution of the vehicle population at month j is lognormal with scale parameter $\ln(j) + 6.85$ and shape parameter 0.72. Thus, the estimate of the probability that a vehicle's mileage is less than or equal to 36,000 miles at month j is

$$\Pr(U \le 36{,}000 | j) = \Phi \left[\frac{\ln(36{,}000) - \ln(j) - 6.85}{0.72} \right].$$

For example, at month 3 the probability estimate is

$$\Pr(U \le 36{,}000 | 3) = \Phi \left[\frac{\ln(36{,}000) - \ln(3) - 6.85}{0.72} \right] = 0.9998.$$

TABLE 11.5 Mechanical Assembly Warranty Data

TIS (months)	TTF (months)											Total	Sales Volume
	1	2	3	4	5	6	7	8	9	10	11		
1	2											2	12,571
2	2	0										2	13,057
3	1	3	3									7	20,806
4	3	2	5	4								14	18,165
5	3	5	4	3	6							21	16,462
6	1	3	5	3	7	4						23	13,430
7	1	1	3	5	4	3	5					22	16,165
8	2	0	1	2	5	4	5	6				25	15,191
9	0	2	1	2	2	3	5	4	4			23	11,971
10	2	0	3	4	4	5	3	4	2	3		30	5,958
11	1	0	1	1	2	2	1	0	1	0	0	9	2,868
Total	18	16	26	24	30	21	19	14	7	3	0	178	146,645

TABLE 11.6 Estimation of Cumulative Hazard Rates

j	$r_{.j}$	N_j	$\Pr(U \le 36{,}000|j)$	n_j	$\hat{h}_j$	$\hat{H}_j$
1	18	146,645	1.0000	146,645	0.000123	0.000123
2	16	134,058	1.0000	134,055	0.000119	0.000242
3	26	120,986	0.9998	120,961	0.000215	0.000457
4	24	100,161	0.9991	100,074	0.000240	0.000697
5	30	81,986	0.9976	81,791	0.000367	0.001064
6	21	65,516	0.9949	65,181	0.000322	0.001386
7	19	52,087	0.9907	51,604	0.000368	0.001754
8	14	35,926	0.9850	35,386	0.000396	0.002150
9	7	20,745	0.9776	20,279	0.000345	0.002495
10	3	8,790	0.9685	8,513	0.000352	0.002847
11	0	2,859	0.9579	2,739	0.000000	0.002847

The estimates of the probability for $j = 1, 2, \ldots, 11$ are given in Table 11.6.

The estimate of the number of the surviving vehicles whose mileage is less than or equal to 36,000 miles at month j is

$$n_j = \Pr(U \le 36{,}000|j)N_j.$$

For example, at month 3 the estimate of the number of such vehicles is

$$n_3 = \Pr(U \le 36{,}000|3)N_3 = 0.9998 \times 120{,}986 = 120{,}961.$$

$n_j (j = 1, 2, \ldots, 11)$ are calculated and shown in Table 11.6.

The estimate of the hazard rate at month j is calculated from (11.26). For example, the hazard rate at month 3 is estimated as

$$\hat{h}_3 = \frac{26}{120,961} = 0.000215 \text{ failures per month.}$$

The estimates of the hazard rate for $j = 1, 2, \ldots, 11$ are given in Table 11.6.

Then the cumulative hazard rate up to j months is computed using (11.28). For example, the cumulative hazard rate up to three months is estimated as

$$\hat{H}_3 = \sum_{i=0}^{3} \hat{h}_i = 0 + 0.000123 + 0.000119 + 0.000215 = 0.000457.$$

The estimates of the cumulative hazard rate for $j = 1, 2, \ldots, 11$ are computed and summarized in Table 11.6.

To estimate the marginal life distribution, (11.19) through (11.22) are fitted to the data points $(j, \hat{H}_j)$, where $j = 1, 2, \ldots, 11$. Linear regression analysis indicates that the Weibull distribution provides the best fit and yields the estimates $\hat{\beta} = 1.415$ and $\hat{\alpha} = 645.9$ months. The Weibull fit is plotted in Figure 11.7. The pdf of the marginal life distribution is

$$f_T(t) = \frac{t^{0.415}}{6.693 \times 10^3} \exp\left[-\left(\frac{t}{645.9}\right)^{1.415}\right].$$

The pdf will be used in Example 11.6. The reliability of the mechanical assembly at the end of the warranty time (36 months) is estimated as

$$\hat{R}(36) = \exp\left[-\left(\frac{36}{645.9}\right)^{1.415}\right] = 0.9833.$$

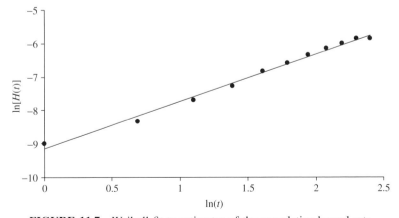

FIGURE 11.7   Weibull fit to estimates of the cumulative hazard rate

11.5.4 Estimation of Conditional Usage-to-Failure Distribution

To utilize (11.10) to calculate the two-dimensional reliability, we need to estimate the conditional pdf $f_{M|T}(m)$. The pdf is determined by the usages to failure of all failed products and has little dependence on the surviving units. In other words, calculation of $f_{M|T}(m)$ should take into account the failed products that exceed the warranty usage limit u_0 and may ignore the survivals.

It is important to differentiate between $f_{M|T}(m)$ and $f_{U|T}(u)$. The former is the conditional pdf of the usage to failure; it describes the conditional life distribution. The latter is the conditional pdf of the usage; it models the distribution of the accumulated usage at a given time. If the failure mode of concern does not tend to occur on high or low usage accumulators, the two conditional pdf's may be the same. To detect failure tendency, we can plot the usages to failure and the mean of usage distribution at various times in service, as shown in Figure 11.8. The maximum time in service in the plot should not be greater than time t_2 before which the warranty dropout fraction is negligible (assumption 2 in Section 11.5.2). If the number of dots above the mean is approximately equal to that below the mean at all time periods, there is some evidence that the failures are independent of usage accumulation (Davis, 1999). If this is the case, we should further test the hypothesis that the usage distribution is equal to the usage-to-failure distribution at each time period. If the hypothesis is not rejected, we have $f_{M|T}(m) = f_{U|T}(u)$. Since $f_{U|T}(u)$ is known, (11.10) is ready for calculating the two-dimensional reliability.

In Figure 11.8, if the numbers of dots above and below the mean usage are apparently unequal, it is evident that the failure is dependent on usage. Then $f_{M|T}(m)$ should be estimated from the usage-to-failure data. In the case where the maximum time period k is less than or equal to t_2, the products exceeding the warranty usage limit are minimal. Then the warranty claim data are approximately complete. Hence, the warranty data can be applied directly to estimate $f_{M|T}(m)$. The calculation yields estimates of the location and scale parameters of the distributions at different time periods. Subsequent regression analysis then establishes the relationships between the location parameter and time, denoted $\mu_m(t)$, and the scale parameter and time, denoted $\sigma_m(t)$.

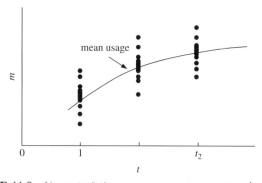

FIGURE 11.8 Usage to failure versus mean usage at various times

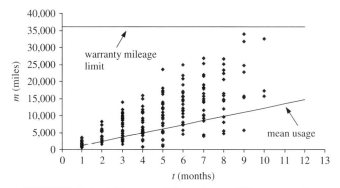

FIGURE 11.9   Mileage-to-failure data at different months

When $k > t_2$, the warranty dropout fraction is no longer negligible, and esti-
mation of $f_{M|T}(m)$ should take into account all failed products, including those
exceeding the warranty usage limit. The estimation is relatively complicated but
can be done by using the sequential regression method described earlier for
modeling usage accumulation. When applying this method, the usage data are
replaced by the usage-to-failure data, and the calculation process remains the
same. The analysis yields $\mu_m(t)$ and $\sigma_m(t)$, in contrast to $\mu_u(t)$ and $\sigma_u(t)$ from
the usage accumulation modeling.

Example 11.5   Refer to Examples 11.3 and 11.4. When the sport utility vehicles
failed and claimed for warranty repair, the mileages and times at failure were
recorded. Figure 11.9 plots the mileages to failure at each month in service up
to 10 months. Table 11.6 shows the number of claims in each month. Estimate
the pdf of the mileage to failure M conditional on the month in service T.

SOLUTION   Because the vehicles had only 11 months in service at the time of
data analysis, the warranty dropout fraction is negligible. It is reasonable to consider that Figure 11.9 includes all failed vehicles. To detect the tendency of failure
occurrence, we plot the mean usage of the vehicle population in Figure 11.9.
Here the mean usage at month t is $\exp[6.85 + \ln(t) + 0.5 \times 0.72^2] = 1223.2t$.
Figure 11.9 shows that the failures tended to occur on high mileage accumula-
tors. Hence, $f_{M|T}(m)$ is not equal to $f_{U|T}(u)$. In this case, $f_{M|T}(m)$ is calculated
directly from the warranty data.

Since there are no warranty repairs in month 11, and only 3 repairs in month
10, the calculation of $f_{M|T}(m)$ uses the data for the first nine months. The
mileage-to-failure data in each of the nine months are adequately fitted by the
Weibull distribution, as shown in Figure 11.10. The Weibull fits approximately
parallel, indicating a common shape parameter to all months. The maximum
likelihood estimates of the Weibull characteristic life $\hat{\alpha}_m$ for the nine months are
calculated and plotted versus time in Figure 11.11. The plot suggests a linear
relationship: $\alpha_m = bt$, where b is a slope. The slope can be estimated using the

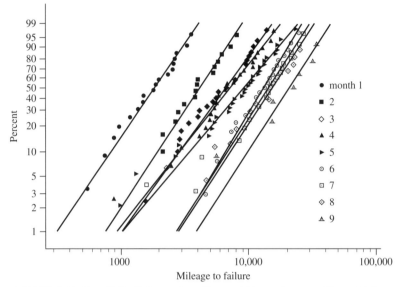

FIGURE 11.10   Weibull fits to the miles to failure of sport utility vehicles

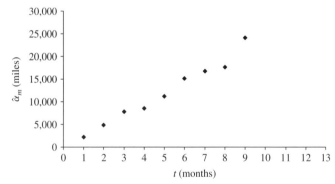

FIGURE 11.11   Estimates of Weibull characteristic life at different months

least squares method. Here the maximum likelihood method is used to get better estimates. From (7.59), the log likelihood function can be written as

$$L(b, \beta) = \sum_{j=1}^{9} \sum_{i=1}^{r_{.j}} \left[\ln(\beta) - \beta \ln(bt_j) + (\beta - 1) \ln(m_{ij}) - \left(\frac{m_{ij}}{bt_j} \right)^{\beta} \right],$$

where β is the common shape parameter, $r_{.j}$ the total number of first failures at month j (see Table 11.6 for values), m_{ij} the ith mileage to failure at month j, and $t_j = j$. Substituting the data into the equation above and then maximizing the log likelihood gives $\hat{\beta} = 2.269$ and $\hat{b} = 2337$. Hence, the conditional pdf of

M at a given month T is

$$f_{M|T}(m) = \frac{m^{1.269}}{1.939 \times 10^7 \times t^{2.269}} \exp\left[-\left(\frac{m}{2337 \times t}\right)^{2.269}\right].$$

This pdf will be used in Example 11.6.

11.5.5 Estimation of Two-Dimensional Reliability and Failure Quantities

After $f_{M|T}(m)$ and $f_T(t)$ are estimated, we may compute the two-dimensional reliability and the probabilities of failure using the formulas given in Section 11.5.1. The calculations involve numerical integrations and need commercial software or small computer programs.

In addition to the reliability and probabilities of failure, failure quantities are often of interest. The number of first failures by the warranty usage limit u_0 and time limit t_0 is

$$N_1(u_0, t_0) = F_{M,T}(u_0, t_0) \sum_{i=1}^{k} n_i, \qquad (11.29)$$

where $F_{M,T}(u_0, t_0)$ is the two-dimensional probability of failure and is calculated from (11.2). These failures occur in region I of Figure 11.3, and the resulting repairs are covered by warranty.

The number of first failures occurring within the warranty time limit t_0 and outside the usage limit u_0 is

$$N_2(u_0, t_0) = \Pr(M > u_0, T \leq t_0) \sum_{i=1}^{k} n_i, \qquad (11.30)$$

where the probability is computed from (11.4). These failures occur in region II of Figure 11.3 and are not reimbursed by the manufacturers.

The number of first failures occurring outside the warranty time limit t_0 and within the usage limit u_0 is

$$N_3(u_0, t_0) = \Pr(M \leq u_0, T > t_0) \sum_{i=1}^{k} n_i, \qquad (11.31)$$

where the probability is calculated from (11.5). These failures fall in region III of Figure 11.3 and are not eligible for warranty coverage.

The number of first failures occurring outside both the warranty usage and time limits (u_0 and t_0) is

$$N_4(u_0, t_0) = \Pr(M > u_0, T > t_0) \sum_{i=1}^{k} n_i, \qquad (11.32)$$

where the probability equals $R(u_0, t_0)$ and is calculated from (11.3). These failures occur in region IV of Figure 11.3, where both the usage and time exceed the warranty limits.

Example 11.6   For the mechanical assembly of the sport utility vehicles dealt with in Examples 11.4 and 11.5, estimate the two-dimensional reliability at the warranty mileage and time limits ($u_0 = 36{,}000$ miles and $t_0 = 36$ months), and calculate $N_1(u_0, t_0)$, $N_2(u_0, t_0)$, $N_3(u_0, t_0)$, and $N_4(u_0, t_0)$.

SOLUTION   For the mechanical assembly, Examples 11.4 and 11.5 have calculated $f_T(t)$ and $f_{M|T}(m)$, respectively. From (11.10) the joint pdf can be written as

$$f_{M,T}(m, t) = 7.706 \times 10^{-12} \times \frac{m^{1.269}}{t^{1.854}}$$

$$\times \exp\left[-\left(\frac{m}{2337 \times t}\right)^{2.269} - \left(\frac{t}{645.9}\right)^{1.415}\right].$$

The two-dimensional reliability at the warranty mileage and time limits is obtained from (11.3) as

$$R(36{,}000, 36) = \int_{36}^{\infty} \int_{36{,}000}^{\infty} f_{M,T}(m, t)\, dm\, dt.$$

Numerical integration of the equation above gives $R(36{,}000, 36) = 0.9809$. From Example 11.4 the total number of vehicles is 146,645. Then the number of vehicles surviving 36 months and 36,000 miles is given by (11.32) as $N_4(36{,}000, 36) = 0.9809 \times 146{,}645 = 143{,}850$.

Similarly, the probability of failure in region I is $F_{M,T}(36{,}000, 36) = 0.00792$. Then the number of first failures under warranty coverage is $N_1(36{,}000, 36) = 0.00792 \times 146{,}645 = 1162$.

The probability of failure in region II is estimated as $\Pr(M > 36{,}000, T \leq 36) = 0.00884$. Then the number of first failures occurring in 36 months and beyond 36,000 miles is $N_2(36{,}000, 36) = 0.00884 \times 146{,}645 = 1297$.

The probability of failure in region III is estimated as $\Pr(M \leq 36{,}000, T > 36) = 0.00342$. This gives the number of first failures occurring beyond 36 months and within 36,000 miles as $N_3(36{,}000, 36) = 0.00342 \times 146{,}645 = 502$.

From (11.8) the estimate of the marginal reliability of T at 36 months is

$$R_T(36) = 1 - F_T(36) = 1 - 0.00792 - 0.00884 = 0.9832.$$

It is approximately equal to that in Example 11.4.

11.6   WARRANTY REPAIR MODELING

In Sections 11.4 and 11.5 we calculated warranty repairs using reliability estimates. The calculations essentially produce approximate results because the reliability estimation utilizes first failures only. The approximation is adequate when second and subsequent failures in the warranty period are negligible. This situation arises when a product in question is highly reliable or the warranty period is relatively short compared to the product mean life. In practice, however, a product may generate multiple failures from the same problem within the warranty period, and thus the approximation underestimates the true warranty repairs. In this section we consider the good-as-new repair, same-as-old repair and the generalized renewal process, and present warranty repair models allowing for the possibility of multiple failures within the warranty period.

11.6.1 Good-as-New Repair

A good-as-new repair returns a failed product to new condition; that is, the failure rate after repair is the same as that when the product was used initially. This type of repair is equivalent to replacement of the faulty product by a new one identical to the original. Such a repair strategy may be appropriate for a simple product when the product is overhauled completely after failure, or for a complex product whose failure is nearly always due to a critical component. If each failure in the warranty period is claimed and repaired, the repair process is an ordinary renewal process. The expected number of renewals $W(t_0)$ within the warranty period t_0 is

$$W(t_0) = F(t_0) + \int_0^{t_0} W(t_0 - x) f(x) \, dx, \qquad (11.33)$$

where $F(t)$ and $f(t)$ are the cdf and pdf of the product, respectively. The cdf and pdf can be estimated from accelerated life test or warranty data. The estimation from warranty data was presented in Section 11.4. Equation (11.33) is the renewal function and was discussed in Chapter 10; it can be solved using the recursive algorithm described by (10.20).

Example 11.7   The manufacturer of a refrigerator offers a five-year free replacement policy covering the sealed refrigeration system, which consists of a compressor, condenser, evaporator, dryer, and others. A laboratory test shows that the life of the compressor can be modeled using the Weibull distribution with shape parameter 1.37 and characteristic life 1228 months. The compressor designers wanted to estimate the expected number of repairs within the warranty period.

SOLUTION   The repair on the compressor often restores the component to as-new condition and thus is considered as a good-as-new repair. The expected number of repairs within the warranty period of five years (60 months) is given by (11.33) as

$$W(60) = 1 - \exp\left[-\left(\frac{60}{1228}\right)^{1.37}\right]$$

$$+ \int_0^{60} W(60 - x) \frac{x^{0.37}}{12458.94} \exp\left[-\left(\frac{x}{1228}\right)^{1.37}\right] dx.$$

Applying (10.20) to the renewal function yields $W(60) = 0.016$.

11.6.2   Same-as-Old Repair

When a product fails, it may be restored to a condition identical to that immediately before failure; that is, the failure rate after repair is the same as that immediately prior to failure. This type of repair is called a *same-as-old repair*, or *minimal repair*, as described in Chapter 10. This repair strategy is usually appropriate for complex products whose failures are not dominated by certain components. Replacing or repairing one or a few components within the products does not appreciably change the failure rate of the products. This is especially true when the components have an exponential distribution, which possesses the memoryless property. For this type of repair, the number of repairs over the warranty period can be modeled with a nonhomogeneous Poisson process (NHPP) with failure intensity function equal to hazard function. Then the expected number of repairs $W(t_0)$ within the warranty period t_0 is

$$W(t_0) = \int_0^{t_0} h(t)\, dt = \ln \frac{1}{1 - F(t_0)}, \qquad (11.34)$$

where $h(t)$ is the hazard rate and $F(t_0)$ is the associated cdf evaluated at t_0. The cdf or hazard rate may be estimated from accelerated life test or warranty data.

If $F(t_0)$ is very small (because of high reliability or a short warranty period), (11.34) can be approximated by

$$W(t_0) \approx F(t_0).$$

Example 11.8   For the washing machines studied in Example 11.2, calculate the expected number of repairs during a warranty period of 12 months for a volume of 22,167 units sold.

SOLUTION   Example 11.2 showed that the life of the washing machines can be adequately fitted with a lognormal distribution with scale parameter 6.03 and shape parameter 1.63, and the probability of failure at the end of warranty period (12 months) is $\hat{F}(12) = 0.0148$. From (11.34), the expected number of repairs within the warranty period is $W(12) = \ln[1/(1 - 0.0148)] = 0.0149$. Note that the value of $W(12)$ is approximately equal to $\hat{F}(12) = 0.0148$, which resulted from the reliability analysis. The expected number of repairs for a volume of 22,167 units is $0.0149 \times 22,167 = 331$.

11.6.3   Generalized Renewal Process

The good-as-new and same-as-old repairs discussed earlier represent the limiting conditions to which a failed product can be restored. In practice, sometimes a repair returns a product to an intermediate condition between these two extremes. Such a repair is often referred to as a *better-than-old-but-worse-than-new repair*. These repair strategies use different modeling processes, as we have seen for the good-as-new and same-as-old cases. Kijima and Sumita (1986) and Kijima (1989) propose a generalized renewal process which treats these repair strategies as special cases. In this subsection we describe briefly the generalized renewal process.

Let V_i and S_i denote, respectively, the virtual age and real age of a product immediately after the ith repair. Here the real age is the elapsed time since a product is put in operation and the virtual age is a fraction of the real age and reflects the condition of a product after a repair. The relationship between the virtual age and real age can be expressed as

$$V_i = qS_i, \tag{11.35}$$

where q is the restoration factor of the ith repair and measures the effectiveness of the repair. If $q = 0$, the virtual age right after the ith repair is zero, meaning that the product is restored to the new condition. Thus, $q = 0$ corresponds to a good-as-new repair. If $q = 1$, the virtual age immediately after the ith repair is equal to the real age, indicating that the product is restored to the same condition as right before failure. Thus, $q = 1$ represents a same-as-old repair. If $0 < q < 1$, the virtual age is between zero and the real age, and thus the repair is a better-than-old-but-worse-than-new repair. In addition, if $q > 1$, the virtual age is greater than the real age. In this case, the product is damaged by the repair to a higher degree than it was right before the respective failure. Such a repair is often called a *worse-than-old repair*.

By using the generalized renewal process, the expected number of repairs $W(t_0)$ within the warranty period t_0 can be written as

$$W(t_0) = \int_0^{t_0} \left[g(\tau|0) + \int_0^\tau w(x)g(\tau - x|x)\,dx \right] d\tau, \tag{11.36}$$

where

$$g(t|x) = \frac{f(t + qx)}{1 - F(qx)}, \quad t, x \geq 0; \qquad w(x) = \frac{dW(x)}{dx};$$

and $f(\cdot)$ and $F(\cdot)$ are the pdf and cdf of the time to first failure distribution. Note that $g(t|0) = f(t)$. Equation (11.36) contains distribution parameters and q, which must be estimated in order to evaluate $W(t_0)$. Kaminskiy and Krivtsov (1998) provide a nonlinear least squares technique for estimating the parameters and a Monte Carlo simulation method for calculating (11.36). Yanez et al. (2002) and Mettas and Zhao (2005) give the maximum likelihood estimates. Kaminskiy and Krivtsov (2000) present an application of the generalized renewal process to warranty repair prediction.

11.7 WARRANTY COST ESTIMATION

In additional to warranty repair counts, manufacturers are often interested in warranty costs. The costs depend on the number of warranty repairs as well as the warranty policy. In this section we estimate the costs for different nonrenewing warranty policies, including free replacement, pro-rata replacement, and combination free and pro-rata replacement.

11.7.1 Warranty Cost Under a Free Replacement Policy

A free replacement policy requires a failure under warranty coverage to be repaired at no cost to the buyer. The manufacturer or seller is responsible for all costs incurred, which include fees for materials, labor, disposal, and others. In general, the cost per repair for the same cause varies from repair to repair. To simplify the calculation, we use the average cost per repair, denoted c_0. Then the expected warranty cost per unit C_w is

$$C_w = c_0 W(t_0), \tag{11.37}$$

where $W(t_0)$ can be calculated from (11.33), (11.34), or (11.36), depending on the repair strategy.

Example 11.9 For the washing machines studied in Examples 11.2 and 11.7, calculate the expected warranty cost for a volume sold of 22,167 units. The average cost per repair is \$155.

SOLUTION   In Example 11.8 we obtained the expected number of repairs per unit as $W(12) = 0.0149$. Then from (11.37), the expected warranty cost per unit is $C_w = 155 \times 0.0149 = \$2.31$. The expected warranty cost for a volume sold of 22,167 units is $2.31 \times 22,167 = \$51,206$.

11.7.2 Warranty Cost Under a Pro-Rata Replacement Policy

When a product is subject to the pro-rata replacement policy and fails within the warranty period, it is repaired or replaced by the manufacturer at a fraction of the repair or replacement cost to the customer. Although the pro-rata replacement policy may conceptually cover repairable products, most products assured by this warranty policy are nonrepairable. For nonrepairable products, to which the following discussion is confined, a warranty repair is actually the replacement of a failed product. Suppose that the purchase price of the product is c_p and the pro-rata value is a linear function of age. Then the warranty cost per unit to the manufacturer can be written as

$$C_w(t) = \begin{cases} c_p \left(1 - \dfrac{t}{t_0} \right), & 0 \le t \le t_0, \\ 0, & t > t_0, \end{cases} \tag{11.38}$$

where t is the life of the product and t_0 is the warranty period. $C_w(t)$ is a function of life and thus is a random variable. Since $C_w(t) = 0$ when $t > t_0$, the expected cost is

$$E[C_w(t)] = \int_0^\infty C_w(t)\, dF(t) = c_p \left[F(t_0) - \frac{\mu(t_0)}{t_0} \right], \tag{11.39}$$

where $F(t)$ is the cdf of the product life and $\mu(t_0)$ is the partial expectation of life over the warranty period, given by

$$\mu(t_0) = \int_0^{t_0} t\, dF(t). \tag{11.40}$$

Example 11.10 The manufacturer of a type of projector lamp offers a pro-rata replacement warranty policy for 12 months from the time of initial purchase. The lamps are sold at \$393 a unit. The lamp life is adequately modeled with a lognormal distribution with scale parameter 8.16 and shape parameter 1.08. A real-world usage study indicates that 95% of the customers use the lamps less than 120 hours a month. Calculate the expected warranty cost per unit for the lamp.

SOLUTION To simplify calculation, we assume that all customers use the lamps 120 hours a month. Then the warranty period of 12 months is equivalent to 1440 hours of continuous use. We have $t_0 = 1440$, $c_p = 393$,

$$F(1440) = \Phi\left[\frac{\ln(1440) - 8.16}{1.08}\right] = 0.2056,$$

$$\mu(1440) = 0.3694 \times \int_0^{1440} \exp\left\{-\frac{[\ln(t) - 8.16]^2}{2.3328}\right\} dt = 179.23.$$

Substituting the data above into (11.39) yields

$$E[C_w(t)] = 393 \times \left(0.2056 - \frac{179.23}{1440}\right) = \$31.89.$$

That is, the manufacturer will pay an expected warranty cost of \$31.89 for each unit it sells. Note that the cost is an approximation because of the variation in customer usage rate.

11.7.3 Warranty Cost Under a Combination Free and Pro-Rata Replacement Policy

As described previously, the combination free and pro-rata replacement policy specifies two warranty periods, denoted t_1 and t_0 ($t_1 < t_0$). If a failure eligible for warranty occurs before t_1 expires, the product is repaired or replaced by the manufacturer free of charge to the customer. When a failure occurs in the interval between t_1 and t_0, the product is repaired or replaced by the manufacturer at a fraction of the repair or replacement cost to the customer. Like the pro-rata replacement policy, the combination policy is usually offered to nonrepairable products. Thus, in this subsection we deal with the nonrepairable case only. Suppose that the purchase price of the product is c_p and the proration is a linear

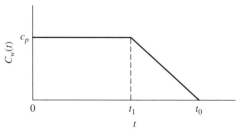

FIGURE 11.12   Warranty cost of a unit under a combination policy

function of age once the pro-rata policy is invoked. Then the warranty cost per unit to the manufacturer can be written as

$$
C_w(t) = \begin{cases} c_p, & 0 \le t \le t_1, \\ \dfrac{c_p(t_0 - t)}{t_0 - t_1}, & t_1 < t \le t_0, \\ 0, & t > t_0, \end{cases} \tag{11.41}
$$

where t is the life of the product. $C_w(t)$ is shown graphically in Figure 11.12.

$C_w(t)$ is a function of life and thus is a random variable. The expectation of the cost is given by

$$
E[C_w(t)] = \int_0^\infty C_w(t)\,dF(t) = \int_0^{t_1} c_p\,dF(t) + \int_{t_1}^{t_0} \frac{c_p(t_0 - t)}{t_0 - t_1}\,dF(t)
$$

$$
= \frac{c_p}{t_0 - t_1}[t_0 F(t_0) - t_1 F(t_1) - \mu(t_0) + \mu(t_1)], \tag{11.42}
$$

where $F(t)$ is the cdf of the product life and $\mu(t_0)$ and $\mu(t_1)$ are the partial expectations of life over t_0 and t_1 as defined in (11.40).

Example 11.11   A type of passenger car battery is sold at \$125 a unit with a combination warranty policy, which offers free replacement in the first 18 months after initial purchase, followed by a linear proration for additional 65 months. The life of the battery is modeled by a Weibull distribution with shape parameter 1.71 and characteristic life 235 months. Estimate the expected warranty cost per unit to the manufacturer of the battery.

SOLUTION   From the data given we have $c_p = 125, t_1 = 18, t_0 = 18 + 65 = 83$,

$$
F(18) = 1 - \exp\left[-\left(\frac{18}{235}\right)^{1.71}\right] = 0.0123, \qquad F(83) = 0.1552,
$$

$$
\mu(18) = \int_0^{18} \frac{t^{1.71}}{6630.25} \exp\left[-\left(\frac{t}{235}\right)^{1.71}\right] dt = 0.1393, \qquad \mu(83) = 7.9744.
$$

Substituting the data above into (11.42) gives

$$E[C_w(t)] = \frac{125}{83 - 18}(83 \times 0.1552 - 18 \times 0.0123 - 7.9744 + 0.1393) = \$9.28.$$

That is, the manufacturer will pay an expected warranty cost of \$9.28 per unit.

11.8 FIELD FAILURE MONITORING

In modern manufacturing, various on-line process control techniques, such as statistical process control (SPC) charts, have been implemented to maintain production processes under control. Such techniques detect process shifts by examining product quality characteristics and are especially effective when patent defects are present. However, process variation may induce latent defects to final products. This type of defect does not appreciably degrade the quality characteristics and thus cannot be discovered during production. Instead, the defective products, if not screened out, are shipped to customers and will fail in a short time in the field. It is vital for a manufacturer to detect such reliability problems and rectify the production process at the earliest time. This may be assisted by monitoring early failures through an analysis of warranty data. In this section we present a simple approach to monitoring field failures by using SPC charts. H. Wu and Meeker (2002) describe a more sophisticated statistical method for early detection of reliability problems from warranty data.

Suppose that n_i units of a product are manufactured in production period i, where $i = 1, 2, \ldots, k$, and $i = 1$ and $i = k$ represent, respectively, the most recent and oldest production times. Often we are interested only in monitoring products under ongoing production. Then the most recent products in the field have aged only one time period. Let r_{i1} denote the number of failures in the first time period among the products made in production period i. Then the probability of failure at the first time period for the products of production period i can be estimated by

$$\hat{p}_i = \frac{r_{i1}}{n_i}. \tag{11.43}$$

This probability is employed as the quality characteristic for generating the SPC charts. Other reliability measures, such as the failure rate at a time of interest (e.g., warranty period) may be used. However, their estimates involve statistical modeling and extrapolation and thus may include large errors.

Suppose that the process is under control over k production periods and the true probability of failure is p. This probability is estimated by the average of the probabilities of failure over k production periods. The estimate is

$$\overline{p} = \frac{\sum_{i=1}^{k} r_{i1}}{\sum_{i=1}^{k} n_i}. \tag{11.44}$$

The occurrence of failures at the first time period can be considered to have a binomial distribution. Thus, a p-chart is appropriate for control of the probability.

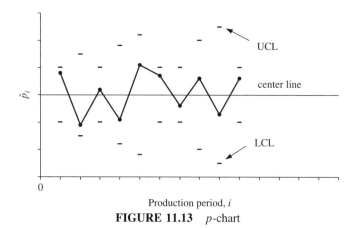

FIGURE 11.13 p-chart

Montgomery (2001a) provides a good description of a p-chart and other control charts. The p-chart is defined by

$$\text{LCL} = \bar{p} - 3\sqrt{\frac{\bar{p}(1-\bar{p})}{n_i}},$$

$$\text{centerline} = \bar{p}, \qquad\qquad (11.45)$$

$$\text{UCL} = \bar{p} + 3\sqrt{\frac{\bar{p}(1-\bar{p})}{n_i}},$$

where UCL stands for *upper control limit* and LCL for *lower control limit*. Figure 11.13 illustrates the concept of the control chart. It is worth noting that the control limits are variable and depend on the volume of each production period. The control limits become constant and form two straight lines when the production volumes are equal.

The actual operation of the control chart consists of computing the probability of failure $\hat{p}_i$ from (11.43) and the corresponding control limits from (11.45) for subsequent production periods, and plotting $\hat{p}_i$ and the control limits on the chart. As long as $\hat{p}_i$ remains within the control limits and the sequence of the plotted points does not display any systematic nonrandom behavior, we can conclude that the infant mortality does not change significantly and the production process is under control. If $\hat{p}_i$ stays outside the control limits, or if the plotted points develop a nonrandom trend, we can conclude that infant mortality has drifted significantly and the process is out of control. In the latter case, investigation should be initiated to determine the assignable causes.

Example 11.12 The manufacturer of a new type of electrical heater wants to detect the unusual infant mortality by monitoring the probability of failure in the first month in service. The heaters have been in production for 10 months. The production volume and the number of failures in the first month in service for each

TABLE 11.7   Data for the Heater Control Chart

Production				With Month 10 Data		Without Month 10 Data	
Month	n_i	r_{i1}	$\hat{p}_i$	LCL	UCL	LCL	UCL
1	9,636	36	0.00374	0.00173	0.00537	0.00160	0.00514
2	9,903	32	0.00323	0.00176	0.00534	0.00162	0.00512
3	10,231	43	0.00420	0.00179	0.00531	0.00165	0.00509
4	13,267	59	0.00445	0.00200	0.00510	0.00186	0.00488
5	23,631	88	0.00372	0.00239	0.00471	0.00224	0.00450
6	30,136	87	0.00289	0.00252	0.00458	0.00237	0.00437
7	32,666	118	0.00361	0.00256	0.00454	0.00241	0.00433
8	23,672	63	0.00266	0.00239	0.00471	0.00224	0.00450
9	20,362	59	0.00290	0.00230	0.00480	0.00215	0.00459
10	18,342	96	0.00523	0.00223	0.00487		

production month are shown in Table 11.7. Develop a p-chart for controlling the probability of failure. In production month 11, 10,325 heaters were manufactured and 58 units failed in the first month. Determine whether the production process was under control in month 11.

SOLUTION   From (11.44), the average probability of failure in the first month in service over 10 months of production is

$$\overline{p} = \frac{36 + 32 + \cdots + 96}{9636 + 9903 + \cdots + 18,342} = 0.00355.$$

Thus, the centerline of the p-chart is 0.00355.

The control limits are calculated from (11.45). For example, the LCL and UCL for production month 1 are

$$LCL = 0.00355 - 3\sqrt{0.00355 \times (1 - 0.00355)/9636} = 0.00173,$$

$$UCL = 0.00355 + 3\sqrt{0.00355 \times (1 - 0.00355)/9636} = 0.00537.$$

The control limits for the 10 production months are calculated in a similar way and summarized in the "With Month 10 Data" columns of Table 11.7.

The probability of failure in the first month in service for each production month is computed using (11.43). For example, for production month 1, we have $\hat{p}_1 = 36/9636 = 0.00374$. The estimates of the probability of failure for the 10 production months are shown in Table 11.7.

The control limits, centerline, and $\hat{p}_i$ ($i = 1, 2, \ldots, 10$) are plotted in Figure 11.14. It is seen that $\hat{p}_{10}$ exceeds the corresponding UCL, indicating that the process was out of control in month 10. As such, we should exclude the month 10 data and revise the control chart accordingly. The new centerline is $\overline{p} = 0.00337$. The control limits are recalculated and shown in the "Without Month 10 Data" columns of Table 11.7. The control chart is plotted in Figure 11.15. On this chart, the control limits for month 10 are LCL = 0.00209 and UCL = 0.00466.

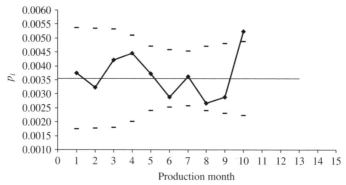

FIGURE 11.14   Control chart including month 10 data

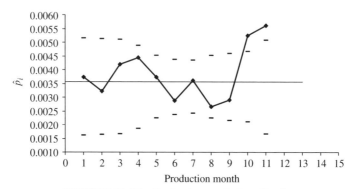

FIGURE 11.15   Revised control chart for $\hat{p}_i$

The probability of failure for month 11 is estimated as $\hat{p}_{11} = 58/10,325 = 0.00562$. The control limits for $\hat{p}_{11}$ are LCL = 0.00166 and UCL = 0.00508. The probability estimates $\hat{p}_{10}$ and $\hat{p}_{11}$ and their control limits are plotted on the revised control chart, as shown in Figure 11.15. It is apparent that the infant mortality started increasing in production month 10, and the trend continued to month 11. Therefore, an investigation into the process should be initiated to determine the assignable cause. This must be followed by appropriate corrective actions.

11.9   WARRANTY COST REDUCTION

Manufacturers now offer more generous warranty packages than ever before in response to intense global competition. The generosity may be helpful in increasing market shares. However, the gain in revenue is often greatly compromised by the pain in warranty expenditure. As a matter of fact, nearly all manufacturers suffer drastic warranty burdens. For example, in recent years, the automotive industry spent some $15 billion annually in the United States alone to reimburse warranty repairs. To maintain and increase profitability, most manufacturers have been taking vigorously action to reduce warranty costs. In general, the action includes design reliability improvement, process variation reduction, stress screening, preventive maintenance, and repair strategy optimization.

Design reliability improvement is the fundamental and most effective approach to warranty cost reduction. This proactive approach allows high reliability to be built into products in the design and development phase and reduces warranty costs by decreasing the number of failures. More important, improved reliability directly boosts customer satisfaction and market shares. As emphasized in this book, design reliability improvement relies on the effective implementation of comprehensive reliability programs, which should start as early as in the product planning stage. In subsequent phases, including product design and development and design verification and process validation, reliability tasks are integrated into, and serve as the essential components of, engineering projects. For example, the robust reliability design technique should be used as an engineering tool to determine optimal levels of the performance characteristics of a product. Such integration enables engineers to do things correctly from the beginning and eliminate any substantive mistake before it propagates. In Chapter 3 we described the development and implementation of effective reliability programs. Reliability programs should be executed to the maximum extent throughout the design cycle. Unfortunately, reliability tasks are sometimes perceived mistakenly as luxury exercises and thus are compromised when time and resources are restricted. The author has occasionally heard the excuse that a design verification test had to use a sample size much smaller than the statistically valid sample because of insufficient test units. There is no doubt that a savings in test units will be offset by an order of magnitude as a result of the number of warranty claims.

Process variation reduction is to minimize production process variation by reducing common causes and preventing the occurrence of special causes. It is an effective approach to the detection and prevention of defects and to improved robustness. Defective products fail in an unexpectedly short time, usually within the warranty period. Thus, the elimination of defects decreases warranty costs directly. Furthermore, improved robustness contributes to an increase in long-term reliability as well as warranty savings.

Because of production process variation and material flaws, latent defects may be built into some products. Such products constitute a substandard population, as described in Chapter 10. If not removed, the defects will manifest themselves as early failures in the field, and thus inevitably incur spending for warranties. Defective products can be reduced or eliminated by stress screening before being shipped to customers. Although this approach is reactive compared to design reliability improvement and process variation reduction, it is effective in preventing early failures and is economically justifiable in most applications. Indeed, screening is the last and a frequent measure that a manufacturer can take proactively to lessen the number of field failures. If a screening strategy is developed to achieve warranty cost objectives, the screen stress and duration should be correlated to the number of failures within the warranty period. Kar and Nachlas (1997) have studied coordinated warranty and screening strategies.

Once products are sold to customers, preventive maintenance on a regular basis plays a critical role in alleviating performance degradation and reducing catastrophic failures in the field. Manufacturers know more about their products than any others, and thus usually provide customers with a guide for appropriate preventive maintenance. Indeed, many manufacturers require their customers to follow the guide and will void the warranty coverage if customers fail to do so. An automobile is a good example. New vehicles always come with a scheduled maintenance guide which specifies maintenance items and frequencies: for example, changing the engine oil every 5000 miles. It is the vehicle owner's responsibility to make sure that all of the scheduled maintenance is performed. The automakers will not reimburse for repairs due to neglect or inadequate maintenance.

The optimization of repair strategy is to minimize the cost per repair without compromising the quality of the repair. This approach requires diagnosing a failure correctly and efficiently, fixing the root cause in the shortest time at the least cost, and testing the repaired product no more or less than necessary. It is important to fix the problem right the first time; doing so minimizes repeat repairs during the warranty period. An optimal repair strategy should be supplied by the manufacturer and not depend heavily on the experience and skills of individual repair providers.

PROBLEMS

11.1 Describe the purposes of warranty analysis. How can warranty analysis help reduce the life cycle cost of a product?

11.2 What are the elements of a warranty policy? Explain the following warranty policies:

(a) Free replacement.

(b) Pro-rata replacement.

(c) Combination free and pro-rata replacement.

(d) Renewing.

(e) Nonrenewing.

(d) Two-dimensional.

11.3 For a warranty repair, what types of data should be recorded in a warranty database? Discuss the use of the data and describe the steps for warranty data mining.

11.4 Explain the limitations of warranty data and how they affect the estimation of product reliability and warranty cost.

11.5 Refer to Example 11.2. If the washing machine warranty data were analyzed two months earlier, what would be the reliability estimate and number of upcoming failures? Compare the results with those in the example.

TABLE 11.8   Electronic Module Warranty Data

TIS (months)	TTF (months)												Sales Volume
	1	2	3	4	5	6	7	8	9	10	11	12	
1	1												836
2	1	2											2063
3	2	0	1										2328
4	1	2	2	1									2677
5	2	1	2	1	2								3367
6	1	2	3	2	2	1							3541
7	2	2	3	3	2	2	3						3936
8	1	2	1	2	1	3	2	2					3693
9	0	2	2	1	0	2	3	1	2				2838
10	2	3	2	1	2	1	2	1	2	2			2362
11	1	1	2	1	0	1	2	1	2	1	1		2056
12	1	1	0	2	1	1	0	1	0	1	1	1	1876

11.6 Refer to Example 11.4. Suppose that the warranty dropout due to the accumulated mileage exceeding 36,000 miles is negligible in the first 11 months. Estimate the life distribution of the mechanical assembly, and calculate the reliability at the end of the warranty period. Compare the results with those in the example.

11.7 An electronic module installed in a luxury car is warranted for 48 months and 48,000 miles, whichever comes first. The mileage accumulation rate of the car can be modeled with a lognormal distribution with scale parameter 7.37 and shape parameter 1.13. The vehicles have a maximum time in service (also called the *maturity*) of 12 months, during which the repeat repairs are negligible. The sales volumes and failure data are shown in Table 11.8. Suppose that the mileage to failure distribution each month is the same as the usage distribution that month.

(a) Calculate the fractions of warranty dropout at the end of 12 and 48 months, respectively.

(b) Determine the month at which the fraction of warranty dropout is 10%.

(c) What would be the warranty mileage limit if the manufacturer wanted 50% of the vehicles to be out of warranty coverage at the end of 48 months?

(d) Calculate the hazard rate estimates $\hat{h}_j$ for $j = 1, 2, \ldots, 12$.

(e) Compute the cumulative hazard rate estimates $\hat{H}_j$ for $j = 1, 2, \ldots, 12$.

(f) Estimate the marginal life distribution of the electronic module.

(g) Estimate the reliability and number of first failures by the end of 48 months.

(h) Write down the joint pdf of the time and mileage to failure.

(i) Estimate the probability of failure at 48 months and 48,000 miles.

(j) How many vehicles will fail at least once due to the module problem in 48 months and 48,000 miles?

(k) Estimate the reliability at 48 months and 48,000 miles.

(l) How many vehicles will survive 48 months and 48,000 miles without the module failing?

(m) Estimate the number of first failures that occur within 48 months not covered by the warranty policy.

11.8 The life (in months) of a battery installed in a laptop computer has the lognormal distribution with scale parameter 3.95 and shape parameter 0.63 under normal usage. When the battery fails, it is replaced with a new one from the same population at a cost of $126. The battery is warranted for one year under a free replacement policy.

(a) Estimate the probability of failure at the end of the warranty period.

(b) Calculate the expected number of repairs per unit within the warranty period.

(c) Compute the expected warranty cost for 1000 computers sold.

11.9 An LCD (liquid-crystal display) television set is assumed to have an exponential distribution with a failure rate of 0.00087 failures per month under average usage. Upon failure, the TV set is repaired to operational conditions, and the repair can be viewed as a same-as-old repair. The TV set is warranted for 12 months. Calculate the expected number of repairs per unit within the warranty period.

11.10 The manufacturer of a water heater currently provides a linear pro-rata warranty coverage over 36 months for the product, which is sold at $58 a unit. The life of the heater has a Weibull distribution with shape parameter 2.23 and characteristic life 183 months. To increase customer satisfaction and market shares, the manufacturer plans to offer a combination free and pro-rata replacement warranty, where the free replacement policy is proposed to cover the first 12 months, followed by 24 months of linear pro-rata replacement.

(a) Calculate the expected warranty cost per unit under the current warranty policy.

(b) Compute the expected warranty cost per unit under the proposed combination warranty policy.

(c) What is the incremental expected warranty cost per unit due to the change in warranty policy?

(d) What additional information is required for the manufacturer to make the decision?

11.11 Derive the centerline and control limits given by (11.45).

11.12 A multi-CD (compact disk) player had been in production for 12 months. The manufacturer wants to establish a control chart using the warranty

TABLE 11.9 Monthly Production and Failure Data

	Production Month											
	1	2	3	4	5	6	7	8	9	10	11	12
Volume	6963	7316	7216	7753	8342	8515	8047	8623	8806	8628	8236	7837
Number of failures	15	18	12	17	13	15	19	26	16	14	21	14

data of the first month in service to detect unusual infant mortality. The production volume and number of failures in the first month in service for each monthly production period are shown in Table 11.9.

(a) Develop a p-chart.

(b) Make comments on the control limits.

(c) Can the variable control limits be approximated by two straight lines? How?

(d) In production month 13, 7638 units were made and 13 of them failed in the first month in service. Determine whether the process was under control in that month.

APPENDIX

ORTHOGONAL ARRAYS, LINEAR GRAPHS, AND INTERACTION TABLES*

$L_4 (2^3)$

$L_4 (2^3)$ Orthogonal Array

Run No.	Column		
	1	2	3
1	0	0	0
2	0	1	1
3	1	0	1
4	1	1	0

Linear Graph for L_4

1 ●————● 2
　　3

---

*The material in this appendix is reproduced with permission from Dr. Genichi Taguchi with assistance from the American Supplier Institute, Inc. More orthogonal arrays, linear graphs, and interaction tables may be found in Taguchi et al. (1987, 2005).

$L_8 (2^7)$

$L_8 (2^7)$ Orthogonal Array

Run No.	Column 1	2	3	4	5	6	7
1	0	0	0	0	0	0	0
2	0	0	0	1	1	1	1
3	0	1	1	0	0	1	1
4	0	1	1	1	1	0	0
5	1	0	1	0	1	0	1
6	1	0	1	1	0	1	0
7	1	1	0	0	1	1	0
8	1	1	0	1	0	0	1

Interaction Table for L_8

Column	Column 1	2	3	4	5	6	7
1	(1)	3	2	5	4	7	6
2		(2)	1	6	7	4	5
3			(3)	7	6	5	4
4				(4)	1	2	3
5					(5)	3	2
6						(6)	1
7							(7)

Linear Graphs for L_8

(1)

(2)

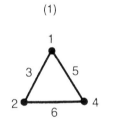

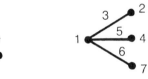

$L_9 (3^4)$

$L_9 (3^4)$ Orthogonal Array

Run No.	Column 1	2	3	4
1	0	0	0	0
2	0	1	1	1
3	0	2	2	2
4	1	0	1	2
5	1	1	2	0
6	1	2	0	1
7	2	0	2	1
8	2	1	0	2
9	2	2	1	0

Linear Graphs for L_9

1 ●———● 2
   3,4

$L_{12}(2^{11})$

$L_{12}(2^{11})$ Orthogonal Array

Run No.	Column										
	1	2	3	4	5	6	7	8	9	10	11
1	0	0	0	0	0	0	0	0	0	0	0
2	0	0	0	0	0	1	1	1	1	1	1
3	0	0	1	1	1	0	0	0	1	1	1
4	0	1	0	1	1	0	1	1	0	0	1
5	0	1	1	0	1	1	0	1	0	1	0
6	0	1	1	1	0	1	1	0	1	0	0
7	1	0	1	1	0	0	1	1	0	1	0
8	1	0	1	0	1	1	1	0	0	0	1
9	1	0	0	1	1	1	0	1	1	0	0
10	1	1	1	0	0	0	0	1	1	0	1
11	1	1	0	1	0	1	0	0	0	1	1
12	1	1	0	0	1	0	1	0	1	1	0

Note: The interaction between any two columns is confounded partially with the remaining nine columns. Do not use this array if the interactions must be estimated.

$L_{16}(2^{15})$

$L_{16}(2^{15})$ Orthogonal Array

Run No.	Column														
	1	2	3	4	5	6	7	8	9	10	11	12	13	14	15
1	0	0	0	0	0	0	0	0	0	0	0	0	0	0	0
2	0	0	0	0	0	0	0	1	1	1	1	1	1	1	1
3	0	0	0	1	1	1	1	0	0	0	0	1	1	1	1
4	0	0	0	1	1	1	1	1	1	1	1	0	0	0	0
5	0	1	1	0	0	1	1	0	0	1	1	0	0	1	1
6	0	1	1	0	0	1	1	1	1	0	0	1	1	0	0
7	0	1	1	1	1	0	0	0	0	1	1	1	1	0	0
8	0	1	1	1	1	0	0	1	1	0	0	0	0	1	1
9	1	0	1	0	1	0	1	0	1	0	1	0	1	0	1
10	1	0	1	0	1	0	1	1	0	1	0	1	0	1	0
11	1	0	1	1	0	1	0	0	1	0	1	1	0	1	0
12	1	0	1	1	0	1	0	1	0	1	0	0	1	0	1
13	1	1	0	0	1	1	0	0	1	1	0	0	1	1	0
14	1	1	0	0	1	1	0	1	0	0	1	1	0	0	1
15	1	1	0	1	0	0	1	0	1	1	0	1	0	0	1
16	1	1	0	1	0	0	1	1	0	0	1	0	1	1	0

Interaction Table for L_{16} (2^{15})

Column	1	2	3	4	5	6	7	8	9	10	11	12	13	14	15
1	(1)	3	2	5	4	7	6	9	8	11	10	13	12	15	14
2		(2)	1	6	7	4	5	10	11	8	9	14	15	12	13
3			(3)	7	6	5	4	11	10	9	8	15	14	13	12
4				(4)	1	2	3	12	13	14	15	8	9	10	11
5					(5)	3	2	13	12	15	14	9	8	11	10
6						(6)	1	14	15	12	13	10	11	8	9
7							(7)	15	14	13	12	11	10	9	8
8								(8)	1	2	3	4	5	6	7
9									(9)	3	2	5	4	7	6
10										(10)	1	6	7	4	5
11											(11)	7	6	5	4
12												(12)	1	2	3
13													(13)	3	2
14														(14)	1
15															(15)

Linear Graphs for L_{16}

(1)

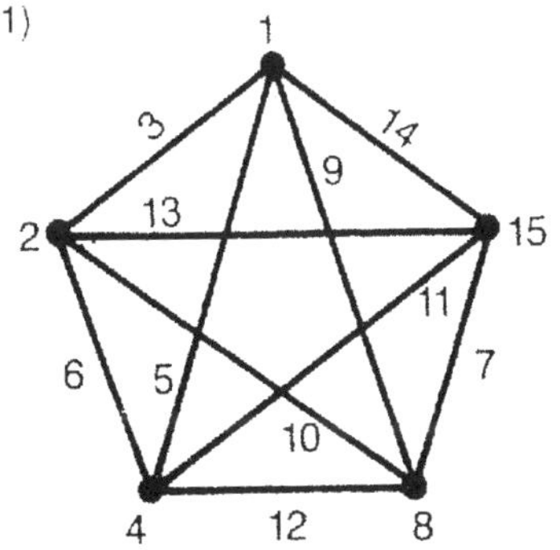

(2)

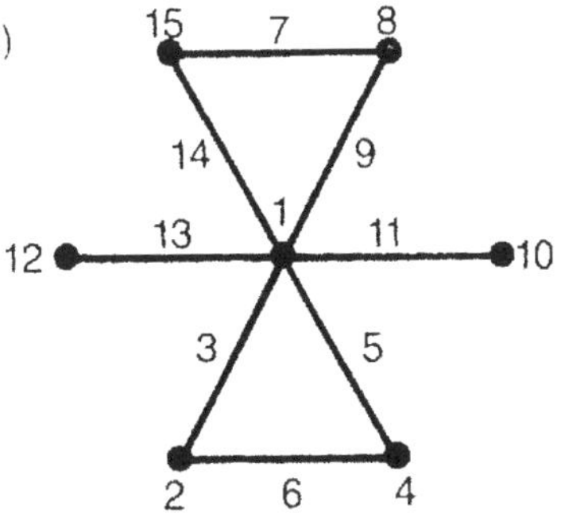

(3)

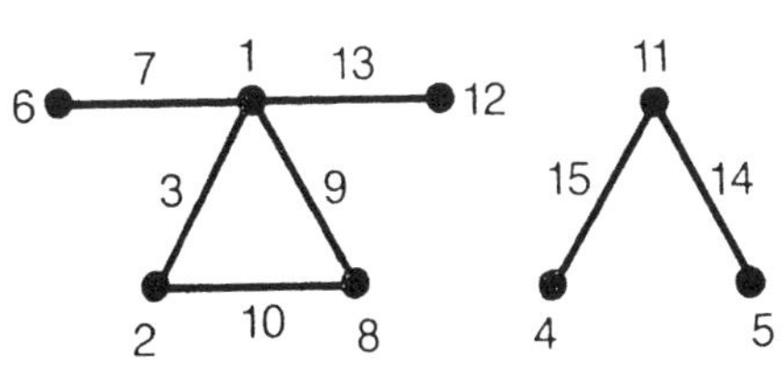

(4)

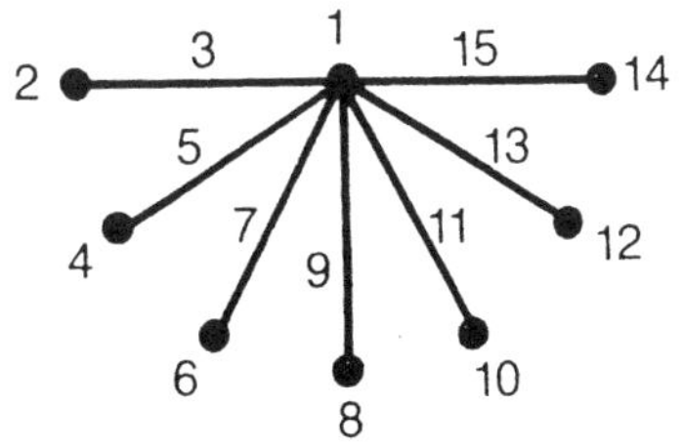

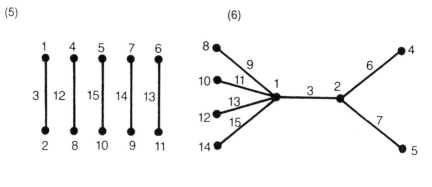

$$L'_{16} (4^5)$$

L'_{16} (4⁵) Orthogonal Array

Run	Column				
No.	1	2	3	4	5
1	0	0	0	0	0
2	0	1	1	1	1
3	0	2	2	2	2
4	0	3	3	3	3
5	1	0	1	2	3
6	1	1	0	3	2
7	1	2	3	0	1
8	1	3	2	1	0
9	2	0	2	3	1
10	2	1	3	2	0
11	2	2	0	1	3
12	2	3	1	0	2
13	3	0	3	1	2
14	3	1	2	0	3
15	3	2	1	3	0
16	3	3	0	2	1

Note: To estimate the interaction between columns 1 and 2, all other columns must be kept empty.

Linear Graph for L'_{16}

1 ●———● 2
   3,4,5

L_{18} $(2^1 \times 3^7)$

L_{18} $(2^1 \times 3^7)$ Orthogonal Array

Run No.	Column							
	1	2	3	4	5	6	7	8
1	0	0	0	0	0	0	0	0
2	0	0	1	1	1	1	1	1
3	0	0	2	2	2	2	2	2
4	0	1	0	0	1	1	2	2
5	0	1	1	1	2	2	0	0
6	0	1	2	2	0	0	1	1
7	0	2	0	1	0	2	1	2
8	0	2	1	2	1	0	2	0
9	0	2	2	0	2	1	0	1
10	1	0	0	2	2	1	1	0
11	1	0	1	0	0	2	2	1
12	1	0	2	1	1	0	0	2
13	1	1	0	1	2	0	2	1
14	1	1	1	2	0	1	0	2
15	1	1	2	0	1	2	1	0
16	1	2	0	2	1	2	0	1
17	1	2	1	0	2	0	1	2
18	1	2	2	1	0	1	2	0

Note: Interaction between columns 1 and 2 is orthogonal to all columns and hence can be estimated without sacrificing any column. The interaction can be estimated from the 2-way table of columns 1 and 2. Columns 1 and 2 can be combined to form a 6-level column. Interactions between any other pair of columns is confounded partially with the remaining columns.

Linear Graph for L_{18}

$L_{25}(5^6)$

$L_{25}(5^6)$ Orthogonal Array

Run No.	Column					
	1	**2**	**3**	**4**	**5**	**6**
1	0	0	0	0	0	0
2	0	1	1	1	1	1
3	0	2	2	2	2	2
4	0	3	3	3	3	3
5	0	4	4	4	4	4
6	1	0	1	2	3	4
7	1	1	2	3	4	0
8	1	2	3	4	0	1
9	1	3	4	0	1	2
10	1	4	0	1	2	3
11	2	0	2	4	1	3
12	2	1	3	0	2	4
13	2	2	4	1	3	0
14	2	3	0	2	4	1
15	2	4	1	3	0	2
16	3	0	3	1	4	2
17	3	1	4	2	0	3
18	3	2	0	3	1	4
19	3	3	1	4	2	0
20	3	4	2	0	3	1
21	4	0	4	3	2	1
22	4	1	0	4	3	2
23	4	2	1	0	4	3
24	4	3	2	1	0	4
25	4	4	3	2	1	0

Note: To estimate the interaction between columns 1 and 2, all other columns must be kept empty.

Linear Graph for L_{25}

3,4,5,6

1 ●————————● 2

$L_{27}(3^{13})$

$L_{27}(3^{13})$ Orthogonal Array

Run No.	Column												
	1	2	3	4	5	6	7	8	9	10	11	12	13
1	0	0	0	0	0	0	0	0	0	0	0	0	0
2	0	0	0	0	1	1	1	1	1	1	1	1	1
3	0	0	0	0	2	2	2	2	2	2	2	2	2
4	0	1	1	1	0	0	0	1	1	1	2	2	2
5	0	1	1	1	1	1	1	2	2	2	0	0	0
6	0	1	1	1	2	2	2	0	0	0	1	1	1
7	0	2	2	2	0	0	0	2	2	2	1	1	1
8	0	2	2	2	1	1	1	0	0	0	2	2	2
9	0	2	2	2	2	2	2	1	1	1	0	0	0
10	1	0	1	2	0	1	2	0	1	2	0	1	2
11	1	0	1	2	1	2	0	1	2	0	1	2	0
12	1	0	1	2	2	0	1	2	0	1	2	0	1
13	1	1	2	0	0	1	2	1	2	0	2	0	1
14	1	1	2	0	1	2	0	2	0	1	0	1	2
15	1	1	2	0	2	0	1	0	1	2	1	2	0
16	1	2	0	1	0	1	2	2	0	1	1	2	0
17	1	2	0	1	1	2	0	0	1	2	2	0	1
18	1	2	0	1	2	0	1	1	2	0	0	1	2
19	2	0	2	1	0	2	1	0	2	1	0	2	1
20	2	0	2	1	1	0	2	1	0	2	1	0	2
21	2	0	2	1	2	1	0	2	1	0	2	1	0
22	2	1	0	2	0	2	1	1	0	2	2	1	0
23	2	1	0	2	1	0	2	2	1	0	0	2	1
24	2	1	0	2	2	1	0	0	2	1	1	0	2
25	2	2	1	0	0	2	1	2	1	0	1	0	2
26	2	2	1	0	1	0	2	0	2	1	2	1	0
27	2	2	1	0	2	1	0	1	0	2	0	2	1

Interaction Table for L_{27} (3^{13})

Column	1	2	3	4	5	6	7	8	9	10	11	12	13
1	(1)	3 4	2 4	2 3	6 7	5 7	5 6	9 10	8 10	8 9	12 13	11 13	11 12
2		(2)	1 4	1 3	8 11	9 12	10 13	5 11	6 12	7 13	5 8	6 9	7 10
3			(3)	1 2	9 13	10 11	8 12	7 12	5 13	6 11	6 10	7 8	5 9
4				(4)	10 12	8 13	9 11	6 13	7 11	5 12	7 9	5 10	6 8
5					(5)	1 7	1 6	2 11	3 13	4 12	2 8	4 10	3 9
6						(6)	1 5	4 13	2 12	3 11	3 10	2 9	4 8
7							(7)	3 12	4 11	2 13	4 9	3 8	2 10
8								(8)	1 10	1 9	2 5	3 7	4 6
9									(9)	1 8	4 7	2 6	3 5
10										(10)	3 6	4 5	2 7
11											(11)	1 13	1 12
12												(12)	1 11
13													(13)

Linear Graph for L_{27}

REFERENCES

Adams, V., and Askenazi, A. (1998), *Building Better Products with Finite Element Analysis*, OnWord Press, Santa Fe, NM.

Aggarwal, K. K. (1993), *Reliability Engineering*, Kluwer Academic, Norwell, MA.

AGREE (1957), *Reliability of Military Electronic Equipment*, Office of the Assistant Secretary of Defense Research and Engineering, Advisory Group of Reliability of Electronic Equipment, Washington, DC.

Ahmad, M., and Sheikh, A. K. (1984), Bernstein reliability model: derivation and estimation of parameters, *Reliability Engineering*, vol. 8, no. 3, pp. 131–148.

Akao, Y. (1990), *Quality Function Deployment*, Productivity Press, Cambridge, MA.

Allmen, C. R., and Lu, M. W. (1994), A reduced sampling approach for reliability verification, *Quality and Reliability Engineering International*, vol. 10, no. 1, pp. 71–77.

Al-Shareef, H., and Dimos, D. (1996), Accelerated life-time testing and resistance degradation of thin-film decoupling capacitors, *Proc. IEEE International Symposium on Applications of Ferroelectrics*, vol. 1, pp. 421–425.

Amagai, M. (1999), Chip scale package (CSP) solder joint reliability and modeling, *Microelectronics Reliability*, vol. 39, no. 4, pp. 463–477.

Ames, A. E., Mattucci, N., MacDonald, S., Szonyi, G., and Hawkins, D. M. (1997), Quality loss functions for optimization across multiple response surfaces, *Journal of Quality Technology*, vol. 29, no. 3, pp. 339–346.

ANSI/ASQ (2003a), *Sampling Procedures and Tables for Inspection by Attributes*, ANSI/ASQ Z1.4–2003, American Society for Quality, Milwaukee, WI, www.asq.org.

Life Cycle Reliability Engineering, by Guangbin Yang
Copyright © 2007 John Wiley & Sons, Inc.

———(2003b), *Sampling Procedures and Tables for Inspection by Variables for Percent Nonconforming*, ANSI/ASQ Z1.9–2003, American Society for Quality, Milwaukee, WI, www.asq.org.

Armacost, R. L., Componation, P. J., and Swart, W. W. (1994), An AHP framework for prioritizing customer requirement in QFD: an industrialized housing application, *IIE Transactions*, vol. 26, no. 4, pp. 72–78.

Bai, D. S., and Yun, H. J. (1996), Accelerated life tests for products of unequal size, *IEEE Transactions on Reliability*, vol. 45, no. 4, pp. 611–618.

Bain, L. J., and Engelhardt, M. (1991), *Statistical Analysis of Reliability and Life-Testing Models: Theory and Methods*, (2nd ed.), Marcel Dekker, New York.

Barlow, R. E., and Proschan, F. (1974), *Importance of System Components and Fault Tree Analysis*, ORC-74-3, Operations Research Center, University of California, Berkeley, CA.

Basaran, C., Tang, H., and Nie, S. (2004), Experimental damage mechanics of microelectronic solder joints under fatigue loading, *Mechanics of Materials*, vol. 36, no. 11, pp. 1111–1121.

Baxter, L., Scheuer, E., McConaloguo, D., and Blischke, W. (1982), On the tabulation of the renewal function, *Technometrics*, vol. 24, no. 2, pp. 151–156.

Bazaraa, M. S., Sherali, H. D., and Shetty, C. M. (1993), *Nonlinear Programming: Theory and Algorithms*, (2nd ed.), Wiley, Hoboken, NJ.

Becker, B., and Ruth, P. P. (1998), Highly accelerated life testing for the 1210 digital ruggedized display, *Proc. SPIE*, International Society for Optical Engineering, pp. 337–345.

Bertsekas, D. P. (1996), *Constrained Optimization and Lagrange Multiplier Methods*, Academic Press, San Diego, CA.

Bhushan, B. (2002), *Introduction to Tribology*, Wiley, Hoboken, NJ.

Birnbaum, Z. W. (1969), On the importance of different components in a multicomponent system, in *Multivariate Analysis—II*, P. R. Krishnaiah, Ed., Academic Press, New York, pp. 581–592.

Black, J. R. (1969), Electromigration: a brief survey and some recent results, *IEEE Transactions on Electronic Devices*, vol. ED-16, no. 4, pp. 338–347.

Blischke, W. R., and Murthy, D. N. P. (1994), *Warranty Cost Analysis*, Marcel Dekker, New York.

——— Eds. (1996), *Product Warranty Handbook*, Marcel Dekker, New York.

———(2000), *Reliability: Modeling, Prediction, and Optimization*, Wiley, Hoboken, NJ.

Boland, P. J., and El-Neweihi, E. (1995), Measures of component importance in reliability theory, *Computers and Operations Research*, vol. 22, no. 4, pp. 455–463.

Bossert, J. L. (1991), *Quality Function Deployment*, ASQC Quality Press, Milwaukee, WI.

Bouissou, M. (1996), An ordering heuristic for building a decision diagram from a fault-tree, *Proc. Reliability and Maintainability Symposium*, pp. 208–214.

Boulanger, M., and Escobar, L. A. (1994), Experimental design for a class of accelerated degradation tests, *Technometrics*, vol. 36, no. 3, pp. 260–272.

Bowles, J. B. (2003), An assessment of RPN prioritization in failure modes effects and criticality analysis, *Proc. Reliability and Maintainability Symposium*, pp. 380–386.

Bowles, J. B., and Pelaez, C. E. (1995), Fuzzy logic prioritization of failures in a system failure mode, effects and criticality analysis, *Reliability Engineering and System Safety*, vol. 50, no. 2, pp. 203–213.

Box, G. (1988), Signal-to-noise ratios, performance criteria, and transformation, *Technometrics*, vol. 30, no. 1, pp. 1–17.

Brooks, A. S. (1974), The Weibull distribution: effect of length and conductor size of test cables, *Electra*, vol. 33, pp. 49–61.

Broussely, M., Herreyre, S., Biensan, P., Kasztejna, P., Nechev, K., and Staniewicz, R. J. (2001), Aging mechanism in Li ion cells and calendar life predictions, *Journal of Power Sources*, vol. 97-98, pp. 13–21.

Bruce, G., and Launsby, R. G. (2003), *Design for Six Sigma*, McGraw-Hill, New York.

Carot, V., and Sanz, J. (2000), Criticality and sensitivity analysis of the components of a system, *Reliability Engineering and System Safety*, vol. 68, no. 2, pp. 147–152.

Chan, H. A., and Englert, P. J., Eds. (2001), *Accelerated Stress Testing Handbook: Guide for Achieving Quality Products*, IEEE Press, Piscataway, NJ.

Chao, A., and Hwang, L. C. (1987), A modified Monte Carlo technique for confidence limits of system reliability using pass–fail data, *IEEE Transactions on Reliability*, vol. R-36, no. 1, pp. 109–112.

Chao, L. P., and Ishii, K. (2003), Design process error-proofing: failure modes and effects analysis of the design process, *Proc. ASME Design Engineering Technical Conference*, vol. 3, pp. 127–136.

Chen, M. R. (2001), Robust design for VLSI process and device, *Proc. 6th International Workshop on Statistical Metrology*, pp. 7–16.

Chi, D. H., and Kuo, W. (1989), Burn-in optimization under reliability and capacity restrictions, *IEEE Transactions on Reliability*, vol. 38, no. 2, pp. 193–198.

Chiao, C. H., and Hamada, M. (2001), Analyzing experiments with degradation data for improving reliability and for achieving robust reliability, *Quality and Reliability Engineering International*, vol. 17, no. 5, pp. 333–344.

Coffin, L. F., Jr. (1954), A study of the effects of cyclic thermal stresses on a ductile metal, *Transactions of ASME*, vol. 76, no. 6, pp. 931–950.

Coit, D. W. (1997), System-reliability confidence-intervals for complex-systems with estimated component-reliability, *IEEE Transactions on Reliability*, vol. 46, no. 4, pp. 487–493.

Condra, L. W. (2001), *Reliability Improvement with Design of Experiments*, (2nd ed.), Marcel Dekker, New York.

Cory, A. R. (2000), Improved reliability prediction through reduced-stress temperature cycling, *Proc. 38th IEEE International Reliability Physics Symposium*, pp. 231–236.

Corzo, O., and Gomez, E. R. (2004), Optimization of osmotic dehydration of cantaloupe using desired function methodology, *Journal of Food Engineering*, vol. 64, no. 2, pp. 213–219.

Cox, D. R. (1962), *Renewal Theory*, Wiley, Hoboken, NJ.

——— (1972), Regression models and life tables (with discussion), *Journal of the Royal Statistical Society*, ser. B, vol. 34, pp. 187–220.

Creveling, C. M. (1997), *Tolerance Design: A Handbook for Developing Optimal Specifications*, Addison-Wesley, Reading, MA.

Croes, K., De Ceuninck, W., De Schepper, L., and Tielemans, L. (1998), Bimodal failure behaviour of metal film resistors, *Quality and Reliability Engineering International*, vol. 14, no. 2, pp. 87–90.

Crowder, M. J., Kimber, A. C., Smith, R. L., and Sweeting, T. J. (1991), *Statistical Analysis of Reliability Data*, Chapman & Hall, London.

Dabbas, R. M., Fowler, J. W., Rollier, D. A., and McCarville, D. (2003), Multiple response optimization using mixture-designed experiments and desirability functions in semi-conductor scheduling, *International Journal of Production Research*, vol. 41, no. 5, pp. 939–961.

Davis, T. P. (1999), A simple method for estimating the joint failure time and failure mileage distribution from automobile warranty data, *Ford Technical Journal*, vol. 2, no. 6, Report FTJ-1999-0048.

Deely, J. J., and Keats, J. B. (1994), Bayes stopping rules for reliability testing with the exponential distribution, *IEEE Transactions on Reliability*, vol. 43, no. 2, pp. 288–293.

Del Casttillo, E., Montgomery, D. C., and McCarville, D. R. (1996), Modified desirability functions for multiple response optimization, *Journal of Quality Technology*, vol. 28, no. 3, pp. 337–345.

Derringer, G., and Suich, R. (1980), Simultaneous optimization of several response variables, *Journal of Quality Technology*, vol. 12, no. 4, pp. 214–219.

Dhillon, B. S. (1999), *Design Reliability: Application and Fundamentals*, CRC Press, Boca Raton, FL.

Dieter, G. E. (2000), *Engineering Design: A Materials and Processing Approach*, McGraw-Hill, New York.

Dimaria, D. J., and Stathis, J. H. (1999), Non-Arrhenius temperature dependence of reliability in ultrathin silicon dioxide films, *Applied Physics Letters*, vol. 74, no. 12, pp. 1752–1754.

Dugan, J. B. (2003), Fault-tree analysis of computer-based systems, tutorial at the Reliability and Maintainability Symposium.

Eliashberg, J., Singpurwalla, N. D., and Wilson, S. P. (1997), Calculating the reserve for a time and usage indexed warranty, *Management Science*, vol. 43, no. 7, pp. 966–975.

Elsayed, E. A. (1996), *Reliability Engineering*, Addison Wesley Longman, Reading, MA.

Ersland, P., Jen, H. R., and Yang, X. (2004), Lifetime acceleration model for HAST tests of a pHEMT process, *Microelectronics Reliability*, vol. 44, no. 7, pp. 1039–1045.

Evans, R. A. (2000), Editorial: Populations and hazard rates, *IEEE Transactions on Reliability*, vol. 49, no. 3, p. 250 (first published in May 1971).

Farnum, N. R., and Booth, P. (1997), Uniqueness of maximum likelihood estimators of the 2-parameter Weibull distribution, *IEEE Transactions on Reliability*, vol. 46, no. 4, pp. 523–525.

Feilat, E. A., Grzybowski, S., and Knight, P. (2000), Accelerated aging of high voltage encapsulated transformers for electronics applications, *Proc. IEEE International Conference on Properties and Applications of Dielectric Materials*, vol. 1, pp. 209–212.

Fleetwood, D. M., Meisenheimer, T. L., and Scofield, J. H. (1994), $1/f$ noise and radiation effects in MOS devices, *IEEE Transactions on Electron Devices*, vol. 41, no. 11, pp. 1953–1964.

Franceschini, F., and Galetto, M. (2001), A new approach for evaluation of risk priorities of failure modes in FMEA, *International Journal of Production Research*, vol. 39, no. 13, pp. 2991–3002.

Fussell, J. B. (1975), How to hand-calculate system reliability and safety characteristics, *IEEE Transactions on Reliability*, vol. R-24, no. 3, pp. 169–174.

Garg, A., and Kalagnanam, J. (1998), Approximation for the renewal function, *IEEE Transactions on Reliability*, vol. 47, no. 1, pp. 66–72.

Gertsbakh, I. B. (1982), Confidence limits for highly reliable coherent systems with exponentially distributed component life, *Journal of the American Statistical Association*, vol. 77, no. 379, pp. 673–678.

———(1989), *Statistical Reliability Theory*, Marcel Dekker, New York.

Ghaffarian, R. (2000), Accelerated thermal cycling and failure mechanisms for BGA and CSP assemblies, *Transactions of ASME: Journal of Electronic Packaging*, vol. 122, no. 4, pp. 335–340.

Gillen, K. T., Bernstein, R., and Derzon, D. K. (2005), Evidence of non-Arrhenius behavior from laboratory aging and 24-year field aging of polychloroprene rubber materials, *Polymer Degradation and Stability*, vol. 87, no. 1, pp. 57–67.

Gitlow, H. S., and Levine, D. M. (2005), *Six Sigma for Green Belts and Champions: Foundations, DMAIC, Tools, Cases, and Certification*, Pearson/Prentice Hall, Upper Saddle River, NJ.

Gnedenko, B., Pavlov, I., and Ushakov, I. (1999), in *Statistical Reliability Engineering*, Chakravarty, S., Ed., Wiley, Hoboken, NJ.

Goddard, P. L. (1993), Validating the safety of real time control systems using FMEA, *Proc. Reliability and Maintainability Symposium*, pp. 227–230.

———(2000), Software FMEA techniques, *Proc. Reliability and Maintainability Symposium*, pp. 118–123.

Guida, M., and Pulcini, G. (2002), Automotive reliability inference based on past data and technical knowledge, *Reliability Engineering and System Safety*, vol. 76, no. 2, pp. 129–137.

Hallberg, O., and Peck, D. S. (1991), Recent humidity acceleration, a base for testing standards, *Quality and Reliability Engineering International*, vol. 7, no. 3, pp. 169–180.

Han, J., and Kamber, M. (2000), *Data Mining: Concepts and Techniques*, Morgan Kaufmann, San Francisco, CA.

Harris, T. A. (2001), *Rolling Bearing Analysis*, (4th ed.), Wiley, Hoboken, NJ.

Harter, H. L., and Moore, A. H. (1976), An evaluation of exponential and Weibull test plans, *IEEE Transactions on Reliability*, vol. R-25, no. 2, pp. 100–104.

Hauck, D. J., and Keats, J. B. (1997), Robustness of the exponential sequential probability ratio test (SPRT) when weibull distributed failures are transformed using a 'known' shape parameter, *Microelectronics Reliability*, vol. 37, no. 12, pp. 1835–1840.

Henderson, T., and Tutt, M. (1997), Screening for early and rapid degradation in GaAs/AlGaAs HBTs, *Proc. 35th IEEE International Reliability Physics Symposium*, pp. 253–260.

Henley, E. J., and Kumamoto, H. (1992), *Probabilistic Risk Assessment: Reliability Engineering, Design, and Analysis*, IEEE Press, Piscataway, NJ.

Hines, W. W., Montgomery, D. C., Goldsman, D. M., and Borror, C. M. (2002), *Probability and Statistics in Engineering*, (4th ed.), Wiley, Hoboken, NJ.

Hirose, H. (1999), Bias correction for the maximum likelihood estimates in the two-parameter Weibull distribution, *IEEE Transactions on Dielectrics and Electrical Insulation*, vol. 6, no. 1, pp. 66–68.

Hobbs, G. K. (2000), *Accelerated Reliability Engineering: HALT and HASS*, Wiley, Chichester, West Sussex, England.

Hwang, F. K. (2001), A new index of component importance, *Operations Research Letters*, vol. 28, no. 2, pp. 75–79.

IEC (1985), *Analysis Techniques for System Reliability: Procedure for Failure Mode and Effects Analysis (FMEA)*, IEC 60812, International Electromechanical Commission, Geneva, www.iec.ch.

―――(1998, 2000), *Functional Safety of Electrical/Electronic/Programmable Electronic Safety-Related Systems*, IEC 61508, International Electromechanical Commission, Geneva, www.iec.ch.

IEEE Reliability Society (2006), http://www.ieee.org/portal/site/relsoc.

Ireson, W. G., Coombs, C. F., and Moss, R. Y. (1996), *Handbook of Reliability Engineering and Management*, McGraw-Hill, New York.

Jeang, A. (1995), Economic tolerance design for quality, *Quality and Reliability Engineering International*, vol. 11, no. 2, pp. 113–121.

Jensen, F. (1995), *Electronic Component Reliability*, Wiley, Chichester, West Sussex, England.

Jensen, F., and Petersen, N. E. (1982), *Burn-in: An Engineering Approach to the Design and Analysis of Burn-in Procedures*, Wiley, Chichester, West Sussex, England.

Jiang, G., Purnell, K., Mobley, P., and Shulman, J. (2003), Accelerated life tests and in-vivo test of 3Y-TZP ceramics, *Proc. Materials and Processes for Medical Devices Conference*, ASM International, pp. 477–482.

Johnson, R. A. (1998), *Applied Multivariate Statistical Analysis*, Prentice Hall, Upper Saddle River, NJ.

Joseph, V. R., and Wu, C. F. J. (2004), Failure amplification method: an information maximization approach to categorical response optimization (with discussion), *Technometrics*, vol. 46, no. 1, pp. 1–12.

Jung, M., and Bai, D. S. (2006), Analysis of field data under two-dimensional warranty, *Reliability Engineering and System Safety*, to appear.

Kalbfleisch, J. D., Lawless, J. F., and Robinson, J. A. (1991), Method for the analysis and prediction of warranty claims, *Technometrics*, vol. 33, no. 3, pp. 273–285.

Kalkanis, G., and Rosso, E. (1989), Inverse power law for the lifetime of a Mylar–polyurethane laminated dc hv insulating structure, *Nuclear Instruments and Methods in Physics Research, Series A: Accelerators, Spectrometers, Detectors and Associated Equipment*, vol. 281, no. 3, pp. 489–496.

Kaminskiy, M. P., and Krivtsov, V. V. (1998), A Monte Carlo approach to repairable system reliability analysis, *Proc. Probabilistic Safety Assessment and Management*, International Association for PSAM, pp. 1063–1068.

―――(2000), G-renewal process as a model for statistical warranty claim prediction, *Proc. Reliability and Maintainability Symposium*, pp. 276–280.

Kaplan, E. L., and Meier, P. (1958), Nonparametric estimation from incomplete observations, *Journal of the American Statistical Association*, vol. 54, pp. 457–481.

Kapur, K. C., and Lamberson, L. R. (1977), *Reliability in Engineering Design*, Wiley, Hoboken, NJ.

Kar, T. R., and Nachlas, J. A. (1997), Coordinated warranty and burn-in strategies, *IEEE Transactions on Reliability*, vol. 46, no. 4, pp. 512–518.

Kececioglu, D. B. (1991), *Reliability Engineering Handbook*, Vol. 1, Prentice Hall, Upper Saddle River, NJ.

———(1994), *Reliability and Life Testing Handbook*, Vol. 2, Prentice Hall, Upper Saddle River, NJ.

Kececioglu, D. B., and Sun, F. B. (1995), *Environmental Stress Screening: Its Quantification, Optimization and Management*, Prentice Hall, Upper Saddle River, NJ.

Kielpinski, T. L., and Nelson, W. B. (1975), Optimum censored accelerated life tests for normal and lognormal life distributions, *IEEE Transactions on Reliability*, vol. R-24, no. 5, pp. 310–320.

Kijima, M. (1989), Some results for repairable systems with general repair, *Journal of Applied Probability*, vol. 26, pp. 89–102.

Kijima, M., and Sumita, N. (1986), A useful generalization of renewal theory: counting process governed by non-negative Markovian increments, *Journal of Applied Probability*, vol. 23, pp. 71–88.

Kim, H. G., and Rao, B. M. (2000), Expected warranty cost of two-attribute free-replacement warranties based on a bivariate exponential distribution, *Computers and Industrial Engineering*, vol. 38, no. 4, pp. 425–434.

Kim, K. O., and Kuo, W. (2005), Some considerations on system burn-in, *IEEE Transactions on Reliability*, vol. 54, no. 2, pp. 207–214.

Kim, T., and Kuo, W. (1998), Optimal burn-in decision making, *Quality and Reliability Engineering International*, vol. 14, no. 6, pp. 417–423.

Kleyner, A., Bhagath, S., Gasparini, M., Robinson, J., and Bender, M. (1997), Bayesian techniques to reduce the sample size in automotive electronics attribute testing, *Microelectronics Reliability*, vol. 37, no. 6, pp. 879–883.

Krivtsov, V., and Frankstein, M. (2004), Nonparametric estimation of marginal failure distributions from dually censored automotive data, *Proc. Reliability and Maintainability Symposium*, pp. 86–89.

Kuo, W., and Zuo, M. J. (2002), *Optimal Reliability Modeling: Principles and Applications*, Wiley, Hoboken, NJ.

Kuo, W., Chien, W. T., and Kim, T. (1998), *Reliability, Yield, and Stress Burn-in*, Kluwer Academic, Norwell, MA.

Kuo, W., Prasad, V. R., Tillman, F. A., and Hwang, C. L. (2001), *Optimal Reliability Design: Fundamentals and Applications*, Cambridge University Press, Cambridge.

Lambert, H. E. (1975), Fault trees for decision making in system analysis, Ph.D. dissertation, University of California, Livermore, CA.

Lawless, J. F. (2002), *Statistical Models and Methods for Lifetime Data*, 2nd ed., Wiley, Hoboken, NJ.

Lawless, J. F., Hu, J., and Cao, J. (1995), Methods for the estimation of failure distributions and rates from automobile warranty data, *Life Data Analysis*, vol. 1, no. 3, pp. 227–240.

Lee, B. (2004), Sequential Bayesian bit error rate measurement, *IEEE Transactions on Instrumentation and Measurement*, vol. 53, no. 4, pp. 947–954.

Lee, C. L. (2000), Tolerance design for products with correlated characteristics, *Mechanism and Machine Theory*, vol. 35, no. 12, pp. 1675–1687.

Levin, M. A., and Kalal T. T. (2003), *Improving Product Reliability: Strategies and Implementation*, Wiley, Hoboken, NJ.

Lewis, E. E. (1987), *Introduction to Reliability Engineering*, Wiley, Hoboken, NJ.

Li, Q., and Kececioglu, D. B. (2003), Optimal design of accelerated degradation tests, *International Journal of Materials and Product Technology*, vol. 20, no. 1–3, pp. 73–90.

Li, R. S. (2004), *Failure Mechanisms of Ball Grid Array Packages Under Vibration and Thermal Loading*, SAE Technical Paper Series 2004-01-1686, Society of Automotive Engineers, Warrendale, PA.

Lomnicki, Z. (1996), A note on the Weibull renewal process, *Biometrics*, vol. 53, no. 3–4, pp. 375–381.

Lu, J. C., Park, J., and Yang, Q. (1997), Statistical inference of a time-to-failure distribution derived from linear degradation data, *Technometrics*, vol. 39, no. 4, pp. 391–400.

Lu, M. W. (1998), Automotive reliability prediction based on early field failure warranty data, *Quality and Reliability Engineering International*, vol. 14, no. 2, pp. 103–108.

Lu, M. W., and Rudy, R. J. (2000), Reliability test target development, *Proc. Reliability and Maintainability Symposium*, pp. 77–81.

Manian, R., Coppit, D. W., Sullivan, K. J., and Dugan, J. B. (1999), Bridging the gap between systems and dynamic fault tree models, *Proc. Reliability and Maintainability Symposium*, pp. 105–111.

Mann, N. R. (1974), Approximately optimum confidence bounds on series and parallel system reliability for systems with binomial subsystem data, *IEEE Transactions on Reliability*, vol. R-23, no. 5, pp. 295–304.

Manson, S. S. (1966), *Thermal Stress and Low Cycle Fatigue*, McGraw-Hill, New York.

Marseguerra, M., Zio, E., and Cipollone, M. (2003), Designing optimal degradation tests via multi-objective generic algorithms, *Reliability Engineering and System Safety*, vol. 79, no. 1, pp. 87–94.

Martz, H. F., and Waller, R. A. (1982), *Bayesian Reliability Analysis*, Wiley, Hoboken, NJ.

Meeker, W. Q. (1984), A comparison of accelerated life test plans for Weibull and lognormal distributions and Type I censoring, *Technometrics*, vol. 26, no. 2, pp. 157–171.

Meeker, W. Q., and Escobar, L. A. (1995), Planning accelerated life tests with two or more experimental factors, *Technometrics*, vol. 37, no. 4, pp. 411–427.

—— (1998), *Statistical Methods for Reliability Data*, Wiley, Hoboken, NJ.

Meeker, W. Q., and Hahn, G. J. (1985), *How to Plan an Accelerated Life Test: Some Practical Guidelines*, volume of ASQC Basic References in Quality Control: Statistical Techniques, American Society for Quality, Milwaukee, WI, www.asq.org.

Meeker, W. Q., and Nelson, W. B. (1975), Optimum accelerated life-tests for the Weibull and extreme value distributions, *IEEE Transactions on Reliability*, vol. R-24, no. 5, pp. 321–332.

Meng, F. C. (1996), Comparing the importance of system components by some structural characteristics, *IEEE Transactions on Reliability*, vol. 45, no. 1, pp. 59–65.

—— (2000), Relationships of Fussell–Vesely and Birnbaum importance to structural importance in coherent systems, *Reliability Engineering and System Safety*, vol. 67, no. 1, pp. 55–60.

Menon, R., Tong, L. H., Liu, Z., and Ibrahim, Y. (2002), Robust design of a spindle motor: a case study, *Reliability Engineering and System Safety*, vol. 75, no. 3, pp. 313–319.

Meshkat, L., Dugan, J. B., and Andrews J. D. (2000), Analysis of safety systems with on-demand and dynamic failure modes, *Proc. Reliability and Maintainability Symposium*, pp. 14–21.

Mettas, A. (2000), Reliability allocation and optimization for complex systems, *Proc. Reliability and Maintainability Symposium*, pp. 216–221.

Mettas, A., and Zhao, W. (2005), Modeling and analysis of repairable systems with general repair, *Proc. Reliability and Maintainability Symposium*, pp. 176–182.

Misra, K. B. (1992), *Reliability Analysis and Prediction: A Methodology Oriented Treatment*, Elsevier, Amsterdam, The Netherlands.

Misra, R. B., and Vyas, B. M. (2003), Cost effective accelerated testing, *Proc. Reliability and Maintainability Symposium*, pp. 106–110.

Mogilevsky, B. M., and Shirn, G. (1988), Accelerated life tests of ceramic capacitors, *IEEE Transactions on Components, Hybrids and Manufacturing Technology*, vol. 11, no. 4, pp. 351–357.

Mok, Y. L., and Xie, M. (1996), Planning and optimizing environmental stress screening, *Proc. Reliability and Maintainability Symposium*, pp. 191–198.

Montanari, G. C., Pattini, G., and Simoni, L. (1988), Electrical endurance of EPR insulated cable models, *Conference Record of the IEEE International Symposium on Electrical Insulation*, pp. 196–199.

Montgomery, D. C. (2001a), *Introduction to Statistical Quality Control*, Wiley, Hoboken, NJ.

———(2001b), *Design and Analysis of Experiments*, 5th ed., Wiley, Hoboken, NJ.

Moore, A. H., Harter, H. L., and Sneed, R. C. (1980), Comparison of Monte Carlo techniques for obtaining system reliability confidence limits, *IEEE Transactions on Reliability*, vol. R-29, no. 4, pp. 178–191.

Murphy, K. E., Carter, C. M., and Brown, S. O. (2002), The exponential distribution: the good, the bad and the ugly: a practical guide to its implementation, *Proc. Reliability and Maintainability Symposium*, pp. 550–555.

Murthy, D. N. P., and Blischke, W. R. (2005), *Warranty Management and Product Manufacture*, Springer-Verlag, New York.

Myers, R. H., and Montgomery, D. C. (2002), *Response Surface Methodology: Process and Product Optimization Using Designed Experiments*, 2nd ed., Wiley, Hoboken, NJ.

Nachlas, J. A. (1986), A general model for age acceleration during thermal cycling, *Quality and Reliability Engineering International*, vol. 2, no. 1, pp. 3–6.

Naderman, J., and Rongen, R. T. H. (1999), Thermal resistance degradation of surface mounted power devices during thermal cycling, *Microelectronics Reliability*, vol. 39, no. 1, pp. 123–132.

Nair, V. N. (1992), Taguchi's parameter design: a panel discussion, *Technometrics*, vol. 34, no. 2, pp. 127–161.

Nair, V. N., Taam, W., and Ye, K. (2002), Analysis of functional responses from robust design studies, *Journal of Quality Technology*, vol. 34, no. 4, pp. 355–371.

Natvig, B. (1979), A suggestion of a new measure of importance of system components, *Stochastic Processes and Their Applications*, vol. 9, pp. 319–330.

Nelson, W. B. (1972), Theory and application of hazard plotting for censored failure data, *Technometrics*, vol. 14, no. 4, pp. 945–966. (Reprinted in *Technometrics*, vol. 42, no. 1, pp. 12–25.)

———(1982), *Applied Life Data Analysis*, Wiley, Hoboken, NJ.

———(1985), Weibull analysis of reliability data with few or no failures, *Journal of Quality Technology*, vol. 17, no. 3, pp. 140–146.

———(1990), *Accelerated Testing: Statistical Models, Test Plans, and Data Analysis*, Wiley, Hoboken, NJ.

———(2003), *Recurrent Events Data Analysis for Product Repairs, Diseases Recurrences, and Other Applications*, ASA and SIAM, Philadelphia, PA, www.siam.org.

———(2004), paperback edition of Nelson (1990) with updated descriptions of software, Wiley, Hoboken, NJ.

———(2005), A bibliography of accelerated test plans, *IEEE Transactions on Reliability*, vol. 54, no. 2, pp. 194–197, and no. 3, pp. 370–373. Request a searchable Word file from WNconsult@aol.com.

Nelson, W. B., and Kielpinski, T. J. (1976), Theory for optimum censored accelerated life tests for normal and lognormal life distributions, *Technometrics*, vol. 18, no. 1, pp. 105–114.

Nelson, W. B., and Meeker, W. Q. (1978), Theory for optimum accelerated censored life tests for Weibull and extreme value distributions, *Technometrics*, vol. 20, no. 2, pp. 171–177.

Neufeldt, V., and Guralnik, D. B., Eds. (1997), *Webster's New World College Dictionary*, 3rd ed., Macmillan, New York.

Nielsen, O. A. (1997), *An Introduction to Integration and Measure Theory*, Wiley, Hoboken, NJ.

Norris, K. C., and Landzberg, A. H. (1969), Reliability of controlled collapse interconnections, *IBM Journal of Research and Development*, vol. 13, pp. 266–271.

O'Connor, P. D. T. (2001), *Test Engineering: A Concise Guide to Cost-Effective Design, Development and Manufacture*, Wiley, Chichester, West Sussex, England.

———(2002), *Practical Reliability Engineering*, 4th ed., Wiley, Chichester, West Sussex, England.

Oraee, H. (2000), Quantative approach to estimate the life expectancy of motor insulation systems, *IEEE Transactions on Dielectrics and Electrical Insulation*, vol. 7, no. 6, pp. 790–796.

Ozarin, N. W. (2004), Failure modes and effects analysis during design of computer software, *Proc. Reliability and Maintainability Symposium*, pp. 201–206.

Park, J. I., and Yum, B. J. (1997), Optimal design of accelerated degradation tests for estimating mean lifetime at the use condition, *Engineering Optimization*, vol. 28, no. 3, pp. 199–230.

———(1999), Comparisons of optimal accelerated test plans for estimating quantiles of lifetime distribution at the use condition, *Engineering Optimization*, vol. 31, no. 1–3, pp. 301–328.

Peck, D. S. (1986), Comprehensive model for humidity testing correlation, *Proc. 24th IEEE International Reliability Physics Symposium*, pp. 44–50.

Phadke, M. S., and Smith, L. R. (2004), Improving engine control reliability through software optimization, *Proc. Reliability and Maintainability Symposium*, pp. 634–640.

Pham, H., Ed. (2003), *Handbook of Reliability Engineering*, Springer-Verlag, London.

Pignatiello, J. J. (1993), Strategies for robust multiresponse quality engineering, *IIE Transactions*, vol. 25, no. 3, pp. 5–15.

Pillay, A., and Wang, J. (2003), Modified failure mode and effects analysis using approximate reasoning, *Reliability Engineering and System Safety*, vol. 79, no. 1, pp. 69–85.

Pohl, E. A., and Dietrich, D. L. (1995a), Environmental stress screening strategies for complex systems: a 3-level mixed distribution model, *Microelectronics Reliability*, vol. 35, no. 4, pp. 637–656.

———(1995b), Environmental stress screening strategies for multi-component systems with Weibull failure-times and imperfect failure detection, *Proc. Reliability and Maintainability Symposium*, pp. 223–231.

Popinceanu, N. G., Gafitanu, M. D., Cretu, S. S., Diaconescu, E. N., and Hostiuc, L. T. (1977), Rolling bearing fatigue life and EHL theory, *Wear*, vol. 45, no. 1, pp. 17–32.

Pyzdek, T. (2003), *The Six Sigma Project Planner: A Step-by-Step Guide to Leading a Six Sigma Project Through DMAIC*, McGraw-Hill, New York.

Qiao, H., and Tsokos, C. P. (1994), Parameter estimation of the Weibull probability distribution, *Mathematics and Computers in Simulation*, vol. 37, no. 1, pp. 47–55.

Rahe, D. (2000), The HASS development process, *Proc. Reliability and Maintainability Symposium*, pp. 389–394.

Rai, B., and Singh, N. (2004), Modeling and analysis of automobile warranty data in presence of bias due to customer-rush near warranty expiration limit, *Reliability Engineering and System Safety*, vol. 86, no. 1, pp. 83–94.

Rantanen, K., and Domb, E. (2002), *Simplified TRIZ: New Problem-Solving Applications for Engineers and Manufacturing Professionals*, CRC Press/St. Lucie Press, Boca Raton, FL.

Rauzy, A. (1993), New algorithms for fault tree analysis, *Reliability Engineering and System Safety*, vol. 40, no. 3, pp. 203–211.

Reddy, R. K., and Dietrich, D. L. (1994), A 2-level environmental-stress-screening (ESS) model: a mixed-distribution approach, *IEEE Transactions on Reliability*, vol. 43, no. 1, pp. 85–90.

ReVelle, J. B., Moran, J. W., and Cox, C. A. (1998), *The QFD Handbook*, Wiley, Hoboken, NJ.

Robinson, T. J., Borror, C. M., and Myers, R. H. (2004), Robust parameter design: a review, *Quality and Reliability Engineering International*, vol. 20, no. 1, pp. 81–101.

Romano, D., Varetto, M., and Vicario, G. (2004), Multiresponse robust design: a general framework based on combined array, *Journal of Quality Technology*, vol. 36, no. 1, pp. 27–37.

Ross, P. J. (1996), *Taguchi Techniques for Quality Engineering: Loss Function, Orthogonal Experiments, Parameter and Tolerance Design*, McGraw-Hill, New York.

Ross, R. (1994), Formulas to describe the bias and standard deviation of the ML-estimated Weibull shape parameter, *IEEE Transactions on Dielectrics and Electrical Insulation*, vol. 1, no. 2, pp. 247–253.

———(1996), Bias and standard deviation due to Weibull parameter estimation for small data sets, *IEEE Transactions. on Dielectrics and Electrical Insulation*, vol. 3, no. 1, pp. 28–42.

Ryoichi, F. (2003), Application of Taguchi's methods to aero-engine engineering development, *IHI Engineering Review*, vol. 36, no. 3, pp. 168–172.

SAE (2000), *Recommended Failure Modes and Effects Analysis (FMEA) Practices for Non-automobile Applications (draft)*, SAE ARP 5580, Society of Automotive Engineers, Warrendale, PA, www.sae.org.

———(2002), *Potential Failure Mode and Effects Analysis in Design (Design FMEA), Potential Failure Mode and Effects Analysis in Manufacturing and Assembly Processes (Process FMEA), and Potential Failure Mode and Effects Analysis for Machinery (Machinery FMEA)*, SAE J1739, Society of Automotive Engineers, Warrendale, PA, www.sae.org.

Seber, G. A., and Wild, C. J. (2003), *Nonlinear Regression*, Wiley, Hoboken, NJ.

Segal, V., Nattrass, D., Raj, K., and Leonard, D. (1999), Accelerated thermal aging of petroleum-based ferrofluids, *Journal of Magnetism and Magnetic Materials*, vol. 201, pp. 70–72.

Sergent, J. E., and Krum, A. (1998), *Thermal Management Handbook for Electronic Assemblies*, McGraw-Hill, New York.

Sharma, K. K., and Rana, R. S. (1993), Bayesian sequential reliability test plans for a series system, *Microelectronics Reliability*, vol. 33, no. 4, pp. 463–465.

Sheu, S. H., and Chien, Y. H. (2004), Minimizing cost-functions related to both burn-in and field operation under a generalized model, *IEEE Transactions on Reliability*, vol. 53, no. 3, pp. 435–439.

———(2005), Optimal burn-in time to minimize the cost for general repairable products sold under warranty, *European Journal of Operational Research*, vol. 163, no. 2, pp. 445–461.

Shohji, I., Mori, H., and Orii, Y. (2004), Solder joint reliability evaluation of chip scale package using a modified Coffin–Manson equation, *Microelectronics Reliability*, vol. 44, no. 2, pp. 269–274.

Silverman, M. (1998), Summary of HALT and HASS results at an accelerated reliability test center, *Proc. Reliability and Maintainability Symposium*, pp. 30–36.

Sinnamon, R. M., and Andrews, J. D. (1996), Fault tree analysis and binary decision diagram, *Proc. Reliability and Maintainability Symposium*, pp. 215–222.

———(1997a), Improved efficiency in qualitative fault tree analysis, *Quality and Reliability Engineering International*, vol. 13, no. 5, pp. 293–298.

———(1997b), New approaches to evaluating fault trees, *Reliability Engineering and System Safety*, vol. 58, no. 2, pp. 89–96.

Smeitink, E., and R. Dekker (1990), A simple approximation to the renewal function, *IEEE Transactions on Reliability*, vol. 39, no. 1, pp. 71–75.

Smith, C. L., and Womack, J. B. (2004), Raytheon assessment of PRISM® as a field failure prediction tool, *Proc. Reliability and Maintainability Symposium*, pp. 37–42.

Sohn, S. Y., and Jang, J. S. (2001), Acceptance sampling based on reliability degradation data, *Reliability Engineering and System Safety*, vol. 73, no. 1, pp. 67–72.

Stachowiak, G. W., and Batchelor, A. W. (2000), *Engineering Tribology*, 2nd ed., Butterworth-Heineman, Woburn, MA.

Stamatis, D. H. (2004), *Six Sigma Fundamentals: A Complete Guide to the System, Methods and Tools*, Productivity Press, New York.

Steinberg, D. S. (2000), *Vibration Analysis for Electronic Equipment*, 3rd ed., Wiley, Hoboken, NJ.

Strifas, N., Vaughan, C., and Ruzzene, M. (2002), Accelerated reliability: thermal and mechanical fatigue solder joints methodologies, *Proc. 40th IEEE International Reliability Physics Symposium*, pp. 144–147.

Strutt, J. E., and Hall, P. L., Eds. (2003), *Global Vehicle Reliability: Prediction and Optimization Techniques*, Professional Engineering Publishing, London.

Suh, N. P. (2001), *Axiomatic Design: Advances and Applications*, Oxford University Press, New York.

Sumikawa, M., Sato, T., Yoshioka, C., and Nukii, T. (2001), Reliability of soldered joints in CSPs of various designs and mounting conditions, *IEEE Transactions on Components and Packaging Technologies*, vol. 24, no. 2, pp. 293–299.

Taguchi, G. (1986), *Introduction to Quality Engineering*, Asian Productivity Organization, Tokyo.

———(1987), *System of Experimental Design*, Unipub/Kraus International, New York.

———(2000), *Robust Engineering*, McGraw-Hill, New York.

Taguchi, G., Konishi, S., Wu, Y., and Taguchi, S. (1987), *Orthogonal Arrays and Linear Graphs*, ASI Press, Dearborn, MI.

Taguchi, G., Chowdhury, S., and Wu, Y. (2005), *Taguchis Quality Engineering Handbook*, Wiley, Hoboken, NJ.

Tamai, T., Miyagawa, K., and Furukawa, M. (1997), Effect of switching rate on contact failure from contact resistance of micro relay under environment containing silicone vapor, *Proc. 43rd IEEE Holm Conference on Electrical Contacts*, pp. 333–339.

Tang, L. C., and Xu, K. (2005), A multiple objective framework for planning accelerated life tests, *IEEE Transactions on Reliability*, vol. 54, no. 1, pp. 58–63.

Tang, L. C., and Yang, G. (2002), Planning multiple levels constant stress accelerated life tests, *Proc. Reliability and Maintainability Symposium*, pp. 338–342.

Tang, L. C., Yang, G., and Xie, M. (2004), Planning of step-stress accelerated degradation test, *Proc. Reliability and Maintainability Symposium*, pp. 287–292.

Tanner, D. M., Walraven, J. A., Mani S. S., and Swanson, S. E. (2002), Pin-joint design effect on the reliability of a polysilicon microengine, *Proc. 40th IEEE International Reliability Physics Symposium*, pp. 122–129.

Teng, S. Y., and Brillhart, M. (2002), Reliability assessment of a high CBGA for high availability systems, *Proc. IEEE Electronic Components and Technology Conference*, pp. 611–616.

Thoman, D. R., Bain, L. J., and Antle, C. E. (1969), Inferences on the parameters of the Weibull distribution, *Technometrics*, vol. 11, no. 3, pp. 445–460.

Tian, X. (2002), Comprehensive review of estimating system-reliability confidence-limits from component-test data, *Proc. Reliability and Maintainability Symposium*, pp. 56–60.

Tijms, H. (1994), *Stochastic Models: An Algorithmic Approach*, Wiley, Hoboken, NJ.

Tseng, S. T., Hamada, M., and Chiao, C. H. (1995), Using degradation data to improve fluorescent lamp reliability, *Journal of Quality Technology*, vol. 27, no. 4, pp. 363–369.

Tsui, K. L. (1999), Robust design optimization for multiple characteristic problems, *International Journal of Production Research*, vol. 37, no. 2, pp. 433–445.

Tu, S. Y., Jean, M. D., Wang, J. T., and Wu, C. S. (2006), A robust design in hard-facing using a plasma transfer arc, *International Journal of Advanced Manufacturing Technology*, vol. 27, no. 9–10, pp. 889–896.

U.S. DoD (1984), *Procedures for Performing a Failure Mode, Effects and Criticality Analysis*, MIL-STD-1629A, U.S. Department of Defense, Washington, DC.

————(1993), *Environmental Stress Screening (ESS) of Electronic Equipment*, MIL-HDBK-344A, U.S. Department of Defense, Washington, DC.

————(1995), *Reliability Prediction of Electronic Equipment*, MIL-HDBK-217F, U.S. Department of Defense, Washington, DC.

————(1996), *Handbook for Reliability Test Methods, Plans, and Environments for Engineering: Development, Qualification, and Production*, MIL-HDBK-781, U.S. Department of Defense, Washington, DC.

————(1998), *Electronic Reliability Design Handbook*, MIL-HDBK-338B, U.S. Department of Defense, Washington, DC.

————(2000), *Environmental Engineering Considerations and Laboratory Tests*, MIL-STD-810F, U.S. Department of Defense, Washington, DC.

————(2002), *Test Method Standard for Electronic and Electrical Component Parts*, MIL-STD-202G, U.S. Department of Defense, Washington, DC.

————(2004), *Test Method Standard for Microcircuits*, MIL-STD-883F, U.S. Department of Defense, Washington, DC.

Ushakov, I. E., Ed. (1996), *Handbook of Reliability Engineering*, Wiley, Hoboken, NJ.

Usher J. M., Roy, U., and Parsaei, H. R. (1998), *Integrated Product and Process Development*, Wiley, Hoboken, NJ.

Vesely, W. E. (1970), A time dependent methodology for fault tree evaluation, *Nuclear Engineering and Design*, vol. 13, no. 2, pp. 337–360.

Vesely, W. E., Goldberg, F. F., Roberts, N. H., and Haasl, D. F. (1981), *Fault Tree Handbook*, U.S. Nuclear Regulatory Commission, Washington, DC.

Vining, G. G. (1998), A compromise approach to multiresponse optimization, *Journal of Quality Technology*, vol. 30, no. 4, pp. 309–313.

Vlahinos, A. (2002), Robust design of a catalytic converter with material and manufacturing variations, SAE Series SAE-2002-01-2888, *Presented at the Powertrain and Fluid Systems Conference and Exhibition*, www.sae.org.

Vollertsen, R. P., and Wu, E. Y. (2004), Voltage acceleration and t63.2 of 1.6–10 nm gate oxides, *Microelectronics Reliability*, vol. 44, no. 6, pp. 909–916.

Wang, C. J. (1991), Sample size determination of bogey tests without failures, *Quality and Reliability Engineering International*, vol. 7, no. 1, pp. 35–38.

Wang, F. K., and Keats, J. B. (2004), Operating characteristic curve for the exponential Bayes-truncated test, *Quality and Reliability Engineering International*, vol. 20, no. 4, pp. 337–342.

Wang, W., and Dragomir-Daescu, D. (2002), Reliability qualification of induction motors: accelerated degradation testing approach, *Proc. Reliability and Maintainability Symposium*, pp. 325–331.

Wang, W., and Jiang, M. (2004), Generalized decomposition method for complex systems, *Proc. Reliability and Maintainability Symposium*, pp. 12–17.

Wang, W., and Loman J. (2002), Reliability/availability of k-out-of-n system with M cold standby units, *Proc. Reliability and Maintainability Symposium*, pp. 450–455.

Wang, Y., Yam, R. C. M., Zuo, M. J., and Tse, P. (2001), A comprehensive reliability allocation method for design of CNC lathes, *Reliability Engineering and System Safety*, vol. 72, no. 3, pp. 247–252.

Wen, L. C., and Ross, R. G., Jr. (1995), Comparison of LCC solder joint life predictions with experimental data, *Journal of Electronic Packaging*, vol. 117, no. 2, pp. 109–115.

Whitesitt, J. E. (1995), *Boolean Algebra and Its Applications*, Dover Publications, New York.

Willits, C. J., Dietz, D. C., and Moore, A. H. (1997), Series-system reliability-estimation using very small binomial samples, *IEEE Transactions on Reliability*, vol. 46, no. 2, pp. 296–302.

Wu, C. C., and Tang, G. R. (1998), Tolerance design for products with asymmetric quality losses, *International Journal of Production Research*, vol. 36, no. 9, pp. 2529–2541.

Wu, C. F. J., and Hamada, M. (2000), *Experiments: Planning, Analysis, and Parameter Design Optimization*, Wiley, Hoboken, NJ.

Wu, C. L., and Su, C. T. (2002), Determination of the optimal burn-in time and cost using an environmental stress approach: a case study in switch mode rectifier, *Reliability Engineering and System Safety*, vol. 76, no. 1, pp. 53–61.

Wu, H., and Meeker, W. Q. (2002), Early detection of reliability problems using information from warranty databases, *Technometrics*, vol. 44, no. 2, pp. 120–133.

Wu, S. J., and Chang, C. T. (2002), Optimal design of degradation tests in presence of cost constraint, *Reliability and System Safety*, vol. 76, no. 2, pp. 109–115.

Xie, M. (1989), On the solution of renewal-type integral equation, *Communications in Statistics*, vol. B18, no. 1, pp. 281–293.

Yan, L., and English, J. R. (1997), Economic cost modeling of environmental-stress-screening and burn-in, *IEEE Transactions on Reliability*, vol. 46, no. 2, pp. 275–282.

Yanez, M., Joglar, F., and Modarres, M. (2002), Generalized renewal process for analysis of repairable systems with limited failure experience, *Reliability Engineering and System Safety*, vol. 77, no. 2, pp. 167–180.

Yang, G. (1994), Optimum constant-stress accelerated life-test plans, *IEEE Transactions on Reliability*, vol. 43, no. 4, pp. 575–581.

———(2002), Environmental-stress-screening using degradation measurements, *IEEE Transactions on Reliability*, vol. 51, no. 3, pp. 288–293.

———(2005), Accelerated life tests at higher usage rate, *IEEE Transactions on Reliability*, vol. 54, no. 1, pp. 53–57.

Yang, G., and Jin, L. (1994), Best compromise test plans for Weibull distributions with different censoring times, *Quality and Reliability Engineering International*, vol. 10, no. 5, pp. 411–415.

Yang, G., and Yang, K. (2002), Accelerated degradation-tests with tightened critical values, *IEEE Transactions on Reliability*, vol. 51, no. 4, pp. 463–468.

Yang, G., and Zaghati, Z. (2002), Two-dimensional reliability modeling from warranty data, *Proc. Reliability and Maintainability Symposium*, pp. 272–278.

———(2003), Robust reliability design of diagnostic systems, *Proc. Reliability and Maintainability Symposium*, pp. 35–39.

———(2004), Reliability and robustness assessment of diagnostic systems from warranty data, *Proc. Reliability and Maintainability Symposium*, pp. 146–150.

————(2006), Accelerated life tests at higher usage rates: a case study, *Proc. Reliability and Maintainability Symposium*, pp 313–317.

————and Kapadia, J. (2005), A sigmoid process for modeling warranty repairs, *Proc. of Reliability and Maintainability Symposium*, pp. 326–330.

Yang, K., and El-Haik, B. (2003), *Design for Six Sigma: A Roadmap for Product Development*, McGraw-Hill, New York.

Yang, K., and Xue, J. (1996), Continuous state reliability analysis, *Proc. Reliability and Maintainability Symposium*, pp. 251–257.

Yang, K., and Yang, G. (1998), Robust reliability design using environmental stress testing, *Quality and Reliability Engineering International*, vol. 14, no. 6, pp. 409–416.

Yang, S., Kobza, J., and Nachlas, J. (2000), Bivariate failure modeling, *Proc. Reliability and Maintainability Symposium*, pp. 281–287.

Yassine, A. M., Nariman, H. E., McBride, M., Uzer, M., and Olasupo K. R. (2000), Time dependent breakdown of ultrathin gate oxide, *IEEE Transactions on Electron Devices*, vol. 47, no. 7, pp. 1416–1420.

Ye, N., Ed. (2003), *The Handbook of Data Mining*, Lawrence Erlbaum Associates, Mahwah, NJ.

Yeo, C., Mhaisalka, S., and Pang, H. (1996), Experimental study of solder joint reliability in a 256 pin, 0.4 mm pitch PQFP, *Journal of Electronic Manufacturing*, vol. 6, no. 2, pp. 67–78.

Young, D., and Christou, A. (1994), Failure mechanism models for electromigration, *IEEE Transactions on Reliability*, vol. 43, no. 2, pp. 186–192.

Yu, H. F. (1999), Designing a degradation experiment, *Naval Research Logistics*, vol. 46, no. 6, pp. 689–706.

————(2003), Designing an accelerated degradation experiment by optimizing the estimation of the percentile, *Quality and Reliability Engineering International*, vol. 19, no. 3, pp. 197–214.

Yu, H. F., and Chiao, C. H. (2002), An optimal designed degradation experiment for reliability improvement, *IEEE Transactions on Reliability*, vol. 51, no. 4, pp. 427–433.

INDEX

Linearly increasing hazard distribution:

$$h(t) = \lambda t$$

$$f(t) = \lambda t \, e^{-\frac{\lambda t^2}{2}}$$

$$F(t) = 1 - e^{-\lambda t^2/2}$$

$$E(t) = \sqrt{\pi/2\lambda}$$

$$Var(t) = 2/\lambda \left(1 - \pi/4\right)$$

$$R(t) = \frac{n_s(t)}{n_0} = \frac{n_0 - n_f(t)}{n_0}$$

$$f(t) = \frac{n_f}{n_0 \Delta t}$$

$$h(t) = \frac{n_f}{n_s(\Delta t)}$$

Mean Square Deviation

$$MSD = (y - m_y)^2$$

$$L = K * MSD$$

$$MSD = \sigma_y^2 + (\mu_y - m_y)^2$$

$$\alpha = \frac{m_y}{\mu_y} = Target/mean$$

$$m_y = \mu_y \alpha$$

If $m_y = \mu_y$ $\quad \alpha = 1$ $\quad$ ⎤
$$MSD = \sigma_y^2 \qquad \qquad $$ ⎦ NTB

If $m_y = 0$ $\quad \alpha = 0$ $\quad$ ⎤
$$MSD = \sigma_y^2 + \mu_y^2 \qquad$$ ⎦ STB

If m_y is larger
$\quad \alpha = 1.5 \quad MTB = \sigma_y^2 + .25\mu_y^2$ ⎤
$\quad \alpha = 2 \qquad MTB = \sigma_y^2 + \mu_y^2$ ⎦ LTB

- Probability Density Function (PDF)
  - → denotes Failure distribution for all time
  - → Absolute Failure Speed

$$\int_{-\infty}^{\infty} f(t)\,dt = 1 \quad \text{(Everything Fails Eventually)}$$

$f(t) \geq 0$ large value $\Rightarrow$ high failure rate over small time period

Commutative Distribution Function (CDF)
  - → Probability that a product will Fail by a specified time
  - → $F(t)$

Reliability = $R(t) = 1 - F(t)$
  - → Survival Function
  - → Probability of failure during time period $[t_1, t_2]$

$$= \int_{t_1}^{t_2} f(t)\,dt = R(t_1) - R(t_2) = F(t_2) - F(t_1) = F(t + \Delta t) - F(t$$

$P(\text{fail in } \Delta t \mid \text{Survive up to } t) = \left[F(t - \Delta t) - F(t) \right] / R(t)$

Instantenous Failure Rate (Hazard Rate) $h(t)$

$$h(t) = f(t) \cdot 1/R(t)$$
  - → Relative failure speed (rate) (failures per time)
  - → Bathtub Curve
  - → DFR, CFR, IFR

$f(t) = (n_f(t))/(n_0 \Delta t)$ $h(t) = (n_f(t)) / (n_s(t) \Delta t)$

$R(t) = f(t)/h(t) = \dfrac{n_s(t)}{n_0}$

Commutative Hazard Function — $H(t)$ — CHF

$$H(t) = \int_{-\infty}^{\infty} h(t)\,dt = -\ln(R(t))$$
$$R(t) = \exp(-H(t)) \approx 1 - H(t)$$

$t_p = t_{.1} \Rightarrow 10\% \text{ Failed}$

$t_p = F^{-1}(P)$ inverse of $F(p)$

Mean Time To Failure — MTTF — Expected Life — $E(t)$.

$$E(t) = \int_0^{\infty} t\,f(t)\,dt = \int_0^{\infty} R(t)$$

Variance — Var(T) — speed of life distribution

$$\text{Var}(T) = \int_0^{\infty} \left[t - E(t) \right]^2 f(t)\,dt \Rightarrow \int_0^{\infty} t^2 f(t)\,dt - (MTTF)^2$$

Standard Deviation — σ

$$\sigma = \sqrt{\text{Var}(T)}$$